TRIAZOLES: 1,2,3

This is the thirty-ninth volume in the series

THE CHEMISTRY OF HETEROCYCLIC COMPOUNDS

THE CHEMISTRY OF HETEROCYCLIC COMPOUNDS

A SERIES OF MONOGRAPHS

ARNOLD WEISSBERGER and EDWARD C. TAYLOR

Editors

TRIAZOLES: 1,2,3

K. Thomas Finley

STATE UNIVERSITY OF NEW YORK

BROCKPORT, NEW YORK

Volume Editor

John A. Montgomery

SOUTHERN RESEARCH INSTITUTE

BIRMINGHAM, ALABAMA

AN INTERSCIENCE ® PUBLICATION

JOHN WILEY & SONS

New York · Chichester · Brisbane · Toronto

An Interscience ® Publication

***Library of Congress Cataloging in Publication Data*:**

Finley, Kay Thomas, 1934–
Triazoles-1,2,3.

(The Chemistry of heterocyclic compounds; v. 39 ISSN 0069-3154)
Includes bibliographical references and index.
1. Triazoles. I. Montgomery, John A. II. Title.
QD401.F505 547'.593 80-13323
ISBN 0-471-07827-1

Printed in the United States of America

10 9 8 7 6 5 4 3 2 1

The Chemistry of Heterocyclic Compounds

The chemistry of heterocyclic compounds is one of the most complex branches of organic chemistry. It is equally interesting for its theoretical implications, for the diversity of its synthetic procedures, and for the physiological and industrial significance of heterocyclic compounds.

A field of such importance and intrinsic difficulty should be made as readily accessible as possible, and the lack of a modern detailed and comprehensive presentation of heterocyclic chemistry is therefore keenly felt. It is the intention of the present series to fill this gap by expert presentations of the various branches of heterocyclic chemistry. The subdivisions have been designed to cover the field in its entirety by monographs which reflect the importance and the interrelations of the various compounds, and accommodate the specific interests of the authors.

In order to continue to make heterocyclic chemistry as readily accessible as possible, new editions are planned for those areas where the respective volumes in the first edition have become obsolete by overwhelming progress. If, however, the changes are not too great so that the first editions can be brought up-to-date by supplementary volumes, supplements to the respective volumes will be published in the first edition.

ARNOLD WEISSBERGER

Research Laboratories
Eastman Kodak Company
Rochester, New York

EDWARD C. TAYLOR

Princeton University
Princeton, New Jersey

Preface

In 1950 Benson and Savell published an excellent review of the *v*-triazole ring system (Fig. 1.1) and exclaimed that through 1947 more than 1400 such

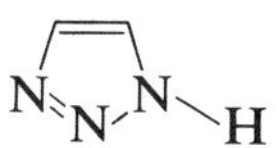

Figure **1.1**

compounds had been described in the literature![1] It had been 60 years since Hans von Pechmann, assistant to Baeyer and later professor at Tübingen, prepared and correctly formulated the 1,2,3-triazole ring both in the parent compound and in substituted derivatives.[2] In the past seven years *Chemical Abstracts* has reported studies involving well over 9000 mono cyclic 1,2,3-triazoles!

The synthesis and chemistry of the 1,2,3-triazoles have been reviewed several times since 1950,[3-6] and these sources should be studied in conjunction with this volume. *The Azoles,* by Schofield, Grimmett, and Keene,[6] is especially important in that it quite successfully places the 1,2,3-triazoles in the context of related heteroaromatic nitrogen compounds.

With so many recent and well-written reviews, the desirability of another requires a specific answer. My purpose in this and succeeding sections is essentially complete coverage of syntheses through *Chemical Abstracts* 1976. The reader should be able to determine if a given compound has been prepared and the best method yet described for that preparation. The major share of space in this book is devoted to comprehensive tables of all monocyclic 1,2,3-triazoles with appropriate references. Only those references are cited that provide useful alternative methods of syntheses. The narrative complements the tables by bringing together, in as brief a manner as possible, those synthetic methods of greatest utility. The objectives of the narrative are to (1) unite closely related examples that are dispersed by the alphabetical arrangement of the tables, and (2) to call attention to those promising areas ripe for further study.

Please notice that a few compound types that have been made in very large numbers and have especially cumbersome names are located in a separate section for ease of searching. Chapter 17 should be consulted for these parent structures rather than the chapter related to their normal structural assignment.

K. THOMAS FINLEY

Brockport, New York
May 1980

References

1. F. R. Benson and W. L. Savell, *Chem. Revs.*, **46,** 1 (1950).
2. H. von Pechmann, *Chem. Ber.*, **21,** 2756 (1888).
3. E. Hoggarth, "*Triazoles,*" in E. H. Rodd, Ed., *Chemistry of Carbon Compounds*, Vol. 4A, Elsevier, London, 1957, Chapter 6.
4. J. H. Boyer, "Monocyclic Triazoles and Benzotriazoles," in R. C. Elderfield, Ed., *Heterocyclic Compounds*, Vol. 7, Wiley, New York, 1961, Chapter 5.
5. T. L. Gilchrist and G. E. Gymer, "1,2,3-Triazoles," in A. R. Katritzky and A. J. Boulton, Eds., *Advances in Heterocyclic Chemistry*, Vol. 16, Academic Press, New York, 1974, Chapter 2.
6. K. Schofield, M. R. Grimmett, and B. R. T. Keene, *Heteroaromatic Nitrogen Compounds: The Azoles*, Cambridge University Press, Cambridge, England, 1976.

Contents

TRIAZOLES: 1,2,3

This is the thirty-ninth volume in the series

THE CHEMISTRY OF HETEROCYCLIC COMPOUNDS

CHAPTER 1

Alkyl- or Aryl-Monosubstituted 1,2,3-Triazoles

1.1. 1-ALKYL- OR ARYL-SUBSTITUTED 1*H*-1,2,3-TRIAZOLES

The nineteenth century story of 1,2,3-triazoles has been well told,[1] and Wiley has thoughtfully outlined the weaknesses of synthetic efforts during the first 80 years.[2] Along with Hussung and Moffat, he described what appears to be the first satisfactory method for preparing the parent compound (**1.1-1**) on a half mole or larger scale (Eq. 1). The overall conversion obtained is 70%. Gold has modified this synthesis, improving the yield to 79% and reducing the time required from one week to a half day.[3]

$$HO_2CC{\equiv}CCO_2H + PhCH_2N_3 \longrightarrow \text{[4,5-bis}(CO_2H)\text{-1-}CH_2Ph\text{-1,2,3-triazole]} \xrightarrow[\text{vacuum}]{\text{heat}} \text{[1-}CH_2Ph\text{-1,2,3-triazole]} \xrightarrow[H_2]{Pd/C} \mathbf{1.1\text{-}1} \qquad (1)$$

The direct addition of hydrazoic acid to acetylenes with electron-withdrawing groups[4] can produce comparable yields of **1.1-1** (Eq. 2), but the safety problems are quite real.[5] Acetylene itself can be converted to **1.1-1** in high yield through reaction with sodium azide under carefully controlled conditions (Eq. 3).[6]

$$HC{\equiv}CCO_2H + HN_3 \longrightarrow \text{[}CO_2H\text{-1,2,3-triazole (NH)]} \longrightarrow \mathbf{1.1\text{-}1} \qquad (2)$$

$$HC{\equiv}CH + Na^+N_3^- \xrightarrow{H_2SO_4} \mathbf{1.1\text{-}1} \qquad (3)$$

The addition of trimethylsilylazide to the carbon-carbon triple bond (Eq. 4) is attractive both for safety and for the ease of removing the silyl group.[7]

With acetylene itself excellent yields have been reported.[8] The thermal rearrangement of the trimethylsilyl group in **1.1-2** is well established. The presence of strong electron-withdrawing groups facilitates the reaction. The decarboxylation product (**1.1-3**) has been obtained quantitatively in dioxane.[9]

$$Me_3SiN_3 + R^1C{\equiv}CR^2 \longrightarrow \mathbf{1.1\text{-}2}\ (2\text{-}SiMe_3\text{-}4\text{-}R^1\text{-}5\text{-}R^2\text{-}1,2,3\text{-triazole}) \xrightarrow{MeOH} \mathbf{1.1\text{-}3} \longrightarrow \mathbf{1.1\text{-}1} \quad (4)$$

1.1-3
$R^1, R^2 = CN, CO_2H, CO_2R, H$

The safest and perhaps most generally applicable synthesis of 1-alkyl-1,2,3-triazoles relies on the nucleophilic character of the heterocyclic nitrogen atoms of **1.1-1** (Eq. 5). Gold has shown[3] that with equimolar reactants, **1.1-4** and **1.1-5** are formed in a ratio of 4:1 whereas an excess of **1.1-1** (or its silver salt) produces a lower conversion and a greater excess of **1.1-4**.

$$\mathbf{1.1\text{-}1} + RBr \xrightarrow[Na^+\ ^-OMe]{MeOH} \mathbf{1.1\text{-}4}\ (1\text{-R-1,2,3-triazole}) + \mathbf{1.1\text{-}5}\ (2\text{-R-1,2,3-triazole}) \quad (5)$$

R = Pr, 2-propenyl

A variety of functionally substituted alkyl chains can be attached to the 1,2,3-triazole nitrogens using Gold's method (Eqs. 6 and 7).[3] Notice that either substitution or addition can occur, given the appropriate reactant.

$$\mathbf{1.1\text{-}1} + ClCH_2CH_2CN \xrightarrow{Na^+\ ^-OMe} 1\text{-}(CH_2)_2CN\text{-1,2,3-triazole}\ (71\%) + 2\text{-}(CH_2)_2CN\text{-1,2,3-triazole}\ (26\%) \quad (6)$$

$$\mathbf{1.1\text{-}1} + CH_2{=}CHNO_2 \xrightarrow[Raney\ Ni]{H_2} 1\text{-}(CH_2)_2NH_2\text{-1,2,3-triazole}\ (68\%) \quad (7)$$

Hubert[10] has prepared a variety of 1-arylmethyl-1,2,3-triazoles (**1.1-6**) (Ar = Cl_5C_6, 2,5-$Me_2C_6H_3$, 4-Cl, 4-MeO and 4-$NO_2C_6H_4$) in acceptable yields (Eq. 8).

$$HC{\equiv}CH + ArCH_2N_3 \xrightarrow{Me_2C{=}O} \text{1-}(CH_2Ar)\text{-1,2,3-triazole} \quad (8)$$

1.1-6

The *N*-aryl substituent played a key role in the birth of triazole chemistry,[1] and the significance of such compounds continues at a high level after nearly a century. El Khadem and his collaborators recognized the possibility of using the aryl-triazoles as intermediates in the preparation of: (1) sulfa drug analogs and (2) azo dyes for cellulose. They have made detailed studies of such compounds, beginning with a significant improvement (Eq. 9) in von Pechmann's original method.[11,12]

$$\begin{array}{l}CH{=}NNHPh\\ |\\ CH{=}NNHPh\end{array} + CuSO_4 \xrightarrow[\text{reflux}]{H_2O} \text{2-Ph-1,2,3-triazole} \quad (9)$$

A fundamentally different method was developed to provide the 1-aryl-1,2,3-triazoles in fair-to-good yields (Eq. 10).[13]

$$2ArN_3 + Na^{+}\,{}^{-}OEt \longrightarrow \text{1-Ar-1,2,3-triazole} \quad (10)$$

The pressures, temperatures, and (often) long reaction times required in the direct addition of azides to acetylene has led to extensive studies of alternatives such as the addition to enol esters (Eq. 11),[14] ethers (Eq. 12),[14] and enamines (Eq. 13).[15]

In some cases the intermediate triazolines are formed with regiospecificity. The exceptions are enamines that can tautomerize. Aromatization can occur spontaneously (e.g., **1.1-7**), or it may require stronger heating or treatment with acid or base (e.g., **1.1-9**, **1.1-10**). The yields are often quite good and the conditions fairly mild.

Nucleophilic substitution has also been applied frequently as a synthesis of *N*-aryl triazoles.[16,17] As we should expect, this method is useful largely with nitrated fluorobenzenes. Both the solvent and the addition sequence appear to be significant in determining the product isomer distribution (Eqs. 14 to 16).[17]

Huisgen and his collaborators[18] have also prepared 1-phenyl-1,2,3-triazole in excellent yield from the azide and an active methylene substrate (Eq. 17).

ArN_3 + $AcOCH{=}CH_2$ → [**1.1-7**, 5-OAc-4,5-dihydro-1-Ar-1,2,3-triazole] → **1.1-8** (1-Ar-1,2,3-triazole) (11)

ArN_3 + $BuOCH{=}CH_2$ → **1.1-9** (5-OBu-4,5-dihydro-1-Ar-1,2,3-triazole) → **1.1-8** (12)

ArN_3 + $CH_2{=}CH{-}N$(morpholino) → **1.1-10** (5-morpholino-4,5-dihydro-1-Ar-1,2,3-triazole) $\xrightarrow{KOH}$ **1.1-8** (13)

1.1-1 + 2-fluoro-1,3,5-trinitrobenzene (F, O_2N, NO_2, NO_2) → 1-(2,4,6-trinitrophenyl)-1,2,3-triazole + 2-(2,4,6-trinitrophenyl)-1,2,3-triazole

Conditions	1-isomer	2-isomer	
$O(CH_2CH_2)_2O$ or $HCONMe_2$	95%	0%	(14)
Me_2SO[a]	53%	41%	(15)
Me_2SO[b]	21%	75%	(16)

[a] **1.1-1** refluxed for several hours before the fluoride is added.
[b] **1.1-1** is added to the dissolved fluoride.

$HC(=O)CH_2CO_2Et + PhN_3 \xrightarrow[EtOH]{Na^+\ ^-OEt}$ 1-Ph-1,2,3-triazole-5-CO_2H $\xrightarrow{heat}$ 1-Ph-1,2,3-triazole (17)

REFERENCES

1. **44:** 3979h
2. **50:** 15517e
3. **64:** 2082g
4. **55:** 17626g
5. **61:** 14674f
6. **70:** 87698a
7. **63:** 4324c
8. **68:** 13061z
9. **83:** 97148d
10. **73:** 3865d
11. **53:** 344f
12. **55:** 27070e
13. **69:** 10402w
14. **62:** 16248a
15. **67:** 64365p
16. **68:** 48845k
17. **74:** 99950x
18. **64:** 5075e

1.2. 2-ALKYL- OR ARYL-SUBSTITUTED 2*H*-1,2,3-TRIAZOLES

The preparation 2-alkyl- and substituted alkyl-2*H*-1,2,3-triazoles has been achieved using Gold's nucleophilic substitution reaction,[19] and although the product distribution is unfavorable for simple alkyls (Eq. 18), proper choice of reactant and subsequent chemistry can make this a most useful preparative approach (Eqs. 19 and 20).

NH-triazole + RBr $\xrightarrow[Na^{+}\ ^{-}OMe]{MeOH}$ 1-R-triazole + 2-R-triazole (18)

R = Pr, 2-propenyl

NH-triazole:

+ $BrCH_2CO_2Et$ $\xrightarrow{Na^{+}\ ^{-}OMe}$ 1-(CH_2CO_2Et)-triazole, 58% + 2-(CH_2CO_2Et)-triazole, 39% (19)

+ $CH_2{=}CHCN$ $\xrightarrow{\text{Triton B}}$ 2-($(CH_2)_2CN$)-triazole, 99% (20)

Nearly thirty years ago Riebsomer made an extensive study of the introduction of functional groups in 2-phenyl-2*H*-1,2,3-triazole.[20] As Equation 21 indicates, the well-known utility of the aryl nitro group was exploited and offers many other possibilities for further development. Lynch has continued the study of intermediate **1.2-1**, cleared up some earlier misunderstanding, and contributed several useful new routes (Eq. 22).[21,22] The yields in these reactions are generally adequate, and the rest could likely be improved with study. An obvious extension of both of these methods involves the nucleophilic substitution of fluoronitroaryl systems (Eq. 23).[23,24]

As indicated in Section 1.1, this method allows rather good control of the product distribution obtained.

1.2-1 (21)

Y = NO_2, NH_2 NHAc, $As(O)(OH)_2$, $NHSO_2C_6H_4NHAc$-4, SO_2Cl, SO_2NH_2 SO_2NHPh

(22)

Ar = Ph, 4-$NO_2C_6H_4$ Ar′ = 4-Br, 4-Ph, 4-$NO_2C_6H_4$

(23)

20% 62%

The most extensive and continuing study of 2-aryl-2*H*-1,2,3-triazole chemistry is that of El Khadem and his collaborators.[25] In addition to their reinvestigation of von Pechmann's original work (Eq. 24), they have developed important methods of introducing functional groups (Eq. 25)[26] and have greatly expanded our knowledge of the synthesis of 2-aryl-2*H*-1,2,3-triazoles from sugars (Eq. 26).[27] The range of substituents available is impressive, and the yields are generally excellent.

(24)

Ar = 4-MeC_6H_4, $CuSO_4$; $KMnO_4$; Ar = Ph, Br_2

(25)

CH=NNHAr
C=NNHAr
HOCH
HCOH
HCOH
CH_2OH

two steps → HO_2C (triazole) N—Ar $\xrightarrow{-CO_2}$ (triazole) N—Ar (26)

REFERENCES

19. **64:** 2082g
20. **43:** 2619i
21. **58:** 6819f
22. **59:** 10053c
23. **74:** 99950x
24. **67:** 64310s
25. **55:** 27070e
26. **55:** 3562f
27. **61:** 5738c

1.3. 4-ALKYL- OR ARYL-SUBSTITUTED *v*-TRIAZOLES

The preparation of a wide variety of 4-alkyl-1,2,3-triazoles (**1.3-1**) has been accomplished by the direct addition of hydrazoic acid[28] or trimethylsilylazide[29,30] to various terminal alkynes (Eq. 27). In general, the yields are good, but the lower homologs are better made either through the silyl derivative or by decarboxylation (Eq. 28).[31] A novel approach has been applied by Haszeldine and his collaborators[30] to the synthesis of the 4-(trifluoromethyl) (**1.3-2**) substituent (Eq. 29). The yield, based on unrecovered starting material, is near quantitative, and this method clearly deserves further development.

RC≡CH { $+HN_3$ → **1.3-1** ; $+Me_3SiN_3$ → ($SiMe_3$-triazole) $\xrightarrow{H_2O}$ **1.3-1** } (27)

$MeC{\equiv}CCO_2H + HN_3$ → (Me, CO_2H, N—H triazole) → (Me, N—H triazole) (28)

$$Me_3SiCHN_2 + CF_3CN \longrightarrow \text{4-}F_3C\text{-1-}SiMe_3\text{-1,2,3-triazole} \longrightarrow \text{4-}F_3C\text{-1H-1,2,3-triazole}\ (\mathbf{1.3\text{-}2}) \quad (29)$$

In an attempt to study triazole analogs of the α-amino acids as antimetabolites, Sheehan and Robinson[32] prepared alanine containing such a substituent (**1.3-3**) along with a series of 4-substituted 1,2,3-triazole intermediates. The best route (Eq. 30) was discovered after several alternatives were investigated.

$$\text{4-CHO-1H-1,2,3-triazole} + PhCONHCH_2CO_2H \xrightarrow[Et_3N]{Ac_2O} \text{1-Ac-triazol-4-yl-}CH{=}\text{(2-Ph-oxazol-5-one)} \xrightarrow[Pt_2O]{H_2}$$

$$\text{1H-triazol-4-yl-}CH_2CH(NHC(=O)Ph)CO_2H \xrightarrow{HCl} \text{1H-triazol-4-yl-}CH_2CH(NH_2)CO_2H\ (\mathbf{1.3\text{-}3}) \quad (30)$$

An impressive array of methods has been applied to the preparation of 4-phenyl-1,2,3-triazole (**1.3-4**): the standard trimethylsilylazide (Eq. 29),[29,33] sodium azide (Eq. 31),[34] and hydrazoic acid[28] additions. But, in addition to these well-documented techniques: tosylazide addition (Eq. 32)[35] intramolecular cyclization (Eq. 33),[34] triphenylphosphonium salt addition (Eq. 34),[36] addition to styrenes bearing strong electron-withdrawing β-groups

$$PhC{\equiv}CH + Na^+N_3^- \xrightarrow{Me_2SO} \text{4-Ph-1H-1,2,3-triazole}\ (\mathbf{1.3\text{-}4}) \quad (31)$$

$$PhC{\equiv}CH + p\text{-}MeC_6H_4SO_2N_3 \longrightarrow \text{4-Ph-1-Ts-1,2,3-triazole} \xrightarrow{H_2SO_4} \mathbf{1.3\text{-}4} \quad (32)$$

$$Ph(N_3)C{=}CH_2 \xrightarrow{K/Al_2O_3} \mathbf{1.3\text{-}4} \quad (33)$$

(Eq. 35),[37] and the cyclization of α,β-bishydrazones (Eq. 36)[38] have all been employed. With the exceptions of Equations 33 and 35 (CN), the yields of these varied methods are uniformly good to excellent. The first (Eq. 27) and last three (Eqs. 34 to 36) methods have all been employed with variously substituted phenyl acetylenes and have generally produced excellent yields.[39–41] The effect of aryl substituents on the yield of triazole in Equation 35 and the rather surprising by-products (which are also very much influenced by the substituent present) have been reported and strongly suggest some interesting chemistry to be studied.[37,40]

$$\mathrm{PhC{\equiv}CX + Ph_3P \longrightarrow PhC{\equiv}CPPh_3^+X^- \xrightarrow[\mathrm{HCONMe_2}]{\mathrm{Na^+N_3^-}}} \quad (34)$$

Ph, PPh_3^+ (triazole anion, N–N–N ring) $\xrightarrow[\mathrm{H_2O}]{\mathrm{Na^+OH^-}}$ **1.3-4**

$$\mathrm{PhCH{=}CH(NO_2\ or\ CN) + Na^+N_3^- \xrightarrow{\mathrm{Me_2SO}}} \textbf{1.3-4} \quad (35)$$

(Ph and H on one carbon, H and NO_2 or CN on the other carbon)

$$\mathrm{PhC(=O)CHBr_2 + N_2H_4 \longrightarrow PhC(=NNH_2)CH{=}NNH_2 \xrightarrow[\mathrm{MnO_2}]{\mathrm{HgO\ or}}}$$

4-Ph-1,2,3-triazole with NH_2 on N $\xrightarrow{\mathrm{HNO_2}}$ **1.3-4** (36)

REFERENCES

28. **49:** 3949b
29. **68:** 13061z
30. **78:** 111412r
31. **50:** 9392f
32. **43:** 6621d
33. **60:** 5537c
34. **73:** 25364s
35. **64:** 5075e
36. **79:** 66254y
37. **75:** 140767d
38. **68:** 59506b
39. **80:** 108452q
40. **77:** 100955y
41. **76:** 85756t

TABLE 1. ALKYL- OR ARYL-MONOSUBSTITUTED 1,2,3-TRIAZOLES

Compound	Reference
1.1. 1-Alkyl- or Aryl-Substituted 1*H*-1,2,3-Triazoles	
v-1,2,3-triazole	**50:** 15517e **55:** 17626g **64:** 2082g **70:** 87698a
v-1,2,3-triazole-1*d*	**82:** 56932r
1-[2-[*N*-[3-acetamido-4-[(3′-chloro-4′-nitrophenyl)azo]-*N*-ethylamino]]ethyl]-	**69:** P28576r
1-acetamide, *N*-[2-chloro-5-[[1-oxo-2-(3-pentadecyl-phenoxy)butyl]amino]phenyl]-α-(2,2-dimethyl-1-oxo-propyl)-4,5-dihydro-4,4-dimethyl-5-oxo-	**83:** P155710u
1-(acetamidomethyl)-	**24:** 3215[2]
1-[(3-acetamido)phenyl]-	**67:** P100995e
1-[(4-acetamido)phenyl]	**67:** P100995e
1-(4-amino-3-chlorophenyl)-	**67:** P100995e
1-(5-amino-2,3-dihydro-3,3-dimethyl-2-hydroxy-naphtho[1,2-*b*:4,3*b*′]difuran-4-yl)-	**74:** 13031p
1-(3-amino-4-ethoxyphenyl)-	**67:** P100995e
1-(2-aminoethyl)-	**64:** 2082g
1-(2-aminophenyl)-	**67:** 64311t
1-(3-aminophenyl)-	**67:** P100995e
1-(4-aminophenyl)-	**81:** 62940n **67:** P100995e
1-(3-aminopropyl)-	**64:** 2082g
1-(2-azidophenyl)-	**67:** 64311t
1-(2-benzimidazolyl)-	**74:** 53669j
1-(2-benzothiazolyl)-	**74:** 53669j
1-[(1*H*-benzotriazol-1-yl)methyl]-	**73:** 3865d
1-benzyl-	**24:** 3232[1] **50:** 15517e **84:** 159507b
1-[2-(1-benzylbenzimidazolyl)]-	**74:** 53669j
1-[(4-benzylamino)-3-oxo-2-phenyl-2*H*-pyridazin-5-yl]-	**77:** 88450v
4,5-bis(trimethylsilyl)-	**60:** 5537c
4,5-bis(trimethylsilyl)-1-phenyl-	**60:** 5537c
1-(4-bromo-2,6-dinitrophenyl)-	**60:** 4132g
1-(4-bromophenyl)-	**69:** 10402w **77:** 126516f
1-[2-(*N*-butylanilino)ethyl]-	**69:** P28576r
1-[(4-carboxyanilino)methyl]-	**50:** 8612h
1-(3-carboxy-4-hydroxyphenyl)-	**81:** P63478e
1-(2-chloro-4,6-dinitrophenyl)-	**60:** 4132g
1-(4-chloro-2,6-dinitrophenyl)-	**60:** 4132g
1-(4-chloro-2-methyl-3-oxo-2*H*-pyridazin-5-yl)-	**78:** 72035y
1-(3-chloro-4-nitrophenyl)-	**67:** P100995e
1-[2-[4-[(2-chloro-4-nitrophenyl)-azo]-*N*-ethylanilino]ethyl]-	**69:** P28576r
1-(4-chloro-3-oxo-2-phenyl-2*H*-pyridazin-5-yl)-	**74:** 125600e
1-(4-chloro-3-oxo-2*H*-pyridazin-5-yl)-	**78:** 72035y
1-(4-chlorophenyl)-	**69:** 10402w
1-[2-(4-chlorophenyl)-1,3-dioxo-2-indanyl]-	**73:** 56037s
1-[(4-chlorophenyl)methyl]-	**73:** 3865d
1-[2-[[4-[3-(4-chlorophenyl)-2-pyrazolin-1-yl]phenyl]-sulfonyl]ethyl]-	**72:** P91491m
1-(2-cyanoethyl)-	**64:** 2082g

TABLE 1 (*Continued*)

Compound	Reference
1.1. 1-Alkyl- or Aryl-Substituted 1*H*-1,2,3-Triazoles (*Continued*)	
1-(cyanomethyl)-	**64:** 2082g
1-[(1-cyano-2-phenyl)ethenyl]-	**64:** 2082g
1-(2,4-dibromophenyl)-	**19:** 476[3]
1-[(4,5-dichloro-4-cyclopenten-1,3-dion-2-yliden)-methyl]-	**78:** P124250m
1-(2,4-dichlorophenyl)-	**19:** 476[3]
1-(2,5-dichlorophenyl)-	**19:** 476[3]
1-[2-[4-[(2,4-dichlorophenyl)azo]-*N*-ethylanilino]ethyl]-	**69:** P28576r
1-[2-[4-[(2,4-dichlorophenyl)azo]-*N*-ethyl-3-methyl-anilino]ethyl]-	**69:** P28576r
1-[2,3-dihydro-4-(*N*-isopropylacetamido)-3-oxo-2-phenyl-5-pyridazinyl]-	**77:** 88450v
1-[2,3-dihydro-6-(*N*-isopropylbenzamido)-3-oxo-2-phenyl-5-pyridazinyl]-	**77:** 88450v
1-[(2,4-dimethylphenyl)methyl]-	**73:** 3865d
1-[2-(3,5-dinitrobenzamido)ethyl]-	**84:** P121664n
1-(2,4-dinitrophenyl)-	**68:** 48845k
1-[1,3-dioxo-2-(4-methoxyphenyl)-2-indanyl]-	**73:** 56037s
1-(1,3-dioxo-2-phenyl-2-indanyl)-	**73:** 56037s
1-(1,3-diphenyl-3-oxopropyl)-	**49:** 13227f
1-[2,6-di-(2-pyridyl)-4-hydroxypyrimidin-5-yl]-	**69:** P87008e
-ethanoic acid	**64:** 2082g
-ethanoic acid, ethyl ester	**64:** 2082g
-ethanoic acid, hydrazide	**64:** 2082g
1-ethenyl-	**80:** 82834p
1-[2-ethenyloxy)ethyl]-	**84:** P59482w
1-[(ethenyloxy)methyl]-	**84:** P59482w
1-[3-(ethenyloxy)propyl]-	**84:** P59482w
1-[2-(*N*-ethylanilino)ethyl]-	**69:** P28576r
1-[2-[*N*-ethyl-(3-methylanilino)]ethyl]-	**69:** P28576r
1-[5-(2-formylfuranyl)]-	**78:** 43170h
1-(4-formylphenyl)-	**78:** P124250m
1-[2-(3-hydroxy-2-phenazineacetamido)ethyl]-, 5,10-dioxide	**72:** P55499m
1-[(1-hydroxyphthalid-1-yl)phenylmethyl]-	**73:** 56037s
1-[(1-hydroxyphthalid-1-yl)phenylmethyl]-, acetate ester	**73:** 56037s
1-(6-imidazo[1,2-*b*]pyridazinyl)-	**79:** 126424r
1-(2-iodo-4-nitrophenyl)-	**60:** 4132g
1-(4-iodo-2-nitrophenyl)-	**60:** 4132g
1-(4-isocyanatophenyl)-	**72:** P100715s
1-[4-(isopropylamino)-3-oxo-2-phenyl-2*H*-pyridazin-5-yl]-	**77:** 88450v
1-(5-isoxazolyl)-	**26:** 1606[3]
1-[(4-methoxyphenyl)methyl]-	**73:** 3865d
1-methyl-	**50:** 9392f **55:** 17626g
1-[2-(1-methylbenzimidazolyl)]-	**74:** 53669j
1-(1-methyl-1-nitroethyl)-	**52:** 7292g
1-(9-methyl-4-oxo-4*H*-pyrido[1,2-*a*]pyrimidin-2-yl)-	**74:** 53669j

TABLE 1 (*Continued*)

Compound	Reference
1.1. 1-Alkyl- or Aryl-Substituted 1*H*-1,2,3-Triazoles (*Continued*)	
1-(2-methylphenyl)-	**19:** 476[3]
1-(4-methylphenyl)-	**69:** 10402w
1-[[4-(2-methylphenyl)amino]-3-oxo-2-phenyl-2*H*-pyridazin-5-yl]-	**77:** 88450v
1-[6-(3-methylpyridazinyl)]-	**74:** 53669j
1-(3-methyl-2,4,6-trinitrophenyl)-	**60:** 4132g
1-[4-(4-morpholinyl)-3-oxo-2-phenyl-2*H*-pyridazin-5-yl]-	**77:** 88450v
1-(2-nitrophenyl)-	**67:** 64311t **68:** 48845k
1-(3-nitrophenyl)-	**67:** P100995e
1-(4-nitrophenyl)-	**69:** 10402w **67:** 64365p **68:** 48845k
1-[[4-(2-nitrophenyl)amino]-3-oxo-2-phenyl-2*H*-pyridazin-5-yl]-	**77:** 88450v
1-(3-oxo-1-butenyl)-	**70:** 72424j
1-[2-(4-oxo-9-methyl-4*H*-pyrido[1,2-*a*]pyrimidinyl]-	**74:** 53669j
1-[3-oxo-2-phenyl-4-(1-pyrrolidinyl)-2*H*-pyridazin-5-yl]-	**77:** 88450v
1-[3-oxo-2-phenyl-4-(phenylamino)-2*H*-pyridazin-5-yl]-	**77:** 88450v
1-(4-oxo-4*H*-pyrido[1,2-*a*]pyrimidin-2-yl)-	**85:** 123849w
1-[(2,3,4,5,6-pentachlorophenyl)methyl]-	**73:** 3865d
1-phenyl-	**58:** 2447a **69:** 10402w
1-[(1-phenyl)-3-oxobutyl]-	**49:** 13227f
1-(phenylthioxomethyl)-	**79:** 78696g
1-[2-(phthalimido)ethyl]-	**64:** 2082g
1-(phthalimidomethyl)-	**73:** 3865d
-1-propanoic acid	**49:** 13227f
-2-propanoic acid	**24:** 3215[2]
1-(2-propenyl)-	**64:** 2082g
1-propyl-	**64:** 2082g
1-(2-pyrimidinyl)-	**74:** 53669j
1-β-D-ribofuranosyl-	**78:** 84715h
1-(succinimidomethyl)-	**73:** 3865d
1-(2,3,5-tri-*O*-acetyl-β-D-ribofuranosyl)-	**78:** 84715h
1-(2,4,6-trinitrophenyl)-	**68:** 48845k **74:** 99950x
1-(trimethylsilyl)-	**79:** 78696g
1.2. 2-Alkyl- or Aryl-Substituted 2*H*-1,2,3-Triazoles	
2-(4-acetamidophenyl)-	**43:** 2619i
2-[4-(4-acetamidobenzenesulfonamido)phenyl]-	**43:** 2619i
2-(2-aminoethyl)-	**64:** 2082g
2-(4-aminophenyl)-	**43:** 2619i **81:** 62940n
2-[4-(4-aminobenzenesulfonamido)phenyl]-	**43:** 2619i
2-(3-aminopropyl)-	**64:** 2082g
2-(4-arsonophenyl)-	**43:** 2619i
2-(4-benzamidophenyl)-	**43:** 2619i
2-(4-biphenylyl)-	**58:** 6819f
2-(4-bromophenyl)-	**21:** 2690[5] **59:** 10053c
2-(3-carboxy-4-hydroxyphenyl)-	**81:** P63478e
2-(3-carboxyphenyl)-	**55:** 27070e

TABLE 1 (*Continued*)

Compound	Reference
1.2 2-Alkyl- or Aryl-Substituted 2*H*-1,2,3-Triazoles (*Continued*)	
2-(4-carboxyphenyl)-	**21:** 2690[5] **55:** 3562f
2-(5-chloro-2-hydroxyphenyl)-	**71:** P124445j
2-(2-chlorophenyl)-	**84:** 135554p
2-(3-chlorophenyl)-	**61:** 5738c
2-(4-chlorophenyl)-	**61:** 5738c
2-[(4-chlorosulfo)phenyl]-	**43:** 2619i
2-(2-cyanoethyl)-	**64:** 2082g
2-[1-(2,6-dichlorophenyl)-2-nitroethyl]-	**81:** P91539u
-2-ethanoic acid	**64:** 2082g
-2-ethanoic acid, ethyl ester	**64:** 2082g
-2-ethanoic acid, methyl ester	**24:** 3215[2]
2-[4-(2-hydroxynaphth-1-yl)azo]phenyl-	**43:** 2619j
2-(2-hydroxyphenyl)-	**71:** P124445j
2-[(4-methoxycarbonyl)phenyl]-	**21:** 2690[5]
2-(2-methoxyphenyl)-	**71:** P124445j
2-methyl-	**50:** 9392f **55:** 17626g **68:** 48845k
2-(3-methylphenyl)-	**55:** 27070e
2-(4-methylphenyl)-	**55:** 3562f
2-(2-nitrophenyl)-	**67:** 64310s
2-(4-nitrophenyl)-	**81:** 62940n
2-phenyl-	**43:** 2619i **84:** 135554p
2-[4-(*N*-phenylsulfamoyl)phenyl]-	**43:** 2619i
2-(phenylthioxomethyl)-	**79:** 78696g
2-[2-(*N*-phthalimido)ethyl]-	**64:** 2082g
-2-propanoic acid	**64:** 2082g
-2-(2-propenyl)-	**64:** 2082g
2-propyl-	**64:** 2082g
2-β-D-ribofuranosyl-	**78:** 84715h
2-(4-sulfamoylbenzene)-	**43:** 2619i
4-(tetramethylphosphindiamidoylidyne)-2-phenyl-	**82:** P156328s
2-(2,3,5-tri-*O*-acetyl-β-D-ribofuranosyl)-	**78:** 84715h
2-(trimethylsilyl)-	**68:** 13061z
2-(2,4,5-trinitrophenyl)-	**74:** 99950x
1.3. 4-Alkyl- or Aryl-Substituted *v*-Triazoles	
-4-alanine	**43:** 6621d
-4-alanine, α-benzamido-	**43:** 6621d
-4-alanine, α-benzamido- (azlactone)	**43:** 6621d
-4-alanine, *N*-benzoyl	**43:** 6621d
4-(2-acetamidoethyl)-	**43:** 6621d
4-(2-aminoethyl)-	**43:** 6621d
4-benzyl-	**77:** 126517g
4-(1,1′-biphenyl-4-yl)-	**76:** 85756t
4-[bis[4-(*N*,*N*-dimethylamino)phenyl]methyl]-	**56:** 12876h
4-(3-bromophenyl)-	**76:** 85756t
4-(4-bromophenyl)-	**75:** 140767d **76:** 85756t **79:** 66254y **85:** 177330p

TABLE 1 (*Continued*)

Compound	Reference
1.3. 4-Alkyl- or Aryl-Substituted *v*-Triazoles (*Continued*)	
4-butyl-	**49:** 3949b
4-[4-(*tert*-butyl)phenyl]-	**76:** 85756t
4-butyl-2-(trimethylsilyl)-	**65:** 15414d
4-(1-butynyl)-	**65:** 15414d
4-(1-butynyl)-2-(trimethylsilyl)-	**65:** 15414d
4-(chlorodiphenylmethyl)-, monohydrochloride	**80:** 133389g
4-(3-chlorophenyl)-	**76:** 85756t
4-(4-chlorophenyl)-	**75:** 140767d **79:** 66254y **80:** 108452q **84:** 164688r
5-(4-chlorophenyl)-4-(triphenylphosphonio)-, bromide	**80:** 108452q
5-(4-chlorophenyl)-4-(triphenylphosphonio)-, hydroxide, inner salt	**79:** 66254y
5-(4-chlorophenyl)-4-(triphenylphosphoranylidene)-	**79:** 66254y
4-(2-cyanoethenyl)-	**85:** 177330p
4-(2-cyanoethenyl)-, (Z)-	**74:** 64224m
4-(2-cyanoethyl)-	**43:** 6621d
4-(4-cyanophenyl)-	**83:** 95918f
4-(4-cyclohexylphenyl)-	**78:** 111412r
4-decyl-	**49:** 3949b
4-(3,5-dichlorophenyl)-	**83:** P10089p
4-(diethoxyphosphinyl)-5-phenyl-	**80:** 108452q
4-[4-(*N*,*N*-dimethylamino)phenyl]-	**84:** 164688r **85:** 177330p
4-(2,4-dinitrophenyl)-	**74:** 99950x
-4-ethanamine	**81:** P154677k
4-ethyl-	**49:** 3949b
4-(4-ethylphenyl)-	**76:** 85756t
4-(4-fluorophenyl)-	**76:** 85756t
4-heptyl-	**49:** 3949b
4-hexyl-	**49:** 3949b
4-isobutyl-	**73:** P27318d
4-isopentyl-	**49:** 3949b
4-isopropyl-	**72:** 110456x
4-[(2-isopropylamino)ethyl]-	**43:** 6621d
-4-methanol, α,α-bis[(4-*N*,*N*-dimethylamino)phenyl]-	**56:** 12876h
-4-methanol, α,α-diphenyl-	**80:** 133389g
4-(4-methoxyphenyl)-	**76:** 85756t **77:** 100955y **84:** 164688r
4-methyl-	**50:** 9392f **68:** 13061z
4-methyl-2-(trimethylsilyl)-	**68:** 13061z
4-(3-methylphenyl)-	**83:** P10089p
4-(4-methylphenyl)-	**80:** 108452q **84:** 164688r
4-(2-naphthyl)-	**68:** 59506b **76:** 85756t
4-[2-(5-nitro-2-furanyl)ethenyl]-	**77:** 101470y
4-(3-nitrophenyl)-	**83:** 95918f
4-(4-nitrophenyl)-	**74:** 99950x **80:** 108452q **84:** 164688r
4-nonyl-	**49:** 3949b
4-octyl-	**49:** 3949b

TABLE 1 (*Continued*)

Compound	Reference
1.3. 4-Alkyl- or Aryl-Substituted *v*-Triazoles (*Continued*)	
4-pentyl-	**49:** 3949b
4-phenyl-	**60:** 5537c **79:** 66254y **80:** 108452q **84:** 164688r
4-phenyl-5*d*-	**76:** 85761r
4-phenyl-2,5-bis(trimethylsilyl)-	**60:** 5537c
4-phenyl-2-(tributylstannyl)-	**78:** 29956n
4-phenyl-1-(trimethylsilyl)-(?)	**60:** 5537b **68:** 13061z(?)
4-phenyl-5-(trimethylsilyl)-	**60:** 5537c
4-phenyl-2-(trimethylstannyl)-	**78:** 29956n
5-phenyl-4-(triphenylphosphoranylidene)-	**79:** 66254y
5-phenyl-4-(triphenylphosphonio)-, bromide	**80:** 108452q
5-phenyl-4-(triphenylphosphonio)ylide-	**79:** 66254y
4-propyl-	**49:** 3949b
4-propyl-2-(trimethylsilyl)-	**65:** 15414d
-4-pyruvic acid, oxime	**43:** 6621d
4-(rhodaninylmethylene)-	**43:** 6621d
4-(1*H*-tetrazol-5-yl)-	**71:** 13067t
4-α-thiopyruvic acid	**78:** 111412r
4-(trifluoromethyl)-	**78:** 111412r
4-(trifluoromethyl)-2-(trimethylsilyl)-	**78:** 111412r
4-(triphenylphosphonio)-, hydroxide, inner salt	**84:** 17493v

REFERENCES

19: 476[3]	F. D. Chattaway, F. L. Garton, and G. D. Parkes, *J. Chem. Soc.*, **125,** 1980 (1924).
21: 2690[5]	K. Fries, *Justus Liebigs Ann. Chem.*, **454,** 121 (1927).
24: 3215[2]	T. Curtius and W. Klavehn, *J. Prakt. Chem.*, **125,** 498 (1930).
24: 3232[1]	T. Curtius and K. Raschig, *J. Prakt. Chem.*, **125,** 466 (1930).
26: 1606[3]	A. Quilico, *Gazz. Chim. Ital.*, **61,** 759 (1931).
43: 2619i	J. L. Riebsomer, *J. Org. Chem.*, **13,** 815 (1948).
43: 6621d	J. C. Sheehan and C. A. Robinson, *J. Amer. Chem. Soc.*, **71,** 1436 (1949).
44: 3979h	F. R. Benson and W. L. Savell, *Chem. Revs.*, **46,** 1 (1950).
44: 4430g	K. Alder and W. Trimborn, *Justus Liebigs Ann. Chem.*, **566,** 58 (1950).
49: 3948e	E. Mugnaini and P. Grünanger, *Atti Accad. Naz. Lincei., Rend., Cl. Sci. Fis. Mat. Nat.*, **14,** 275 (1953).
49: 3949b	L. W. Hartzel and F. R. Benson, *J. Amer. Chem. Soc.*, **76,** 667 (1954).
49: 10274b	R. H. Wiley and J. Moffat, *J. Amer. Chem. Soc.*, **77,** 1703 (1955).
49: 13227f	R. H. Wiley, N. R. Smith, D. M. Johnson, and J. Moffat, *J. Amer. Chem. Soc.*, **76,** 4933 (1954).
50: 8612h	J. J. Licari, L. W. Hartzel, G. Dougherty, and F. R. Benson, *J. Amer. Chem. Soc.*, **77,** 5386 (1955).
50: 9392f	R. Hüttel and G. Welzel, *Justus Liebigs Ann. Chem.*, **593,** 207 (1955).
50: 15517e	R. H. Wiley, K. F. Hussung, and J. Moffat, *J. Org. Chem.*, **21,** 190 (1956).

52: 7292g S. Maffei and G. F. Bettinetti, *Ann. Chim.* (*Rome*), **47,** 1286 (1957).
53: 344f H. El Khadem and Z. M. El-Shafei, *J. Chem. Soc.*, 3117 (1958).
55: 3562f H. El Khadem, Z. M. El-Shafei, and Y. S. Mohammed, *J. Chem. Soc.*, 3993 (1960).
55: 17626g C. Pedersen, *Acta Chem. Scand.*, **13,** 888 (1959).
55: 27070e H. El Khadem, Z. M. El-Shafei, and M. H. Meshreki, *J. Chem. Soc.*, 2957 (1961).
56: 12876h R. Hüttel, W. Schwarz, J. Miller, and F. Wunsch, *Chem. Ber.*, **95,** 222 (1962).
58: 2447a T. Kindt-Larsen and C. Pedersen, *Acta Chem. Scand.*, **16,** 1800 (1962).
58: 6819f B. M. Lynch and T-L. Chan, *Can. J. Chem.*, **41,** 274 (1963).
59: 10053c B. M. Lynch, *Can. J. Chem.*, **41,** 2380 (1963).
59: P15420d R. Raue and H. Gold, Belgian Patent, 621,482 (1962).
60: 4132g D. S. Deorha, V. K. Mahesh, and S. K. Mukerji, *J. Indian Chem. Soc.*, **40,** 901 (1963).
60: 5537b L. Birkofer, A. Ritter, and P. Richter, *Chem. Ber.*, **96,** 2750 (1963).
60: 5537c L. Birkofer, A. Ritter, and H. Uhlenbrauck, *Chem. Ber.*, **96,** 3280 (1963).
61: 5738c H. El Khadem, M. H. Meshreki, and G. H. Labib, *J. Chem. Soc.*, 2306 (1964).
61: 14674f J. H. Boyer, R. Moriarty, B. de B. Darwen, and P. A. S. Smith, *Chem. Eng. News*, **42,** (31), 6 (1964).
62: 16248a R. Huisgen, L. Möbius, and G. Szeimies, *Chem. Ber.*, **98,** 1138 (1965).
63: 4324c L. Birkofer and A. Ritter, *Angew. Chem. Int. Ed.*, **4,** 417 (1965).
64: 2082g H. Gold, *Justus Liebigs Ann. Chem.*, **688,** 205 (1965).
64: P11215h Farbenfabriken Bayer A.-G., Dutch Patent, 6,505,524 (1965).
64: 5075e R. Huisgen, R. Knorr, L. Möbius, and G. Szeimies, *Chem. Ber.*, **98,** 4014 (1965).
65: 15414d L. Birkofer and P. Wegner, *Chem. Ber.*, **99,** 2512 (1966).
67: 64310s R. A. Carboni, J. C. Kauer, W. R. Hatchard, and R. J. Harder, *J. Amer. Chem. Soc.*, **89,** 2626 (1967).
67: 64311t J. C. Kauer and R. A. Carboni, *J. Amer. Chem. Soc.*, **89,** 2633 (1967).
67: 64365p P. Ferruti, D. Pocar, and G. Bianchetti, *Gazz. Chim. Ital.*, **97,** 109 (1967).
67: P100995e Farbenfabriken Bayer A.-G., Dutch Patent, 6,610,038 (1967).
68: 13061z L. Birkofer and P. Wegner, *Chem. Ber.*, **100,** 3485 (1967).
68: 48845k J. Elguero, E. Gonzalez, and R. Jacquier, *Bull. Soc. Chim. Fr.*, 2998 (1967).
68: 59506b S. Hauptmann, H. Wilde, and K. Moser, *Tetrahedron Lett.*, 3295 (1967).
68: P106086b Hickson and Welch, Ltd., Dutch Patent, 6,613,363 (1967).
69: 10402w H. El Khadem, H. A. R. Mansour, and M. H. Meshreki, *J. Chem. Soc.*, (*C*), 1329 (1968).
69: P28576r G. Wolfrum and H. Gold, British Patent, 1,111,236 (1968).
69: P87008e H. J. Kabbe and E. Eiter, German Patent, 1,271,116 (1968).
70: 11634f S. Hoffmann, S. Kreissl, and E. Mühle, *Z. Chem.*, **8,** 381 (1968).
70: 72424j S. Hoffmann, *Z. Chem.*, **9,** 25 (1969).
70: 87698a A. Catino, *Ann. Chim.* (*Rome*), **58,** 1507 (1968).
71: 2032b S. Petersen, W. Gauss, H. Kiehne, and L. Jühling, *Z. Krebsforsch.*, **72,** 162 (1969).
71: 13067t C. Arnold, Jr. and D. N. Thatacher, *J. Org. Chem.*, **34,** 1141 (1969).
71: 70272e S. McLean and D. M. Findlay, *Tetrahedron Lett.*, 2219 (1969).
71: P124445j Geigy, A.-G., French Patent, 1,559,131 (1969).
72: P79056r J. C. Kauer, U.S. Patent, 3,489,761 (1970).

72: P91491m	Farbenfabriken Bayer A.-G., French Patent, 2,001,069 (1969).
72: P100715s	S. Petersen, H. Striegler, H. Gold, and A. Haberkorn, South African Patent, 69 01,244 (1969).
72: 110456x	C. G. Overberger and P. S. Yuen, *J. Amer. Chem. Soc.*, **92,** 1667 (1970).
73: 3865d	A. J. Hubert, *Bull. Soc. Chim. Belges*, **79,** 195 (1970).
73: 25364s	F. P. Woerner and H. Reimlinger, *Chem. Ber.*, **103,** 1908 (1970).
73: P27318d	S. H. Patinkin, U.S. Patent, 3,511,623 (1970).
73: 56037s	N. Bruvele and E. Gudriniece, *Latv. PSR Zinat. Akad. Vestis, Kim. Ser.*, 198 (1970).
74: 53669j	A. J. Hubert and H. Reimlinger, *Chem. Ber.*, **103,** 3811 (1970).
74: 64224m	N. S. Zefirov and N. K. Chapovskaya, *Zh. Org. Khim.* (English Translation), **6,** 2605 (1970).
74: P75496c	K. H. Buechel, W. Draber, R. R. Schmidt, E. Regel, and L. Eue, German Patent, 1,935,292 (1971).
74: 99950x	P. N. Neuman, *J. Heterocycl. Chem.*, **8,** 51 (1971).
74: P112048f	G. Jaeger and K. H. Buechel, German Patent, 1,940,626 (1971).
74: 125600e	E. Gudriniece and V. Urbans, *Latv. PSR Zinat. Akad. Vestis, Kim. Ser.*, 82 (1971).
74: P125697s	W. Draber and K. H. Buechel, German Patent 1,940,627 (1971).
74: P125698t	K. H. Buechel and W. Draber, German Patent, 1,940,628 (1971).
75: P36045f	P. Scheiner, U.S. Patent, 3,579,531 (1971).
75: P110997m	J. Schroeder and C. W. Schellhammer, German Patent, 1,955,068 (1971).
75: 140767d	N. S. Zefirov, N. K. Chapovskaya, and V. V. Kolesnikov, *Chem. Commun.*, 1001 (1971).
76: 72454t	M. Begtrup, *Acta Chem. Scand.*, **25,** 3500 (1971).
76: 85756t	S. Hauptmann, H. Wilde, and K. Moser, *J. Prakt. Chem.*, **313,** 882 (1971).
76: 85761r	R. Selvarajan and J. H. Boyer, *J. Heterocycl. Chem.*, **9,** 87 (1972).
76: 126875b	D. Pocar, R. Stradi, and L. M. Rossi, *J. Chem. Soc., Perkin Trans.*, **I,** 769 (1972).
77: P7315e	A. Dorlars and H. Gold, German Patent, 2,040,189 (1972).
77: 88450v	V. Urbans, E. Gudriniece, and T. V. Kotenko, *Latv. PSR Zinat. Akad. Vestis, Kim. Ser.*, 353 (1972).
77: 100955y	N. S. Zefirov, N. K. Chapovskaya, U. R. Apsalon, and V. V. Kolesnikov, *Zh. Org. Khim.* (English Translation), **8,** 1355 (1972).
77: 101470y	I. Hirao and Y. Kato, *Bull. Chem. Soc. Jap.*, **45,** 2055 (1972).
77: 126516f	H. Cardoen, S. Toppet, G. Smets, and G. L'abbé, *J. Heterocycl. Chem.*, **9,** 971 (1972).
78: 29956n	S. Kogima, T. Itano, N. Mihara, K. Sisido, and T. Isida, *J. Organometal. Chem.*, **44,** 117 (1972).
78: 43170h	F. Lieb and K. Eiter, *Justus Liebigs Ann. Chem.*, **761,** 130 (1972).
78: 72035y	V. Urbans, E. Gudriniece, and L. A. Khoroshanskaya, *Latv. PSR Zinat. Akad. Vestis, Kim. Ser.*, 712 (1972).
78: 84715h	F. A. Lehmkuhl, J. T. Witkowski, and R. K. Robins, *J. Heterocycl. Chem.*, **9,** 1195 (1972).
78: 111412r	J. M. Crossman, R. N. Haszeldine, and A. E. Tipping, *J. Chem. Soc., Dalton Trans.*, 483 (1973).
78: P124250m	K. Grohe, H. Kaspers, and H. Scheinpflug, German Patent, 2,140,737 (1973).
78: 135769p	D. M. Findlay, M. L. Roy, and S. McLean, *Can. J. Chem.*, **50,** 3186 (1972).
79: 66254y	Y. Tanaka and S. I. Miller, *J. Org. Chem.*, **38,** 2708 (1973).
79: 78696g	W. Walter and M. Rodke, *Justus Liebigs Ann. Chem.*, 636 (1973).

79: P78815v E. Aufderhaar and A. Meyer, German Patent, 2,263,878 (1973).

79: 126424r S. Polanc, B. Stanovnik, and M. Tišler, *J. Heterocycl. Chem.*, **10,** 565 (1973).

80: 82834p O. V. Voishcheva, V. D. Galkin, B. I. Mikhant'ev, and G. V. Shatalov, *Izv. Vyssh. Ucheb. Zaved., Khim. Khim. Tekhnol.*, **16,** 1913 (1973).

80: 108452q Y. Tanaka, S. R. Velen, and S. I. Miller, *Tetrahedron*, **29,** 3271 (1973).

80: 133389g E. M. Burgess and J. P. Sanchez, *J. Org. Chem.*, **39,** 940 (1974).

81: 13832v B. I. Mikhant'ev, G. V. Shatalov, and V. D. Galkin, *Monomery Vysokomol. Soedin.*, 82 (1973).

81: 62940n P. Bouchet, C. Coquelet, and J. Elguero, *J. Chem. Soc., Perkin Trans.*, **2,** 449 (1974).

81: P63478e L. H. Sarett and W. V. Ruyle, South African Patent, 72 01,129 (1973).

81: P63684u G. G. Skovortsova, E. S. Domnina, D. G. Kim, and L. P. Makhno, Russian Patent, 420,629 (1974).

81: P91539u U. Petersen, S. Petersen, H. Scheinpflug, and B. Hamburger, German Patent, 2,260,704 (1974).

81: P154677k K. Sato, Japanese Patent, 74 02,267 (1974).

82: 16641p D. N. Reinhoudt and C. G. Kouwenhoven, *Tetrahedron Lett.*, 2163 (1974).

82: 56932r G. O. Sørensen, L. Nygaard, and M. Begtrup, *J. Chem. Soc., Chem. Commun.*, 605 (1974).

83: 95918f H. Wilde, S. Hauptmann, and E. Kleinpeter, *Z. Chem.*, **15,** 155 (1975).

83: 97148d B. I. Mikhant'ev, G. V. Shatalov, and V. D. Galkin, *Monomery Vysokomol. Soedin.*, 6 (1974).

83: P99223y H. Aebli, F. Fleck, and H. Schmid, Swiss Patent, 562,228 (1975).

83: P155710u W. Fujimatsu, S. Sato, T. Kojima, S. Itoh, and T. Endo, German Patent, 2,433,812 (1975).

83: 206290s K. H. Buechel, H. Gold, P. E. Frohberger, and H. Kaspers, German Patent, 2,407,305 (1975).

84: 17493v C. Ivancsics and E. Zbiral, *Monatsh. Chem.*, **106,** 839 (1975).

84: P59482w V. S. Sukhinin and V. G. Mikhailov, Russian Patent 487,074 (1975).

84: P121664n H. J. Kabbe and H. Otten, German Patent, 2,428,673 (1976).

84: 164688r N. S. Zefirov, N. K. Chapovskaya, and U. R. Apsalon, *Zh. Org. Khim.*, **12,** 148 (1976).

85: 123849w S. Polanc, B. Stanovnik, and M. Tišler, *J. Org. Chem.*, **41,** 3152 (1976).

85: 177330p N. S. Zefirov, N. K. Chapovskaya, and S. S. Trach, *Tezisy Dokl.-Simp. Khim. Tekhnol. Geterotsikl. Soedin. Goryuch. Iskop., 2nd* ed., Vol. 36 (1973).

CHAPTER 2

Alkyl- or Aryl-Disubstituted 1,2,3-Triazoles

2.1. 1,4-ALKYL- OR ARYL-DISUBSTITUTED 1*H*-1,2,3-TRIAZOLES

The traditional synthetic method—addition of azides to terminal acetylenes—has been used extensively and often gives a good yield of 1,4-disubstituted product. Moulin[1] carried out a large number of such reactions (Eq. 1) and found a high degree of regiospecificity (except with (**2.1-1**, R = R′ = Ph where the yield of each isomer is poor). This particular problem was overcome by Munk and Kim[2] using an enamine (**2.1-2**) (Eq. 2). They also prepared the 5-methyl analog of **2.1-3** with an overall yield of 45%. Experiments with the opposite isomer of **2.1-2** demonstrated the same highly regiospecific product formation[2]. Kirmse and Horner[3] greatly improved the yield of this reaction, but they also found a preference for 1,5-product (**2.1-3**; 1,4 = 42.5% and 1,5 = 51.5%).

$$RN_3 + R'C{\equiv}CH \longrightarrow \text{(1,4-disubstituted triazole, 1-R, 4-R')} \qquad (1)$$

2.1-1

R = $PhCH_2$, Ph, 4-MeOPh, 4-NO_2Ph
R′ = R''_2COH, Ph

$$\text{2.1-2 (piperidino-CH=CHPh)} + PhN_3 \longrightarrow \text{triazoline (91\%)} \xrightarrow[MeOH]{K^+OH^-} \text{2.1-3 (95\%)} \qquad (2)$$

2.1-2 91% 95% **2.1-3**

Most studies[4–7] of azide addition have shown a pattern of approximately equal yields of 1,4- and 1,5-disubstitution; however, there are a few intriguing exceptions. For example, 2-nitroethylazide produced a 3:1 excess of 1,4-addition (**2.1-4** over **2.1-5**) (Eq. 3).[7] Tišler and his collaborators[8] found an even greater preference for 1,4-addition with a series of heteroaromatic azides (Eq. 4). An especially interesting case comes from the much-studied field of 1,2,3-triazoles involving sugar residues where García-López[4] found that with 2-*O*-trichloroacetyl-tri-*O*-acetyl-α-D-glucopyranosylazide, the trichloroacetyl group is lost and only 1,4-product is formed. In another study[9] members of this laboratory reported that azidomethyl ethers and sulfides produce only fair yields of mixtures except in three cases (Eq. 5).

$$O_2NCH_2CH_2N_3 + PhC{\equiv}CH \longrightarrow \text{2.1-4 (3 pts.)} + \text{2.1-5 (1 pt.)} \quad (3)$$

Ph; N N N; $CH_2CH_2NO_2$; 3 pts.; **2.1-4**

Ph; N N N; $CH_2CH_2NO_2$; 1 pt.; **2.1-5**

$$\text{Ar-}N_3 + PhC{\equiv}CH \longrightarrow \text{(5 pts.)} + \text{(1 pt.)} \quad (4)$$

N N N N N; N_3; Ar (and related heterocycles); Ph; N N N; Ar; 5 pts.; Ph; N N N; Ar; 1 pt.

$$R{-}X{-}CH_2N_3 + PhC{\equiv}CH \longrightarrow \text{(triazole)}{-}CH_2XR \quad (5)$$

R = Ph, Me; X = O, S; O

A variety of other methods has been applied to the synthesis of 1,4-disubstituted 1,2,3-triazoles; for example, the method developed by Sheehan[10] has been modified by Hüttel and his collaborators (Eq. 6).[11] This modification allows the conversion, in high yield, of the formyl group (**2.1-6**) into various functionally substituted side chains (Eq. 7).

CHO O H N N N N Ph + S S ⟶ CH= O H N S S N N N Ph

2.1-6 82%

⟶ various derivatives (6)

CHO N N N R + $PhCONHCH_2CO_2H$ $\xrightarrow[Ac_2O]{Na^+ \ ^-OAc}$ CH= O O N Ph N N N R (7)

R = Me, Ph Ph = 75%

The anion of *N*-nitrosomethylethylamine (**2.1-7**) reacts with nitriles to form products of structure **2.1-8** in fair-to-good yield (Eq. 8).[12] *N*-methyl-1,2,3-triazoles can be prepared by alkylation although both 1- and 2-methyl products are formed (Eq. 9).[13] The reaction of benzonitrile with diazomethane in the presence of trialkyl-aluminum catalysts provides a modest yield of **2.1-9**.[13]

$RCN + Me\bar{C}HN(NO)Me$ ⟶ R N N N Me (8)

2.1-7 **2.1-8**

R = Ph, 4-MePh, β-$C_{10}H_7$ 40–70%

Ph N N N H + Me_2SO_4 ⟶ Ph N N N Me + Ph N N N Me (9)

2.1-9

62% 38%

The acid catalyzed condensation of *N,N*-dimethylaniline with the formyl group (Eq. 10) produces an excellent yield of the methanol derivative **2.1-10**.[14] The generality of this reaction should be examined.

CHO NMe$_2$ OH

Me–N N N + (benzene) $\xrightarrow{H^+}$ Me–N N N –C–(C$_6$H$_4$–NMe$_2$)$_2$ (10)

2.1-10
83%

A large number of 3-propenyl derivatives (**2.1-11**) may be obtained in high yield from the reaction of 3-acetamidopyridine with cyanogen bromide and anilines followed by diazotization (Eq. 11).[15] Similar results have been reported in the diazotization of *N*-alkylpyridinium salts (Eq. 12).[16]

NHAc (pyridine, N) + $ArNH_2 + BrCN \longrightarrow$ NHAc, NHPh, HN^+–Ar, Br^-

$\xrightarrow[HBr]{Na^+\ ^-NO_2}$ OHC–CH=CH–(triazole N N N)–Ar (11)

2.1-11

Ar = Ph, 4-MePh, 4-Et_2NPh, 4-NO_2Ph, etc. Yields = 60–80% (lower for 4-ROPh)

(pyridinium N^+–R, Br^-) $\xrightarrow[HBr]{Na^+\ ^-NO_2}$ OHCCH=CH–(triazole N N N)–R (12)

R = Me, Et, $CH_2CH{=}CH_2$, $PhCH_2$

REFERENCES

1. **46:** 8651g
2. **61:** 2926b
3. **53:** 5253d
4. **72:** 3718n
5. **79:** 66254y
6. **83:** 79201n
7. **83:** 193181e
8. **83:** 97213w
9. **72:** 43576w
10. **45:** 9037f
11. **49:** 6241h
12. **78:** 72017u
13. **67:** 108605y
14. **56:** 12876h
15. **64:** 3522g
16. **66:** 37839u

2.2. 1,5-ALKYL- OR ARYL-DISUBSTITUTED 1*H*-1,2,3-TRIAZOLES

The 1,5-addition product is often a synthetically interesting by-product of reactions producing 1,4-disubstituted 1,2,3-triazoles. An example of this, which also illustrates the importance of electronic effects in these reactions, has been provided by Tsypin and his collaborators (Eq. 13).[17] These results should be compared with Equation 3 in Section 2.1.

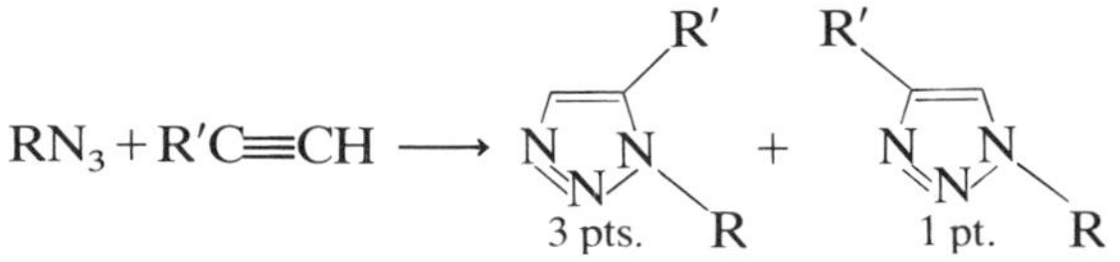

$$RN_3 + R'C{\equiv}CH \longrightarrow \text{(1-R-4-R'-1,2,3-triazole) 3 pts.} + \text{(1-R-5-R'-1,2,3-triazole) 1 pt.} \quad (13)$$

$R = MeCHNO_2, O_2NCH_2CH_2$ $\quad R' = Ph, HOCH$

Olsen[18] has reported that the addition of benzyl azide to acetone forms a modest yield of 1-benzyl-5-methyl-1*H*-1,2,3-triazole (**2.2-1**) along with the trisubstituted product **2.2-2** (Eq. 14). He was able to isolate the intermediate 5-hydroxy-1,2,3-triazole (**2.2-3**).

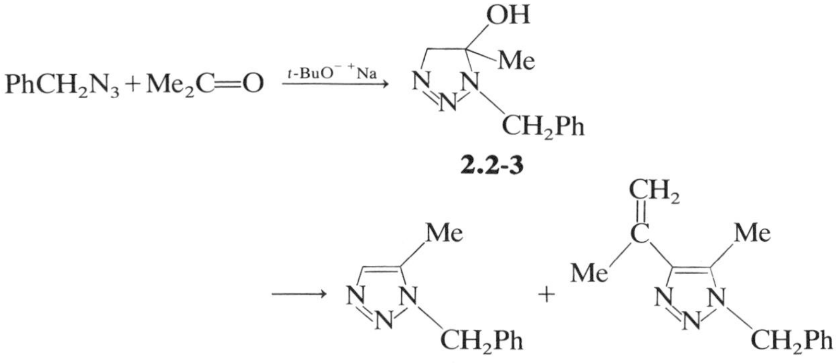

$$PhCH_2N_3 + Me_2C{=}O \xrightarrow{t\text{-BuO}^- \, {}^+Na} \text{(5-OH, 5-Me, 1-}CH_2Ph\text{ triazoline)}$$

2.2-3

$$\longrightarrow \text{(5-Me, 1-}CH_2Ph\text{ triazole)} + \text{(4-}C(Me){=}CH_2\text{, 5-Me, 1-}CH_2Ph\text{ triazole)} \quad (14)$$

2.2-1 37% **2.2-2** 40%

L'abbé and his collaborators have also demonstrated the importance of electronic effects in the synthesis of 1-vinyl-1,2,3-triazoles by adding azides to either acetylenes (Eq. 15) or active methylene compounds (Eq. 16).[19,20] In both cases iodoalkyl azides may be added, with comparable results, to produce precursors of **2.2-4** and **2.2-5**. This very productive group has pioneered the excellent general method for 1,5-disubstituted-1,2,3-triazoles involving phosphorus ylides (Eq. 17).[21] The high yields and wide range of substituents employed makes this method most attractive. Product structures were demonstrated by the Dimroth addition and decarboxylation (Eq. 16).

N_3–CH=CH–Bu-*t* + PhC≡CH ⟶ **2.2-4** (70%) + **2.2-5** (30%) (15)

N_3–CH=CH–Bu-*t* + $RCOCH_2CO_2Et$ $\xrightarrow[-CO_2]{H_2O}$ **2.2-4** (16)

R = Me, Ph; 60–84%

In a closely related study[22] the same high degree of regiospecificity was obtained although the yields varied greatly (Eq. 18). Other laboratories have contributed to the development of 1,2,3-triazole synthesis through phosphorus-containing intermediates (Eqs. 19,20).[23,24] The subsequent hydrolysis of **2.2-7** was not reported in this particular study, but it should be possible in high yield.

Ph_3P=CHCOR + ArN_3 ⟶ [Ph_3P^+, O^-, H, R, N, N, N, Ar] ⟶ R, N, N, N, Ar (17)

R = Me, Ph, 4-NO_2Ph   Ar = Ph, 4-NO_2Ph, 4-BrPh, 4-MeOPh

$N_3(R^1)C=C(R^3)R^2$ + $R^5COC(R^4)=PPh_3$ ⟶ **2.2-6** (18)

R^1	R^2	R^3	R^4	R^5	%
H	H	PhCO	H	Me	98
H	H	PhCO	H	Ph	95
PhCO	H	Me	H	Me	20
PhCO	H	Me	H	Ph	15

$PhCOCH_2PO(OEt)_2 + ArN_3 \longrightarrow$ O, $(EtO)_2P$, O^-, H, Ph, N, N, N, Ar $\longrightarrow$ Ph, N, N, N, Ar (19)

Ar = 4-ClPh 36%
4-NO_2Ph 64%

$R_2POC{\equiv}CMe + PhN_3 \longrightarrow$ O, R_2P, Me, N, N, N, Ph (20)

R = EtO, *n*-PrO, Cl, Et, Ph

2.2-7
47–61%

The use of organometallics shows promise in the preparation of 1,5-disubstituted 1,2,3-triazoles. A preliminary announcement[25] of the use of Iotsich complexes followed by two detailed studies[26,27] illustrates the rather broad range of possible application (Eqs. 21,22). The yields of **2.2-8** bearing an alkynyl group are good to excellent except with R = Me or Ph.[27] The Russian scientists engaged in these studies have reported experiments with alkynyl lithium salts[28] in which only 4-triazeno products are obtained with certain azides (Eq. 23).

R, N, N, N, Ar

2.2-8
70–90%

$RC{\equiv}CMgBr + ArN_3$ (21)

$\xrightarrow{CO_2}$ HO_2C, R, N, N, N, Ar $\xrightarrow{-CO_2}$ **2.2-8**

R = Ph, $CH_2{=}CH$, $MeCH{=}CH$, 1-cyclohexenyl
Ar = Ph, 4-BrPh, 3- and 4-NO_2Ph

$RC{\equiv}CC{\equiv}CMgBr + R'N_3 \longrightarrow$ **2.2-8** R = $C{\equiv}C$(H or Et) (22)

R = H, Et R′ = Me, Ph, 4-BrPh, 4-NO_2Ph

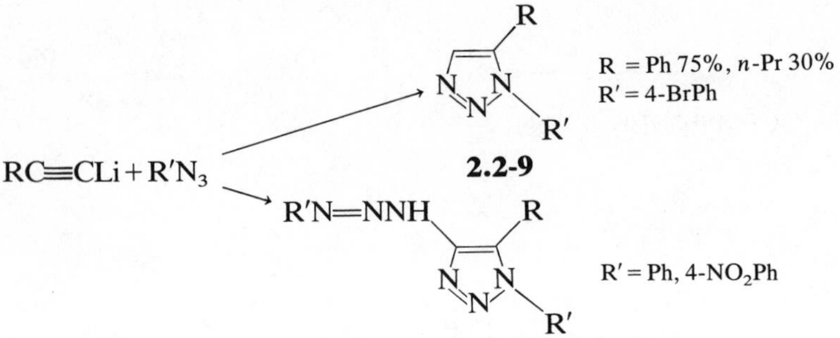

R

N N

N R'

R = Ph 75%, n-Pr 30%
R' = 4-BrPh

$RC{\equiv}CLi + R'N_3$ **2.2-9** (23)

$R'N{=}NNH$ R

N N

N R'

R' = Ph, 4-NO_2Ph

The addition of azides to enamines (Eqs. 24,25)[29,30] and enol ethers (Eq. 26)[31] followed by elimination from the relatively unstable triazolines (*e.g.*, **2.2-10**) represents an important advance. Simplicity and generally excellent yields make these methods worthy of further study.

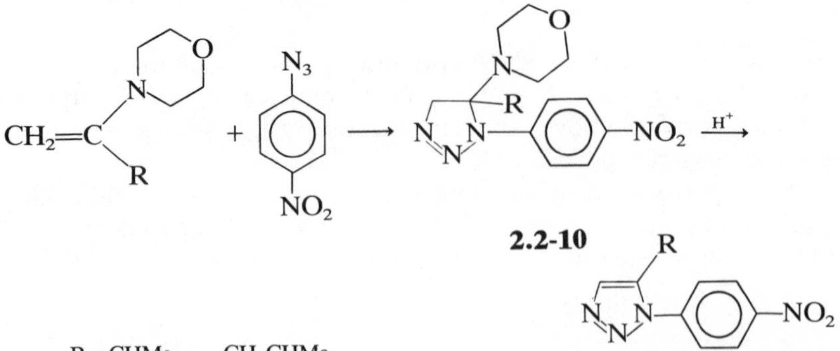

O

N N_3 O

N

R

$CH_2{=}C$ + ⟶ N N ─ ─NO_2 $\xrightarrow{H^+}$

R N

NO_2

2.2-10 R

N N ─ ─NO_2 (24)

N

R = $CHMe_2$ CH_2CHMe_2
% **2.2-11** = 70 90

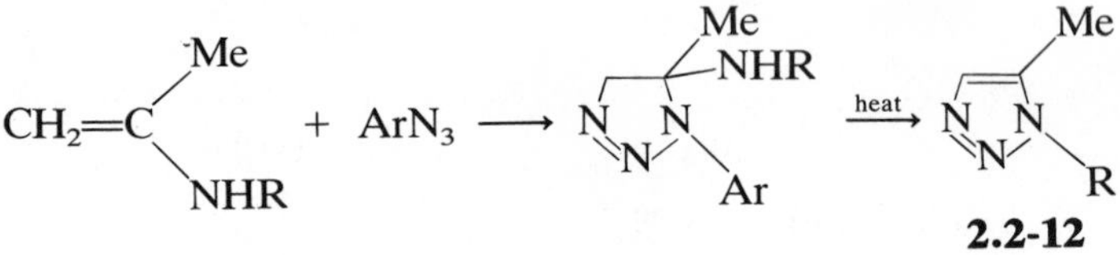

2.2-11

Me Me

Me NHR

$CH_2{=}C$ + ArN_3 ⟶ N N $\xrightarrow{heat}$ N N (25)

N N

NHR Ar R

2.2-12

R	= Me	Et	n-Pr	n-Bu	$PhCH_2$	$Ph(CH_2)_2$	c-C_6H_{11}
% **2.2-12**	= 30	56	60	85	78	43	62

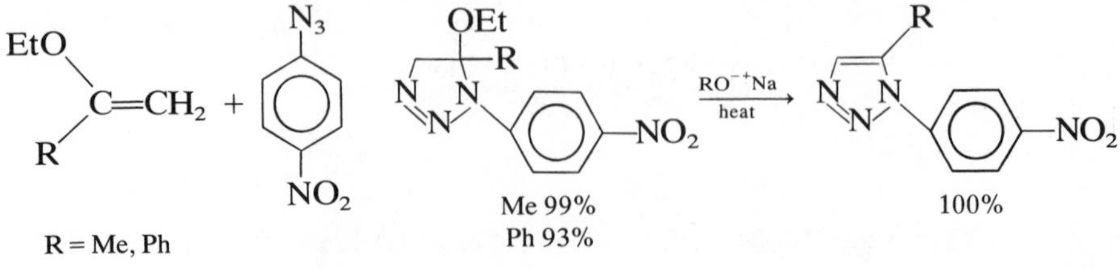

N_3 OEt R

EtO R

$C{=}CH_2$ + N N $\xrightarrow[heat]{RO^-\,^+Na}$ N N

N ─NO_2 N ─NO_2

R

NO_2 Me 99% 100%

R = Me, Ph Ph 93%

(26)

An especially promising approach to the 1,5-diaryl-1,2,3-triazoles was suggested some years ago, but apparently it has not yet been pursued.[32] The addition of phenyl azide to β-nitrostyrenes (Eq. 27) produces a poor yield of triazolines (**2.2-13**), but the subsequent reduction and elimination are essentially quantitative.

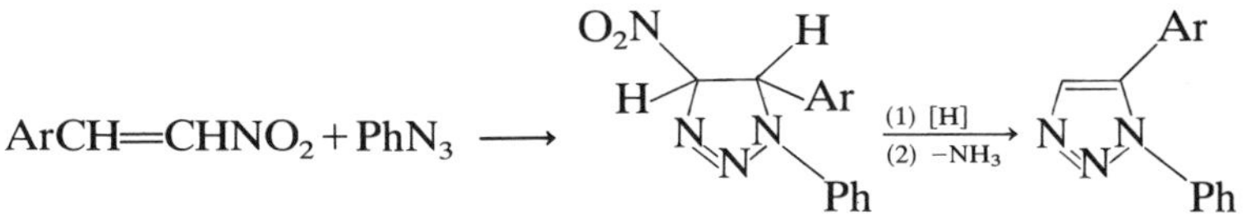

$$ArCH{=}CHNO_2 + PhN_3 \longrightarrow \textbf{2.2-13} \xrightarrow[(2)\ -NH_3]{(1)\ [H]} \quad (27)$$

Ar =	2-BrPh	4-BrPh	2-ClPh	4-ClPh	2-HOPh
% **2.2-13** =	33	60	59	31	29

Similar problems of variable yield exist in the Dimroth addition of aryl azides and ethyl acetoacetate followed by decarboxylation (Eq. 28).[33] The relationship between yields of **2.2-14** and **2.2-15** (*e.g.*, Ph = 30% : 85%; 4-NO_2Ph = 77% : 35%) is a lead that deserves to be followed.

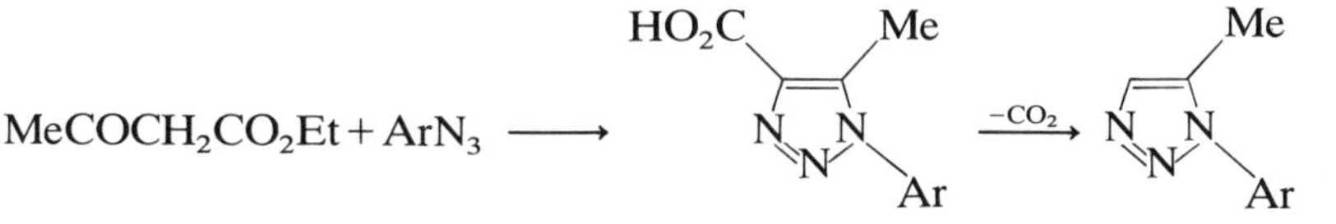

$$MeCOCH_2CO_2Et + ArN_3 \longrightarrow \textbf{2.2-14} \xrightarrow{-CO_2} \textbf{2.2-15} \quad (28)$$

Ar =	Ph	4-BrPh	4-ClPh	4-IPh	4-NO_2Ph
% **2.2-14** =	30	54	80	75	77
% **2.2-15** =	85	58	50	40	35

Pocar[34] and Crandall[35,36] and their collaborators have reported reactions (Eqs. 29 to 31) that may be either isolated examples or the basis of exciting new methods. Yields of **2.2-16** were generally much less than the example shown and **2.2-17** was reported only as the major product.

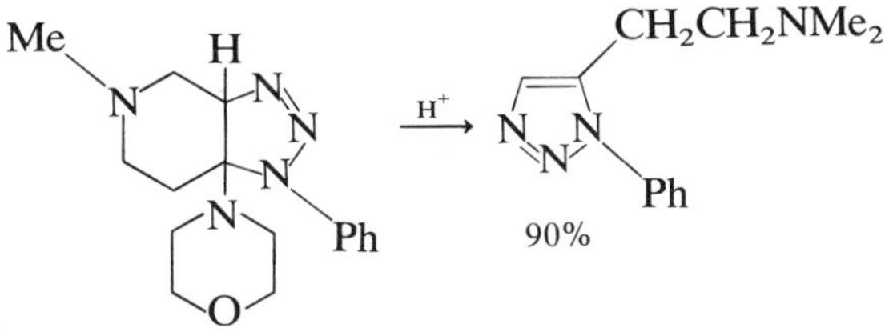

$$\xrightarrow{H^+} \quad 90\% \quad (29)$$

R″, R′, CO_2R (methylenecyclopropane) $=CH_2 + PhN_3 \longrightarrow$ (R″)(R′)$CHCHCO_2R$-substituted triazole, N-Ph (30)

2.2-16

R = Me, Et, R′ = H, Me
R″ = H, CO_2Me

91% R′H, R″CO_2Me

H_2C= (aziridine) $NCHMe_2 + PhN_3 \longrightarrow$ (spiro aziridine–triazoline, Me_2CH, N–Ph) $\longrightarrow$

$CH_2NHCHMe_2$-substituted triazole, N-Ph (31)

2.2-17

REFERENCES

17. **83:** 193181e	18. **80:** 47912r	19. **75:** 19622z	20. **72:** 121447w
21. **74:** 141988t	22. **77:** 164609w	23. **79:** 78884s	24. **79:** 126568r
25. **64:** 9713h	26. **67:** 100071a	27. **68:** 87243g	28. **68:** 105100q
29. **63:** 11551c	30. **67:** 82171b	31. **62:** 164248b	32. **66:** 10887k
33. **69:** 10402w	34. **61:** 3096e	35. **83:** 79160x	36. **83:** 79201m

2.3. 2,4-ALKYL- OR ARYL-DISUBSTITUTED 2*H*-1,2,3-TRIAZOLES

A large amount of effort has been expended on the conversion of osazones (**2.3-1**, D-glucose) to osatriazoles (**2.3-2**, D-glucose) (Eq. 32). The method, pioneered by Hann and Hudson,[37] has been developed extensively, notably by Hardegger[38] and El Khadem.[39] The range of 2-aryl-1,2,3-triazoles containing a sugar residue in the 4-position is very broad, but the basic ring-formation step has remained substantially the same.

Using glucose as the starting material and modifying the 2- and 4-substituents, Riebsomer and his students[40,41] produced a number of new 1,2,3-triazoles based mostly on the chemistry of the 4-formyl group (Eq. 33). The yield of **2.3-3** is nearly quantitative, and its conversion to various derivatives proceeds in good to excellent yield. An alternative, high-yield route to these products is found in the lithium aluminum hydride reduction of the 4-methyl ester (Eq. 34).[42]

CH=NNHPh
C=NNHPh
HO—C—H
H—C—OH
H—C—OH
CH_2OH
2.3-1

$\xrightarrow[H_2O]{CuSO_4}$

N=N—Ph (triazole ring)
HO—C—H
H—C—OH
H—C—OH
CH_2OH
2.3-2

(32)

2.3-2 $\xrightarrow{Na^+ \ ^-IO_4}$ CHO, N N, N, Ph **2.3-3** $\longrightarrow$ Y, N N, N, Ph (33)

$Y = Me, CH_2OH, CH_2Br, CH_2CN$

CO_2Me, N N, N, Ph $\xrightarrow{LiAlH_4}$ CH_2OH, N N, N, Ph $\longrightarrow$ Derivatives as in Eq. 33 (34)

The analogous reactions of α,β-bisarylhydrazones (**2.3-4**) to 1,2,3-triazoles lacking a sugar substituent (Eq. 35) has been reported,[43] but in general the yields are poor to fair. A few exceptions, such as 4-Ph-2-(4-IPh) 80% and 4-Ph-2-(4-BrPh) 60%, indicate the need for further examination. Another reason why this route shows potential is the reported improved preparation of the starting material **2.3-4** (Eq. 36).[44]

$Ar^1C{=}NNHAr^2$
$CH{=}NNHAr^2$
2.3-4

$\xrightarrow[EtOH]{CuCl_2}$ Ar^1, N N, N, Ar^2 (35)

$Ar^1 = Ph, 4\text{-}BrPh$
$Ar^2 = Ph, 4\text{-}XPh, 4\text{-}NO_2Ph$

X, X, X, X (tetrahalobenzodioxole) O, O —C(=O)Ar + $PhNHNH_2$ $\longrightarrow$ **2.3-4** $Ar^2 = Ph$ (36)

$X = Cl, Br \quad Ar = Ph, 4\text{-}MePh, 4\text{-}MeOPh, 4\text{-}ClPh, \beta\text{-}C_{10}H_7$

The synthesis of *N*-methyl-1,2,3-triazoles from benzonitrile and diazomethane in the presence of trialkylaluminums (Eq. 37) can produce moderate yields when the proper choice of catalysts and reaction conditions is made. The direct preparation of **2.3-5** with dimethyl sulfate has also been reported in comparable yield.[45]

$$PhCN + CH_2N_2 \xrightarrow{ICH_2AlEt_2} \text{2.3-5 (4-Ph-2-Me-1,2,3-triazole), 43\%} \quad (37)$$

Tanaka and Miller[46] have found that 1,2,3-triazole ylides (**2.3-6**) are excellent nucleophiles (Eq. 38).

$$\text{2.3-6} \xrightarrow{RX} \text{[2-R-4-Ar-5-}PPh_3^+X^-\text{-triazole]} \xrightarrow{OH^-} \text{2.3-7} \quad (38)$$

RX	= EtI	$EtO_2CCH_2CH_2Cl$	2,4-$(NO_2)_2PhBr$
% **2.3-7**	= 71	100	67

REFERENCES

37. **38:** 3622[3]
38. **42:** 1209h
39. **58:** 11454d
40. **43:** 2619b
41. **46:** 6123e
42. **56:** 5943d
43. **69:** 76826u
44. **76:** 153646j
45. **67:** 108605y
46. **79:** 66254y

2.4. 4,5-ALKYL- OR ARYL-DISUBSTITUTED *v*-TRIAZOLES

The early study of Sheehan and Robinson[47] in which they sought 1,2,3-triazole analogs of histamine (*e.g.*, **2.4-1**), included some 5-phenyl intermediates (Eq. 39). The yields are generally excellent.

Organometallic azides add smoothly to acetylenes (Eqs. 40,41), and the resulting 2-trialkylsilicon (**2.4-2**)[48] or tin (**2.4-3**)[49] derivatives hydrolyze in near quantative yield. The addition of acetylazide to triphenylphosphorane methylenes (**2.4-4**) followed by hydrolysis (Eq. 42)[50] is another attractive route. Yields of 60% and greater were reported for a variety of substituents on **2.4-5**.

Ph CHO N N N H + S N O S H ⟶ Ph CH N N N H S N O S H

several steps ⟶ Ph $CH_2CH_2NH_2$ N N N H (39)

2.4-1

R R′ N N N $SiMe_3$ (40)

Me_3SiN_3

H^+

$RC{\equiv}CR'$

2.4-2

R R′ N N N H

n-Bu_3SnN_3

R, R′ = Me, Et, Ph

R R′ N N N $Sn(Bu\text{-}n)_3$

H^+ (41)

2.4-3

R^1 R^2 R^1 R^2

COR^1

$Ph_3P{=}C$ + $MeCON_3$ ⟶ N N N Ac $\xrightarrow{OH^-}$ N N N H (42)

R^2

2.4-4 **2.4-5**

The preparation of 4-aryl-5-heteroaromatic-1,2,3-triazoles in excellent yield can be accomplished by the direct addition of the azide ion to appropriate styryl sulfones (Eq. 43).[51] Except for Ph and 9-anthryl (50 and 60%), the yields are 80 to 90%.

Ar Y

SO_2Ph

$ArCH{=}C$ + $Na^+\ ^-N_3$ $\xrightarrow{HCONMe_2}$ N N N H (43)

Y

2.4-6

Ar = Ph, 4-MeOPh, 4-ClPh, 4-Me_2NPh, 3-pyridyl, 3,4,5-$(MeO)_3$Ph, 9-anthryl

Y = (4-methyl-6-methyl-2-pyranone), (5,7-dichlorobenzoxazol-2-yl), (5-phenyl-1,3,4-oxadiazol-2-yl)

$$\mathrm{ArC{\equiv}CCO_2Me + Na^+ \, ^-N_3 \longrightarrow}$$ 4-Ar-5-CO_2Me-1H-1,2,3-triazole $\xrightarrow{\text{PhLi}}$ 4-Ar-5-(CPh_2OH)-1,2,3-triazole (**2.4-7**)

Ar = Ph, 4-ClPh

4-ClPh 92%

2.4-7 4-ClPh 96%, Ph 88% (44)

The conversion of a carbomethoxy group to various hydroxymethyl derivatives (**2.4-7**) in very high yield has been reported by Burgess and Sanchez (Eqs. 44,45).[52] Of special interest is the high yield of the fulvene (**2.4-8**) and the product composition reported for **2.4-9** and **2.4-10**.

2.4-7 (Ar = Ph) $\xrightarrow{SOCl_2}$ 4-Ph-5-(CPh_2Cl)-triazole $NH_2^+Cl^-$ (90%)

- $\xrightarrow[\text{Et}_3\text{N,THF}]{\text{piperidine}}$ 4-Ph-5-(CPh_2-piperidino)-1H-triazole, 95%
- $\xrightarrow[\text{Et}_3\text{N,THF}]{\text{MeOH}}$ 4-Ph-5-(CPh_2OMe)-1H-triazole, 94%
- $\xrightarrow[\text{C}_6\text{H}_6\text{,THF}]{\text{Et}_3\text{N}}$ 4-Ph-5-(=CPh_2)-triazole, **2.4-8**, 98% (45)

2.4-9 (Ph, CPh_2OH, intramolecular H-bond to N1–H) 35%

2.4-10 (Ph, CPh_2OH, 2H-triazole) 53%

REFERENCES

47. **45:** 9037f 48. **65:** 15414d 49. **78:** 29956n 50. **71:** 112865h
51. **80:** 3439n 52. **80:** 133389g

TABLE 2. ALKYL- OR ARYL-DISUBSTITUTED 1,2,3-TRIAZOLES

Compound	Reference
2.1. 1,4-Alkyl- or Aryl-Disubstituted 1*H*-1,2,3-Triazoles	
-4-acetaldehyde, 1-phenyl-, oxime	**49:** 6242c
1-(2-acetamido-2-deoxy-β-D-glucopyranosyl)-4-phenyl-	**72:** 3718n
1-(2-acetamido-2-deoxy-β-D-glucopyranosyl)-4-phenyl-, 3′,4′,6′-triacetate ester	**72:** 3718n
-1-acetic acid, 4(or 5)-(1-hydroxycyclohexyl)-, ethyl ester	**60:** 15859d
-1-acetic acid, 4(or 5)-(1-hydroxyethyl)-, ethyl ester	**60:** 15859d
-1-acetic acid, 4-(hydroxy-1-methylethyl)-, ethyl ester	**60:** 15859d
-1-acetic acid, 4-phenyl-	**80:** 108451p
-1-acetic acid, 4(or 5)-phenyl-	**60:** 15859d
-1-acetic acid, 4-phenyl-, ethyl ester	**80:** 108451p
-1-acetic acid, 4(or 5)-phenyl-, ethyl ester	**60:** 15859d
-1-acetonitrile, 4-(methoxymethyl)-α-(4-methoxy-salicylidene)-	**59:** P15420h
-4-acetonitrile, 1-methyl-	**49:** 6242e
-4-acetonitrile, 1-phenyl-	**45:** 9038c
-4-acrolein, 1–allyl-	**66:** 37839u
-4-acrolein, 1-benzyl-	**66:** 37839u
-4-acrolein, 1-benzyl, phenylhydrazone	**66:** 37839u
-4-acrolein, α-bromo-1-phenyl-	**64:** 3523f
-4-acrolein, α-bromo-1-phenyl-, (4-nitrophenyl)-hydrazone	**64:** 3523f
-4-acrolein, 1-(4-bromophenyl)-	**64:** 3523e
-4-acrolein, 1-(4-bromophenyl)-, phenylhydrazone	**64:** 3523e
-4-acrolein, 1-(2,4-dinitrophenyl)-	**64:** 3523f
-4-acrolein, 1-ethyl-	**66:** 37839u
-4-acrolein, 1-ethyl-, phenylhydrazone	**66:** 37839u
-4-acrolein, 1-(4-methoxyphenyl)-	**64:** 3523e
-4-acrolein, 1-methyl-, (E)-	**75:** 63702z
-4-acrolein, 1- methyl-, phenylhydrazone	**66:** 37839u
-4-acrolein, 1-(4-methylphenyl)-	**64:** 3523e
-4-acrolein, 1-[4-[4-(methylphenyl)sulfinyl]phenyl]-	**64:** 3523f
-4-acrolein, 1-(1-naphthyl)-	**64:** 3523e
-4-acrolein, 1-(2-naphthyl)-	**64:** 3523e
-4-acrolein, 1-(3-nitrophenyl)-	**64:** 3523e
-4-acrolein, 1-(4-nitrophenyl)-	**64:** 3523e
-4-acrolein, 1-phenyl-, (2,4-dinitrophenyl)hydrazone	**64:** 3523e
-4-acrolein, 1-phenyl-, (4-nitrophenyl)hydrazone	**64:** 3523e
-4-acrolein, 1-phenyl-, oxime	**64:** 3523e

TABLE 2 (*Continued*)

Compound	Reference
2.1. 1,4-Alkyl- or Aryl-Disubstituted 1*H*-1,2,3-Triazoles (*Continued*)	
-4-acrolein, 1-phenyl-, phenylhydrazone	**64:** 3523e
-4-acrylic acid, α-benzamido-1-phenyl-	**49:** 6242b
-4-acrylic acid, 1-benzyl-	**50:** 4129c
-4-acrylic acid, α-cyano-1-methyl, ethyl ester	**67:** 118072m
-4-acrylic acid, 1-methyl-	**50:** 4129c, **66:** 37839u
-4-acrylic acid, 1-phenyl-	**64:** 3523g
-4-alanine, *N*-benzoyl-1-phenyl-	**49:** 6242b
-4-alanine, 1-methyl-	**49:** 6242b
-4-alanine, 1-phenyl-	**49:** 6241i
4-(2-aminoethyl)-1-methyl-	**49:** 6242d
4-(2-aminoethyl)-1-phenyl-	**45:** 9038c, **49:** 6242b
4-(2-amino-4-thiazolyl)-1-phenyl-	**50:** 4924e
1-(4-anilino-3-oxo-2-phenyl-2*H*-pyridazin-5-yl)-4-(chloromethyl)-	**72:** 55344g
4-[1-*C*-(R-D-arabinitolyl)]-1-phenyl-, 1,2,3,4,5-pentaacetate ester	**79:** 66696u
4-[(D-*arabino*-1,2,3,4,5-pentahydroxypentyl)-2,3:4,5-bis-*O*-(1-methylethylidene)-1-*C*-]-1-phenyl-, (R)-	**84:** 180505j
4-[(D-*arabino*-1,2,3,4,5-pentahydroxypentyl)-2,3:4,5-bis-*O*-(1-methylethylidene)-1-*C*-]-1-phenyl-, (S)-	**84:** 180505j
4-[(D-*arabino*-1,2,3,4,5-pentahydroxypentyl)-2,3:4,5-bis-*O*-(1-methylethylidene)-1-*C*-]-1-phenyl-, 1-acetate ester, (R)-	**84:** 180505j
4-[(D-*arabino*-1,2,3,4,5-pentahydroxypentyl)-2,3:4,5-bis-*O*-(1-methylethylidene)-1-*C*-]-1-phenyl-, 1-acetate ester, (S)-	**84:** 180505j
4-(D-*arabino*-1,2,3,4,5-pentahydroxypentyl-1-*C*)-1-phenyl-, (R)-	**84:** 180505j
4-(D-*arabino*-1,2,3,4,5-pentahydroxypentyl-1-*C*)-1-phenyl-, (S)-	**84:** 180505j
4-(D-*arabino*-1,2,3,4,5-pentahydroxypentyl-1-*C*)-1-phenyl-, 1,2,3,4,5-pentaacetate ester, (S)-	**84:** 180505j
4-(azidomethyl)-1-(4-azido-3-oxo-2-phenyl-2*H*-pyridazin-5-yl)-	**72:** 66882g
1-(4-azido-3-oxo-2-phenyl-2*H*-pyridazin-5-yl)-4-(chloromethyl)-	**72:** 66882g
4-[(2-azidophenoxy)methyl]-1-[2-(2-propynloxy)phenyl]-	**84:** 105510d
1-(1*H*-benzimidazol-2-yl)-4(or 5)-(1-hydroxycyclohexyl)-	**79:** P78815v
1-(2-benzothiazolyl)-4(or 5)-(bromomethyl)-	**79:** P78815v
1-(2-benzothiazolyl)-4(or 5)-(1-hydroxycyclohexyl)-	**79:** P78815v
1-(2-benzothiazolyl)-4(or 5)-[*N*-(2-hydroxyphenyl)-*N*-(methylamino)methyl]-, methylcarbamate ester	**79:** P78815v
1-(2-benzothiazolyl)-4(or 5)-pentyl-	**79:** P78815v
1-(2-benzothiazolyl)-4(or 5)-phenyl-	**79:** P78815v
1-[4-(benzylamino)-3-oxo-2-phenyl-2*H*-pyridazin-5-yl]-4-[(benzylamino)methyl]-	**72:** 55344g
1-[4-(benzylamino)-3-oxo-2-phenyl-2*H*-pyridazin-5-yl]-4-(chloromethyl)-	**72:** 55344g, **76:** 153690u

TABLE 2 (*Continued*)

Compound	Reference
2.1. 1,4-Alkyl- or Aryl-Disubstituted 1*H*-1,2,3-Triazoles (*Continued*)	
1-[4-(benzylamino)-3-oxo-2-phenyl-2*H*-pyridazin-5-yl]-4-[(dimethylamino)methyl]-	**72:** 55344g
1-[2-[(4,6-*O*-benzylidene-2-deoxy-α-D-altro-pyranosidyl)]methyl]-4-phenyl-	**75:** 88891y
1-benzyl-4-isopropenyl-	**46:** 8651g
1-benzyl-4-isopropyl-	**46:** 8652b
1-benzyl-4(or 5)-methyl-	**60:** 15859d
1-benzyl-4-phenyl-	**51:** 13854i, **53:** 5253d
1-[(benzylsulfonyl)methyl]-4-phenyl-	**72:** 43576w
1-[(benzylsulfonyl)methyl]-4(or 5)-phenyl-	**51:** 13855f
1-[(benzylthio)methyl]-4-phenyl-	**72:** 43576w
1-benzyl-4-[D-*threo*-(1,2,3-trihydroxypropyl)]-	**75:** 88891y
1-benzyl-4-(2,3,5-*tris*-*O*-benzyl-β-D-ribofuranosyl)-	**77:** 34803u, **82:** 43677m
4-[bis[4-(dimethylamino)phenyl]methyl]-1-methyl-	**56:** 12877g
1-[5-bromo-1-(2-cyanoethyl)-6-oxo-6*H*-pyridazin-4-yl]-4-(chloromethyl)-	**72:** 121473b
1-[5-bromo-1-(2-cyanoethyl)-6-oxo-6*H*-pyridazin-4-yl]-4-(hydroxymethyl)-	**70:** 3996k, **72:** 121473b
1-[4-bromo-2-(2-hydroxyethyl)-3-oxo-2*H*-pyridazin-5-yl]-4-(hydroxymethyl)-	**70:** 3996k, **72:** 121473b
1(4-bromo-2-methyl-3-oxo-2*H*-pyridazin-5-yl)-4-(chloromethyl)-	**72:** 121473b
1-(4-bromo-2-methyl-3-oxo-2*H*-pyridazin-5-yl)-4-(hydroxymethyl)-	**70:** 3996k, **72:** 121473b
4-(bromomethyl)-1-phenyl-	**84:** 105510d
1-(4-bromo-3-oxo-2-phenyl-2*H*-pyridazin-5-yl)-4-[(*tert*-butylamino)methyl]-	**72:** 55344g
1-(4-bromo-3-oxo-2-phenyl-2*H*-pyridazin-5-yl)-4-(chloromethyl)-	**72:** 121473b
1-(4-bromo-3-oxo-2-phenyl-2*H*-pyridazin-5-yl)-4-(hydroxymethyl)-	**70:** 3996k, **72:** 65613h, **72:** 121473b
1-(4-bromo-3-oxo-2-phenyl-2*H*-pyridazin-5-yl)-4-(iodomethyl)-	**72:** 121473b
1-(4-bromo-3-oxo-2-phenyl-2*H*-pyridazin-5-yl)-4-[(methylamino)methyl]-	**72:** 55344g
1-[(5-bromo-6-oxo-1-phenyl-6*H*-pyridazin-4-yl)-4-(methylamino)-, triethylammonium iodide	**72:** 55344g
1-(4-bromo-3-oxo-2-phenyl-2*H*-pyridazin-5-yl)-4-(*aci*-nitromethyl)-	**72:** 66882g
1-(4-bromo-3-oxo-2*H*-pyridazin-5-yl)-4-(chloromethyl)-	**72:** 121473b
1-(4-bromo-3-oxo-2*H*-pyridazin-5-yl)-4-(hydroxymethyl)-	**70:** 3996k, **72:** 65613h, **72:** 121473b
1-(5-bromo-6-oxo-1-phenyl-6*H*-pyridazin-4-yl)-4-[(thiocyanato)methyl]-	**72:** 66882g
1-(4-bromophenyl)-4-methyl-	**69:** 10402w
4-(2-bromophenyl)-1-phenyl-	**66:** 10887k
4-(4-bromophenyl)-1-phenyl-	**66:** 10887k
1-(4-bromophenyl)-4-vinyl-	**77:** 126516f

TABLE 2 (*Continued*)

Compound	Reference
2.1. 1,4-Alkyl- or Aryl-Disubstituted 1*H*-1,2,3-Triazoles (*Continued*)	
-2-butanedioic acid, 1-phenyl-, dimethyl ester	**83:** 79160x
-1-butanoic acid, 4-phenyl-	**84:** P164786w, **85:** P123926u
1-[4-(butylamino)-3-oxo-2-phenyl-2*H*-pyridazin-5-yl]-4-(hydroxymethyl)-	**74:** 125600e, **76:** 153690u
1-butyl-4(or 5)-ethyl-	**60:** 15859d
1-butyl-4(or 5)-methyl-	**60:** 15859d
1-*tert*-butyl-4-phenyl-	**78:** 72017u
4-[(5-chloro-2-hydroxyphenyl)methyl]-1-(2,4,6-trinitrophenyl)-	**84:** 90080n
1-(4-chloro-2-methyl-3-oxo-2*H*-pyridazin-5-yl)-4-(hydroxymethyl)-	**78:** 72035y
1-(4-chloro-2-methyl-3-oxo-2*H*-pyridazin-5-yl)-4-phenyl-	**78:** 72035y
4-(chloromethyl)-1-(4-chloro-3-oxo-2-phenyl-2*H*-pyridazin-5-yl)-	**76:** 153690u
4-(chloromethyl)-1-[4-(methylamino)-3-oxo-2-phenyl-2*H*-pyridazin-5-yl]-	**72:** 55344g
4-(chloromethyl)-1-[4-[(3-methylphenyl)amino]-3-oxo-2-phenyl-2*H*-pyridazin-5-yl]-	**72:** 55344g
4-(chloromethyl)-1-[4-[(4-methylphenyl)amino]-3-oxo-2-phenyl-2*H*-pyridazin-5-yl]-	**72:** 55344g
4-(chloromethyl)-1-[4-(4-morpholinyl)-3-oxo-2-phenyl-2*H*-pyridazin-5-yl]-	**72:** 55344g
4-(chloromethyl)-1-(3-oxo-2-phenyl-4-piperidinyl-2*H*-pyridazin-5-yl]-	**72:** 55344g
4-(chloromethyl)-1-[3-oxo-2-phenyl-4-(1-pyrrolidinyl)-2*H*-pyridazin-5-yl]-	**72:** 55344g
1-(4-chloro-3-oxo-2-phenyl-2*H*-pyridazin-5-yl)-4-(hydroxymethyl)-	**74:** 125600e
1-(4-chloro-3-oxo-2*H*-pyridazin-5-yl)-4-(hydroxymethyl)-	**78:** 72035y
1-(4-chloro-3-oxo-2*H*-pyridazin-5-yl)-4-phenyl-	**78:** 72035y
1-(4-chlorophenyl)-4-methyl-	**69:** 10402w
1-(4-chlorophenyl)-4-[2-(4-morpholinyl)ethyl]-	**76:** 126879f
1-(4-chlorophenyl)-4-(4-morpholinylmethyl)-	**76:** 126875b
4-(2-chlorophenyl)-1-phenyl-	**66:** 10887k
4-(4-chlorophenyl)-1-phenyl-	**66:** 10887k
1-(4-chlorophenyl)-4-[1-(1-piperidinyl)ethyl]-	**76:** 126879f
1-(4-chlorophenyl)-4-(1-piperidinylmethyl)-	**76:** 126875b
1-[4-(cyclohexylamino)-3-oxo-2-phenyl-2*H*-pyridazin-5-yl]-4-(hydroxymethyl)-	**76:** 153690u
1-[6-(6-deoxy-1,2:3,4-di-*O*-isopropylidene-α-D-galactopyranosyl)]-4-phenyl-	**75:** 88891y
4-[(3,5-dichloro-2-hydroxyphenyl)methyl]-1-(2,4,6-trinitrophenyl)-	**84:** 90080n
1-[4-(diethylamino)-3-oxo-2-phenyl-2*H*-pyridazin-5-yl]-4-(chloromethyl)-	**72:** 55344g
1-(3,4-dihydro-1*H*-2-benzopyran-1-yl)-4-phenyl-	**81:** 91486z

TABLE 2 (*Continued*)

Compound	Reference
2.1. 1,4-Alkyl- or Aryl-Disubstituted 1*H*-1,2,3-Triazoles (*Continued*)	
1-(3,4-dihydro-1*H*-2-benzothiopyran-1-yl)-4-phenyl-	**81:** 91486z
1-(3,4-dihydro-1*H*-2-benzothiopyran-1-yl)-4-phenyl-, S,S-dioxide	**81:** 91486z
4-[6-*C*-(1,2:3,4-di-*O*-isopropylidene-D-glycero-α-D-galactohexopyranosyl)]-1-phenyl-	**75:** 88891y
1,4-dimethyl-	**85:** 77337c
1-[4-(dimethylamino)-3-oxo-2-phenyl-2*H*-pyridazin-5-yl]-4-(chloromethyl)-	**72:** 55344g
1-[4-(dimethylamino)-3-oxo-2-phenyl-2*H*-pyridazin-5-yl]-4-[(dimethylamino)methyl]-	**72:** 55344g
1-[4-(dimethylamino)-3-oxo-2-Phenyl-2*H*-pyridazin 5-yl]-4-(piperidinylmethyl)-	**72:** 55344g
4-[(1,3-dimethyl-2-oxo-3,5-cyclohexadien-1-yl)methyl]-1-(2,4,6-trinitrophenyl)-	**84:** 90080n
1-(3,3-dimethyl-1-butenyl)-4-phenyl-	**75:** 19622z
1-(2,5-dimethylphenyl)-4-isopropyl-	**22:** 3411[1]
1-(2,5-dimethylphenyl)-4-methyl-	**22:** 3411[1]
1-(2,5-dimethylphenyl)-4-phenyl-	**22:** 3411[2]
1-(1,3-dioxo-2-phenyl-2-indanyl)-4-(azidomethyl)-	**73:** 98912d
1-(1,3-dioxo-2-phenyl-2-indanyl)-4-(chloromethyl)-	**73:** 98912d
1-(1,3-dioxo-2-phenyl-2-indanyl)-4-(hydroxymethyl)-	**73:** 98912d
1-(1,3-dioxo-2-phenyl-2-indanyl)-4-(iodomethyl)-	**73:** 98912d
1-(1,3-dioxo-2-phenyl-2-indanyl)-4-[(thiocyanato)-methyl]-	**73:** 98912d
1,4-diphenyl-	**82:** 154936q
1,4-diphenyl-, -4-^{13}C	**82:** 154936q
1,4-diphenyl-5-(trimethylsilyl)-	**60:** 5538a
4-(dipiperidinylmethyl)-1-phenyl-	**49:** 6242c
-1-ethanol, 4(or 5)-ethyl-	**60:** 15859d
-1-ethanol, 4(or 5)-(1-hydroxycyclohexyl)-	**60:** 15859d
-1-ethanol, 4(or 5)-(1-hydroxyethyl)-	**60:** 15859d
-1-ethanol, 4-(hydroxymethyl)-	**60:** 15859d, **83:** 193181e
-1-ethanol, 4(or 5)-(3-hydroxy-3-methyl-1-butynyl)-	**60:** 15859d
-1-ethanol, 4-(1-hydroxy-1-methylethyl)-	**60:** 15859d
-1-ethanol, 4-(1-hydroxy-2-methylpropyl)-	**62:** P10443h
-1-ethanol, 4-phenyl-	**83:** 193181e
-1-ethanol, 4(or 5)-phenyl-	**60:** 15859d
1-[4-(ethenyloxy)butyl]-4-phenyl-	**84:** P59482w
1-[2-(ethenyloxy)ethyl]-4-phenyl-	**84:** P59482w
1-[(ethenyloxy)methyl]-4-phenyl-	**84:** P59482w
1-[3-(ethenyloxy)propyl]-4-phenyl-	**84:** P59482w
1-(ethoxymethyl)-4(or 5)-phenyl-	**51:** 13855f
1-ethyl-4-phenyl-	**51:** 13854i
1-[(ethylthio)methyl]-4(or 5)-phenyl-	**51:** 13855f
1-β-D-galactopyranosyl-4-phenyl-	**54:** 12125c, **72:** 3718n
1-β-D-galactopyranosyl-4-phenyl-, 2′,3′,4′,6′-tetraacetate ester	**72:** 3718n
1-α-D-glucopyranosyl-4-phenyl-	**72:** 3718n

TABLE 2 (*Continued*)

Compound	Reference
2.1. 1,4-Alkyl- or Aryl-Disubstituted 1*H*-1,2,3-Triazoles (*Continued*)	
1-β-D-glucopyranosyl-4-phenyl-	**54:** 12125b, **72:** 3718n
1-α-D-glucopyranosyl-4-phenyl-, 3′,4′,6′-triacetate ester	**72:** 3718n
1-β-D-glucopyranosyl-4-phenyl-, 2′,3′,4′,6′-tetraacetate ester	**72:** 3718n
4(or 5)-(1-hydroxycyclohexyl)-1-(6-isopropyl-4-methyl-2-pyrimidinyl)-	**79:** P78815v
1-[4-[(3-hydroxyphenyl)amino]-3-oxo-2-phenyl-2*H*-pyridazin-5-yl]-4-(hydroxymethyl)-	**76:** 153690u
4-[(2-hydroxyphenyl)methyl]-1-(2,4,6-trinitrophenyl)-	**84:** 90080n
4-[(2-hydroxyphenyl)methyl]-1-(2,4,6-trinitrophenyl)-, acetate ester	**84:** 90080n
4-(2-hydroxyphenyl)-1-phenyl-	**66:** 10887k
1-(imidazo[1,2-*b*]pyridazin-6-yl)-4-phenyl-	**83:** 97213w
1-(2-iodo-2,3-dimethylbutyl)-4-phenyl-	**75:** 19622z
1-[4-(isobutylamino)-3-oxo-2-phenyl-2*H*-pyridazin-5-yl]-4-(chloromethyl)-	**72:** 55344g
1-[4-(isobutylamino)-3-oxo-2-phenyl-2*H*-pyridazin-5-yl]-4-[(isobutylamino)methyl]-	**72:** 55344g
4(or 5)-isopropenyl-1-[(4-methylphenyl)methyl]-	**60:** 15859d
4-isopropenyl-1-phenyl-	**46:** 8652b
1-[4-(isopropylamino)-3-oxo-2-phenyl-2*H*-pyridazin-5-yl]-4-(chloromethyl)-	**76:** 153690u
1-[4-(isopropylamino)-3-oxo-2-phenyl-2*H*-pyridazin-5-yl]-4-(hydroxymethyl)-	**76:** 153690u
1-isopropyl-4-phenyl-	**78:** 72017u
4-isopropyl-1-phenyl-	**46:** 8652b
-4(or 5)-methanamine, 1-(1*H*-benzimidazol-2-yl)-*N*,*N*-dimethyl-	**79:** P78815v
-4(or 5)-methanamine, 1-(2-benzothiazolyl)-*N*,*N*-dimethyl-	**79:** P78815v
-4-methanamine, 1-(4-chlorophenyl)-*N*-methyl-*N*-phenyl-	**76:** 126875b
-4(or 5)-methanamine, *N*,*N*-diethyl-1-[4-(ethylamino)-6-(isopropylamino)-1,3,5-triazin-2-yl]-	**79:** P78815v
-4-methanamine, *N*,*N*-diethyl-1-(4-nitrophenyl)-	**76:** 126875b
-4-methanamine, *N*-isopropyl-1-phenyl-	**83:** 79201m
-4-methanamine, *N*-methyl-1-(4-nitrophenyl)-*N*-phenyl-	**76:** 126875b
-4-methanol, 1-(3-aminophenyl)-α,α-dimethyl-	**53:** 3199i
-4-methanol, 1-(4-aminophenyl)-α,α-dimethyl-	**46:** 8651h
-4(or 5)-methanol, 1-(1*H*-benzimidazol-2-yl)-	**79:** P78815v
-4(or 5)-methanol, 1-(1*H*-benzimidazol-2-yl)-α,α-dimethyl-	**79:** P78815v
-4(or 5)-methanol, 1-(1*H*-benzimidazol-2-yl)-α-pentyl-	**79:** P78815v
-4(or 5)-methanol, 1-(1*H*-benzimidazol-2-yl)-α-phenyl-	**79:** P78815v
-4(or 5)-methanol, 1-(2-benzothiazolyl)-	**79:** P78815v
-4(or 5)-methanol, 1-(2-benzothiazolyl)-α,α-dimethyl-	**79:** P78815v
-4(or 5)-methanol, 1-(2-benzothiazolyl)-α-ethyl-α-methyl-	**79:** P78815v

TABLE 2 (*Continued*)

Compound	Reference
2.1. 1,4-Alkyl- or Aryl-Disubstituted 1*H*-1,2,3-Triazoles (*Continued*)	
-4(or 5)-methanol, 1-(2-benzothiazolyl)-α-methyl-	**79:** P78815v
-4(or 5)-methanol, 1-(2-benzothiazolyl)-α-pentyl-	**79:** P78815v
-4(or 5)-methanol, 1-(2-benzothiazolyl)-α-phenyl-	**79:** P78815v
-4(or 5)-methanol, 1-(2-benzoxazolyl)-	**79:** P78815v
-4-methanol, 1-benzyl-	**46:** 8651g, **73:** 98912d
-4-methanol, 1-benzyl-α,α-dimethyl-	**46:** 8651f
-4-methanol, 1-benzyl-α-hexyl-	**46:** 8651g
-4(or 5)-methanol, 1-benzyl-α-methyl-	**60:** 15859d
-4-methanol, α,α-bis[4-(dimethylamino)phenyl]-1-methyl-	**56:** 12877h
-4(or 5)-methanol, 1-butyl-	**60:** 15859d
-4(or 5)-methanol, 1-butyl-α,α-dimethyl-	**60:** 15859d
-4(or 5)-methanol, 1-butyl-α-methyl-	**60:** 15859d
-4(or 5)-methanol, 1-[6-(*sec*-butylamino)-4-(ethylamino)-1,3,5-triazin-2-yl]-	**79:** P78815v
-4-methanol, 1-[1-(2-carboxybenzoyl)phenylmethyl]-	**73:** 98912d
-4(or 5)-methanol, α-(3,4-dihydro-2,6-dimethyl-2*H*-pyran-2-yl)-1-(2-ethoxyethyl)-	**60:** 15859d
-4(or 5)-methanol, α,α-dimethyl-1-(6-isopropyl-4-methyl-2-pyrimidinyl)-	**79:** P78815v
-4(or 5)-methanol, α,α-dimethyl-1-[4-(methylphenyl)-methyl]-	**60:** 15859d
-4-methanol, α,α-dimethyl-1-(3-nitrophenyl)-	**53:** 3199h
-4-methanol, α,α-dimethyl-1-(4-nitrophenyl)-	**46:** 8651h
-4-methanol, α,α-dimethyl-1-phenyl-	**46:** 8651h
-4-methanol, α,α-dimethyl-1-phenyl-, acetate ester	**46:** 8651h
-4-methanol, α,α-dimethyl-1-[4-(triphenylplumbyl)butyl]-	**67:** 82249h
-4-methanol, α,α-dimethyl-1-[5-(triphenylplumbyl)pentyl]-	**67:** 82249h
-4-methanol, α,α-dimethyl-1-[3-(triphenylplumbyl)propyl]-	**67:** 82249h
-4(or 5)-methanol, 1-(2-ethoxyethyl)-	**60:** 15859d
-4(or 5)-methanol, 1-(2-ethoxyethyl)-α-methyl-	**60:** 15859d
-4(or 5)-methanol, 1-[4-(ethylamino)-6-(isopropyl-amino)-1,3,5-triazin-2-yl]-α,α-dimethyl-	**79:** P78815v
-4(or 5)-methanol, 1-[4-(ethylamino)-6-(isopropyl-amino)-1,3,5-triazin-2-yl]-α-phenyl-	**79:** P78815v
-4-methanol, 1-[4-(2-hydroxy-1-naphthylazo)phenyl]-α,α-dimethyl-	**46:** 8651h
-4-methanol, 1-(1-hydroxyphthalid-1-yl)phenylmethyl)-	**73:** 98912d
-4(or 5)-methanol, 1-(6-isopropyl-4-methyl-2-pyrimidinyl)-	**79:** P78815v
-4(or 5)-methanol, 1-(6-isopropyl-4-methyl-2-pyrimidinyl)-α-phenyl-	**79:** P78815v
-4-methanol, 1-[(4-methylphenyl)methyl]-α,α-dimethyl-	**46:** 8651g
-4-methanol, α-methyl-1-(4-nitrophenyl)-(?)	**52:** P2086a
-4-methanol, 1-[4-(4-morpholinyl)-3-oxo-2-phenyl-2*H*-pyridazin-5-yl]-	**76:** 153690u
-4(or 5)-methanol, 1-(6-nitro-2-benzothiazolyl)-	**79:** P78815v

TABLE 2 (*Continued*)

Compound	Reference
2.1. 1,4-Alkyl- or Aryl-Disubstituted 1*H*-1,2,3-Triazoles (*Continued*)	
-4-methanol, 1-(1-nitroethyl)-	**83:** 193181e
-4-methanol, 1-(2-nitroethyl)-	**83:** 193181e
-4-methanol, 1-(4-nitrophenyl)-	**58:** 12561b
4-methanol, 1-[4-[(2-nitrophenyl)amino]-3-oxo-2-phenyl-2*H*-pyridazin-5-yl]-	**76:** 153690u
4-methanol, 1-[4-[(4-nitrophenyl)amino]-3-oxo-2-phenyl-2*H*-pyridazin-5-yl]-	**76:** 153690u
4-methanol, 1-[3-oxo-2-phenyl-4-(phenylamino)-2*H*-pyridazin-5-yl]-	**76:** 153690u
4-methanol, 1-[3-oxo-2-phenyl-4-(1-piperidinyl)-2*H*-pyridazin-5-yl]-	**76:** 153690u
4-methanol, 1-[3-oxo-2-phenyl-4-(1-pyrrolidinyl)-2*H*-pyridazin-5-yl]-	**76:** 153690u
-4(or 5)-methanol, 1-pentyl-	**60:** 15859d
-4-methanol, 1-pentyl-	**83:** 193181e
-4-methanol, 1-phenyl-	**37:** 119[4], **46:** 8651h, **48:** 2685b
-4-methanol, 1-phenyl-, benzoate ester	**49:** 3948e
-4-methanol, 1-(4-pyridyl)-, l′-oxide	**55:** 27338f
-4-methanol, 1-(4-quinolyl)-	**55:** 27338e
1-(methoxymethyl)-4-phenyl-	**72:** 43576w
1-methyl-4-(4-methylphenyl)-	**78:** 72017u
4-methyl-1-(4-methylphenyl)-	**69:** 10402w
1-methyl-4-(2-naphthyl)-	**78:** 72017u
4-methyl-1-(4-nitrophenyl)-	**69:** 10402w
1-methyl-4-phenyl-	**67:** 108605y, **72:** 43576w, **78:** 72017u
4-methyl-1-phenyl-	**61:** 2926c, **69:** 10402w
1-[(methylsulfonyl)methyl]-4-phenyl-	**72:** 43576w
1-[(methylthio)methyl]-4-phenyl-	**72:** 43576w
4-(4-morpholinylmethyl)-1-(4-nitrophenyl)-	**76:** 126875b
4(or 5)-(4-morpholinylmethyl)-1(or 2)-[2-(5-nitro-2-furanyl)ethenyl]-	**77:** 101470y
1-(6-nitro-2-benzothiazolyl)-4(or 5)-phenyl-	**79:** P78815v
4-(2-nitroethenyl)-1-phenyl-	**49:** 6242c
4-(2-nitroethenyl)-1-propyl-	**62:** 15262a
1-(1-nitroethyl)-4-phenyl-	**83:** 193181e
1-(2-nitroethyl)-4-phenyl-	**83:** 193181e
1-(4-nitrophenyl)-4-phenyl-	**62:** 16248b, **70:** 19987u
1-(4-nitrophenyl)-4-(4-pyrrolidinylmethyl)-	**76:** 126875b
4-[(2-oxo-1,3,4,5,6-pentachloro-3,5-cyclohexadien-1-yl)methyl]-1-(2,4,6-trinitrophenyl)-	**84:** 90080n
1-(3-oxo-2-phenyl-4-piperidinyl-2*H*-pyridazin-5-yl)-4-(piperidinylmethyl)-	**72:** 55344g
4-[(2-oxo-1,3,5-trichloro-3,5-cyclohexadien-1-yl) methyl]-1-(2,4,6-trinitrophenyl)-	**84:** 90080n
1-(pentafluorophenyl)-4-phenyl-	**81:** 135588r
1-pentyl-4-phenyl-	**83:** 193181e

TABLE 2 (*Continued*)

Compound	Reference
2.1. 1,4-Alkyl- or Aryl-Disubstituted 1*H*-1,2,3-Triazoles (*Continued*)	
1-(phenoxymethyl)-4-phenyl-	**72:** 43576w
4-(phenoxymethyl)-1-phenyl-	**84:** 105510d
1-(1-phenylethenyl)-4-phenyl-	**82:** 156182q
1-phenyl-4-[2-*N*-phenylformimidoyl)ethenyl]-	**64:** 3523e
4-phenyl-1-[(phenylsulfonyl)methyl]-	**72:** 43576w
4(or 5)-phenyl-1-[(phenylsulfonyl)methyl]-	**51:** 13855f
4-phenyl-1-[(phenylthio)methyl]-	**72:** 43576w
4(or 5)-phenyl-1-[(phenylthio)methyl]-	**51:** 13855f
4(or 5)-phenyl-1-(piperidinylmethyl)-	**52:** 13727b
4-phenyl-1-β-D-ribofuranosyl-	**72:** 3718n
4-phenyl-1-β-D-ribofuranosyl-, 2′,3′,5′-tribenzoate ester	**72:** 3718n
4-phenyl-1-styryl-, (E)-	**71:** 38476p
4-phenyl-1-(tetrazolo[1,5-*b*]pyridazin-6-yl)-	**83:** 97213w
1-phenyl-4-[D-*threo*-(1,2,3-trihydroxypropyl)]-	**75:** 88891y
1-phenyl-4-[D-*threo*-(1,2,3-trihydroxypropyl)]-, triacetate ester	**75:** 88891y
4-phenyl-1-β-D-xylopyranosyl-, 2′,3′,4′-triacetate ester	**75:** 88891y
-1-propanoic acid, 4-[3-(acetylamino)phenyl]-	**84:** P164786w
-1-propanoic acid, 4-(3-aminophenyl)-	**84:** P164786w
-1-propanoic acid, 4-(3-bromophenyl)-	**84:** P164786w
-1-propanoic acid, 4-(3-chlorophenyl)-	**84:** P164786w
-4-propanoic acid, 1-(3-chlorophenyl)-	**84:** P164786w
-1-propanoic acid, 4-(3-ethoxyphenyl)-	**84:** P164786w
-1-propanoic acid, 4-(3-fluorophenyl)-	**84:** P164786w
-4-propanoic acid, 1-(3-fluorophenyl)-	**84:** P164786w
-4-propanoic acid, 1-methyl-	**73:** 98878x, **75:** 63702z
-4-propanoic-α,β-d_2 acid, 1-methyl-, methyl ester	**73:** 98878x, **75:** 63702z
-4-propanoic acid, 1-methyl-, methyl ester	**73:** 98878x, **75:** 63702z
-4-propanoic acid, α-methyl-1-phenyl, ethyl ester	**83:** 79160x
-1-propanoic acid, β-methyl-4-phenyl-	**83:** P10089p
-1-propanoic acid, 4-(3-methylphenyl)-	**84:** P164786w
-1-propanoic acid, 4-(3-nitrophenyl)-	**84:** P164786w
-1-propanoic acid, 4-(3-nitrophenyl)-, isopropyl ester	**84:** P164786w
-4-propanoic acid, β-oxo-1-phenyl-	**37:** 5405[1], **79:** 66254y
-1-propanoic acid, 4-phenyl-	**80:** 108451p
-1-propanoic acid, 4-phenyl-, ethyl ester	**80:** 108451p
-4-propanoic acid, 1-phenyl-	**83:** 79160x
-4-propanoic acid, 1-phenyl-, ethyl ester	**83:** 79160x
-4-propanoic acid, 1-phenyl-, methyl ester	**83:** 79160x
-4-propanol, 1-phenyl-	**64:** 3523f, **83:**79160x
-4-propanol, 1-phenyl-, benzoate ester	**64:** 3523f
-4-prop-2-enal, 1-phenyl-	**84:** P164786w
-4-prop-2-enoic acid, 1-(3-fluorophenyl)-	**85:** P123926u
-4-prop-2-enoic acid, 1-phenyl-	**83:** 10089p
-4-pyruvic acid, 1-methyl-	**49:** 6242e

TABLE 2 (*Continued*)

Compound	Reference
2.1. 1,4-Alkyl- or Aryl-Disubstituted 1*H*-1,2,3-Triazoles (*Continued*)	
-4-pyruvic acid, 1-methyl-, oxime	**49:** 6242e
-4-pyruvic acid, 1-phenyl-, oxime	**45:** 9038b
1-(2,3,5,6-tetrafluoro-4-pyridyl)-4-phenyl-	**78:** 58205x
-4-α-thiopyruvic acid, 1-phenyl-	**45:** 9038b
2.2. 1,5-Alkyl- or Aryl-Disubstituted 1*H*-1,2,3-Triazoles	
1-(2-acetamido-2-deoxy-β-D-glucopyranosyl)-5-phenyl-	**72:** 3718n
1-(2-acetamido-2-deoxy-β-D-glucopyranosyl)-5-phenyl-, 3′,4′,6′-triacetate ester	**72:** 3718n
-1-acetic acid, 5-phenyl-	**80:** 108451p
-1-acetic acid, 5-phenyl-, ethyl ester	**80:** 108451p
5-(2-aminoethyl)-1-phenyl-	**61:** 3097e
1-(4-aminophenyl)-5-phenyl-	**67:** 100071a
5-[(D-*arabino*-1,2,3,4,5-pentahydroxypentyl)-2,3:4,5-bis-*O*-(1-methylethylidene)-1-*C*-]-1-phenyl-, (R)-	**84:** 180505j
5-[(D-*arabino*-1,2,3,4,5-pentahydroxypentyl)-2,3:4,5-bis-*O*-(1-methylethylidene)-1-*C*-]-1-phenyl-, (S)-	**84:** 180505j
5-[(D-*arabino*-1,2,3,4,5-pentahydroxypentyl)-2,3:4,5-bis-*O*-(1-methylethylidene)-1-*C*-]-1-phenyl-, 1-acetate ester, (R)-	**84:** 180505j
5-[(D-*arabino*-1,2,3,4,5-pentahydroxypentyl)-2,3:4,5-bis-*O*-(1-methylethylidene)-1-*C*-]-1-phenyl-, 1-acetate ester, (S)-	**84:** 180505j
5-[(D-*arabino*-1,2,3,4,5-pentahydroxypentyl)-2,3:4,5-bis-*O*-(1-methylethylidene)-1-*C*-]-1-phenyl-, 1-benzoate ester, (R)-	**84:** 180505j
5-(D-*arabino*-1,2,3,4,5-pentahydroxypentyl-1-*C*)-1-phenyl-, 1,2,3,4,5-pentaacetate ester, (R)-	**84:** 180505j
1-(2-azido-1,5-diphenyl-3-oxo-1,4-pentadien-4-yl)-5-phenyl-, (Z,Z)-	**77:** 164609w
1-[1-benzoyl-2-(3-nitrophenyl)ethenyl]-5-methyl-, (Z)-	**77:** 164609w
1-[1-benzoyl-2-(3-nitrophenyl)ethenyl]-5-(3-nitrophenyl)-, (Z)	**77:** 164609w
1-[1-benzoyl-2-(4-nitrophenyl)ethenyl]-5-(4-nitrophenyl)-, (Z),	**77:** 164609w
1-[1-benzoyl-2-(4-nitrophenyl)ethenyl]-5-phenyl-, (Z)-	**77:** 164609w
1-(1-benzoyl-2-phenylethenyl)-5-methyl-, (Z)-	**77:** 164609w
1-(1-benzoyl-2-phenylethenyl)-5-(4-nitrophenyl)-, (Z)-	**77:** 164609w
1-(1-benzoyl-2-phenylethenyl)-5-phenyl-, (Z)-	**77:** 164609w
1-(1-benzoyl-1-propenyl)-5-methyl-, (Z),	**77:** 164609w
1-(1-benzoyl-1-propenyl)-5-phenyl-, (Z)-	**77:** 164609w
5-[2-(benzylamino)ethyl]-1-phenyl-	**61:** 3097d
1-benzyl-5-*tert*-butyl-	**67:** 82171b, **80:** 47912r
1-benzyl-5-methyl-	**67:** 82171b, **80:** 47912r, **84:** 30978b
5-[2-(benzylmethylamino)ethyl]-1-phenyl-	**61:** 3097d

TABLE 2 (*Continued*)

Compound	Reference
2.2. 1,5-Alkyl- or Aryl-Disubstituted 1*H*-1,2,3-Triazoles (*Continued*)	
1-benzyl-5-phenyl-	**46:** 8651i, **51:** 13854i, **53:** 5253d, **80:** 47912r, **84:** 30978b
1-benzyl-5-(2,3,5-*tris*-*O*-benzyl-β-D-ribofuranosyl)-	**77:** 34803u, **82:** 43677m
1-[(benzylsulfonyl)methyl]-5-phenyl-	**72:** 43576w
1-[(benzylthio)methyl]-5-phenyl-	**72:** 43576w
1,5-bis(4-nitrophenyl)-	**74:** 141988t. **79:** 31992k
5-(bromomethyl)-1-phenyl-	**52:** 17246i
1-(4-bromophenyl)-5-(1-butynl)-	**68:** 87243g
1-(4-bromophenyl)-5-(diphenylmethyl)-	**70:** 115081q, **83:** 114297w
1-(4-bromophenyl)-5-ethynyl-	**68:** 87243g
1-(4-bromophenyl)-5-methyl-	**69:** 10402w, **79:** 31992k
1-(4-bromophenyl)-5-(4-nitrophenyl)-	**79:** 31992k
1-(4-bromophenyl)-5-phenyl-	**68:** 105100q, **74:** 141988t, **79:** 31992k
1-(4-bromophenyl)-5-propenyl-	**67:** 100071a
1-(4-bromophenyl)-5-propyl-	**68:** 105100q
1-(4-bromophenyl)-5-vinyl-	**67:** 82171b
1-butyl-5-methyl-	**67:** 82171b
1-butyl-5-phenyl-	**67:** 82170a
1-butyl-5-propyl-	**53:** 16120e
5-(1-butynyl)-1-methyl-	**68:** 87243g
5-(1-butynyl)-1-(3-nitrophenyl)-	**68:** 87243g
5-(1-butynyl)-1-(4-nitrophenyl)-	**68:** 87243g
1-(2-chloro-1-oxo-2,4,6-cycloheptatrien-5-yl)-5-methyl-	**76:** 140726k
1-(4-chlorophenyl)-5-(diphenylmethyl)-	**83:** 114297w
1-(2-chlorophenyl)-5-methyl-	**53:** 16120e
1-(4-chlorophenyl)-5-methyl-	**69:** 10402w
1-(4-chloro-3-oxo-2-phenyl-2*H*-pyridazin-5-yl)-5-phenyl-	**74:** 125600e
1-(4-chlorophenyl)-5-phenyl-	**79:** 78884s
1-(4-chlorophenyl)-5-phenyl-4-phosphono-, monomethyl ester	**79:** 78884s
5-(1-cyclohexen-1-yl)-1-methyl-	**67:** 100071a
5-(1-cyclohexen-1-yl)-1-(4-nitrophenyl)-	**67:** 100071a
5-(1-cyclohexen-1-yl)-1-phenyl-	**67:** 100071a
1-cyclohexyl-5-methyl-	**67:** 82171b
1-(3,7-dibromo-2-hydroxy-1-oxo-2,4,6-cyclo-heptatrien-5-yl)-5-methyl-	**76:** 140726k
1-(3,7-dibromo-2-hydroxy-1-oxo-2,4,6-cyclo-heptatrien-5-yl)-5-phenyl-	**76:** 140726k
1-(2,4-dichlorophenyl)-5-methyl-	**53:** 18946h
1-(2,5-dichlorophenyl)-5-methyl-	**53:** 18946h
4-(diethylphosphinyl)-5-methyl-1-phenyl-	**79:** 126568r
1-(3,4-dihydro-1*H*-2-benzothiopyran-1-yl)-5-phenyl-, S,S-dioxide	**81:** 91486z
1,5-dimethyl-	**67:** 82171b

TABLE 2 (*Continued*)

Compound	Reference
2.2. 1,5-Alkyl- or Aryl-Disubstituted 1*H*-1,2,3-Triazoles (*Continued*)	
5-[2-(dimethylamino)ethyl]-1-phenyl-	**61:** 3097a
1-(1,2-dimethyl-6-benzimidazolyl)-5-methyl-	**74:** 48048b
1-(3,3-dimethyl-1-butenyl)-5-methyl-, (E)-	**72:** 121447w
1-(3,3-dimethyl-1-butenyl)-5-phenyl-	**75:** 19622z
1-(3,3-dimethyl-1-butenyl)-5-phenyl-, (E)-	**72:** 121447w
1-(1,3-dioxo-2-phenyl-2-indanyl)-5-(hydroxymethyl)-	**73:** 98912d
1,5-diphenyl-	**46:** 8651i, **53:** 5253d, **61:** 2926c, **64:** 9713h, **66:** 10887k, **67:** 100071a, **79:** 31992k, **82:** 154936q
-5-^{13}C, 1,5-diphenyl-	**82:** 154936q
5-(diphenylmethyl)-1-(4-methylphenyl)-	**70:** 115081q **83:** 114297w
5-(diphenylmethyl)-1-phenyl-	**83:** 114297w
4-(diphenylphosphinyl)-5-methyl-1-phenyl-	**79:** 126568r
1,5-diphenyl-4-(trimethylsilyl)-	**60:** 5537h
-1-ethanol, 5-(hydroxymethyl)-	**83:** 193181e
-1-ethanol, 5-phenyl-	**83:** 193181e
1-ethyl-5-methyl-	**67:** 82171b
5-ethyl-1-phenyl-	**61:** 3097b, **67:** 100071a
5-ethyl-1-propyl-	**67:** 82170a
5-ethynyl-1-(4-nitrophenyl)-	**68:** 87243g
5-ethynyl-1-phenyl-	**68:** 87243g
1-(3-fluorophenyl)-5-phenyl-	**85:** P21383x
1-β-D-galactopyranosyl-5-phenyl-	**72:** 3718n
1-β-D-glucopyranosyl-5-phenyl-	**72:** 3718n
1-β-D-glucopyranosyl-5-phenyl, 2′,3′,4′,6′-tetra-acetate ester	**72:** 3718n
1-(2-hydroxy-1-oxo-2,4,6-cycloheptatrien-5-yl)-5-methyl-	**70:** P28923x
1-(2-hydroxy-1-oxo-2,4,6-cycloheptatrien-5-yl)-5-phenyl-	**76:** 140726k
1-(2-iodo-3,3-dimethylbutyl)-5-phenyl-	**75:** 19622z
1-(4-iodophenyl)-5-methyl-	**69:** 10402w
5-isobutyl-1-(4-nitrophenyl)-	**58:** 12560h **63:** 11551d
5-isopropenyl-1-methyl-	**69:** P96740q
5-isopropenyl-1-(4-nitrophenyl)-	**67:** 100071a
5-isopropenyl-1-phenyl-	**67:** 100071a
5-isopropyl-1-(4-nitrophenyl)-	**63:** 11551e
5-isopropyl-1-propyl-	**67:** 82170a
1-(5-isoxazolyl)-5-phenyl-	**26:** 1606[4]
-5-methanamine, *N*-isopropyl-1-phenyl-	**83:** 79201m
-5-methanol, 1-(1-nitroethyl)-	**83:** 193181e
-5-methanol, 1-(2-nitroethyl)-	**83:** 193181e
-5-methanol, α-methyl-1-(4-nitrophenyl)-(?)	**52:** P2086a
-5-methanol, 1-(4-nitrophenyl)-	**58:** 12561c
-5-methanol, 1-pentyl-	**83:** 193181e
-5-methanol, 1-phenyl-(?)	**52:** P2085h
1-(4-methoxyphenyl)-5-methyl-	**74:** 141988t, **79:** 31992k

TABLE 2 (*Continued*)

Compound	Reference
2.2. 1,5-Alkyl- or Aryl-Disubstituted 1*H*-1,2,3-Triazoles (*Continued*)	
1-(4-methoxyphenyl)-5-(4-nitrophenyl)-	**74:** 141988t, **79:** 31992k
1-(4-methoxyphenyl)-5-phenyl-	**62:** 16248f, **74:** 141988t, **79:** 31992k
5-[2-(methylamino)ethyl]-1-phenyl-	**61:** 3097c
1-[1-[2-(methylamino)phenyl]-1-methylethyl]-5-(4-nitrophenyl)-	**73:** 14769b
5-methyl-1-(2-methylphenyl)-	**53:** 16120e, **53:** 18946h
1-methyl-5-[[(4-methylphenyl)sulfonyl]methyl]-	**30:** 1027[3]
5-methyl-1-(4-nitrophenyl)-	**58:** 12561a, **62:** 16247g, **69:** 10402w, **79:** 31992k
5-methyl-1-(3-oxo-3-phenyl-1-propenyl)-, (E)-	**77:** 164609w
1-methyl-5-phenyl-	**67:** 100071a, **72:** 43576w
1-methyl-5-phenyl-, monopicrate	**72:** 43576w
5-methyl-1-phenyl-	**69:** 10402w, **79:** 31992k, **80:** 47912r
1-(2-methyl-1-phenyl-5-benzimidazolyl)-5-methyl-	**74:** 48048b
5-methyl-1-(1-phenylethenyl)-	**72:** 121447w, **77:** 164609w
5-methyl-1-(2-phenylethyl)-(?)	**67:** 82171b
5-methyl-1-phenyl-4-phosphono-	**79:** 126568r
5-methyl-1-phenyl-4-phosphono-, dichloride	**79:** 126568r
5-methyl-1-phenyl-4-phosphono-, diethyl ester	**79:** 126568r
1-methyl-5-propenyl-	**67:** 100071a
5-methyl-1-propyl-	**63:** 11552e, **67:** 82171b
1-[(methylsulfonyl)methyl]-5-phenyl-	**72:** 43576w
5-methyl-1-styryl, (E)-	**72:** 121447w
1-[(methylthio)methyl]-5-phenyl-	**72:** 43576w
1-methyl-5-vinyl-	**69:** P96740q
1-(1-nitroethyl)-5-phenyl-	**83:** 193181e
1-(2-nitroethyl)-5-phenyl-	**83:** 193181e
1-(4-nitrophenyl)-5-pentyl-	**68:** 12925x
1-(4-nitrophenyl)-5-phenyl-	**79:** 31992k, **79:** 78884s
5-(4-nitrophenyl)-1-phenyl-	**74:** 141988t, **79:** 31992k
1-(4-nitrophenyl)-5-phenyl-4-phosphono-, monoethyl ester	**79:** 78884s
1-(3-nitrophenyl)-5-propenyl-	**67:** 100071a
1-(4-nitrophenyl)-5-propenyl-	**67:** 100071a
1-(4-nitrophenyl)-5-propyl-	**68:** 12927z
1-(3-nitrophenyl)-5-vinyl-	**67:** 100071a
1-(4-nitrophenyl)-5-vinyl-	**67:** 100071a
1-(3-oxo-3-phenyl-1-propenyl)-5-(4-nitrophenyl)-, (E)-	**77:** 164609w
1-(3-oxo-3-phenyl-1-propenyl)-5-phenyl-, (E)-	**77:** 164609w
1-(pentafluorophenyl)-5-phenyl-	**81:** 135588r
1-pentyl-5-phenyl-	**83:** 193181e
1-phenyl-5-propenyl-	**67:** 100071a
5-phenyl-1-(1-phenylethenyl)-	**72:** 121447w, **77:** 164609w
5-phenyl-1-[(phenylsulfonyl)methyl]-	**72:** 43576w
1-phenyl-5-propyl-	**68:** 105100q
5-phenyl-1-β-D-ribofuranosyl-	**72:** 3718n

TABLE 2 (*Continued*)

Compound	Reference
2.2. 1,5-Alkyl- or Aryl-Disubstituted 1*H*-1,2,3-Triazoles (*Continued*)	
5-phenyl-1-β-D-ribofuranosyl-, 2′,3′,5′-tribenzoate ester	**72:** 3718n
5-phenyl-1-styryl-, (E)-	**71:** 38476p
5-phenyl-1-(2,3,5,6-tetrafluoro-4-pyridinyl)-	**78:** 58205x
5-phenyl-1-tetrazolo[1,5-*b*]pyridazin-6-yl-	**83:** 97213w
1-phenyl-5-vinyl-	**61:** 3097b, **64:** 9713h, **67:** 100071a
-1-propanoic acid, 5-(3-fluorophenyl)-	**85:** P123926u
-1-propanoic acid, 5-phenyl-	**79:** 66254y, **80:** 108451p, **85:** P123926u
-1-propanoic acid, 5-phenyl-, ethyl ester	**80:** 108451p
-5-propanoic acid, 1-phenyl-	**83:** 79160x
-5-propanol, 1-phenyl-	**83:** 79160x
2.3. 2,4-Alkyl- or Aryl-Disubstituted 2*H*-1,2,3-Triazoles	
-4-acetaldehyde, α-(formylmethoxy)-2-phenyl-	**49:** 13901e
-4-acetaldehyde, α-(formylmethoxy)-2-phenyl-, bis(4-nitrophenylhydrazone)	**49:** 13901e
-4-acetamide, α,α-dichloro-*N*,*N*,2-trimethyl-	**74:** 76377b
2-[4-(acetamidomethyl)phenyl]-4-phenyl-	**72:** P100716t
2-(4-acetamidophenyl)-4-(D-*arabino*-1,2,3,4-tetrahydroxybutyl)-, tetraacetate ester	**58:** 11454d
2-(4-acetamidophenyl)-4-methyl-	**46:** 6123i
-4-acetic acid, 2-(3-chloro-4-methylphenyl)-, sodium salt	**81:** P51143d
-4-acetic acid, 2-(3-chlorophenyl)-, sodium salt	**81:** P51143d
-4-acetic acid, 2-(4-methylphenyl)-, sodium salt	**81:** P51143d
-2-acetic acid, 4-phenyl-	**79:** 66254y, **80:** 108451p
-4-acetic acid, α-amino-2-phenyl-	**46:** 6123d
-4-acetic acid, α,α-dichloro-2-methyl-, ethyl ester	**74:** 76377b
-4-acetic acid, 2-phenyl-	**56:** 5943e
-4-acetic acid, 2-phenyl-, methyl ester	**56:** 5943e
-4-acetonitrile, 2-(3-chlorophenyl)-	**81:** P51143d
-4-acetonitrile, α,α-dichloro-2-methyl-	**74:** 76377b
-4-acetonitrile, 2-phenyl-	**46:** 6123e, **56:** 5943d
4-(3-acetoxy-4-chlorotetrahydrofuryl)-2-phenyl-, [2S-(2α, 3β, 4α)]-	**80:** 133719q
4-(2-acetoxyphenyl)-2-phenyl-	**71:** 22079a
4-[2-(6-acetyl-3-oxo-6-azabicyclo[3.1.0]hexyl)]-2-phenyl-, [1R-(1α, 2β, 5α)]-	**77:** 165027s
4-[1-[(acetyloxy)methyl]propyl-2,3-bis(acetyloxy)-β-D-glucopyranosidyl]-2-phenyl-, tetraacetate ester, [1S(R), 2R]-	**76:** 14832d
4-[1-[(acetyloxy)methyl]propyl-2,3-bis(acetyloxy)-β-D-glucopyranosidyl]-2-phenyl-, 2,3,4,6-tetraacetate ester, [1R-[1R*(S*), 2S*]]-	**79:** 66704v

TABLE 2 (*Continued*)

Compound	Reference
2.3. 2,4-Alkyl- or Aryl-Disubstituted 2*H*-1,2,3-Triazoles (*Continued*)	
4-[1-[(acetyloxy)methyl]propyl-2,3-bis(acetyloxy)]4-*O*-(2,3,4,6-tetra-*O*-acetyl-β-D-glucopyranosyl)-β-D-glucopyranosidyl]-2-phenyl-, triacetate ester	**76:** 14832d
4-[1-[(acetyloxy)methyl]propyl-2,3-bis(acetyloxy) *O*-2,3,4,6-tetra-*O*-acetyl-β-D-glucopyranosyl-(1*4)]-, triacetate ester	**76:** 14832d
4-[1-[(acetyloxy)methyl]propyl-2,3-bis(acetyloxy) *O*-2,3,4,6-tetra-*O*-acetyl-β-D-glucopyranosyl-(1*4)-*O*-2,3,6-tri-*O*-acetyl-β-D-glucopyranosyl-(1*4)-*O*-2,3,6-tri-*O*-acetyl-β-D-glucopyranosyl-(1*4)]-, triacetate ester	**76:** 14832d
-4-acrylic acid, 2-phenyl-	**43:** 2619c, **56:** 5943e, **64:** 8276b
-4-acrylic acid, 2-phenyl-, ethyl ester	**43:** 2619c
4-(1-amino-3-oxo-1-butenyl)-2-phenyl-	**83:** P206088g
2-(4-aminophenyl)-4-(D-*arabino*-1,2,3,4-tetrahydroxybutyl)-, tetraacetate ester	**57:** 7353f, **58:** 11454f
2-(4-aminophenyl)-4-(D-*lyxo*-1,2,3,4-tetrahydroxybutyl)-, tetraacetate ester	**59:** 11691a
2-(4-aminophenyl)-4-(L-*xylo*-1,2,3,4-tetrahydroxybutyl)-, tetraacetate ester	**59:** 11691a
2-(4-aminophenyl)-4-methyl-	**46:** 6123i
4-(4-aminophenyl)-2-methyl-	**80:** P49275j
2-[4-(aminosulfonyl)phenyl]-4-(4-formylphenyl)-	**84:** P181616h
4-(D-*arabino*-1,2,3,4-tetrahydroxybutyl)-2-(4-bromo-2-chlorophenyl)-	**63:** 1850g
4-(D-*arabino*-1,2,3,4-tetrahydroxybutyl)-2-(4-bromophenyl)-, 4′-(4-toluenesulfonate) ester	**61:** 5738c
4-(D-*arabino*-1,2,3,4-tetrahydroxybutyl)-2-(4-bromo-2,5-dimethylphenyl)-	**63:** 1850h
4-(D-*arabino*-1,2,3,4-tetrahydroxybutyl)-2-(3-carboxyphenyl)-	**70:** 78274g
4-(D-*arabino*-1,2,3,4-tetrahydroxybutyl)-2-(4-carboxyphenyl)-	**70:** 78274g
4-(D-*arabino*-1,2,3,4-tetrahydroxybutyl)-2-(3-chlorophenyl)-	**70:** 78274g
4-(D-*arabino*-1,2,3,4-tetrahydroxybutyl)-2-(4-chlorophenyl)-	**70:** 78274g
4-(D-*arabino*-1,2,3,4-tetrahydroxybutyl)-2-(2-ethylphenyl)-	**60:** 643h
4-(D-*arabino*-1,2,3,4-tetrahydroxybutyl)-2-(2-ethylphenyl)-, tetraacetate ester	**60:** 643h
4-(D-*arabino*-1,2,3,4-tetrahydroxybutyl)-2-[4-(hydroxyamino)phenyl]-	**61:** 5738c
4-(D-*arabino*-1,2,3,4-tetrahydroxybutyl)-2-[4-(hydroxyamino)phenyl]-, tetraacetate ester	**61:** 5738c
4-(D-*arabino*-1,2,3,4-tetrahydroxybutyl)-2-(2-methylphenyl)-	**54:** 1504d

TABLE 2 (*Continued*)

Compound	Reference
2.3. 2,4-Alkyl- or Aryl-Disubstituted 2*H*-1,2,3-Triazoles (*Continued*)	
4-(D-*arabino*-1,2,3,4-tetrahydroxybutyl)-2-(3-methylphenyl)-	**54:** 1504d, **55:** 27071f, **70:** 78274g
4-(D-*arabino*-1,2,3,4-tetrahydroxybutyl)-2-(4-methylphenyl)-	**53:** 344h, **54:** 9899e, **55:** 27071f, **55:** 3562h, **70:** 78274g
4-(D-*arabino*-1,2,3,4-tetrahydroxybutyl)-2-(4-methylphenyl)-, 4′-(4-toluenesulfonate) ester	**61:** 5738e
4-(D-*arabino*-1,2,3,4-tetrahydroxybutyl)-2-(4-nitrophenyl)-	**57:** 7353e
4-(D-*arabino*-1,2,3,4-tetrahydroxybutyl)-2-phenyl-	**57:** 2297f, **61:** 14759e, **70:** 78274g
4-(D-*arabino*-1,2,3,4-tetrahydroxybutyl)-2-phenyl-, 4′-benzoate ester	**57:** 2297f
4-(D-*arabino*-1,2,3,4-tetrahydroxybutyl)-2-phenyl-, 3,4-diacetate ester	**71:** 81672m
4-(D-*arabino*-1,2,3,4-tetrahydroxybutyl)-2-phenyl-, tetraacetate ester	**61:** 10754g
4-(D-*arabino*-1,2,3,4-tetrahydroxybutyl)-2-phenyl-, 4′-(4-toluenesulfonate) ester	**61:** 5738c
4-(D-*arabino*-1,2,3,4-tetrahydroxybutyl)-2-(4-sulfamoylphenyl)-	**50:** 10009g
4-(L-*arabino*-1,2,3-trihydroxybutyl)-2-phenyl-	**68:** 69231x
4-(4-azidotetrahydro-3-furanol-2-yl)-2-phenyl-, methanesulfonate ester, [2S-(2α, 3β, 4α)]-	**80:** 133719q
4-(4-azidotetrahydro-3-furanol-2-yl)-2-phenyl-, 4-toluenesulfonate ester, [2S-(2α, 3β, 4α)]-	**77:** 165027s, **80:** 133719q
2-(4-benzamidophenyl)-4-methyl-	**46:** 6123i
4-(2-benzoylethenyl)-2-phenyl-	**43:** 2619b
2-benzyl-5-phenyl-4-phosphono-, diethyl ester	**80:** 108452q
2-(4-biphenylyl)-4-phenyl-	**58:** 6819f
4-[1,3-bis[4-(benzotriazol-2-yl)-2-chlorophenyl]-2-propyl]-2-phenyl-	**85:** 5562z
2,4-bis(4-bromophenyl)-	**69:** 76826u
2,4-bis(4-nitrophenyl)-	**69:** 76826u
4-[4′-(bromomethyl)[1,1′-biphenyl]-4-yl]-2-phenyl-	**79:** P147435k
4-(bromomethyl)-2-(4-bromophenyl)-	**66:** 37841p
4-(bromomethyl)-2-(3-chlorophenyl)-	**81:** P51143d
4-(bromomethyl)-2-phenyl-	**46:** 6123e, **56:** 5943d
4-[4-(bromomethyl)phenyl]-2-phenyl-	**84:** P181616h
2-(4-bromophenyl)-4-(D-*galacto*-1,2,3,4,5-pentahydroxypentyl)-	**69:** 106976n
2-(4-bromophenyl)-4-(D-*manno*-1,2,3,4,5-pentahydroxypentyl)-	**69:** 106976n
2-(4-bromophenyl)-4-methyl-	**59:** 13483f, **66:** 37841p
2-(4-bromophenyl)-4-phenyl-	**69:** 76826u
4-(4-bromophenyl)-2-phenyl-	**69:** 76826u
4-(3-bromopropyl)-2-phenyl-	**56:** 5943f
-4-butanenitrile, 2-phenyl-	**56:** 5943g
-4-butanoamide, 2-phenyl-	**56:** 5943g

TABLE 2 (*Continued*)

Compound	Reference
2.3. 2,4-Alkyl- or Aryl-Disubstituted 2*H*-1,2,3-Triazoles (*Continued*)	
-2-butanoic acid, 4-phenyl-	**85:** P123926u
-4-butanoic acid, 2-phenyl-	**56:** 5943g
-4-butanoic acid, 2-phenyl-, methyl ester	**56:** 5943g
4-(1-butyl-4,4-dimethyl-2-imidazolidinyl)-2-phenyl-	**43:** 2619c
4-(1-butyl-4,4-dimethyl-2-imidazolin-2-yl)-2-phenyl-	**43:** 2619e
2-[4-(carboxymethyl)phenyl]-4-phenyl-	**72:** P100716t
2-(4-carboxyphenyl)-4-(4-formylphenyl)-	**84:** P181616h
2-(4-carboxyphenyl)-4-methyl-	**83:** P61732a
2-(4-carboxyphenyl)-4-methyl-, *N*-oxide	**83:** P61732a
2-(3-carboxyphenyl)-4-phenyl-	**72:** P100716t, **74:** P65594u
2-(4-chloro-2-hydroxy-5-phenoxyphenyl)-4-phenyl-	**82:** P4995u
2-(5-chloro-2-hydroxyphenyl)-4-phenyl-	**71:** P124445j
2-(5-chloro-2-methoxyphenyl)-4-phenyl-, 1-oxide	**81:** P79389n
2-[4-(chloromethyl)phenyl]-4-phenyl-	**74:** P65594u
2-(chloro-4-nitrophenyl)-4-methyl-(?)	**59:** 13483f
4-(4-chlorophenyl)-2-(3-cyano-4-methylphenyl)-, *N*-oxide	**83:** P81213k
5-(4-chlorophenyl)-2-ethyl-4-(triphenylphosphonio)-, iodide	**79:** 66254y
4-[*N*-(4-chlorophenyl)formimidoyl]-2-phenyl-	**85:** 5562z
2-(4-chlorophenyl)-4-(4-formylphenyl)-	**84:** P181616h
2-[[[4-(2-chlorophenyl)imino]methylene]phenyl]-4-phenyl-	**84:** P181618k
4-[[[4-(2-chlorophenyl)imino]methylene]phenyl]-2-phenyl-	**84:** P181618k
2-(2-chlorophenyl)-4-methyl-	**84:** 135554p
2-(4-chlorophenyl)-4-phenyl-	**69:** 76826u
4-(4-chlorotetrahydro-3-furanol-2-yl)-2-phenyl-, [S-(2α, 3β, 4α)]-	**80:** 133719q
4-[2-(4-cinnolinyl)ethenyl]-2-phenyl-	**49:** 5488g
2-[4-(dichloromethyl)phenyl]-4-phenyl-	**74:** P65594u
2-(2,4-dichlorophenyl)-4-(2-hydroxyphenyl)-	**71:** 22079a
5-(diethoxyphosphinyl)-2-(2-methoxy-5-methylphenyl)-4-methyl-	**77:** P7313c
4-[(diethoxyphosphinyl)methyl]-2-phenyl-	**79:** P106145x
4-[1-(1,2-dihydroxyethyl)-2-(2,2-dimethyl-1,3-dioxolan-4-yl)]-2-phenyl-	**71:** 81672m
4-[1-(1,2-dihydroxyethyl)-2-(2,2-dimethyl-1,3-dioxolan-4-yl)]-2-phenyl-, diacetate ester	**71:** 81672m
4-[1-(1,2-dihydroxyethyl)-2-(2-phenyl-1,3-dioxolan-4-yl)]-2-phenyl-, stereoisomer	**71:** 81672m
4-[1-(1,2-dihydroxyethyl)-2-(2-phenyl-1,3-dioxolan-4-yl)]-2-phenyl-, stereoisomer	**71:** 81672m
4-[1-(1,2-dihydroxyethyl)-2-(2-phenyl-1,3-dioxolan-4-yl)]-2-phenyl-, diacetate ester, stereoisomer	**71:** 81672m
4-[1-(1,2-dihydroxyethyl)-2-(2-phenyl-1,3-dioxolan-4-yl)]-2-phenyl-, diacetate ester, stereoisomer	**71:** 81672m

TABLE 2 (*Continued*)

Compound	Reference
2.3. 2,4-Alkyl- or Aryl-Disubstituted 2*H*-1,2,3-Triazoles (*Continued*)	
4-[4-(1,2-dihydroxyethyl)-2-(2-phenyl-1,3-dioxolan-5-yl)]-2-phenyl-	**71:** 81762m, **77:** 114720e
4-[1-(1,2-dihydroxyethyl)-2-(2-phenyl-1,3-dioxolan-5-yl)]-2-phenyl-, [2S-[2α, 4α(S*), 5β]]-	**77:** 114720e
4-[4-(1,2-dihydroxyethyl)-2-(2-phenyl-1,3-dioxolan-5-yl)]-2-phenyl-, diacetate ester	**71:** 81672m
4-[1-(1,2-dihydroxy-3,4-dimethoxybutyl)]-2-phenyl-	**71:** 81672m
4-[1-(1,2-dihydroxy-3,4-dimethoxybutyl)]-2-phenyl-, diacetate ester	**71:** 81672m
4-[4-(1,2-dihydroxy-3,4-dimethoxybutyl)]-2-phenyl-	**71:** 81672m
4-[4-(1,2-dihydroxy-3,4-dimethoxybutyl)]-2-phenyl-, diacetate ester	**71:** 81672m
4-[1-(1,2-dihydroxy-3,4-epoxybutyl)]-2-phenyl-	**69:** 27660b
4-[1-(1,3-dihydroxy-2-methoxybutyl)]-2-phenyl-, diacetate ester, [1S-(1R*, 2R*, 3R*)]-	**79:** 66704v
4-[[2-(2,3-dihydroxy)propyl]-D-apiofuranosidyl]-	**73:** 31627w
4-[5-(1,2-dimethoxyethyl)-2-phenyl-1,3-dioxolan-4-yl]-2-phenyl-, stereoisomer	**71:** 81672m
4-[5-(1,2-dimethoxyethyl)-2-phenyl-1,3-dioxolan-4-yl]-2-phenyl-, stereoisomer	**71:** 81672m
2,4-dimethyl-	**85:** 77337c
4-[1-[3-(dimethylamino)propyl]piperazinyl]-2-phenyl-, 1:2 compound with maleic acid	**68:** P59611g
4-[2-(2,2-dimethyl-1,3-dioxolan-4-yl)-1,2-dimethoxyethyl]-2-phenyl-	**71:** 81672m
4-[5-(2,2-dimethyl-1,3-dioxolan-4-yl)-2,2-dimethyl-1,3-dioxolan-4-yl]-2-phenyl-	**71:** 81672m
4-[4,4-dimethyl-1-(4-methylphenyl)-2-imidazolidinyl]-2-phenyl-	**43:** 2619c
4-(4,4-dimethyl-1-phenyl-2-imidazolidinyl)-2-phenyl-	**43:** 2619c
4-[2-[6-(2,4-dinitrophenyl)-3-oxa-6-azabicyclo[3.1.0]hexyl]]-2-phenyl-, [1S-(1α, 2β, 5α)]-	**77:** 165027s
2-(2,4-dinitrophenyl)-5-phenyl-4-(triphenylphosphonio)-, bromide	**79:** 66254y
4-(3,6-dioxabicyclo[3.1.0]hex-2-yl)-2-phenyl-, [1R-(1α, 2β, 5α)]	**77:** 165027s, **80:** 133719q
4-(3,6-dioxabicyclo[3.1.0]hex-2-yl)-2-phenyl-, [1S-(1α, 2α, 5α)]-	**77:** 165027s, **80:** 133719q
2-(1,3-dioxo-2-ethyl-1*H*-benz[*de*]isoquinolin-2*H*-6-yl)-4-ethyl-	**82:** P59922y
2,4-diphenyl-	**36:** 2862[9], **69:** 76826u **76:** 153646j, **83:** 96642y
2-[4-(1,3-diphenyl-2-imidazolidinyl)-3-hydroxyphenyl]-4-phenyl-	**84:** 164334x
4-(4-*C*-L-4-(4-*C*-L-erythrosyl)-2-phenyl-	**69:** 27660b
4-(4-*C*-L-erythrosyl)-2-phenyl-, phenylhydrazone	**69:** 27660b
4-(1-L-*erythro*-1,2,3-trihydroxypropyl)-2-phenyl-	**68:** 69231x, **69:** 36377s
2-[2-(1-ethyl-2-oxo-1*H*-benz[*cd*]indol-6-yl) 2*H*-benzotriazol-5-yl]-4-phenyl-	**82:** P87680m

TABLE 2 (*Continued*)

Compound	Reference
2.3. 2,4-Alkyl- or Aryl-Disubstituted 2*H*-1,2,3-Triazoles (*Continued*)	
4-formazyl-2-phenyl-	**49:** 12451h
4-(3,4-furandiol-2-yl)-2-phenyl-, bis(4-toluenesulfonate) ester, [2S-(2α, 3β, 4β)]-	**77:** 165027s, **80:** 133719q
4-(3,4-furandiol-2-yl)-2-phenyl-, dimethanesulfonate ester, [2S-(2α, 3β, 4β)]-	**77:** 165027s, **80:** 133719q
2-(2-fluorophenyl)-4-[1-(1,2,3,4-tetrahydroxybutyl)]-, [1R-(1R*, 2S*, 3R*)]-	**80:** 48283y
2-(4-fluorophenyl)-4-[1-(1,2,3-trihydroxypropyl)]-, triacetate ester, [R-(R*, S*)]-	**77:** 19906m
4-(4-fluorotetrahydro-3-furanol-2-yl)-2-phenyl-, [2S-(2α, 3β, 4α)]-	**80:** 133719q
4-(4-fluorotetrahydro-3-furanol-2-yl)-2-phenyl-, acetate ester, [2S-(2α, 3β, 4α)]-	**80:** 133719q
4-(4-formylphenyl)-2-(2-methoxyphenyl)-	**84:** P181616h
4-(4-formylphenyl)-2-(4-methoxyphenyl)-	**84:** P181616h
4-(4-formylphenyl)-2-[4-(methylsulfonyl)phenyl]-	**84:** P181616h
2-(4-formylphenyl)-4-phenyl-	**84:** P181616h
4-(1-D-*glycero*-1,2-dihydroxyethyl)-2-phenyl-	**69:** 36377s
4-(1-*C*-D-*glycero*-L-glucohexitolyl)-2-phenyl-	**75:** 36556y
4-(1-*C*-D-*glycero*-L-glucohexitolyl)-2-phenyl-, 1,2,3,4,5,6-hexaacetate ester	**75:** 36556y
4-(1-*C*-D-*glycero*-L-glucohexitolyl)-2-phenyl-, 1,2,3,4,5,6-hexabenzoate ester	**75:** 36556y
4-(4-*C*-L-*glycero*-tetrosulosyl)-2-phenyl-, bis(phenylhydrazone)	**69:** 27660b
4-(4-*C*-L-*glycero*-tetrosulosyl)-2-phenyl-, 3,4-diacetate ester, bis(phenylhydrazone)	**69:** 27660b
4-(2-D-*glycero*-1,2,3-trihydroxypropyl)-2-phenyl-	**73:** 25790c
4-[1-(hydroxymethyl)-2,3-(dihydroxypropyl)-α-D-glucopyranosidyl]-2-phenyl-	**68:** 69251d
2-[4-(hydroxymethyl)phenyl]-4-phenyl-	**74:** P65594u
2-(2-hydroxyphenyl)-4-methyl-	**71:** P124445j
2-(2-hydroxyphenyl)-4-phenyl-	**71:** P124445j
2-(3-hydroxyphenyl)-4-phenyl-	**84:** 164334x
4-(2-hydroxyphenyl)-2-phenyl-	**71:** 22079a
4-(iodomethyl)-2-phenyl-	**50:** 16685b
2-(4-iodophenyl)-4-phenyl-	**69:** 76826u
4-(1-isopropyl-4,4-dimethyl-2-imidazolidinyl)-2-phenyl-	**43:** 2619c
4-[(1-isopropyl-4,4-dimethyl-2-imidazolin-2-yl)methyl]-2-phenyl-	**56:** 5943e
4-(1-isopropyl-4,4-dimethyl-2-imidazolin-2-yl)-2-phenyl-	**43:** 2619c
4-(1,2-*O*-isopropylidene-4-*C*-α-D-*xylo*-tetrafuranosyl)-2-phenyl-	**69:** 27660b
4-(D-*lyxo*-1,2,3,4-tetrahydroxybutyl)-2-phenyl-	**61:** 14759e
4-(D-*lyxo*-1,2,3,4-tetrahydroxybutyl)-2-phenyl-, tetraacetate ester	**61:** 10754g
4-(1-L-*lyxo*-1,2,3-trihydroxybutyl)-2-phenyl-	**68:** 69231x

Compound	Reference
2.3. 2,4-Alkyl- or Aryl-Disubstituted 2*H*-1,2,3-Triazoles (*Continued*)	
4-(D-*lyxo*-1,2,3-trihydroxybutyl)-β-D-xylopyranosid-4-yl)-2-phenyl-	**72:** 3666u
4-(D-*lyxo*-1,2,3-trihydroxybutyl)-β-D-xylopyranosid-4-yl)-2-phenyl-, hexaacetate ester	**72:** 3666u
-4-methanol, α-[5-(acetyloxy)-1,3-doxan-4-yl]-2-phenyl-, acetate ester, [4S-[4α(S*), 5α]]-	**77:** 114720e
-4-methanol, α-(aminometyl)-2-phenyl-	**50:** 4923i
-4-methanol, 2-(4-bromo-3-chlorophenyl)-	**63:** 1850h
-4-methanol, 2-(4-bromo-3-methylphenyl)-	**63:** 1850h
-4-methanol, 2-(4-bromophenyl)-	**63:** 1850h
-4-methanol, 2-(3-chlorophenyl)-	**63:** 1850h
-4-methanol, 2-(4-chlorophenyl)-	**63:** 1850h
-4-methanol, α-(ethylaminomethyl)-2-phenyl-	**50:** 4924c
-4-methanol, α-(5-hydroxy-1,3-dioxan-4-yl)-2-phenyl-, [4S-[4α(S*), 5α]]-	**77:** 114720e
-4-methanol, α-(5-hydroxy-2-phenyl-1,3-dioxan-4-yl)-2-phenyl-	**71:** 81672m
-4-methanol, α-(5-hydroxy-2-phenyl-1,3-dioxan-4-yl)-2-phenyl-, diacetate ester	**71:** 81672m
-4-methanol, α-(methylaminomethyl)-2-phenyl-	**50:** 4924h
-4-methanol, α-methyl-2-phenyl-	**43:** 2619d
-4-methanol, 2-(3-methylphenyl)-	**63:** 1850h
-4-methanol, 2-(4-methylphenyl)-	**63:** 1850h
-4-methanol, 2-phenyl-	**43:** 2619b, **56:** 5943d, **63:** 1850h
-4-methanol, 2-phenyl-α-(2-phenyl-1,3-dioxolan-4-yl)-	**64:** 8276b
2-[(methoxycarbonyl)phenyl]-4-phenyl-	**74:** P65594u
4-[methoxy(-5-methoxy-1,3-dioxan-4-yl)methyl]-2-phenyl-, [4S-[4α(S*), 5α]]-	**77:** 114720e
4-(2-methoxyphenyl)-2-phenyl-	**71:** 22079a
4-[1-(4-methoxy-1,2,3-trihydroxybutyl)]-2-phenyl-, triacetate ester, [1R-(1R*, 2R*, 3R*)]-	**76:** 14832d
4-[4-(4-methoxy-1,2,3-trihydroxybutyl)]-2-phenyl-, triacetate ester, [2R-(2R*, 3S*, 4R*)]-	**76:** 14832d
4-methyl-2-(4-nitrophenyl)-	**46:** 6123i, **57:** 15100e
4-methyl-2-(4-nitrophenyl)-, *N*-oxide	**66:** 28250t
2-methyl-4-phenyl-	**67:** 108605y
4-methyl-2-phenyl-	**46:** 6123h, **57:** 15100e **84:** 135554p
4-methyl-2-phenyl-, *N*-oxide	**66:** 28250t
4-methyl-2-(4-methylphenyl)-	**57:** 15100e
2-(4-methylphenyl)-4-phenyl-	**57:** 15100i
4-(4-methylphenyl)-2-phenyl-	**76:** 153646j
2-methyl-4-(trichloromethyl)-	**74:** 76377b
4-(2-nitroethenyl)-2-phenyl-	**62:** 15262a
2-(4-nitrophenyl)-4-phenyl-	**69:** 76826u
4-[2-(3-oxa-6-azabicyclo[3.1.0]hexyl)]-2-phenyl-, [1S-(1α, 2β, 5α)]-	**80:** 133719q

TABLE 2 (*Continued*)

Compound	Reference
2.3. 2,4-Alkyl- or Aryl-Disubstituted 2*H*-1,2,3-Triazoles (*Continued*)	
4-[1-(phenylazo)-4-*C*-L-erythrosyl]-2-phenyl-, phenylhydrazone	**69:** 27660b
2-phenyl-4-[2-phenyl-5-(2-phenyl-1,3-dioxolan-4-yl)-1,3-dioxolan-4-yl]-	**71:** 81672m, **72:** 121865f
2-phenyl-4-[2-phenyl-5-(2-phenyl-1,3-dioxolan-4-yl)-1,3-dioxolan-4-yl]-, stereoisomer	**72:** 121865f
2-phenyl-4-[2-phenyl-5-(2-phenyl-1,3-dioxolan-4-yl)-1,3-dioxolan-4-yl]-, stereoisomer	**72:** 121865f
2-phenyl-4-[2-phenyl-5-(2-phenyl-1,3-dioxolan-4-yl)-1,3-dioxolan-4-yl]-, stereoisomer	**72:** 121865f
2-phenyl-4-(tetrahydro[1,3]dioxino[5,4-*d*]-1,3-dioxin-4-yl)-, [4R-(4α, 4aβ, 8aα)]-	**77:** 114720e
2-phenyl-4-(1,2,3,4-tetrahydroxybutyl)-, [1R-(1R*, 2S*, 3R*)]	**84:** P180230r
2-phenyl-4-(1,2,3-trimethoxypropyl)-, [R-(R*, R*)]-	**76:** 113449r
2-phenyl-4-(D-*xylo*-1,2,3,4-tetrahydroxybutyl)-	**61:** 14759e
2-phenyl-4-(D-*xylo*-1,2,3,4-tetrahydroxybutyl)-, tetraacetate ester	**61:** 10754g
-4-propanamide, 2-phenyl-	**56:** 5943f
-2-propanoic acid, 4-(3-acetamidophenyl)-	**85:** P123926u
-2-propanoic acid, 4-[4-(acetylamino)phenyl]-	**84:** P164786w
-2-propanoic acid, 4-(3-aminophenyl)-	**84:** P164786w
-2-propanoic acid, 4-(3-bromophenyl)-	**85:** P123926u
-2-propanoic acid, 4-(4-bromophenyl)-	**84:** P164786w
-4-propanoic acid, 2-(3-bromophenyl)-	**84:** P164786w
-2-propanoic acid, 4-(3-chlorophenyl)-	**84:** P164786w
-4-propanoic acid, 2-(3-chlorophenyl)-	**83:** P10089p
-2-propanoic acid, 4-(3,5-dichlorophenyl)-	**84:** P164786w
-2-propanoic acid, 4-(3-methylphenyl)-	**85:** P123926u
-2-propanoic acid, 4-(4-methylphenyl)-	**84:** P164786w
-2-propanoic acid, 4-(3-nitrophenyl)-	**84:** P164786w
-2-propanoic acid, 4-phenyl-	**79:** 66254y, **80:** 108451p
-2-propanoic acid, 4-phenyl-, ethyl ester	**79:** 66254y
-4-propanoic acid, 2-phenyl-	**56:** 5943f, **64:** 8276b
-4-propanoic acid, 2-phenyl-, methyl ester	**56:** 5943f
-4-propanol, 2-phenyl-	**56:** 5943f
-4-prop-2-enoic acid, 2-(3-bromophenyl)-	**85:** P123926u
-3-prop-2-enoic acid, 4-(4-chlorophenyl)-, ethyl ester, (E)-	**79:** 66254y
-3-prop-2-enoic acid, 4-(4-chlorophenyl)-, ethyl ester, (Z)-	**79:** 66254y
-4-prop-2-enoic acid, 2-(3-chlorophenyl)-	**83:** P10089p
-3-prop-2-enoic acid, 4-phenyl-, ethyl ester, (E)-	**79:** 66254y
-3-prop-2-enoic acid, 4-phenyl-, ethyl ester, (Z)-	**79:** 66254y
-4-prop-2-enoic acid, 2-phenyl-	**83:** P10089p
4-(tetrahydro-3,4-furandiol-2-yl)-2-phenyl-, [2S-(2α, 3β, 4α)]-	**80:** 133719q

TABLE 2. (*Continued*)

Compound	Reference
2.3. 2,4-Alkyl- or Aryl-Disubstituted 2*H*-1,2,3-Triazoles (*Continued*)	
4-(tetrahydro-3,4-furandiol-2-yl)-2-phenyl-, [2S-(2α, 3β, 4β)]-	**77:** 165027s, **80:** 133719q
4-(tetrahydro-3,4-furandiol-2-yl)-2-phenyl-, 3-(4-toluenesulfonate) ester	**77:** 165027s
4-(1,2,3,4-tetrahydroxybutyl)-2-(4-bromophenyl)-, tetraacetate ester, [1R-(1R*, 2S*, 3R*)]-	**77:** 19906m
4-(1,2,3,4-tetrahydroxybutyl)-2-(4-nitrophenyl)-, tetraacetate ester, [1R-(1R*, 2R*, 3R*)]-	**77:** 19906m
4-(1,2,3,4-tetrahydroxybutyl)-2-(4-nitrophenyl)-, tetraacetate ester, [1R-(1R*, 2S*, 3R*)]-	**77:** 19906m
4-(1,2,3,4-tetrahydroxybutyl)-2-(4-nitrophenyl)-, tetraacetate ester, [1R-(1R*, 2S*, 3S*)]-	**77:** 19906m
4-(1,2,3,4-tetrahydroxybutyl)-2-phenyl-, tetraacetate ester, [1R-(1R*, 2R*, 3R*)]-	**77:** 19906m
4-(1,2,3,4-tetrahydroxybutyl)-2-phenyl-, tetraacetate ester, [1R-(1R*, 2S*, 3S*)]-	**77:** 19906m
4-(D-*threo*-1,2,3-trihydroxypropyl)-2-phenyl-	**68:** 69231x
4-[1-(1,2,3-trihydroxybutyl)]-2-(4-nitrophenyl)-, triacetate ester, [1S-(1R*, 2R*, 3R*)]-	**77:** 19906m
4-[1-(1,2,3-trihydroxybutyl)]-2-phenyl-, 1,2,3-triacetate ester, [1S-(1R*, 2S*, 3S*)]-	**76:** 14832d
4-(1,2,3-trihydroxypropyl)-2-phenyl-, [S-(R*, S*)]-	**76:** 113449r
4-(1,2,3-trihydroxypropyl)-2-phenyl-, triacetate ester, [R-(R*, R*)]-	**79:** 66704v
4-(1,2,3-trihydroxypropyl)-2-phenyl-, 1,2,3-triacetate ester, [S-(R*, R*)]-	**77:** 19906m
4-[1-[1,2,3-*tris*(acetyloxy)-4-α-D-galactopyranosidylbutyl]]-2-phenyl-, tetraacetate ester	**76:** 14832d
4-[1-[1,2,3-*tris*(acetyloxy)-4-β-D-glucopyranosidylbutyl]]-2-phenyl-, tetraacetate ester	**76:** 14832d
4-(L-*xylo*-1,2,3,4-tetrahydroxybutyl)-2-phenyl-	**68:** 69231x
2.4. 4,5-Alkyl- or Aryl-Disubstituted *v*-Triazoles	
4-(2-acetamidoethyl)-5-phenyl-	**45:** 9038a
-4-acetonitrile, 5-phenyl-	**45:** 9037g
4-(2-aminoethyl)-5-phenyl-	**45:** 9037g
5-(9-anthracenyl)-4-(2-benzoxazolyl)-	**80:** P72087h
5-(9-anthracenyl)-4-(5,7-dichloro-2-benzoxazolyl)-	**80:** 3439n
5-(9-anthracenyl)-4-(5-phenyl-1,3,4-oxadiazol-2-yl)-	**80:** 3439n
4-(2-benzoxazolyl)-5-(3,4,5-trimethoxyphenyl)-	**80:** P72087h
4-(2-carboxyphenyl)-5-methyl-	**84:** 59333y
4-(chlorodiphenylmethyl)-5-(4-chlorophenyl)-, monohydrochloride	**80:** 133389g
4-(chlorodiphenylmethyl)-5-phenyl-, monohydrochloride	**80:** 133389g
2-(chloromercurio)-5-phenyl-4-(triphenylphosphonio)-, chloride	**79:** 66254y
5-(4-chlorophenyl)-4-(5,6-dimethyl-2-benzoxazolyl)-	**80:** P72087h

TABLE 2 (*Continued*)

Compound	Reference
2.4. 4,5-Alkyl- or Aryl-Disubstituted *v*-Triazoles (*Continued*)	
5-(4-chlorophenyl)-4-(diphenylmethylene)-	**80:** 133389g
5-(4-chlorophenyl)-4-(2-oxo-2*H*-pyran-6-yl)-	**80:** 3439n
4-cyclobutyl-5-propyl-	**71:** 112865h
5-(5,7-dichloro-2-benzoxazolyl)-4-[(4-dimethylamino)-phenyl]-	**80:** 3439n
4-(5,7-dichloro-2-benzoxazolyl)-5-(4-methoxyphenyl)-	**80:** 3439n
4-(5,7-dichloro-2-benzoxazolyl)-5-(4-nitrophenyl)-	**80:** P72087h
4-(5,7-dichloro-2-benzoxazolyl)-5-(3-pyridyl)-	**80:** 3439n
4,5-diethyl-	**78:** 29956n
4,5-diethyl-, monohydrochloride	**78:** 29956n
4,5-diethyl-2-(tributylstannyl)-	**78:** 29956n
4,5-diethyl-2-(trimethylstannyl)-	**78:** 29956n
4,5-dimethyl-	**50:** 9393b, **51:** 2757b, **52:** 19999c, **79:** 104711e, **68:** 13061z
4-[4-(*N,N*-dimethylamino)phenyl]-5-(5,6-dimethyl-2-benzoxazolyl)-	**80:** P72087h
4-[4-(*N,N*-dimethylamino)phenyl]-5-(5-phenyl-1,3,4-oxadiazol-2-yl)-	**80:** 3439n
5-(4,6-dimethyl-2-benzoxazolyl)-4-[4-(methoxy-carbonyl)phenyl]-	**80:** P72087h
5-(5,6-dimethyl-2-benzoxazolyl)-4-[4-(methoxy-carbonyl)phenyl]-	**80:** P72087h
4-(3,3-dimethyl-2-phenyl-2-oxetanyl)-5-methyl-	**75:** 109587w
4,5-dimethyl-2-[(*N,N,N',N'*-tetramethyldiamido)-phosphino]-	**79:** 104711e
4,5-dimethyl-2-(trimethylsilyl)-	**65:** 15414d
4,5-di-9-octadecenyl-, scandium salt	**73:** P27318d
4,5-diphenyl-	**56:** 1388i, **64:** 14184g, **65:** 15414f, **78:** 29956n, **82:** 156182q
4,5-diphenyl-, silver (1+) salt	**80:** 70667y
4-(diphenylmethylene)-5-phenyl-	**80:** 133389g
4-[(α,α-diphenyl)piperidinylmethyl]-5-phenyl-	**80:** 133389g
4,5-diphenyl-2-(tributylstannyl)-	**78:** 29956n
4,5-diphenyl-2-(trimethylstannyl)-	**78:** 29956n
4,5-dipropyl-	**71:** 112865h
5-ethyl-4-methyl-	**56:** 14021g
4-ethyl-5-phenyl-	**49:** 3910b
4-(9-hydroxy-9*H*-fluoren-9-yl)-5-phenyl-	**80:** 133389g
4-(4-hydroxy-3-methoxyphenyl)-5-(5-nitro-2-benzoxazolyl)-	**80:** P72087h
4-(4-hydroxyphenyl)-5-(5-nitro-2-benzoxazolyl)-	**80:** P72087h
4-(4-hydroxyphenyl)-5-(6-nitro-2-benzoxazolyl)	**80:** P72087h
4-isopropyl-5-methyl-	**71:** 112865h
-4-methanol, 5-(4-chlorophenyl)-α,α-diphenyl-	**80:** 133389g
-4-methanol, α,α,5-triphenyl-	**80:** 133389g
4-[5-(methoxycarbonyl)benzoxazol-2-yl]-5-(4-methoxyphenyl)-	**80:** P72087h

TABLE 2 (*Continued*)

Compound	Reference
2.4. Alkyl- or Aryl-Disubstituted *v*-Triazoles (*Continued*)	
4-[4-(methoxycarbonyl)phenyl]-5-(6-methyl-2-benzoxazolyl)-	**80:** P72087h
4-(methoxydiphenylmethyl)-5-phenyl-	**78:** 71045q, **80:** 133389g
5-(4-methoxyphenyl)-4-(4-methyl-2-oxo-2*H*-pyran-6-yl)-	**80:** 3439n
5-(4-methoxyphenyl)-4-(5-nitro-2-benzoxazolyl)-	**80:** P72087h
5-(4-methoxyphenyl)-4-(6-nitro-2-benzoxazolyl)-	**80:** P72087h
5-(4-methoxyphenyl)-4-(5-phenyl-1,3,4-oxadiazol-2-yl)-	**80:** 3439n
4-methyl-5-(3-methyl-5-isoxazolyl)-	**80:** 108452q
4-methyl-5-(5-methyl-1*H*-pyrazol-3-yl)-	**80:** 108452q
4-(4-methyl-2-oxo-2*H*-pyran-6-yl)-5-phenyl-	**80:** 3439n
4-(4-methyl-2-oxo-2*H*-pyran-6-yl)-5-(3,4,5-trimethoxyphenyl)-	**80:** 3439n
5-methyl-4-(10*H*-phenothiazin-10-yl)-	**80:** 37051v
4-methyl-5-phenyl-	**49:** 3910b, **82:** 156182q
4-(naphth[2,3-*d*]oxazol-2-yl)-5-(1-naphthyl)-	**80:** P72087h
4-(6-nitro-2-benzoxazolyl)-5-(2-thienyl)-	**80:** P72087h
5-(4-nitrophenyl)-4-(5-phenyl-1,3,4-oxadiazol-2-yl)-	**80:** 3439n
4-phenyl-5-(3-phenyl-5-isoxazolyl)-	**80:** 108452q
4-phenyl-5-(5-phenyl-1*H*-pyrazol-3-yl)-	**80:** 108452q
4-phenyl-5-vinyl-	**35:** 2470^2
4-propenyl-5-propyl-	**71:** 112865h
-4-pyruvic acid, 5-phenyl-, oxime	**45:** 9037f
-4-α-thiopyruvic acid, 5-phenyl-	**45:** 9037f

REFERENCES

22: 3411^1	A. Bertho and F. Hölder, *J. Prakt. Chem.*, **119,** 173 (1928).
26: 1606^4	A. Quilico, *Gazz. Chim. Ital.*, **61,** 759 (1931).
30: 1027^3	F. Arndt, H. Scholz, and Frobel, *Justus Liebigs Ann. Chem.*, **521,** 95 (1935).
35: 2470^2	S. G. Fridman and N. M. Lisovs'ka, *Zapiski Inst. Khim., Akad. Nauk URSR, Inst. Khim.*, **6,** 353 (1940).
36: 2862^9	E. Ghigi and T. Pozzo-Balbi, *Gazz. Chim. Ital.*, **71,** 228 (1941).
37: 119^4	R. Hüttel, *Chem. Ber.*, **74B,** 1680 (1941).
37: 5405^1	W. Borsche, H. Hahn, and M. Wagner-Roemmich, *Justus Liebigs Ann. Chem.*, **554,** 15 (1943).
43: 2619b–e	J. L. Riebsomer and G. Sumrell, *J. Org. Chem.*, **13,** 807 (1948).
45: 9037–8f–c	J. C. Sheehan and C. A. Robinson, *J. Amer. Chem. Soc.*, **73,** 1207 (1951).
46: 6123e–i	J. L. Riebsomer and D. A. Stauffer, *J. Org. Chem.*, **16,** 1643 (1951).
46: 8651–2g–b	Fr. Moulin, *Helv. Chim. Acta*, **35,** 167 (1952).
48: 2685b	E. Mugnaini and P. Grünanger, *Atti Accad. Naz. Lincei., Rend., Cl. Sci. Fis. Mat. Nat.*, **14,** 95 (1953).
49: 3910b	R. Meier, *Chem. Ber.*, **86,** 1483 (1953).
49: 3948e	E. Mugnaini and P. Grünanger, *Atti Accad. Naz. Lincei., Rend., Cl. Sci. Fis. Mat. Nat.*, **14,** 275 (1953).

49: 5488g	R. N. Castle and D. B. Cox, *J. Org. Chem.*, **18,** 1706 (1953).
49: 6241–2h–b	R. Hüttel, T. Schneiderhan, H. Hertwig, A. Leuchs, V. Reincke, and J. Miller, *Justus Liebigs Ann. Chem.*, **585,** 115 (1954).
49: 12451h	H. J. Cothrell, D. L. Pain, and R. Slack, *J. Chem. Soc.*, 2968 (1954).
49: 13901e	E. Schreier, G. Stöhr, and E. Hardegger, *Helv. Chim. Acta*, **37,** 574 (1954).
50: 4129c	R. H. Wiley, N. R. Smith, D. M. Johnson, and J. Moffat, *J. Amer. Chem. Soc.*, **77,** 3412 (1955).
50: 4923–4i–e	M. L. Stein and L. D'Antoni, *Il Farmaco (Pavia) Ed. Sci.*, **10,** 235 (1955).
50: 9392–3f–b	R. Hüttel and G. Welzel, *Justus Liebigs Ann. Chem.*, **593,** 207 (1955).
50: 10009g	G. Henseke and H-J Binte, *Chem. Ber.*, **88,** 1167 (1955).
50: 16685b	G. H. Daub and R. N. Castle, *J. Org. Chem.*, **19,** 1571 (1954).
51: 2757b	E. Muller and W. Rundel, *Chem. Ber.*, **89,** 1065 (1956).
51: 13854i	R. Gompper, *Chem. Ber.*, **90,** 382 (1957).
51: 13855c	H. Bohme and D. Morf, *Chem. Ber.*, **90,** 446 (1957).
52: P2085–6h–a	Badische Anilin- und Soda-Fabrik Akt.-Ges., German Patent 818,048 (1951).
52: 13727b	H. Bohme and D. Morf, *Chem. Ber.*, **91,** 660 (1958).
52: 17246a–i	M. C. Ford and D. Mackay, *J. Chem. Soc.*, 1290 (1958).
52: 19999c	J. H. Boyer and L. R. Morgan, Jr., *J. Amer. Chem. Soc.*, **80,** 3012 (1958).
53: 344f	H. El Khadem and Z. M. El-Shafei, *J. Chem. Soc.*, 3117 (1958).
53: 3198–9f–i	A. Dornow and K. Rombusch, *Chem. Ber.*, **91,** 1841 (1958).
53: 5253d	W. Kirmse and L. Horner, *Justus Liebigs Ann. Chem.*, **614,** 1 (1958).
53: 16120d	D. Dal Monte and P. Veggetti, *Boll. Sci. Fac. Chim. Ind. Bologna*, **16,** 1 (1958).
53: 18946h	D. Dal Monte, A. M. Casoni, and M. Fernando, *Gazz. Chim. Ital.*, **88,** 1035 (1958).
54: 1504b	H. El Khadem and Z. M. El-Shafei, *J. Chem. Soc.*, 1655 (1959).
54: 9899c	G. Henseke and M. Winter, *Chem. Ber.*, **93,** 45 (1960).
54: 12125a	F. Micheel and G. Baum, *Chem. Ber.*, **90,** 1595 (1957).
55: 3562f	H. El Khadem, Z. M. El-Shafei, and Y. S. Mohammed, *J. Chem. Soc.* 3993 (1960).
55: 27070e	H. El Khadem, Z. M. El-Shafei, and M. H. Meshreki, *J. Chem. Soc.*, 2957 (1961).
55: 27338b	T. Itai and S. Kamiya, *Chem. Pharm. Bull. (Tokyo)*, **9,** 87 (1961).
56: 1388a	R. Huisgen and M. Seidel, *Chem. Ber.*, **94,** 2509 (1961).
56: 5943d	J. Collier and J. L. Riebsomer, *J. Org. Chem.*, **26,** 1647 (1961).
56: 12876h	R. Hüttel, W. Schwarz, J. Miller, and F. Wunsch, *Chem. Ber.*, **95,** 222 (1962).
56: 14021g	R. Fusco, G. Bianchetti, and D. Pocar, *Gazz. Chim. Ital.*, **91,** 849 (1961).
57: 2296g	H. Simon and J. Steffens, *Chem. Ber.*, **95,** 358 (1962).
57: 7353e	H. El Khadem and M. H. Meshreki, *Nature*, **194,** 373 (1962).
57: 15100b	G. Henseke and G. Mueller, *J. Prakt. Chem.*, **18,** 47 (1962).
58: 6819f	B. M. Lynch and T-L. Chan, *Can. J. Chem.*, **41,** 274 (1963).
58: 11454d	H. El Khadem, Z. M. El-Shafei, and M. H. Meshreki, *J. Chem. Soc.*, 3154 (1962).
58: 12560a	R. Fusco, G. Bianchetti, D. Pocar, and R. Ugo, *Gazz. Chim. Ital.*, **92,** 1040 (1962).
59: 11690g	H. El Khadem, G. H. Labid, and M. H. Meshreki, *J. Chem. Soc.*, 3528 (1963).
59: 13483f	E. Borello and A. Zecchina, *Spectrochim. Acta*, **19,** 1703 (1963).

59: P15420d	R. Raue and H. Gold, Belgian Patent, 621,482 (1962).
60: 643g	B. B. Bishay, H. El Khadem, and Z. M. El-Shafei, *J. Chem. Soc.*, 4980 (1963).
60: 5537b	L. Birkofer, A. Ritter, and P. Richter, *Chem. Ber.*, **96,** 2750 (1963).
60: 15859d	J. A. Durden, Jr., H. A. Stansbury, and W. H. Catlette, *J. Chem. Eng. Data*, **9,** 228 (1964).
61: 2926b	M. E. Munk and Y. K. Kim, *J. Amer. Chem. Soc.*, **86,** 2213 (1964).
61: 3096e	D. Pocar, G. Bianchetti, and P. D. Croce, *Chem. Ber.*, **97,** 1225 (1964).
61: 5738c	H. El Khadem, M. H. Meshreki and G. H. Labib, *J. Chem. Soc.*, 2306 (1964).
61: 10754f	M. L. Wolfrom, H. El Khadem, and H. Alfes, *J. Org. Chem.*, **29,** 2072 (1964).
61: 14759e	K. S. Panwar and J. N. Gaur, *Indian J. Appl. Chem.*, **27,** 13 (1964).
62: P10443g	H. A. Stansbury, Jr., J. A. Durden, Jr., and W. H. Catlette, U.S. Patent, 3,161,651 (1964).
62: 15262a	A. Clitheroe, D. Green, A. B. A. Jansen, P. C. Phillips, and A. W. Rule, *J. Pharm. Pharmacol.*, **17,** 167 (1965).
62: 164248b	R. Huisgen, L. Möebius, and G. Szeimies, *Chem. Ber.*, **98,** 1138 (1965).
63: 1850f	H. El Khadem, Z. M. El-Shafei, and M. H. Meshreki, *J. Chem. Soc.*, 1524 (1965).
63: 11551c	G. Bianchetti, D. Pocar, P. D. Croce, and A. Vigevani, *Chem. Ber.*, **98,** 2715 (1965).
63: 11552a	G. Bianchetti, P. D. Croce, and D. Pocar, *Tetrahedron Lett.*, 2039 (1965).
64: 3522g	W. Koenig, M. Coenen, W. Lorenz, F. Bader, and A. Bassl, *J. Prakt. Chem.*, **30,** 96 (1965).
64: 8276b	H. El Khadem, Z. M. El-Shafei, M. H. Meshreki, and M. A. Shaban, *J. Chem. Soc.*, 91 (1966).
64: 9713h	G. S. Akimova, V. N. Chistokletov, and A. A. Petrov, *Zh. Org. Khim.*, **1,** 2077 (1965).
64: 14184e	J. O. Fournier and J. B. Miller, *J. Heterocycl. Chem.*, **2,** 488 (1965).
65: 15414d	L. Birkofer and P. Wegner, *Chem. Ber.*, **99,** 2512 (1966).
66: 10887k	G. Rembarz, B. Kirchhoff, and G. Dongowski, *J. Prakt. Chem.*, **33,** 199 (1966).
66: 18699g	T. Sasaki and K. Minamoto, *J. Org. Chem.*, **31,** 3914 (1966).
66: 28250t	H. Borello, A. Zecchina, and E. Guglielminotti, *J. Chem. Soc., B*, 1243 (1966).
66: 37839u	W. Koenig, M. Coenen, F. Bahr, B. May, and A. Bassl, *J. Prakt. Chem.*, **33,** 54 (1966).
66: 37841p	G. Henseke and I. Schmeisky, *J. Prakt. Chem.*, **33,** 256 (1966).
67: 82170a	G. Bianchetti, P. D. Croce, D. Pocar, and A. Vigevani, *Gazz. Chim. Ital.*, **97,** 289 (1967).
67: 82171b	G. Bianchetti, D. Pocar, P. D. Croce, and R. Stradi, *Gazz. Chim. Ital.*, **97,** 304 (1967).
67: 82249h	H. Gorth and M. C. Henry, *J. Organometal. Chem.*, **9,** 117 (1967).
67: 100071a	G. S. Akimova, V. N. Chistokletov, and A. A. Petrov, *Zh. Org. Khim.*, **3,** 968 (1967).
67: 108605y	H. Hoberg, *Justus Liebigs Ann. Chem.*, **707,** 147 (1967).
67: 118072m	W. Koenig, M. Coenen, F. Bahr, A. Bassl, and B. May, *J. Prakt. Chem.*, **36,** 202 (1967).
68: 12925x	G. Bianchetti, P. Ferruti, and D. Pocar, *Gazz. Chim. Ital.*, **97,** 579 (1967).
68: 12927z	D. Pocar, G. Bianchetti, and P. Ferruti, *Gazz. Chim. Ital.*, **97,** 597 (1967).

68: 13061z L. Birkofer and P. Wegner, *Chem. Ber.*, **100,** 3485 (1967).

68: P59611g A. S. Tomcufcik and A. M. Hoffman, U.S. Patent, 3,331,830 (1967).

68: 69231x H. S. El Khadem, D. Horton, and T. F. Page, Jr., *J. Org. Chem.*, **33,** 734 (1968).

68: 69251d J. N. BeMiller and D. R. Smith, *Carbohyd. Res.*, **6,** 118 (1968).

68: 87243g G. S. Akimova, V. N. Chistokletov and A. A. Petrov, *Zh. Org. Khim.*, **3,** 2241 (1967).

68: 105100q G. S. Akimova, V. N. Chistokletov, and A. A. Petrov, *Zh. Org. Khim.*, **4,** 389 (1968).

69: 10402w H. El Khadem, H. A. R. Mansour, and M. H. Meshreki, *J. Chem. Soc., C,* 1329 (1968).

69: 27660b G. Hanisch and G. Henseke, *Chem. Ber.*, **101,** 2074 (1968).

69: 36377s G. G. Lyle and M. J. Piazza, *J. Org. Chem.*, **33,** 2478 (1968).

69: 76826u H. El Khadem, M. M. El-Sadik, and M. H. Meshreki, *J. Chem. Soc., C,* 2097 (1968).

69: P96740q G. S. Akimova, V. N. Christokletov, and A. A. Petrov, Russian Patent, 216,737 (1968).

69: 106976n H. El Khadem, M. M. A. Abdel Rahman, and M. A. E. Sallam, *J. Chem. Soc., C,* 2411 (1968).

70: 3996k A. Karklina and E. Gudriniece, *Latv. PSR Zinat Akad. Vestis, Kim. Ser.*, 374 (1968).

70: 19987u M. Regitz, W. Anscheutz, and A. Liedhegener, *Chem. Ber.*, **101,** 3734 (1968).

70: P28923x T. Nozoe, T. Mukai, T. Toda, and H. Horino, Japanese Patent, 68 14,705 (1968).

70: 78274g E. Morita, M. Uemura, and T. Megusu, *Kyushu Kogyo Daigaku Kenkyu Hokoku,* 31 (1968).

70: 115081q M. W. Barker and J. H. Gardner, *J. Heterocycl. Chem.*, **6,** 251 (1969).

71: 22079a H. Wittmann, E. Ziegler, F. Eichenseer, and G. Dworak, *Monatsh. Chem.*, **100,** 602 (1969).

71: 38476p J. H. Boyer, W. E. Krueger, and R. Modler, *J. Org. Chem.*, **34,** 1987 (1969).

71: 81672m D. J. Brecknell and R. M. Carman, *Aust. J. Chem.*, **22,** 1669 (1969).

71: 112865h E. Zbiral and J. Stroh, *Monatsh. Chem.*, **100,** 1438 (1969).

71: P124445j Geigy, A.-G., French Patent, 1,559,131 (1969).

72: 3666u D. Rutherford and N. K. Richtmyer, *Carbohyd. Res.*, **11,** 341 (1969).

72: 3718n M. T. García-López, G. García-Muñoz, J. Iglesias, and R. Madroñero, *J. Heterocycl. Chem.*, **6,** 639 (1969).

72: 43576w G. García-Muñoz, R. Madroñero, M. Rico, and M. C. Saldaña, *J. Heterocycl.* Chem., **6,** 921 (1969).

72: 55344g A. Karklins and E. Gudriniece, *Latv. PSR Zinat. Akad. Vestis, Kim. Ser.*, 579 (1969).

72: 65613h A. Karklina, V. M. Slin'ko and A. Ya. Maslenikova, *Khim. Regul. Rosta Razv. Rast.*, 99 (1969).

72: 66882g A. Karklins and E. Gudriniece, *Latv. PSR Zinat. Akad. Vestis, Kim. Ser.*, 606 (1969).

72: P100716t Farbenfabriken Bayer A.-G., French Patent, 1,577,760 (1969).

72: 121447w G. L'abbé and A. Hassner, *J. Heterocycl. Chem.*, **7,** 361 (1970).

72: 121473b A. Karklina and E. Gudriniece, *Khim. Geterotsikl. Soedin.*, 1127 (1969).

72: 121865f R. M. Carman and J. D. Petty, *Aust. J. Chem.*, **23,** 801 (1970).

73: 14769b M. Regitz and G. Himbert, *Justus Liebigs Ann. Chem.*, **734,** 70 (1970).

73: 25790c P. K. Kindel, *Carbohyd. Res.*, **12,** 466 (1970).

73: 27318d S. H. Patinkin, U.S. Patent, 3,511,623 (1970).

73: 31627w D. A. Hart and P. K. Kindel, *Biochemistry,* **9,** 2190 (1970).

73: 98878x	L. S. Davies and G. Jones, *Tetrahedron Lett.*, 3475 (1970).
73: 98912d	N. Bruvele and E. Gudriniece, *Latv. PSR Zinat. Akad. Vestis, Kim. Ser.*, 368 (1970).
74: 48048b	L. F. Avramenko, Yu. B. Vilenskii, B. M. Ivanov, N. V. Kudryavskays, I. A. Ol'shevskaya, V. Ya. Pochinok, L. I. Skripnik, L. N. Fedorova, and I. P. Fedorova, *Usp. Nauch. Fotogr.* **14,** 12 (1970).
74: P65594u	A. Dorlars and O. Neuner, German Patent, 1,917,740 (1970).
74: 76377b	J. M. Stewart, R. L. Clark, and P. E. Pike, *J. Chem. Eng. Data*, **16,** 98 (1971).
74: 125600e	E. Gudriniece and V. Urbans, *Latv. PSR Zinat. Akad. Vestis, Kim. Ser.*, 82 (1971).
74: 141988t	P. Ykman, G. L'abbé, and G. Smets, *Tetrahedron*, **27,** 845 (1971).
75: 19622z	G. L'abbé and A. Hassner, *Bull. Soc. Chim. Belg.*, **80,** 209 (1971).
75: 36556y	N. K. Richtmyer, *Carbohyd. Res.*, **17,** 401 (1971).
75: 63702z	L. S. Davies and G. Jones, *J. Chem. Soc., C*, 2572 (1971).
75: 88891y	H. El Khadem, D. Horton, and M. H. Meshreki, *Carbohyd. Res.*, **16,** 409 (1971).
75: 109587w	J. Van Thielen, T. Van Thien, and F. C. De Schryver, *Tetrahedron Lett.*, 3031 (1971).
76: 14832d	O. S. Chizhov, N. K. Kochetkov, N. N. Malysheva, A. I. Shiyonok, and V. L. Chashchin, *Org. Mass. Spectrom.*, **5,** 1145 (1971).
76: 113449r	A. Kraus and H. Simon, *Chem. Ber.*, **105,** 954 (1972).
76: 126875b	D. Pocar, R. Stradi, and L. M. Rossi, *J. Chem. Soc., Perkin Trans.* **1,** 769 (1972).
76: 126879f	D. Pocar, R. Stradi, and L. M. Rossi, *J. Chem. Soc., Perkin Trans.* **1,** 619 (1972).
76: 140726k	H. Horino and T. Toda, *Bull. Chem. Soc., Jap.*, **45,** 584 (1972).
76: 153646j	N. Latif and S. A. Meguid, *J. Chem. Soc., Perkin Trans.* **1,** 1095 (1972).
76: 153690u	V. Urbans and E. Gudriniece, *Latv. PSR Zinat. Akad. Vestis, Kim. Ser.*, 107 (1972).
77: P7313c	H. Lind, German Patent, 2,133,012 (1972).
77: 19906m	D. Horton and J. D. Wander, *J. Org. Chem.*, **37,** 1630 (1972).
77: 34803u	J. G. Buchanan, A. R. Edgar, and M. J. Power, *J. Chem. Soc., Chem. Commun.*, 346 (1972).
77: 101470y	I. Hirao and Y. Kato, *Bull. Chem. Soc. Jap.*, **45,** 2055 (1972).
77: 114720e	R. M. Carman, B. C. McKenzie, I. M. Shaw, and G. W. Zerner, *Aust. J. Chem.*, **25,** 1823 (1972).
77: 126516f	H. Cardoen, S. Toppet, G. Smets, and G. L'abbé, *J. Heterocycl. Chem.*, **9,** 971 (1972).
77: 164609w	P. Ykman, G. Mathys, G. L'abbé, and G. Smets, *J. Org. Chem.*, **37,** 3213 (1972).
77: 165027s	F. Nouaille, A. M. Sepulchre, G. Lukacs, and S. D. Gero, *C.R. Acad. Sci., Ser. C*, **275,** 423 (1972).
78: 29956n	S. Kozima, T. Itano, N. Mihara, K. Sisido, and T. Isida, *J. Organometal. Chem.*, **44,** 117 (1972).
78: 58205x	R. E. Banks and G. R. Sparkes, *J. Chem. Soc., Perkin Trans.* **1,** 2964 (1972).
78: 71045q	E. M. Burgess and J. P. Sanchez, *J. Org. Chem.*, **38,** 176 (1973).
78: 72017u	D. Seebach and D. Enders, *Angew. Chem., Int. Ed. Engl.*, **11,** 1102 (1972).
78: 72035y	V. Urbans, E. Gudriniece, and L. A. Khoroshanskays, *Latv. PSR Zinat. Akad. Vestis, Kim. Ser.*, 712 (1972).
79: 31992k	G. L'abbé, *Verh. Kon. Acad. Wetensch., Lett. Schone Kunsten Belg., Kl. Wetensch.*, **34,** 49 pp. (1972).

79: 66254y Y. Tanaka and S. I. Miller, *J. Org. Chem.*, **38,** 2708 (1973).
79: 66696u D. Horton, A. Liav, and S. E. Walker, *Carbohydr. Res.*, **28,** 201 (1973).
79: 66704v O. S. Chizhov, N. N. Malysheva, and N. K. Kochetkov, *Izv. Akad. Nauk SSSR, Ser. Khim.*, 1022 (1973).
79: P78815v E. Aufderhaar and A. Meyer, German Patent, 2,263,878 (1973).
79: 78884s U. Heep, *Justus Liebigs Ann. Chem.*, 578 (1973).
79: 104711e D. Kiessling, K. P. Knackmuss, H. Schicht, F. Winn, and R. Schoellner, *J. Prakt. Chem.*, **315,** 577 (1973).
79: P106145x K. Weber and C. Luethi, German Patent, 2,262,632 (1973).
79: 126568r A. N. Pudovik, N. G. Khusainova, and Z. A. Nasyballina, *Zh. Obshch. Khim.*, **43,** 1683 (1973).
79: P147435k F. Fleck, H. Kittl, and S. Valenti, German Patent, 2,262, 340 (1973).
80: 3439n G. Beck and D. Guenther, *Chem. Ber.*, **106,** 2758 (1973).
80: 37051v G. Himbert and M. Regitz, *Justus Liebigs Ann. Chem.*, 1505 (1973).
80: 47912r C. E. Olsen, *Acta Chem. Scand.*, **27,** 1987 (1973).
80: 48283y H. S. El Khadem and D. L. Swartz, *Carbohydr. Res.*, **30,** 400 (1973).
80: P49275j A. Domergue and R. F. M. Sureau, German Patent, 2,319,828 (1973).
80: 70667y T. L. Gilchrist, G. Gymer, and C. W. Rees, *J. Chem. Soc., Chem. Commun.*, 835 (1973).
80: P72087h G. Beck and G. Dieter, German Patent, 2,212,694 (1973).
80: 108451p Y. Tanaka and S. I. Miller, *Tetrahedron*, **29,** 3285 (1973).
80: 108452q Y. Tanaka, S. R. Velen, and S. I. Miller, *Tetrahedron*, **29,** 3271 (1973).
80: 133389g E. M. Burgess and J. P. Sanchez, *J. Org. Chem.*, **39,** 940 (1974).
80: 133719q F. Nouaille, A. M. Sepulchre, G. Lukacs, A. Kornprobst, and S. D. Gero, *Bull. Soc. Chim. Fr.*, 143 (1974).
81: P51143d H. Schlaepfer, German Patent, 2,329,991 (1974).
81: P79389n R. Kirchmayr, U.S. Patent, 3,816,413 (1974).
81: 91486z H. Boehme and F. Ziegler, *Justus Liebigs Ann. Chem.*, 734 (1974).
81: 135588r R. E. Banks and A. Prakash, *J. Chem. Soc., Perkin Trans.*, **1,** 1365 (1974).
82: P4995u M. Matsuo, K. Koga, S. Hotta, and T. Akamatsu, Japanese Patent, 74 54,281 (1974).
82: 43677m J. G. Buchanan, A. R. Edgar, and M. J. Power, *J. Chem. Soc., Perkin Trans.*, **1,** 1943 (1974).
82: P59922y A. Dorlars, German Patent, 2,310,123 (1974).
82: P87680m H. Harnisch and A. Brock, German Patent, 2,225,648 (1973).
82: 154936q T. L. Gilchrist, C. W. Rees, and C. Thomas, *J. Chem. Soc., Perkin Trans.*, **1,** 8 (1975).
82: 156182q T. L. Gilchrist, G. E. Gymer, and C. W. Rees, *J. Chem. Soc., Perkin Trans.*, **1,** 1 (1975).
83: P10089p Miles Laboratories, Dutch Patent, 74 07,127 (1974).
83: P61732a A. Dorlars and O. Neuner, German Patent, 2,258,276 (1974).
83: 79160x J. K. Crandall, W. W. Conover, and J. B. Komin, *J. Org. Chem.*, **40,** 2042 (1975).
83: 79201m J. K. Crandall, L. C. Crawley, and J. B. Komin, *J. Org. Chem.*, **40,** 2045 (1975).
83: P81213k F. Fleck, H. Schmid, A. V. Mercer, and R. Paver, Swiss Patent, 561,709 (1975).
83: 96642y K. S. Balachandrin, I. Hiryakhanavar, and M. V. George, *Tetrahedron*, **31,** 1171 (1975).
83: 97213w B. Mihelčič, S. Simonič, B. Stanovnik, and M. Tišler, *Croat. Chem. Acta*, **46,** 275 (1974).
83: 114297w A. Martvon, S. Stankovsky, and J. Svetlik, *Collect. Czech. Chem. Commun.*, **40,** 1199 (1975).

83: 193181e G. I. Tsypin, T. N. Timofeeva, V. V. Mel'nikov, and B. V. Gidaspov, *Zh. Org. Khim.*, **11,** 1395 (1975).
83: P206088g T. Honma and Y. Tada, Japanese Patent, 75 49,271 (1975).
84: 30978b C. E. Olsen, *Acta Chem. Scand., Ser. B*, **29,** 953 (1975).
84: 59333y H. Reimlinger, W. R. F. Lingier, and R. Merenyi, *Chem. Ber.*, **108,** 3794 (1975).
84: P59482w V. S. Sukhinin and V. G. Mikhailov, Russian Patent, 487,074 (1975).
84: 90080n S. Boerresen and J. K. Crandall, *J. Org. Chem.*, **41,** 678 (1976).
84: 105510d O. Tsuge, K. Ueno, and A. Inaba, *Heterocycles*, **4,** 1 (1976).
84: 135554p J. A. C. Allison, H. S. El Khadem, and C. A. Wilson, *J. Heterocycl. Chem.*, **12,** 1275 (1975).
84: 164334x J. Hocker, H. Giesecke, and R. Merten, *Angew. Chem.*, **88,** 151 (1976).
84: P164786w R. T. Buckler, H. E. Hartzler, and S. Hayao, U.S. Publ. Pat. Appl. B, 488,111 (1976).
84: P166274b G. Kormany, G. Kabas, H. Schlaepfer, and A. E. Siegrist, German Patent, 2,535,615 (1976).
84: P180230r M. Tsujimoto, R. Tsukahara, M. Torisu, and I. Okubo, German Patent, 2,538,399 (1976).
84: 180505j D. Horton and A. Liav, *Carbohydr. Res.*, **47,** 81 (1976).
84: P181616h G. Kormany, G. Kabas, H. Schlaepfer, and A. E. Siegrist, German Patent, 2,535,613 (1976).
84: P181617j G. Kormany, G. Kabas, H. Schlaepfer, and A. E. Siegrist, German Patent, 2,535,612 (1976).
84: P181618k G. Kormany, G. Kabas, H. Schaepfer, and A. E. Siegrist, German Patent, 2,535,614 (1976).
85: 5562z V. Coviello and A. E. Siegrist, *Helv. Chim. Acta*, **59,** 802 (1976).
85: P21383x Societa Italina Resine S.p.A., French Patent, 2,268,018 (1975).
85: 77337c M. Begtrup, *J. Chem. Soc., Perkin Trans.*, **2,** 736 (1976).
85: P123926u R. T. Buckler, H. E. Hartzler, S. Hayao, and G. Nichols, U.S. Patent, 3,948,932 (1976).

CHAPTER 3

Alkyl- or Aryl-Trisubstituted 1,2,3-Triazoles

3.1. 1,4,5-ALKYL- OR ARYL-TRISUBSTITUTED 1*H*-1,2,3-TRIAZOLES

Comparatively few 1,2,3-triazoles bearing three carbon substituents have been prepared, and the general methods found in Chapters 1 and 2 have been applied as in the direct addition of azides to acetylenes. Some questions, such as the effect of an *ortho*-substituent in an aryl azide, which can greatly alter the product yield, deserve more attention (Eq. 1).[1] Compound **3.1-1** also represents a successful approach to controling mono- or diaddition to alkadiynes.

Many azides have been added to diphenylacetylene, but the report of 95% yield with benzyl azide offers what may well be optimal conditions.[2] High yields have also been obtained from 1,4-but-2-ynediol (**3.1-2**) and the exciting bisacetal (**3.1-3**) with a variety of azides (Eqs. 2,3).[3] A recent note describing the preparation of fluorinated acetylenes included an example of

$$Me_2C(OH)C{\equiv}CC{\equiv}CC(OH)Me_2 + N_3C_6H_4NO_2 \longrightarrow \text{triazole}: 4\text{-}Me_2C(OH),\ 5\text{-}C{\equiv}CC(OH)Me_2,\ 1\text{-}C_6H_4NO_2 \qquad (1)$$

3.1-1

2-NO_2Ph 10% 4-NO_2Ph 80%

$$HOCH_2C{\equiv}CCH_2OH + PhN_3 \longrightarrow \text{triazole}: 4,5\text{-}(CH_2OH)_2,\ 1\text{-}Ph \qquad (2)$$

3.1-2 73%

a highly regiospecific 1,2,3-triazole formation (Eq. 4).[4] This observation stands in marked contrast to the poor results found by Banks and his collaborators in the addition of perfluoroaromatics (Eq. 5).[5–7]

$$(EtO)_2CHC{\equiv}CCH(OEt)_2 + RN_3 \longrightarrow \text{3.1-4} \quad (3)$$

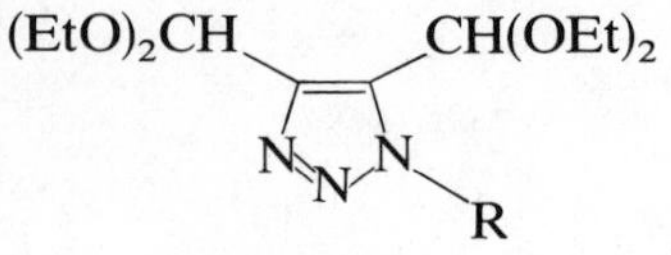

3.1-3 → **3.1-4**: 1-R-4,5-bis[$CH(OEt)_2$]-1,2,3-triazole

R	= Ph	$PhCH_2$	$Me(CH_2)_5$	$Me(CH_2)_{11}$
% **3.1-4**	= 82	78	90	75

$$n\text{-}C_6F_{13}C{\equiv}CPh + PhN_3 \longrightarrow \text{3.1-5} \quad (4)$$

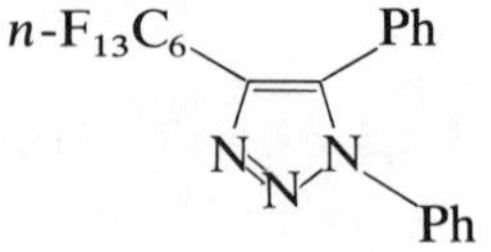

3.1-5 (4-n-$F_{13}C_6$-1,5-diphenyl-1,2,3-triazole), 76%

$$C_6F_5N_3 + PhC{\equiv}CPh \longrightarrow \text{1-pentafluorophenyl-4,5-diphenyl-1,2,3-triazole} \quad (5)$$

27%

4-azidotetrafluoropyridine: 15%

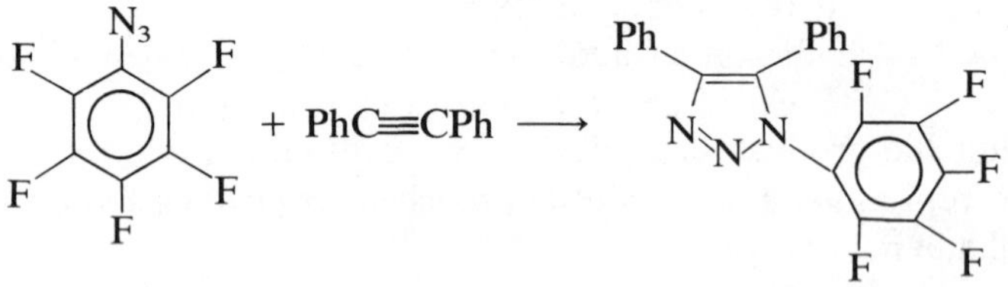

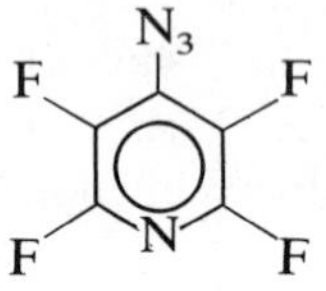

2-azido-3,5,6-trifluoro-4-(trifluoromethyl)pyridine (CF_3, N_3): 4%

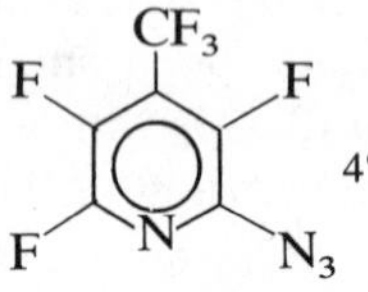

The addition of heteroaromatics takes place in somewhat better, although still only fair, yield (Eq. 6).[8] A radically different approach to the synthesis of similar compounds deserves greater elaboration (Eq. 7).[9] The fact that yields for only the second step are given make the attractiveness of this method hard to evaluate.

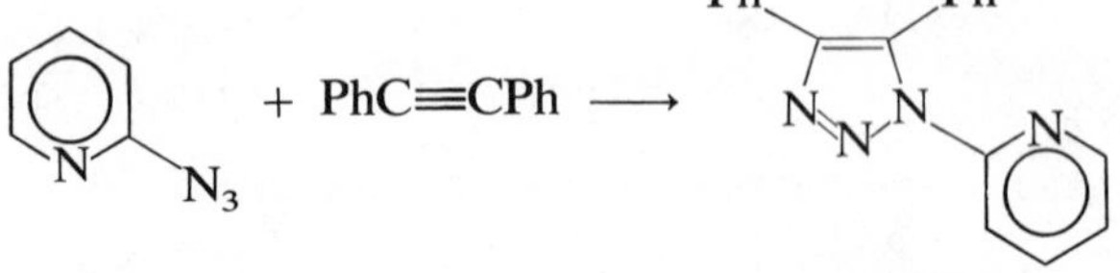

$$\text{2-azidopyridine} + PhC{\equiv}CPh \longrightarrow \text{4,5-diphenyl-1-(2-pyridyl)-1,2,3-triazole} \quad (6)$$

37%

2-azido-4,6-dimethylpyrimidine (Me, Me, N_3): 49%

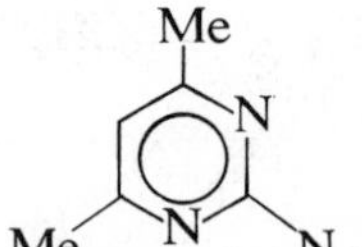

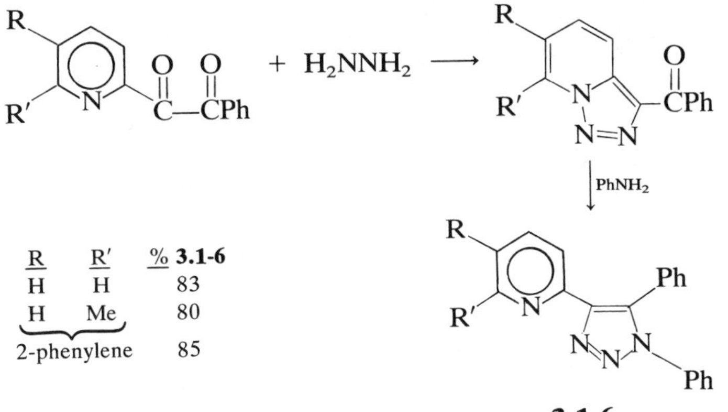

R	R′	% **3.1-6**
H	H	83
H	Me	80
2-phenylene		85

(7)

The formation of relatively unstable amino-1,2,3-triazolines (**3.1-7**) from enamines is another traditional route employed for trisubstituted 1,2,3-triazole synthesis. Munk and Kim demonstrated the great regiospecificity of such reactions (Eq. 8).[10] Pocar and his collaborators have extended the use of enamines, including the following examples (Eqs. 9,10).[11,12] An especially interesting example involves an acid-catalyzed rearrangement (Eq. 11).[13]

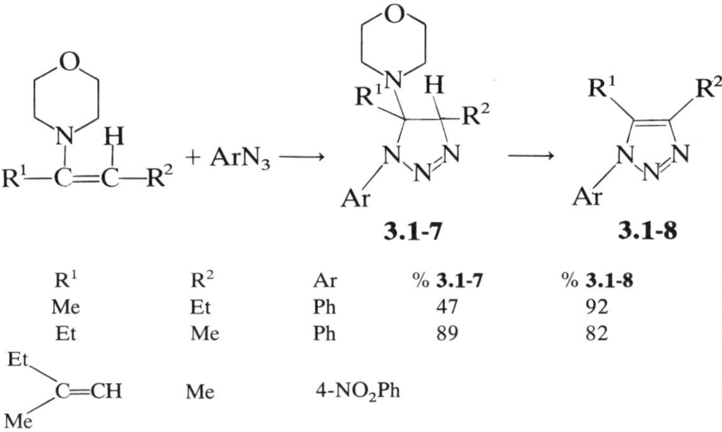

(8–10)

R^1	R^2	Ar	% **3.1-7**	% **3.1-8**	Eq.[ref]
Me	Et	Ph	47	92	8[10]
Et	Me	Ph	89	82	8[10]
Et(Me)C=CH	Me	$4\text{-}NO_2Ph$			9[11]
$PhCH_2$(Me)C=CH	Ph	$4\text{-}NO_2Ph$		80	10[12]
Me	$CHMe_2$	$4\text{-}NO_2Ph$		92	11[13]

Olsen and Pedersen have investigated extensively the reactions of ketones and azides in the presence of potassium *t*-butoxide (Eq. 12).[14] The yield of the intermediate 5-hydroxy-1,2,3-triazolines (**3.1-9**) were generally good, but the dehydration process varied from good ($R'' = PhCH_2$) to poor ($R'' =$ Ph). Acetone gave a fair yield of the 4-(2-propenyl) product (**3.1-10**) along

$$Me_2CHCH{=}C(Me)(\text{morpholino}) + 4\text{-}N_3C_6H_4NO_2 \xrightleftharpoons{CHCl_3} \text{triazolines} \xrightarrow{HOAc} \text{triazoles} \quad (11)$$

CH₂CHMe₂ triazole (1-(4-NO₂Ph)) 90%; Me₂CH / Me triazole (1-(4-NO₂Ph)) 92%

with the 1,5-derivative described in Section 2.2 Chapter 2 (Eq. 13).[15] A later report showed significant improvements in the yields of certain cases (Eq. 14).[16]

$$RCH(Me)COR' + R''N_3 \xrightarrow{K^{+}\,{}^{-}OBu\text{-}t} \mathbf{3.1\text{-}9} \xrightarrow[MeOH]{K^{+}\,{}^{-}OH} \text{triazole} \quad (12)$$

R = H, Me; R′ = Me, Et, Me₂CH; R″ = Ph, PhCH₂, 4-NO₂Ph

$$Me_2C{=}O + PhCH_2N_3 \xrightarrow{K^{+}\,{}^{-}OBu\text{-}t} \mathbf{3.1\text{-}10} + \text{1-benzyl-5-methyltriazole} \quad (13)$$

3.1-10 40%; 37%

$$R^4CH_2COR^5 + R^1N_3 \xrightarrow{\text{two steps}} \text{triazole} \quad (14)$$

R^1 = Me, Ph, PhCH₂
R^4 = Me, Et, Ph
R^5 = Me, Et, t-Bu, Ph

4-Ph-5-Me
1-Me 92%, 1-Ph 79%, 1-PhCH₂ 85%

The direct *N*-alkylation of 4,5-diphenyl-1,2,3-triazole produces mixtures, but it is possible to obtain fair yields of the 1-alkyl isomer using the silver or thallium salt (Eq. 15).[17,18] Alkylation at carbon has been reported and produces surprisingly good yields (Eq. 16).[19]

M = Ag, Tl R = Me, Et, *i*-Pr (15)

R^4, R^5 = H, Ph; (1) *n*-BuLi (2) MeI; 1,5-Ph_2-4-Me 99%; 1,4-Ph_2-5-Me 78% (16)

The decomposition of a sydnone (**3.1-11**) in the presence of diphenylacetylene was shown to involve an intermediate azide and to produce a 1,2,3-triazole in fair yield (Eq. 17).[20]

3.1-11 (c-C_6H_{11}) ⟶ CO_2 + cyclohexyl azide; PhC≡CPh; 37% (17)

Sheehan and Robinson, using the method described in Chapters 1 and 2, prepared 1,5-diphenyl-4-(2-aminoethyl)-1,2,3-triazole in good yield.[21] L'abbé and his collaborators included one trisubstituted example in their study of α-keto phosphorus ylides (Eq. 18).[22]

$CH_2{=}C(N_3)Ph$ + $Ac(Me)C{=}PPh_3$ ⟶ 4,5-dimethyl-1-($C(Ph){=}CH_2$)-1,2,3-triazole 54% (18)

Smith and his collaborators prepared 5-diazomethyl-1,2,3-triazoles (Eq. 19) and studied their carbene chemistry (Eq. 20).[23,24] Results with other monosubstituted benzenes (Y = *i*-Pro, NO_2, NH_2 and NMe_2) did not, for various reasons, contribute to this study. For example isopropoxybenzene

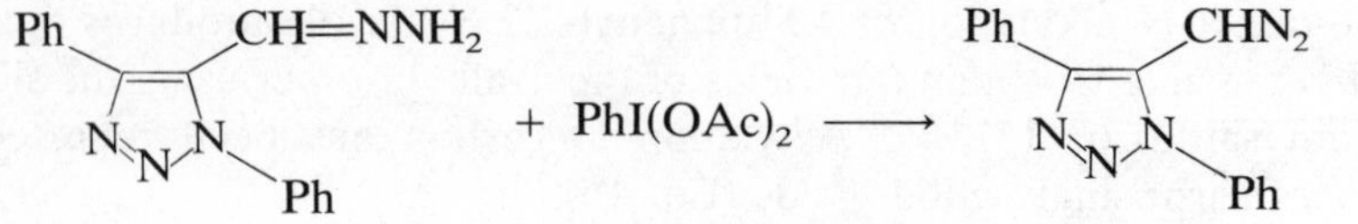

(19)

3.1-12

80%

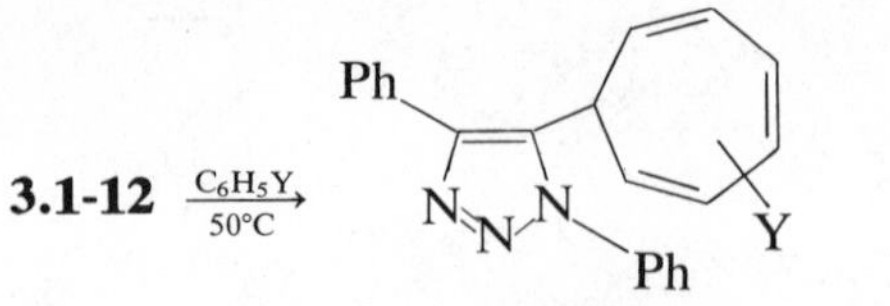

(20)

3.1-13

Y	% **3.1-13** (total)	Isomer distribution		
		Y-3	Y-2	Y-1
H	66	—	—	—
Me	90	50.0	27.3	22.7
MeO	94	34.8	38.3	26.4
F	72	50.0	23.5	26.5

gave a 92% yield, but the isomers could not be accurately resolved, and nitrobenzene reacted poorly (about 10%).[24] In other solvents **3.1-12** gave analogous carbene products in good yields (Eqs. 21,22).[23] The photochemical reactions of **3.1-12** at 10° (where the thermal decomposition is insignificant) showed essentially the same results.[23] The 1,5-diphenyl isomer of **3.1-13** gave 65% of the aldazine (**3.1-14**) and 12 to 30% of **3.1-15** in all solvents (Eq. 23).[23]

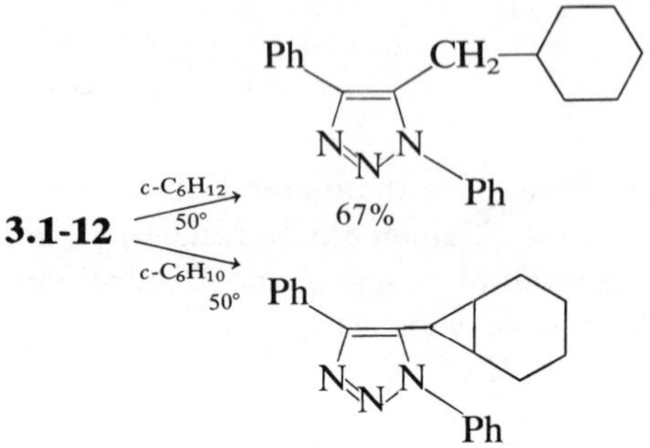

(21)

67%

(22)

about 60%

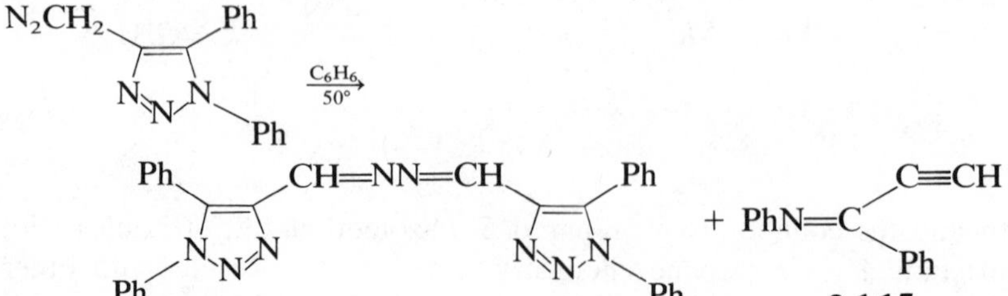

(23)

3.1-14

3.1-15

REFERENCES

1. **53:** 3198f
2. **56:** 1388a
3. **80:** 108454s
4. **83:** 131249z
5. **78:** 58205x
6. **81:** 135588r
7. **82:** 125237g
8. **71:** 81296s
9. **73:** 14768a
10. **61:** 2926b
11. **63:** 11552a
12. **64:** 9717g
13. **63:** 11551c
14. **69:** 96589x
15. **80:** 47912r
16. **80:** 59897c
17. **80:** 70667y
18. **82:** 156182q
19. **75:** 48994c
20. **68:** 68159t
21. **45:** 9037f
22. **77:** 164609w
23. **68:** 78204t
24. **80:** 145179b

3.2. 2,4,5-ALKYL- OR ARYL-TRISUBSTITUTED 2*H*-1,2,3-TRIAZOLES

With only two exceptions the methods developed for the synthesis of these compounds give disappointing yields. Both of these more promising approaches involve either the cyclization of nitrogen derivatives of α,β-dicarbonyl substrates (Eqs. 24,25)[25,26] or the rearrangement of another heterocycle (Eq. 26).[27]

PhC(=NNHPh)C(=NOAc)Me $\xrightarrow[Na_2SO_4]{KMnO_4}$ 4-Ph-5-Me-2-Ph-2*H*-1,2,3-triazole, 50% (24)

MeC(=NOH)C(=NN(Me)Ph)Me $\xrightarrow{Ac_2O}$ 4,5-Me$_2$-2-Ph-2*H*-1,2,3-triazole, 85% (25)

Ph(Ar^{2}HNN=)C-(3-isoxazolyl-5-Ar^1) $\xrightarrow[\text{Cu powder}]{\text{heat}}$ 4-Ph-5-(CH_2COAr^1)-2-Ar^2-2*H*-1,2,3-triazole (26)

$Ar^1 = Ar^2 = Ph$ 50% $Ar^2 = 4\text{-}NO_2Ph$; $Ar^1 = Ph$ 100%, $3\text{-}NO_2Ph$ 90%

Two other methods involving heterocyclic rearrangement have been described: tetrazole (Eq. 27)[28] and syndnone (Eq. 28)[29,30] photochemical conversions. The first reaction was not sensitive to changes in solvent or concentration[31] and the latter was verified by another laboratory.

Ph—(tetrazole: N—N(Me), N=N) $\xrightarrow{h\nu}$ 4,5-diphenyl-2-methyl-1,2,3-triazole (Ph, Ph; N–N(Me)–N) (27)

3.2-1
22–25%

(Ph, O^-; ^+N(R)–N–O sydnone ring) $\xrightarrow{h\nu}$ (Ph, Ph; N–N(R)–N triazole) (28)

3.2-2
R = Ph 24%, 4-MePh, 4-ClPh, c-C_6H_{11} 12%

Direct alkylation of 4,5-diphenyl-1,2,3-triazole gives a mixture of 1- and 2-alkyl products.[32,33]

George and his collaborators have found poor-to-fair results in the oxidation of aldehyde and phenone phenylhydrazones (Eqs. 29,30)[34,35]

ArCH=NNHPh $\xrightarrow{MnO_2}$ (Ar, Ar; N–N(Ph)–N triazole) (29)

Ar = Ph 44%; 4-MePh 15%; 4-MeOPh 7%

Ar(C=NNHPh)–(C=NNHPh)Ar $\xrightarrow{NiO_2}$ Ar(N=NPh)C=C(N=NPh)Ar $\xrightarrow{NiO_2}$ (Ar, Ar; N–N(Ph)–N triazole) (30)

Ar = Ph 13%; 4-MeOPh 21%

REFERENCES

25. **25:** 1827[1] 26. **40:** 7191[4] 27. **74:** 141643b 28. **72:** 43569w
29. **75:** 63700x 30. **78:** 16099k 31. **75:** 88545p 32. **80:** 70667y
33. **82:** 156182q 34. **67:** 43526b 35. **83:** 96642y

TABLE 3. ALKYL- OR ARYL-TRISUBSTITUTED 1,2,3-TRIAZOLES

Compound	Reference
3.1. 1,4,5-Alkyl- or Aryl-Trisubstituted 1*H*-1,2,3-Triazoles	
-1-acetamide, 4-(1–hydroxycyclohexyl)-5-(1-hydroxy-cyclohexylethynyl)-α,α-dimethyl-	**53:** 3199h
-1-acetamide, 5-(3-hydroxy-3-methyl-1-butynyl)-4-(1-hydroxy-1-methylethyl)-α,α-dimethyl-	**53:** 3199e
-1-acetic acid, 4-(1-hydroxycyclohexyl)-5-(1-hydroxy-cyclohexylethynyl)-α,α-dimethyl-, ethyl ester	**53:** 3199g
-4-acetonitrile, 1,5-diphenyl-	**45:** 9038a
1-(2-aminoethyl)-4,5-bis(methoxymethyl)-	**64:** P11216c
4-(2-aminoethyl)-1,5-diphenyl-	**45:** 9038b
1-(1*H*-benzimidazol-2-yl)-4,5-bis[(2-hydroxyethoxy) methyl]-	**79:** P78815v
1-[4-benzo[*f*]quinolin-1-ylmethyleneamino)phenyl]-4,5-dimethyl-	**41:** 1682a
1-(2-benzothiazolyl)-4,5-bis[(2-hydroxyethoxy)methyl]-	**79:** P78815v
1-benzyl-4,5-bis(chloromethyl)-	**58:** 5662d
1-benzyl-4,5-bis(trifluoromethyl)-	**24:** 3232[1], **56:** 1388h
1-benzyl-5-*tert*-butyl-4-methyl-	**80:** 59897c
1-benzyl-5-*tert*-butyl-4-methyl-, 1:1 adduct with 2,4,6-trinitrophenol	**80:** 59897c
1-benzyl-4,5-dimentyl-	**80:** 59897c, **84:** 30978b
1-benzyl-4,5-diphenyl-	**24:** 3232[1], **56:** 1388h, **82:** 156182q
1-benzyl-5-ethyl-4-methyl-, picrate	**69:** 96589x
1-benzyl-5-ethyl-4-methyl-, 1:1 adduct with 2,4,6-trinitrophenol	**80:** 59897c
1-benzyl-4-isopropenyl-5-methyl-	**80:** 47912r
1-benzyl-5-isopropyl-4-methyl-	**84:** 30978b
1-benzyl-4-methyl-5-phenyl-	**80:** 59897c, **84:** 30978b
1-benzyl-5-methyl-4-phenyl-	**80:** 59897c
4,5-bis(diethoxymethyl)-1-phenyl-	**80:** 108454s
5-(*tert*-butoxymethyl)-1,4-diphenyl-	**68:** 78204t
1-butyl-4-ethyl-5-methyl-	**67:** 82170a
5-*tert*-butyl-1,4-dimethyl-	**80:** 59897c, **84:** 30978b
4-butyl-4-(2-methyl-1-heptenyl)-1-(4-nitrophenyl)-	**65:** 7173f
5-*tert*-butyl-4-methyl-1-phenyl-	**80:** 59897c
5-(chloromethyl)-1,4-diphenyl-	**68:** 78204t
5-(2,4,6-cycloheptatrien-1-yl)-1,4-diphenyl-	**68:** 78204t
5-(2,4,6-cycloheptatrien-1-yl)-1,4-diphenyl-, 1:1 adduct with maleic anhydride	**68:** 78204t
1-cyclohexyl-4,5-diphenyl-	**68:** 68159t
5-(cyclohexylmethyl)-1,4-diphenyl-	**68:** 78204t
4-(diazomethyl)-1,5-diphenyl-	**68:** 78204t
5-(diazomethyl)-1,4-diphenyl-	**65:** 20120a, **68:** 78204t, **78:** 84325f, **80:** 145179b
5-(diazophenylmethyl)-1,4-diphenyl-	**78:** 84325f, **80:** 145179b
5-(dibromomethyl)-1,4-diphenyl-	**68:** 78204t
4-(1,2-dichloro-3,3,3-trifluoropropenyl)-1-[4-(dimethylamino)phenyl]-5-(trifluoromethyl)-	**65:** 18484g

TABLE 3 (*Continued*)

Compound	Reference
3.1. 1,4,5-Alkyl- or Aryl-Trisubstituted 1*H*-1,2,3-Triazoles (*Continued*)	
-4,5-dimethanol, 1-(4-arsonophenyl)-	**52:** P2086b
-4,5-dimethanol, 1-(1*H*-benzimidazol-2-yl)-	**79:** P78815v
-4,5-dimethanol, 1-(2-benzothiazolyl)-	**79:** P78815v
-4,5-dimethanol, 1-(2-benzothiazolyl)-α,α'-dimethyl-	**79:** P78815v
-4,5-dimethanol, 1-benzyl-	**55:** 23505h
-4,5-dimethanol, 1-benzyl-$\alpha,\alpha,\alpha',\alpha'$-tetramethyl-	**46:** 8651i, **50:** 4129c
-4,5-dimethanol, 1-butyl-$\alpha,\alpha,\alpha',\alpha'$-tetramethyl-	**60:** 15859d
-4,5-dimethanol, 1-(2-ethoxyethyl)-	**60:** 15859d
-4,5-dimethanol, 1-[4-(ethylamino)-6-(isopropylamino)-1,3,5-triazin-2-yl]-	**79:** P78815v
-4,5-dimethanol, 1-(2-hydroxyethyl)-$\alpha,\alpha,\alpha',\alpha'$-tetramethyl-	**60:** 15859d
-4,5-dimethanol, 1-phenyl-	**48:** 2685b **80:** 108454s
-4,5-dimethanol, 1-phenyl-, dibenzoate ester	**49:** 3948h
-4,5-dimethanol, 1-(4-phenylazophenyl)-	**52:** P2085i
-4,5-dimethanol, 1-phenyl-α,α'-bis(phenylethynyl)-	**82:** 140029j
4-[1-(1,2-dimethylhydrazino)ethyl]-1,5-diphenyl-	**76:** 113139q
4-[1-(1,2-dimethylhydrazino)ethyl]-1,5-diphenyl-, monohydrochloride	**76:** 113139q
4-[(1,2-dimethylhydrazino)methyl]-1,5-diphenyl-	**76:** 113139q
1,4-dimethyl-5-isopropyl-	**84:** 30978b
4,5-dimethyl-1-(4-nitrophenyl)-	**68:** 12925x
1,4-dimethyl-5-phenyl-	**80:** 59897c, **82:** 156182q **84:** 30978b
1,5-dimethyl-4-phenyl-	**80:** 59897c, **82:** 156182q
4,5-dimethyl-1-phenyl-	**80:** 59897c
4,5-dimethyl-1-phenyl-, 1:1 adduct with 2,4,6-trinitrophenol	**80:** 59897c
4,5-dimethyl-1-(1-phenylethenyl)-	**77:** 164609w
4,5-dimethyl-1-propyl-	**67:** 82170a
1-(4,6-dimethylpyrimidin-2-yl)-4,5-diphenyl-	**71:** 81296s
4,5-diphenyl-1-(1-ethoxyethyl)-	**78:** 124549r
4,5-diphenyl-1-ethyl-	**80:** 70667y, **82:** 156182q
1,4-diphenyl-5-(3-fluoro-2,4,6-cycloheptatrien-1-yl)-	**80:** 145179b
1,4-diphenyl-5-(4-fluoro-2,4,6-cycloheptatrien-1-yl)-	**80:** 145179b
4,5-diphenyl-1-[4-(heptafluoro-2-propyl)-3,5,6-trifluoropyridin-2-yl]-	**82:** 125237g
4,5-diphenyl-1-isopropyl-	**82:** 156182q
1,4-diphenyl-5-(2-methoxy-2,4,6-cycloheptatrien-1-yl)-	**80:** 145179b
1,4-diphenyl-5-(3-methoxy-2,4,6-cycloheptatrien-1-yl)-	**80:** 145179b
1,4-diphenyl-5-(4-methoxy-2,4,6-cycloheptatrien-1-yl)-	**80:** 145179b
1,4-diphenyl-5-methyl-	**68:** 78204t, **75:** 48994c **80:** 59897c
1,5-diphenyl-4-methyl-	**75:** 48994c, **80:** 59897c
4,5-diphenyl-1-methyl-	**80:** 70667y, **82:** 156182q
1,4-diphenyl-5-(2-methyl-2,4,6-cycloheptatrien-1-yl)-	**80:** 145179b
1,4-diphenyl-5-(3-methyl-2,4,6-cycloheptatrien-1-yl)-	**80:** 145179b
1,4-diphenyl-5-(4-methyl-2,4,6-cycloheptatrien-1-yl)-	**80:** 145179b

TABLE 3 (*Continued*)

Compound	Reference
3.1. 1,4,5-Alkyl- or Aryl-Trisubstituted 1*H*-1,2,3-Triazoles (*Continued*)	
1,4-diphenyl-5-[(1-methylethoxy)-2,4,6-cyclo-heptatrien-1-yl]-	**80:** 145179b
4,5-diphenyl-1-(1-methyl-1-nitroethyl)-	**52:** 7293a
1,5-diphenyl-4-(6-methyl-2-pyridyl)-	**73:** 14768a
4,5-diphenyl-1-[(methylthio)methyl]-	**72:** 43576w
4,5-diphenyl-1-(3-nitrophenyl)-	**75:** 62695n
1,4-diphenyl-5-(7-norcaryl)-	**68:** 78204t
4,5-diphenyl-1-(pentafluorophenyl)-	**78:** 124171m, **81:** 135588r
1,5-diphenyl-4-(2-pyridyl)-	**73:** 14768a
4,5-diphenyl-1-(2-pyridyl)-	**71:** 81296s
1,5-diphenyl-4-(2-quinolyl)-	**73:** 14768a
1,4-diphenyl-5-(1,2,2,2-tetrachloroethyl)-	**68:** 78204t
4,5-diphenyl-1-(tetrafluoropyridin-4-yl)-	**78:** 58205x
1,4-diphenyl-5-(trichloroethenyl)-	**68:** 78204t
1,5-diphenyl-4-(tridecafluorohexyl)-	**83:** 131249z
4-ethyl-5-methyl-1-(4-nitrophenyl)-	**68:** 12925x, **68:** 12927z
5-ethyl-4-methyl-1-(4-nitrophenyl)-	**56:** 14020b
4-ethyl-5-(2-methyl-1-pentenyl)-1-(4-nitrophenyl)-	**65:** 7173f
4-ethyl-5-methyl-1-phenyl-	**61:** 2926c
5-ethyl-4-methyl-1-phenyl-	**61:** 2926c, **80:** 59897c
5-ethyl-4-methyl-1-phenyl-, 1:1 adduct with 2,4,6-trinitrophenol	**80:** 59897c
5-ethyl-4-methyl-1-propyl-	**67:** 82170a
4-(1-hydrazonoethyl)-1-(4-hydrazino-3-oxo-2-phenyl-2*H*-pyridazin-5-yl)-5-methyl-	**78:** 16120k
4-isopropenyl-5-methyl-1-phenyl-	**80:** 47912r
4-isopropyl-5-methyl-1-(4-nitrophenyl)-	**63:** 11551d
-5-methanamine, α,1,4-triphenyl-	**80:** 145179b
-5-methanimine, α,1,4-triphenyl-	**80:** 145179b
-4-methanol, 1-(4-aminophenyl)-5-(3-hydroxy-3-methyl-1-butynyl)-α,α-dimethyl-	**53:** 3199b
-4-methanol, 1-benzyl-α,α-dimethyl-5-piperidinyl-	**46:** 8652b
-5-methanol, 1,4-diphenyl-	**68:** 78204t
-4-methanol, 5-(3-hydroxy-3-methyl-1-butynyl)-α,α-dimethyl-1-(2-nitrophenyl)-	**53:** 3199a
-4-methanol, 5-(3-hydroxy-3-methyl-1-butynyl)-α,α-dimethyl-1-(3-nitrophenyl)-	**53:** 3199a
-4-methanol, 5-(3-hydroxy-3-methyl-1-butynyl)-α,α-dimethyl-1-(4-nitrophenyl)-	**53:** 9198i
-4-methanol, 5-[3(or 5)-(1-hydroxy-1-methylethyl) pyrazol-4-yl]-α,α-dimethyl-1-(4-nitrophenyl)-	**53:** 3199i
-5-methanol, α,1,4-triphenyl-	**80:** 145179b
5-(2-methyl-1-hexenyl)-1-(4-nitrophenyl)-4-propyl-	**65:** 7173f
4-(methyl-5-(2-methyl-1-butenyl)-1-(4-nitrophenyl)-	**63:** 11552c, **65:** 7137f
5-methyl-1-(4-nitrophenyl)-4-phenyl-	**67:** 82171b
5-methyl-1-[3-oxo-2-phenyl-4-(2-phenylhydrazino)-2*H*-pyradizin-5-yl]-4-[1-(phenylhydrazono)ethyl]	**78:** 16120k
5-(2-methyl-3-phenylpropenyl)-1-(4-nitrophenyl)-4-phenyl-	**64:** 9717g

TABLE 3 (*Continued*)

Compound	Reference
3.1. 1,4,5-Alkyl- or Aryl-Trisubstituted 1*H*-1,2,3-Triazoles (*Continued*)	
4-methyl-5-phenyl-1-propyl-	**67:** 82171b
-1-propanoic acid, 4,5-diphenyl-	**85:** P123926u
-2-propanoic acid, 4,5-diphenyl-	**85:** P123926u
-2-propanoic acid, 4-methyl-5-phenyl-	**85:** P123926u
-4-pyruvic acid, 1,5-diphenyl-, oxime	**45:** 9038a
-4-α-thiopyruvic acid, 1,5-diphenyl-	**45:** 9038a
1,4,5-trimethyl-	**80:** 59897c, **84:** 30978b
1,4,5-triphenyl-	**46:** 8651i
3.2. 2,4,5-Alkyl- or Aryl-Trisubstituted 2*H*-1,2,3-Triazoles	
-4-acetamide, 2-(4-bromophenyl)-5-methyl-	**66:** 37841p
-4-acetamide, 5-methyl-2-phenyl-	**66:** 37841p
2-(4-acetamido-2-hydroxyphenyl)-4,5-diphenyl-	**71:** P124445j
4-(4–acetamidophenyl)-5-(3,4-dimethylphenyl)-2-(2-hydroxyphenyl)-	**71:** P124445j
4-(4-acetamidophenyl)-2-(2-hydroxyphenyl)-5-methyl-	**71:** P124445j
4-(4-acetamidophenyl)-2-(2-hydroxyphenyl)-5-phenyl-	**71:** P124445j
-4-acetic acid, 2-(4-bromophenyl)-5-methyl-	**66:** 37841p
-4-acetic acid, 5-methyl-2-phenyl-	**66:** 37841p
-4-acetonitrile, 2-(4-bromophenyl)-5-methyl-	**66:** 37841p
-4-acetonitrile, 5-methyl-2-phenyl-	**66:** 37841p
2-[3-(acrylamidomethyl)-2-hydroxy-5-methylphenyl]-4,5-diphenyl-	**71:** P124445j
2-(3-allyl-2-hydroxyphenyl)-4,5-diphenyl-	**71:** P124445j
2-[4-(allyloxy)-2-hydroxyphenyl]-4,5-diphenyl-	**71:** P124445j
2-(2-amino-4-arsenosophenyl)-4,5-dimethyl-	**40:** 7191[8]
2-(5-amino-2-chlorophenyl)-4,5-dimethyl-	**80:** P49275j
2-(4-amino-2-chlorophenyl)-4,5-dimethyl-, *N*-oxide	**80:** P49275j
2-(4-amino-2-hydroxyphenyl)-4,5-diphenyl-	**71:** P124445j
2-(2-aminophenyl)-4,5-dimethyl-	**40:** 7191[6]
2-(4-aminophenyl)-4,5-dimethyl-	**40:** 7191[6]
4-(4-aminophenyl)-2-(2-hydroxyphenyl)-5-phenyl-	**71:** P124445j
2-[5-(aminosulfonyl)-*N*-butyl-2-hydroxy-*N*-(2-hydroxyethyl)phenyl]-4,5-bis(4-chlorophenyl)-	**71:** P124445j
2-(4-benzamido-2-hydroxyppenyl)-4,5-diphenyl-	**71:** P124445j
2-[3-(benzamidomethyl)-2-hydroxy-5-methylphenyl]-4,5-diphenyl-	**71:** P124445j
4-[4-(benzamido)phenyl]-2-(2-hydroxyphenyl)-5-phenyl-	**71:** P124445j
4-[4-(benzenesulfonamido)phenyl]-2-(2-hydroxyphenyl)-5-phenyl-	**71:** P124445j
2-[4-(2-benzofuranyl)phenyl]-4-methyl-5-phenyl-	**75:** P22479p
2-[4-(2-benzofuranyl)phenyl]-4-methyl-5-phenyl-, 3-oxide	**75:** P22479p
4-benzoylmethyl-2,5-diphenyl-	**74:** 141643b
4-benzoylmethyl-2,5-diphenyl-, (2,4-dinitrophenyl) hydrazone	**74:** 141643b
4-benzoylmethyl-2-(4-nitrophenyl)-5-phenyl-	**74:** 141643b

TABLE 3 (*Continued*)

Compound	Reference
3.2. 2,4,5-Alkyl- or Aryl-Trisubstituted 2*H*-1,2,3-Triazoles (*Continued*)	
4-[3-(*N*-benzylacetamido)phenyl]-2-(2-hydroxyphenyl)-5-phenyl-	**77:** P115464m
2-(3-benzyl-2-hydroxy-5-methylphenyl)-4,5-diphenyl-	**71:** P124445j
4-benzyl-2-(2-hydroxy-5-methylphenyl)-5-methyl-	**71:** P124445j
2-(5-benzyl-2-hydroxyphenyl)-4,5-diphenyl-	**71:** P124445j
2-(5-benzyl-2-hydroxyphenyl)-4-ethyl-5-phenyl-	**82:** P4995u
4-benzyl-2-(2-hydroxyphenyl)-5-methyl-	**71:** P124445j
2-(5-benzyl-2-hydroxyphenyl)-4-methyl-5-phenyl-	**71:** P124445j
4-benzyl-2-(2-hydroxyphenyl)-5-phenyl-	**71:** P124445j
2-[3-[2-(benzylmethylamino)acetamidomethyl]-2-hydroxy-5-methylphenyl]-4,5-diphenyl-	**71:** P124445j
2-[5-[(*N*-benzyl-*N*-methylamino)carbonyl]-2-hydroxyphenyl]-4,5-bis(4-chlorophenyl)-	**71:** P124445j
2-[5-[(*N*-benzyl-*N*-methylamino)sulfonyl]-2-hydroxyphenyl]-4,5-bis(4-chlorophenyl)-	**71:** P124445j
2-[4-[(benzyloxycarbonyl)amino]-2-hydroxyphenyl]-4,5-diphenyl-	**71:** P124445j
2-[4-(benzyloxy)-2-hydroxyphenyl]-4,5-diphenyl-	**71:** P124445j
2-[4-(benzylsulfonamido)-2-hydroxyphenyl]-4,5-diphenyl-	**71:** P124445j
2-[3-[3-(benzylthio)propanamidomethyl]-2-hydroxy-5-methylphenyl]-4,5-diphenyl-	**71:** P124445j
4-[1,1′-biphenyl]-4-yl-2-(4-carboxyphenyl)-5-methyl-	**83:** P61732a
4-[1,1′-biphenyl]-4-yl-2-[(4-chlorocarbonyl)phenyl]-5-methyl-	**83:** P61732a
4-[1,1′-biphenyl]-4-yl-2-(4-formylphenyl)-5-methyl-	**83:** P61732a
4,5-bis[4-(allyloxy)phenyl]-2-(2-hydroxyphenyl)-	**71:** P124445j
4,5-bis[4-(benzyloxy)phenyl]-2-(2-hydroxyphenyl)-	**71:** P124445j
4,5-bis(4-chlorophenyl)-2-[5-[[*N*-butyl-*N*-(2-hydroxyethyl)amino]carbonyl]-2-hydroxyphenyl]-	**71:** P124445j
4,5-bis(4-chlorophenyl)-2-(5-*tert*-butyl-2-hydroxyphenyl)-	**71:** P124445j
4,5-bis(4-chlorophenyl)-2-(5-chloro-2-hydroxyphenyl)-	**71:** P124445j
4,5-bis(4-chlorophenyl)-2-[5-[(*N*-cyclohexyl-*N*-methylamino)carbonyl]-2-hydroxyphenyl]-	**71:** P124445j
4,5-bis(4-chlorophenyl)-2-[5-[(*N*-cyclohexyl-*N*-methylamino)sulfonyl]-2-hydroxyphenyl]-	**71:** P124445j
4,5-bis(4-chlorophenyl)-2-(3,4-dichloro-2-hydroxyphenyl)-	**71:** P124445j
4,5-bis(4-chlorophenyl)-2-[5-[(*N,N*-diethylamino) carbonyl]-2-hydroxyphenyl]-	**71:** P124445j
4,5-bis(4-chlorophenyl)-2-[5-(*N,N*-diethylamino) sulfonyl]-2-hydroxyphenyl]-	**71:** P124445j
4,5-bis(4-chlorophenyl)-2-(4-hydroxy-1,1′-biphenyl-3-yl)-	**71:** P124445j
4,5-bis(4-chlorophenyl)-2-(2-hydroxy-5-methoxyphenyl)-	**71:** P124445j
4,5-bis(4-chlorophenyl)-2-[2-hydroxy-5-[[3-(methoxy-propyl)amino]carbonyl]phenyl]-	**71:** P124445j

TABLE 3 (*Continued*)

Compound	Reference
3.2. 2,4,5-Alkyl- or Aryl-Trisubstituted 2*H*-1,2,3-Triazoles (*Continued*)	
4,5-bis(4-chlorophenyl)-2-(2-hydroxyphenyl)-	**71:** P124445j
4,5-bis(4-chlorophenyl)-2-[5-[[*N*-(3-methoxypropyl) amino]sulfonyl]-2-hydroxyphenyl]-	**71:** P124445j
4,5-bis(4-chlorophenyl)-2-[5-[(*N*-octylamino) carbonyl]-2-hydroxyphenyl]-	**71:** P124445j
4,5-bis(4-chlorophenyl)-2-[5-[(*N*-octylamino) sulfonyl]-2-hydroxyphenyl]-	**71:** P124445j
4,5-bis(4-chlorophenyl)-2-phenyl-	**73:** 120593z
4,5-bis(3,5-dibromo-4-methoxyphenyl)-2-(4-bromophenyl)-	**68:** 114514e
4,5-bis(4-hydroxyphenyl)-2-(2-hydroxyphenyl)-	**71:** P124445j
4,5-bis(4-hydroxyphenyl)-2-(2-hydroxyphenyl)-, 4,5-diacetate diester	**71:** P124445j
4,5-bis(4-hydroxyphenyl)-2-(2-hydroxyphenyl)-, 4,5-dibenzoate diester	**71:** P124445j
4,5-bis(4-hydroxyphenyl)-2-(2-hydroxyphenyl)-, 4,5-didodecanoate diester	**71:** P124445j
4,5-bis(2-methoxyethyl)-2-(2-hydroxyphenyl)-	**71:** P124445j
4,5-bis(4-methoxyphenyl)-2-(4-bromophenyl)-	**68:** 114514e
4,5-bis(4-methoxyphenyl)-2-(5-chloro-2-hydroxyphenyl)-	**71:** P124445j
4,5-bis(4-methoxyphenyl)-2-(3,5-dichloro-2-hydroxyphenyl)-	**71:** P124445j
4,5-bis(4-methoxyphenyl)-2-(4,5-dimethyl-2-hydroxyphenyl)-	**71:** P124445j
4,5-bis(4-methoxyphenyl)-2-(5-dodecyl-2-hydroxyphenyl)-	**71:** P124445j
4,5-bis(4-methoxyphenyl)-2-(4-hydroxy-1,1′-biphenyl-3-yl)-	**71:** P124445j
4,5-bis(4-methoxyphenyl)-2-(2-hydroxy-5-methoxyphenyl)-	**71:** P124445j
4,5-bis(4-methoxyphenyl)-2-(2-hydroxy-5-methylphenyl)-	**71:** P124445j
4,5-bis(2-methoxyphenyl)-2-(2-hydroxyphenyl)-	**71:** P124445j
4,5-bis(4-methoxyphenyl)-2-(2-hydroxyphenyl)-	**71:** P124445j
4,5-bis(4-methoxyphenyl)-2-(4-iodophenyl)-	**68:** 114514e
4,5-bis(4-methoxyphenyl)-2-(4-methoxy-3-biphenyl)-	**71:** P124445j
4,5-bis(4-methoxyphenyl)-2-(4-methylphenyl)-	**68:** 114514e
4,5-bis(4-methoxyphenyl)-2-phenyl-	**67:** 43526b, **68:** 114514e, **73:** 120593z, **83:** 96642y
4,5-bis[(3,4-methylenedioxy)phenyl]-2-phenyl-	**73:** 120593z
4,5-bis(4-methylphenyl)-2-(5-chloro-2-hydroxyphenyl)-	**71:** P124445j
4,5-bis(4-methylphenyl)-2-(2-hydroxyphenyl)-	**71:** P124445j
4,5-bis(4-methylphenyl)-2-(2-methoxyphenyl)-	**71:** P124445j
4,5-bis(4-methylphenyl)-2-phenyl-	**67:** 43526b, **73:** 120593z
4,5-bis(4-nitrophenyl)-2-phenyl-	**68:** 114514e
4,5-bis[4-(octyloxy)phenyl]-2-(2-hydroxyphenyl)-	**71:** P124445j
4-(bromomethyl)-2-(4-bromophenyl)-5-methyl-	**66:** 37841p
4-[4-(bromomethyl)-3-cyanophenyl]-5-methyl-2-phenyl-	**79:** P147435k

TABLE 3 (*Continued*)

Compound	Reference
3.2. 2,4,5-Alkyl- or Aryl-Trisubstituted 2*H*-1,2,3-Triazoles (*Continued*)	
4-(bromomethyl)-5-methyl-2-phenyl-	**66:** 37841p
4-(bromomethyl)-2-(4-nitrophenyl)-5-methyl-	**66:** 37841p
2-(4-bromophenyl)-4,5-dimethyl-	**66:** 37841p
2-(4-bromophenyl)-4,5-dimethyl-, 1-oxide	**66:** 28250t
2-(4-bromophenyl)-4,5-diphenyl-	**68:** 114514e
2-[3-(butanamidomethyl)-2-hydroxy-5-methylphenyl]-4,5-diphenyl-	**71:** P124445j
4-(3-butanamidophenyl)-2-(2-hydroxyphenyl)-5-phenyl-	**77:** P115464m
2-[3-[(1-butanesulfonamido)phenyl]-4-hydroxyphenyl]-4,5-diphenyl-(?)	**71:** 124445j
2-[3-(2-(butoxyacetamido)methyl]-2-hydroxy-5-methylphenyl]-4,5-diphenyl-	**71:** 124445j
2-(5-*tert*-butyl-2-hydroxyphenyl)-4,5-diphenyl-	**71:** P124445j
4-[3-(butylmethoxyamino)phenyl]-2-(2-hydroxyphenyl)-5-phenyl-	**77:** P115464m
2-[4-(butylsulfonamido)-2-hydroxyphenyl]-4,5-diphenyl-	**71:** P124445j
2-[3-[[[[2-(carboxyethyl)thio]carbamoyl]ethyl] methyl]-2-hydroxy-5-methylphenyl]-4,5-diphenyl-	**71:** P124445j
2-[3-[[[[(carboxymethyl)thio]carbamoyl]ethyl] methyl]-2-hydroxy-5-methylphenyl]-4,5-diphenyl-	**71:** P124445j
2-(4-carboxyphenyl)-4,5-diphenyl-	**77:** P7313c
4-(4-carboxyphenyl)-2-(2-hydroxyphenyl)-5-methyl-	**71:** P124445j
2-(4-carboxyphenyl)-4-methyl-5-(2-naphthalenyl)-	**83:** P61732a
2-(4-carboxyphenyl)-4-methyl-5-phenyl-	**72:** P100716t
2-[3-[2-(chloroacetamido)methyl]-2-hydroxy-5-methylphenyl]-4,5-diphenyl-	**71:** P124445j
2-[5-[[(2-chloroacetamido)phenyl]methyl]-2-hydroxyphenyl]-4-methyl-5-phenyl-(?)	**71:** P124445j
2-[4-(2-chlorobenzamido)-2-hydroxyphenyl]-4,5-diphenyl-	**71:** P124445j
2-[3-(2-chlorobenzamido)methyl]-2-hydroxy-5-methylphenyl]-4,5-diphenyl-	**71:** P124445j
4-[4-(2-chlorobenzamido)phenyl]-2-(2-hydroxyphenyl)-5-phenyl-	**71:** P124445j
2-[4-(5-chloro-2-benzofuranyl)phenyl]-4-methyl-5-phenyl-	**75:** P22479p
2-[(4-chlorocarbonyl)phenyl]-4-methyl-5-(2-naphthalenyl)-	**83:** P61732a
2-(5-chloro-2-hydroxyphenyl)-4,5-diphenyl-	**71:** P124445j
2-(5-chloro-2-hydroxyphenyl)-4-(2-furanyl)-5-methyl-	**71:** P124445j
2-(5-chloro-2-hydroxyphenyl)-4-methyl-5-phenyl-	**71:** P124445j
2-(5-chloro-2-hydroxyphenyl)-4-methyl-5-(2-pyridyl)-	**71:** P124445j
2-(5-chloro-2-hydroxyphenyl)-4-methyl-5-(2-thienyl)-	**71:** P124445j
2-(5-chloro-2-methoxyphenyl)-5-methyl-4-(2-pyridyl)-	**71:** P124445j
2-(5-chloro-2-methoxyphenyl)-5-methyl-4-(2-pyridyl)-, oxide	**71:** P124445j
2-(5-chloro-2-methoxyphenyl)-4-methyl-5-(2-thienyl)-, 3-oxide	**71:** P124445j

TABLE 3 (*Continued*)

Compound	Reference
3.2. 2,4,5-Alkyl- or Aryl-Trisubstituted 2*H*-1,2,3-Triazoles (*Continued*)	
2-[4-(5-chloro-3-methyl-2-benzofuranyl)phenyl]-4-methyl-5-phenyl-	**75:** P22479p
2-[4-(5-chloro-3-phenyl-2-benzofuranyl)phenyl]-4-methyl-5-phenyl-	**75:** P22479p
2-(2-chlorophenyl)-4,5-dimethyl-	**84:** 135554p
2-(4-chlorophenyl)-4,5-dimethyl-	**59:** 13483f
2-(2-chlorophenyl)-4,5-diphenyl-	**84:** 135554p
2-(4-chlorophenyl)-4,5-diphenyl-	**68:** 114514e, **75:** 63700x
2-(4-chlorophenyl)-4-ethyl-5-methyl-	**59:** 13483f
4-(4-chlorophenyl)-2-(2-hydroxyphenyl)-5-[4-(octyloxy)phenyl]-	**71:** P124445j
2-[4-[(4-chlorophenyl)sulfonamido]-2-hydroxyphenyl]-4,5-diphenyl-	**71:** P124445j
2-[4-(5-chloro-3,4,6-trimethyl-2-benzofuranyl) phenyl]-4-methyl-5-phenyl-	**75:** P22479p
4-(3-cyano-4-methylphenyl)-5-methyl-2-phenyl-	**79:** P147435k
2-[4-(2-cyano-2-phenylethenyl]-3-methoxyphenyl]-4,5-dimethyl-	**81:** P51153g
2-[[3-(cyclohexanecarboxamido)methyl]-2-hydroxy-5-methylphenyl]-4,5-diphenyl-	**71:** P124445j
4-[[4-(cyclohexanecarboxamido)phenyl]-2-(2-hydroxy-phenyl)-5-phenyl]-	**71:** P124445j
2-cyclohexyl-4,5-diphenyl-	**78:** 16099k
2-(5-cyclohexyl-2-hydroxyphenyl)-4,5-diphenyl-	**71:** P124445j
2-(5-cyclohexyl-2-hydroxyphenyl)-4-methyl-5-phenyl-	**71:** P124445j
2-[3-[2-(cyclohexylmethylamino)acetamidomethyl]-2-hydroxy-5-methylphenyl]-4,5-diphenyl-	**71:** P124445j
2-[4-[(cyclohexyloxycarbonyl)amino]-2-hydroxy-phenyl]-4,5-diphenyl-	**71:** P124445j
2-[4-(cyclohexylsulfonamido)-2-hydroxyphenyl]-4,5-diphenyl-	**71:** P124445j
2-[3-[3-(cyclohexylthio)propanamidomethyl]-2 hydroxy-5-methylphenyl]-4,5-diphenyl-	**71:** P124445j
4-decyl-2-(2-hydroxyphenyl)-5-phenyl-	**71:** P124445j
4,5-dibenzyl-2-(2-hydroxyphenyl)-	**71:** P124445j
2-[3-[2-(dibutylamino)acetamidomethyl]-2-hydroxy-5-methylphenyl]-4,5-diphenyl-	**71:** P124445j
2-[5-[(*N,N*-dibutylamino)carbonyl]-2-hydroxyphenyl]-4,5-diphenyl-	**77:** P115464m
4,5-dibutyl-2-(2-hydroxyphenyl)-	**71:** P124445j
2-(3,5-di-*tert*-butyl-2-hydroxyphenyl)-4,5-diphenyl-	**71:** P124445j
2-[4-(5,7-dichloro-2-benzofuranyl)phenyl]-4-methyl-5-phenyl-	**75:** P22479p
2-[4-(5,7-dichloro-3-methyl-2-benzofuranyl)phenyl]-4-methyl-5-phenyl-	**75:** P22479p
4-(3,4-dichlorophenyl)-2-(2-hydroxyphenyl)-5-methyl-	**71:** P124445j
4,5-dicyclohexyl-2-(2-hydroxyphenyl)-	**71:** P124445j
2-[5-[(*N,N*-diethylamino)sulfonyl]-2-hydroxyphehyl]-4,5-diphenyl-	**77:** P115464m

TABLE 3 (*Continued*)

Compound	Reference
3.2. 2,4,5-Alkyl- or Aryl-Trisubstituted 2*H*-1,2,3-Triazoles (*Continued*)	
4-[[3-(*N,N*-diethylamino)sulfonyl]phenyl]-2-(2-hydroxy phenyl)-5-phenyl-	**77:** P115464m
4,5-di-2-furanyl-2-(2-hydroxyphenyl)-	**77:** P115464m
2-[1,4-dihydroxy-5-(2*H*-naphtho[1,2-*d*]triazol-2-yl) phenyl]-4,5-dimethyl-	**83:** P132619g
2-(2,4-dihydroxyphenyl)-4,5-diphenyl-	**71:** P124445j
2-(2,4-dihydroxyphenyl)-4,5-diphenyl-, 4-acetate ester	**71:** P124445j
2-(2,4-dihydroxyphenyl)-4,5-diphenyl-, 4-benzoate ester	**71:** P124445j
2-(2,4-dihydroxyphenyl)-4,5-diphenyl-, 4-dodecanoate ester	**71:** P124445j
2-(2,4-dihydroxyphenyl)-4-methyl-5-phenyl-	**71:** P124445j
4-(2,4-dimethoxyphenyl)-2-(2-hydroxyphenyl)-5-phenyl-	**71:** P124445j
4-(4-dimethylaminophenyl)-5-methyl-2-(4-nitrophenyl)-	**25:** 933[1]
2-[4-(3,5-dimethyl-2-benzofuranyl)phenyl]-4-methyl-5-phenyl-	**75:** P22479p
4-(1,3-dimethyl-2-furanyl-1*H*-benzimidazol-2-ylium)-5-methyl-2-phenyl-, methyl sulphate	**84:** P91661q
4,5-dimethyl-2-(2-hydroxyphenyl)-	**71:** P124445j
2-(3,4-dimethyl-2-hydroxyphenyl)-4,5-diphenyl-	**77:** P115464m
2-(3,5-dimethyl-2-hydroxyphenyl)-4,5-diphenyl-	**71:** P124445j
4,5-dimethyl-2-[x[[(methoxycarbonyl)aminomethyl] phenyl]-2-(2-hydroxyphenyl)]-	**77:** P115464m
4,5-dimethyl-2-(2-methoxyphenyl)-	**71:** P124445j
4,5-dimethyl-2-(4-methylphenyl)-, 1-oxide	**66:** 28250t
4,5-dimethyl-2-(2-nitrophenyl)-	**21:** 2133[5], **40:** 7191[5]
4,5-dimethyl-2-(4-nitrophenyl)-, 1-oxide	**66:** 28250t
2,4-dimethyl-5-phenyl-	**82:** 156182q
4,5-dimethyl-2-phenyl-	**40:** 7191[4], **55:** 18756b **84:** 135554p
4,5-dimethyl-2-phenyl-, 1-oxide	**53:** 18947a, **66:** 28250t
2-[3-(1,1-dimethyl-2-propenyl)-2-hydroxyphenyl]-4,5-diphenyl-	**71:** P124445j
2-[3-(1,3-dioxo-1*H*-isoindol-2(2*H*)-yl)methyl]-2-hydroxy-5-methylphenyl]-4,5-diphenyl-	**82:** P17959k
2-(1,3-dioxo-2-methyl-1*H*-benz[*de*]isoqinolin-6(2*H*)-yl)-4-methyl-5-phenyl-	**77:** P90079z
4,5-diphenethyl-2-(2-hydroxyphenyl)-	**71:** P124445j
4,5-diphenyl-2-(4-dodecanamido-2-hydroxyphenyl)-	**71:** P124445j
4,5-diphenyl-2-[3-(dodecanamidomethyl)-2-hydroxy-5-methylphenyl]-	**71:** P124445j
4,5-diphenyl-2-(5-dodecyl-2-hydroxyphenyl)-	**71:** P124445j
4,5-diphenyl-2-[3-[2-(dodecyloxy)acetamidomethyl]-2-hydroxy-5-methylphenyl]-	**71:** P124445j
4,5-diphenyl-2-[3-[(ethoxycarbonyl)aminomethyl]-2-hydroxy-5-methylphenyl]-	**71:** P124445j
4,5-diphenyl-2-[3-[(ethoxycarbonyl)ethylaminomethyl]-2-hydroxy-5-methylphenyl]-	**71:** P124445j

TABLE 3 (*Continued*)

Compound	Reference
3.2. 2,4,5-Alkyl- or Aryl-Trisubstituted 2*H*-1,2,3-Triazoles (*Continued*)	
4,5-diphenyl-2-ethyl-	**82:** 156182q
4,5-diphenyl-2-(4-hydroxy-1,1′-biphenyl-3-yl)-	**71:** P124445j
4,5-diphenyl-2-[3-[2-hydroxy-3-[(hydroxyethyl)thio] propanamidomethyl]-5-methylphenyl]-	**71:** P124445j
4,5-diphenyl-2-(2-hydroxy-3-isobutylphenyl)-	**71:** P124445j
4,5-diphenyl-2-[2-hydroxy-4-(4-methoxybenzamido) phenyl]-	**71:** P124445j
4,5-diphenyl-2-(2-hydroxy-4-methoxyphenyl)-	**71:** P124445j
4,5-diphenyl-2-(2-hydroxy-5-methoxyphenyl)-	**71:** P124445j
4,5-diphenyl-2-[2-hydroxy-4-[(4-methoxyphenyl) sulfonamido]phenyl]-	**71:** P124445j
4,5-diphenyl-2-[2-hydroxy-4-(*N*-methylbenzamido) phenyl]-	**77:** P115464m
4,5-diphenyl-2-[2-hydroxy-4-(3-methylbenzamido) phenyl]-	**71:** P124445j
4,5-diphenyl-2-[2-hydroxy-5-methyl-3-[(4-methylbenz-amido)methyl]phenyl]-	**71:** P124445j
4,5-diphenyl-2-[2-hydroxy-5-methyl-3-(octadecanamido-methyl)phenyl]-	**71:** P124445j
4,5-diphenyl-2-(2-hydroxy-5-methyl-3-octadecylphenyl)-	**71:** P124445j
4,5-diphenyl-2-(2-hydroxy-5-methyl-3-octanoylphenyl)-	**71:** P124445j
4,5-diphenyl-2-(2-hydroxy-5-methyl-3-octylphenyl)-	**71:** P124445j
4,5-diphenyl-2-[2-hydroxy-5-methyl-3-[2-(octylthio) acetamidomethyl]phenyl]-	**71:** P124445j
4,5-diphenyl-2-[2-hydroxy-5-methyl-3-[[2-(octylthio) propanamido]methyl]phenyl]-	**77:** P115464m
4,5-diphenyl-2-[2-hydroxy-5-methyl-3-(oxazolidin-2-oneylmethyl)phenyl]-	**71:** P124445j
4,5-diphenyl-2-[2-hydroxy-5-methyl-3-[(2-oxo-2*H*-azepin-1-yl)methyl]]-	**71:** P124445j
4,5-diphenyl-2-(2-hydroxy-5-methylphenyl)-	**71:** P124445j
4,5-diphenyl-2-[2-hydroxy-5-methyl-3-[2-(phenylacet-amido)methyl]phenyl]-	**71:** P124445j
4,5-diphenyl-2-[2-hydroxy-4-[(4-methylphenyl)sulfon-amido]phenyl]-	**71:** P124445j
4,5-diphenyl-2-[2-hydroxy-5-methyl-3-[2-(phenylthio) acetamidomethyl]phenyl]-	**71:** P124445j
4,5-diphenyl-2-[2-hydroxy-5-methyl-3-[(2-propenamido) methyl]phenyl]-	**71:** P124445j
4,5-diphenyl-2-[2-hydroxy-3-(2-methyl-2-propenyl) phenyl]-	**71:** P124445j
4,5-diphenyl-2-[2-hydroxy-5-methyl-3-(1-pyrrolidin-2-oneylmethyl)phenyl]-	**71:** P124445j
4,5-diphenyl-2-[2-hydroxy-4-(methylsulfonamido)phenyl]-	**71:** P124445j
4,5-diphenyl-2-[2-hydroxy-5-(4-morpholinylcarbonyl) phenyl]-	**77:** P115464m
4,5-diphenyl-2-(2-hydroxy-4-nitrophenyl)-	**71:** P124445j
4,5-diphenyl-2-[2-hydroxy-4-(octadecanamido)phenyl]-	**71:** P124445j

TABLE 3 (*Continued*)

Compound	Reference
3.2. 2,4,5-Alkyl- or Aryl-Trisubstituted 2*H*-1,2,3-Triazoles (*Continued*)	
4,5-diphenyl-2-(2-hydroxy-5-octadecylphenyl)-	**71:** P124445j
4,5-diphenyl-2-(2-hydroxy-5-octanoylphenyl)-	**71:** P124445j
4,5-diphenyl-2-[2-hydroxy-4-[(octyloxycarbonyl) amino]phenyl]-	**71:** P124445j
4,5-diphenyl-2-[2-hydroxy-5-(octyloxy)phenyl]-	**71:** P124445j
4,5-diphenyl-2-(2-hydroxy-5-octylphenyl)-	**71:** P124445j
4,5-diphenyl-2-(2-hydroxy-3-*tert*-pentylphenyl)-	**71:** P124445j
4,5-diphenyl-2-[2-hydroxy-4-[(phenoxycarbonyl) amino]phenyl]-	**71:** P124445j
4,5-diphenyl-2-(2-hydroxy-5-phenoxyphenyl)-	**82:** P4995u
4,5-diphenyl-2-(2-hydroxyphenyl)-	**71:** P124445j
4,5-diphenyl-2-(2-hydroxyphenyl)-, acetate ester	**71:** P124445j
4,5-diphenyl-2-[2-hydroxy-3-(1-phenyl-2-propenyl) phenyl]-	**71:** P124445j
4,5-diphenyl-2-[2-hydroxy-3-(1-phenylpropyl)phenyl]-	**71:** P124445j
4,5-diphenyl-2-[2-hydroxy-4-(phenylsulfonamido)-phenyl]-	**71:** P124445j
4,5-diphenyl-2-[2-hydroxy-5-(1-piperidinylcarbonyl) phenyl]-	**77:** P115464m
4,5-diphenyl-2-(2-hydroxy-3-propylphenyl)-	**71:** P124445j
4,5-diphenyl-2-(4-hydroxy-3-sulfophenyl)-	**71:** P124445j
4,5-diphenyl-2-(2-hydroxy-5,6,7,8-tetrahydronaphth-1-yl)-	**82:** P17965j
4,5-diphenyl-2-[2-hydroxy-5-(1,1,3,3-tetramethyl-butyl)phenyl]-	**71:** P124445j
2-[4-(1,3-diphenyl-2-imidazolidinyl)-3-hydroxy-phenyl]-4-ethyl-5-methyl-	**84:** 164334x
2-[4-(1,3-diphenyl-2-imidazolidinyl)-3-hydroxy-phenyl]-4-methyl-5-phenyl-	**84:** 164334x
4,5-diphenyl-2-(4-iodophenyl)-	**68:** 114514e
4,5-diphenyl-2-isopropyl-	**82:** 156182q
4,5-diphenyl-2-[4-[(methoxycarbonyl)amino]-2-hydroxyphenyl]-	**71:** P124445j
4,5-diphenyl-2-(2-methoxy-5-methylphenyl)-	**71:** P124445j
4,5-diphenyl-2-(2-methoxyphenyl)-	**71:** P124445j
4,5-diphenyl-2-(2-methoxyphenyl)-, 1-oxide	**71:** P124445j
2,5-diphenyl-4-methyl-, 1-oxide	**78:** 97571h
4,5-diphenyl-2-methyl-	**72:** 43569w, **82:** 156182q
4,5-diphenyl-2-(2-methylphenyl)-	**68:** 114514e
4,5-diphenyl-2-(4-methylphenyl)-	**68:** 114514e, **75:** 63700x **75:** 88545p
4,5-diphenyl-2-[3-[3-(octylthio)propanamidomethyl]-2-hydroxy-5-methylphenyl]-	**71:** P124445j
4,5-di-3-pyridyl-2-(2-hydroxyphenyl)-	**77:** P115464m
4-[4-(dodecanamido)phenyl]-2-(2-hydroxyphenyl)-5-phenyl-	**71:** P124445j
4-[4-(dodecycloxy)phenyl]-2-(2-hydroxyphenyl)-5-phenyl-	**71:** P124445j

TABLE 3 (*Continued*)

Compound	Reference
3.2. 2,4,5-Alkyl- or Aryl-Trisubstituted 2*H*-1,2,3-Triazoles (*Continued*)	
4-(4-dodecylphenyl)-2-(2-hydroxyphenyl)-5-phenyl-	**71:** P124445j
-4-ethanesulfonic acid, 2-(4-cyanophenyl)-5-methyl-β-oxo-, sodium salt	**84:** P181615g
-4-ethanesulfonic acid, α,α'-1,2-ethenediylbis-[(3-sulfo-4,1-phenylene)-2-hydrazinyl-1-ylidene]bis[5-methyl-α-oxo-2-phenyl]-, tetrasodium salt	**84:** P181615g
-4-ethanesulfonic acid, 2-(4-methylphenyl)-5-methyl-α-oxo-, sodium salt	**84:** P181615g
-4-ethanesulfonic acid, 5-methyl-α-oxo-2-phenyl-, sodium salt	**84:** P181615g
-4-ethanol, 2-(2-hydroxyphenyl)-5-phenyl-	**71:** P124445j
4-[4-(ethoxycarbonyl)aminophenyl]-2-(2-hydroxy-phenyl)-5-phenyl-	**71:** P124445j
4-[1-(ethoxycarbonyl)methyl]-2-furanyl-1*H*-benz-imidazol-2-yl]-5-methyl-2-phenyl-	**84:** P91661q
2-(4-ethoxy-2-hydroxyphenyl)-4-methyl-5-phenyl-	**71:** P124445j
4-ethyl-2-[2-hydroxy-5-(4-morpholinylsulfonyl)-phenyl]-5-methyl-	**77:** P115464m
4-ethyl-2-(3-hydroxyphenyl)-5-methyl-	**84:** 164334x
4-ethyl-5-methyl-2-(4-nitrophenyl)-	**59:** 13483f
4-ethyl-5-methyl-2-(4-nitrophenyl)-, 1-oxide	**66:** 28250t
4-ethyl-5-methyl-2-(4-nitrophenyl)-, 3-oxide	**66:** 28250t
4-ethyl-5-methyl-2-phenyl-	**59:** 13483f
4-ethyl-5-methyl-2-phenyl, 1-oxide	**66:** 28250t
4-ethyl-5-methyl-2-phenyl-, 3-oxide	**66:** 28250t
4-(4-ethylphenyl)-2-(2-hydroxyphenyl)-5-phenyl-	**71:** P124445j
4-(4-ethylphenyl)-2-(2-hydroxyphenyl)-5-phenyl-, 3-oxide	P124445j
4-(4-ethylphenyl)-2-(2-methoxyphenyl)-5-phenyl-	**71:** P124445j
4-(5-formylfuran-2-yl)-5-methyl-2-phenyl-	**84:** P91661q
2-(4-formyl-2-hydroxyphenyl)-4-methyl-5-phenyl-	**82:** P172618s
2-(4-formyl-2-methoxyphenyl)-4-methyl-5-phenyl-	**82:** P172618s
2-(4-formylphenyl)-4-methyl-5-(2-naphthalenyl)-	**83:** P61732a
4-(2-furanyl-1*H*-benzimidazol-2-yl)-5-methyl-2-phenyl-	**84:** P91661q
4-(2-furanyl)-2-(2-hydroxy-5,6,7,8-tetrahydronaphth-1-yl)-5-octyl-	**82:** P17965j
2-(4-hydroxy-1,1′-biphenyl-3-yl)-4-methyl-5-phenyl-	**71:** P124445j
2-(2-hydroxy-4-methoxyphenyl)-4-methyl-5-phenyl-	**71:** P124445j
2-(2-hydroxy-5-methoxyphenyl)-4-methyl-5-phenyl-	**71:** P124445j
4-(2-hydroxy-4-methylphenyl)-2-(2-hydroxyphenyl)-5-phenyl-	**71:** P124445j
4-[3-(5-hydroxy-3-methyl-1-phenylpyrazol-4-yl)-1-(3-methyl-5-oxo-1-phenyl-2-pyrazolin-4-yl)-2-propenyl idene]-5-methyl-2-(4-nitrophenyl)-	**74:** P77409a
4-[3-(5-hydroxy-3-methyl-1-phenylpyrazol-4-yl)-3-(3-methyl-5-oxo-1-phenyl-2-pyrazolin-4-yl)-2-propenyl-idene]-5-methyl-2-(4-nitrophenyl)-	**72:** P80363v
2-(2-hydroxyphenyl)-4-(4-hydroxyphenyl)-5-methyl-	**71:** P124445j

TABLE 3 (*Continued*)

Compound	Reference
3.2. 2,4,5-Alkyl- or Aryl-Trisubstituted 2*H*-1,2,3-Triazoles (*Continued*)	
2-(2-hydroxyphenyl)-4-[4-(methanesulfonamido) phenyl]-5-phenyl-	**71:** P124445j
2-(2-hydroxyphenyl)-4-[4-(2-methoxybenzamido)phenyl]-5-phenyl	
2-(2-hydroxyphenyl)-4-[4-[1-[[methoxycarbonyl) amino]methyl]propyl]phenyl]-5-phenyl-	**77:** P115464m
2-(2-hydroxyphenyl)-4-(4-methoxyphenyl)-5-methyl-	**71:** P124445j
2-(2-hydroxyphenyl)-4-(4-methoxyphenyl)-5-(4-methylphenyl)-	**71:** P124445j
2-(2-hydroxyphenyl)-4-methyl-5-(4-methylphenyl)-	**71:** P124445j
2-(2-hydroxyphenyl)-4-methyl-5-(3-nitrophenyl)-	**71:** P124445j
2-(2-hydroxyphenyl)-4-methyl-5-pentyl-	**71:** P124445j
2-(2-hydroxyphenyl)-4-methyl-5-phenyl-	**71:** P124445j
2-(2-hydroxyphenyl)-4-methyl-5-phenyl, 3-oxide	**71:** P124445j
2-(3-hydroxyphenyl)-4-methyl-5-phenyl-	**84:** 164334x
2-(2-hydroxyphenyl)-4-methyl-5-(3-thienyl)-	**77:** P115464m
2-(2-hydroxyphenyl)-4-[2-(methylthio)ethyl]-5-phenyl-	**71:** P124445j
2-(2-hydroxyphenyl)-5-methyl-4-[4-(4-toluenesulfon-amido)phenyl]-	**71:** P124445j
2-(2-hydroxyphenyl)-4-(4-nitrophenyl)-5-phenyl-	**71:** P124445j
2-(2-hydroxyphenyl)-4-(4-octylphenyl)-5-phenyl-	**71:** P124445j
2-(2-hydroxyphenyl)-5-phenyl-4-[3-(1-piperidinyl-sulfonyl)phenyl]-	**77:** P115464m
2-(2-hydroxyphenyl)-4-phenyl-5-propyl-	**71:** P124445j
2-[2-hydroxy-4-(1-piperidinylsulfonyl)phenyl]-4-phenyl-5-propyl-	**77:** P115464m
2-[2-hydroxy-3-(2-propenyl)phenyl]-4-methyl-5-phenyl-	**77:** P115464m
2-[2-hydroxy-5-(1,1,3,3-tetramethylbutyl)phenyl]-4-methyl-5-phenyl-	**71:** P124445j
-4-methanol, α-benzyl-5-methyl-2-phenyl-α-(1-piperidinyl)methyl-	**79:** P18725f
-4-methanol, 2-(4-bromophenyl)-5-methyl-α-[2-(1-piperidinyl)ethyl]-, monohydrochloride	**75:** 62010k
-4-methanol, 2-(4-bromophenyl)-5-methyl-α-phenyl-α-[2-(1-piperidinyl)ethyl]-, hydrochloride	**79:** P18725f
-4-methanol, 2-(4-chlorophenyl)-α,5-dimethyl-α-[2-(4-morpholinyl)ethyl]-, hydrochloride	**79:** P18725f
-4-methanol, 2-(3-chlorophenyl)-5-methyl-α-[2-(1-piperidinyl)ethyl]-	**79:** P126505t
-4-methanol, 2-(4-chlorophenyl)-5-methyl-α-[2-(1-piperidinyl)ethyl]-, monohydrochloride	**75:** 62010k
-4-methanol, α-[2-(dimethylamino)ethyl]-5-methyl-2-phenyl-	**79:** P126505t
-4-methanol, α-[2-(dimethylamino)ethyl]-5-methyl-2-phenyl-, ethanedioate salt	**79:** P126505t
-4-methanol, 2-(4-methoxyphenyl)-5-methyl-α-[2-(1-piperidinyl)ethyl]-, monohydrochloride	**79:** P126505t

TABLE 3 (*Continued*)

Compound	Reference
3.2. 2,4,5-Alkyl- or Aryl-Trisubstituted 2*H*-1,2,3-Triazoles (*Continued*)	
-4-methanol, 5-methyl-2-(4-nitrophenyl-α-[2-(1-piperidinyl)ethyl]-	**79:** P126505t
-4-methanol, 5-methyl-2-(4-nitrophenyl)-α-[2-(1-piperidinyl)ethyl]-, monohydrochloride	**75:** 62010k
-4-methanol, 5-methyl-2-phenyl-	**66:** 37841p
-4-methanol, 5-methyl-2-phenyl-α-[2-(1-piperidinyl)ethyl]-, monohydrochloride	**79:** P126505t
-4-methanol, 5-methyl-2-phenyl-α-[(1-piperidinyl)methyl]-, monohydrochloride	**75:** 62010k
2-[4-(6-methoxy-3-methyl-2-benzofuranyl)phenyl]-4-methyl-5-phenyl-	**75:** P22479p
2-(6-methoxy-3-methylphenyl)-4-methyl-5-phenyl-	**71:** P124445j
2-[4-(5-methoxy-3-phenyl-2-benzofuranyl)phenyl]-4-methyl-5-phenyl-	**75:** P22479p
2-(2-methoxyphenyl)-4-methyl-5-(3-nitrophenyl)-	**71:** P124445j
2-(2-methoxyphenyl)-4-methyl-5-phenyl-	**71:** P124445j
2-(2-methoxyphenyl)-4-methyl-5-phenyl-, 3-oxide	**71:** P124445j
4-methyl-2,5-diphenyl-	**25:** 1827[8]
4-methyl-2-[4-(3-methyl-2-benzofuranyl)phenyl]-5-phenyl-	**75:** P22479p
4-methyl-5-(1-methyl-2-oxopropyl)-2-phenyl-	**42:** 2970c
4-methyl-5-(1-methyl-2-oxopropyl)-2-phenyl-, oxime	**42:** 2970c
4-methyl-5-(1-methyl-2-oxopropyl)-2-phenyl-, semicarbazone	**42:** 2970c
4-methyl-5-(2-methylphenyl)-2-phenyl-	**25:** 1828[1]
4-methyl-5-(2-oxopropyl)-2-phenyl-	**35:** 3638[2]
5-methyl-2-(4-nitrophenyl)-4-(2-oxopropyl)-	**68:** 2862k
5-methyl-2-(4-nitrophenyl)-4-(2-oxopropyl)-, phenylhydrazone	**68:** 2862k
4-methyl-5-phenyl-2-[4-(3-phenyl-2-benzofuranyl)phenyl]-	**75:** P22470p
4-methyl-5-phenyl-2-(3-sulfophenyl)-, sodium salt	**81:** P38957c
-2-propanoic acid, 2-(4-chlorophenyl)-5-methyl-α-oxo-, ethyl ester	**84:** P181615g
-2-propanoic acid, 4,5-diphenyl-	**84:** P164786w
-2-propanoic acid, 4,5-diphenyl-, methyl ester	**84:** P164786w
-4-propanoic acid, 2-(2-hydroxyphenyl)-5-phenyl-, ethyl ester	**71:** P124445j
-2-propanoic acid, 5-methyl-α-oxo-2-phenyl-, ethyl ester	**84:** P181615g
-2-propanoic acid, 4-methyl-5-phenyl-	**84:** P164786w
2-[3-[(α-toluenesulfonamido)phenyl]-4-hydroxyphenyl]-4,5-diphenyl-(?)	**71:** P124445j
2,4,5-triphenyl-	**47:** 2154a, **49:** 5372gh, **57:** 15100h, **73:** 120593z, **75:** 63700x, **83:** 96642y **84:** 135554p
2,4,5-triphenyl-, 1-oxide	**66:** 28250t
2,4,5-*tris*(4-bromophenyl)-	**68:** 114514e

REFERENCES

21: 2133^{5} P. C. Guha and M. K. De, *Quart. J. Indian Chem. Soc.*, **3,** 41 (1926).

24: 3232^{1} T. Curtius and K. Raschig, *J. Prakt. Chem.*, **125,** 466 (1930).

25: 933^{1} A. Quilico and M. Freri, *Gazz. Chem. Ital.*, **60**, 606 (1930).

25: $1827–8^{1}$ M. Gallotti, G. Barro, and L. Salto, *Gazz. Chem. Ital.*, **60,** 866 (1930).

35: 3638^{2} T. Ajello and Sl Cusmano, *Gazz. Chem. Ital.*, **70,** 770 (1940).

40: 7191^{4-8} R. F. Coles and C. F. Hamilton, *J. Amer. Chem. Soc.*, **68,** 1799 (1946).

41: 1682a R. E. Benson and C. S. Hamilton, *J. Amer. Chem. Soc.*, **68,** 2644 (1946).

42: 2970c T. Ajello and B. Tornetta, *Gazz. Chem. Ital.*, **77,** 332 (1947).

45: 9037–8f–c J. C. Sheehan and C. A. Robinson, *J. Amer. Chem. Soc.*, **73,** 1207 (1951).

46: 8651–2i–b Fr. Moulin, *Helv. Chim. Acta*, **35,** 167 (1952).

47: 2154a Al. V. Spasov, D. Elenkov, and St. Robev, *Balgarska. Akad. Nauk., Otdel. Geol.-Geograf. Khim. Nauk., Izvest. Khim. Inst.*, **1,** 229 (1951).

48: 2685b E. Mugnaini and P. Grünanger, *Atti Accad. Naz. Lincei., Rend., Cl. Sci. Fis. Mat. Nat.*, **14,** 95 (1953).

49: 3948h E. Mugnaini and P. Grünanger, *Atti Accad. Naz. Lincei., Rend., Cl. Sci. Fis. Mat. Nat.*, **14,** 275 (1953).

49: 5372g,h A. Spasov and S. Robev, *Bull. inst. chim. acad. bulgare sci.*, **2,** 3 (1953).

50: 4129c R. H. Wiley, N. R. Smith, D. M. Johnson, and J. Moffat, *J. Amer. Chem. Soc.*, **77,** 3412 (1955).

52: P2085–6i–b Badische Anilin- und Soda-Fabrik Akt.-Ges., German Patent, 818,048 (1951).

52: 7293a S. Maffei and G. F. Bettinetti, *Ann. Chim. (Rome)*, **47,** 1286 (1957).

53: 3198–9f–i A. Dornow and K. Rombusch, *Chem. Ber.*, **91,** 1841 (1958).

53: 18947a D. Dal Monte A. M. Casoni, and M. Fernando, *Gazz. Chim. Ital.*, **88,** 1035 (1958).

55: 18756b J. Hádáček, *Spisy prirodovedecke fak. univ. Brne*, **417,** 373 (1960).

55: 23505h E. Profft and W. Georgi, *Justus Liebigs Ann. Chem.*, **643,** 136 (1961).

56: 1388a R. Huisgen, J. Sauer, and M. Seidel, *Chem. Ber.*, **94,** 2503 (1961).

56: 14020b R. Fusco, G. Blanchetti, and D. Pocar, *Gazz. Chem. Ital.*, **91,** 849 (1961).

57: 15100h G. Henseke and G. Mueller, *J. Prakt. Chem.*, **18,** 47 (1962).

58: 5662d J. Moffat and J. L. Rutter, *J. Chem. Eng. Data.*, **7,** 542 (1962).

59: 13483f E. Borello and A. Zecchina, *Spectrochim. Acta*, **19,** 1703 (1963).

60: 15859d J. A. Durden, Jr., H. A. Stansbury, and W. H. Catlette, *J. Chem. Eng. Data*, **9,** 228 (1964).

61: 2926b M. E. Munk and Y. K. Kim, *J. Amer. Chem. Soc.*, **86,** 2213 (1964).

63: 11551c G. Bianchetti, D. Pocar, P. D. Croce, and A. Vigevani, *Chem. Ber.*, **98,** 2715 (1965).

63: 11552a G. Bianchetti, P. D. Croce, and D. Pocar, *Tetrahedron Lett.*, 2039 (1965).

64: 9717g D. Pocar, G. Bianchetti, and P. D. Croce, *Gazz. Chim. Ital.*, **95,** 1220 (1965).

64: P11216c Farbenfabriken Bayer A.-G., Dutch Patent, 6,505,524 (1965).

65: 18484g W. P. Norris and W. G. Finnegan, *J. Org. Chem.*, **31,** 3292 (1966).

65: 20120a E. Fahr, K. Doeppert, and F. Scheckenbach, *Justus Liebigs Ann. Chem.*, **696,** 136 (1966).

65: 7173f G. Bianchetti, P. D. Croce, and D. Pocar, *Rend. Ist. Lombardo Sci. Lettere, A*, **99,** 259 (1965).

66: 28250t H. Borello, A. Zecchina, and E. Guglielminotti, *J. Chem. Soc., B*, 1243 (1966).

66: 37841p G. Henseke and I. Schmeisky, *J. Prakt. Chem.*, **33,** 256 (1966).

67: 43526b I. Bhatnagar and M. V. George, *J. Org. Chem.*, **32,** 2252 (1967).

67: 82170a G. Bianchetti, P. D. Croce, D. Pocar, and A. Vigevani, *Gazz. Chim. Ital.*, **97,** 289 (1967).

67: 82171b G. Bianchetti, D. Pocar, P. D. Croce, and R. Stradi, *Gazz. Chim. Ital.*, **97,** 304 (1967).

68: 2862k A. J. Boulton, A. R. Katritzky, and A. M. Hamid, *J. Chem. Soc., C,* 2005 (1967).

68: 12925x G. Bianchetti, P. Ferruti, and D. Pocar, *Gazz. Chim. Ital.*, **97,** 579 (1967).

68: 12927z D. Pocar, G. Bianchetti, and P. Ferruti, *Gazz. Chim. Ital.*, **97,** 597 (1967).

68: 68159t R. Huisgen, H. Gotthardt, and R. Grashey, *Chem. Ber.*, **101,** 535 (1968).

68: 78204t P. A. S. Smith and J. G. Wirth, *J. Org. Chem.*, **33,** 1145 (1968).

68: 114514e H. El Khadem, Z. M. El-Shafei, and M. M. Hashem, *J. Chem. Soc., C,* 949 (1968).

69: 96589x C. E. Olsen and C. Pedersen, *Tetrahedron Lett.*, 3805 (1968).

71: 81296s R. Huisgen, K. von Fraunberg, and H. J. Sturm, *Tetrahedron Lett.*, 2589 (1969).

71: P124445j Geigy, A.-G., French Patent, 1,559,131 (1969).

72: 43569w R. R. Fraser, Gurudata, and K. E. Haque, *J. Org. Chem.*, **34,** 4118 (1969).

72: 43576w G. García-Muñoz, R. Madroñero, M. Rico, and M. C. Saldaña, *J. Heterocycl. Chem.*, **6,** 921 (1969).

72: P80363v E. Foerster and B. Hirsch, German Patent, 1,800,581 (1969).

72: P100716t Farbenfabriken Bayer A.-G., French Patent, 1,577,760 (1969).

73: 14768a B. Eistert and E. Endres, *Justus Liebigs Ann. Chem.*, **734,** 56 (1970).

73: 120593z A. Spasov and B. Chemishev, *Dokl. Bolq. Akad. Nauk.*, **23,** 791 (1970).

74: P77409a Veb Filmfabrik Wolfen, French Patent, 1,592,402 (1970).

74: 141643b T. Sasaki, T. Yoshioka, and Y. Suzuki, *Bull. Chem. Soc. Jap.*, **44,** 185 (1971).

75: P22479p G. Kabas and H. Schlaepfer, German Patent, 2,031,819 (1971).

75: 48994c R. Raap, *Can. J. Chem.*, **49,** 1792 (1971).

75: 62010k I. Gruebner, W. Klinger, and H. Ankermann, *Arch. Int. Pharmacodyn. Ther.*, **191,** 37 (1971).

75: 62695n F. Compernolle and M. Dekeirel, *Org. Mass Spectrom.*, **5,** 427 (1971).

75: 63700x Y. Huseya, A. Chinone, and M. Ohta, *Bull. Chem. Soc. Jap.*, **44,** 1667 (1971).

75: 88545p M. Maerky, H. J. Huasen, and H. Schmid, *Helv. Chim. Acta*, **54,** 1275 (1971).

76: 72461t D. J. Anderson, T. L. Gilchrist, G. E. Gymer, and C. W. Rees, *J. Chem. Soc., D,* 1518 (1971).

76: 113139q R. Jacquier, C. Pellier, C. Petrus, and F. Petrus, *Bull. Soc. Chim. Fr.*, 4078 (1971).

77: P7313c H. Lind, German Patent, 2,133,012 (1972).

77: P90079z T. Noguchi, K. Tsukamoto, K. Isogami, and H. Hojo, Japanese Patent, 71 33,148 (1971).

77: P115464m R. Kirchmayr, H. J. Heller, and J. Rody, Swiss Patent, 522,432 (1972).

77: 164609w P. Ykman, G. Mathys, G. L'abbé and G. Smets, *J. Org. Chem.*, **37,** 3213 (1972).

78: 16099k Y. Huseya, A. Chinone, and M. Ohta, *Bull. Chem. Soc. Jap.*, **45,** 3202 (1972).

78: 16120k V. V. Solov'eva and E. Gudriniece, *Latv. PSR Zinat. Akad. Vestis, Khim. Ser.*, 572 (1972).

78: 58205x R. E. Banks and G. R. Sparkes, *J. Chem. Soc., Perkin Trans.* **1,** 2964 (1972).

78: 84325f — E. M. Bruckmann, *University Microfilms*, 72-29,008, 210 pp (1972).
78: 97571h — S. A. Amitina and L. B. Volodarskii, *Khim. Geterotsikl. Soedin.*, 131 (1973).
78: 111219h — T. L. Gilchrist, G. E. Gymer, and C. W. Rees, *J. Chem. Soc., Perkin Trans.* **1,** 555 (1973).
78: 124171m — E. R. Banks and A. Prakash, *Tetrahedron Lett.*, 99 (1973).
78: 124549r — C. W. Rees and A. A. Sale, *J. Chem. Soc., Perkin Trans.* **1,** 545 (1973).
79: P18725f — B. Hirsch, D. Lohmann, and H. Froemmel, East German Patent, 94,010 (1972).
79: P78815v — E. Aufderhaar and A. Meyer, German Patent, 2,263,878 (1973).
79: P126505t — B. Hirsch, D. Lohmann, and H. Froemmel, East German Patent, 92,924 (1972).
79: P147435k — F. Fleck, H. Kittl, and S. Valenti, German Patent, 2,262,340 (1973).
80: 47912r — C. E. Olsen, *Acta Chem. Scand.*, **27,** 1987 (1973).
80: P49275j — A. Domergue and R. F. M. Sureau, German Patent, 2,319,828 (1973).
80: 59897c — C. E. Olsen, *Acta Chem. Scand.*, **27,** 2983 (1973).
80: 70667y — T. L. Gilchrist, G. E. Gymer, and C. W. Rees, *J. Chem. Soc., Chem. Commun.*, 835 (1973).
80: 108454s — W. Winter and E. Mueller, *Chem. Ber.*, **107,** 715 (1974).
80: 145179b — P. A. S. Smith and E. M. Bruckmann, *J. Org. Chem.*, **39,** 1047 (1974).
81: P38957c — H. Gold and U. Claussen, German Patent 2,242,784 (1974).
81: P51153g — T. Yanagisawa, Japanese Patent, 73 37,324 (1973).
81: 135588r — R. E. Banks and A. Prakash, *J. Chem. Soc., Perkin Trans.* **1,** 1365 (1974).
82: P4995u — M. Matsuo, K. Koga, S. Hotta, and T. Akamatsu, Japanese Patent, 74 54,281 (1974).
82: P17959k — M. Matsuo, K. Koga, S. Hotta, and T. Akamatsu, Japanese Patent, 74 63,681 (1974).
82: P17965j — M. Matsuo, K. Koga, S. Hotta, and T. Akamatsu, Japanese Patent, 74 61,068 (1974).
82: 125237g — R. E. Banks and A. Prakash, *J. Chem. Soc., Perkin Trans.* **1,** 2479 (1974).
82: 140029j — E. Mueller and W. Winter, *Justus Liebigs Ann. Chem.*, 1876 (1975).
82: 156182q — T. L. Gilchrist, G. E. Gymer, and C. W. Rees, *J. Chem. Soc., Perkin Trans.*, **1,** 1 (1975).
82: P172618s — J. Schroeder, German Patent, 2,335,218 (1975).
83: P10089p — Miles Laboratories, Dutch Patent, 74 07,127 (1974).
83: P61732a — A. Dorlars and O. Neuner, German Patent, 2,258,276 (1974).
83: 96642y — K. S. Balachandrin, I. Hiryakhanavar, and M. V. George, *Tetrahedron*, **31,** 1171 (1975).
83: 131249z — R. J. DePasquale, C. D. Padgett, and R. W. Rosser, *J. Org. Chem.*, **40,** 810 (1975).
83: P132619g — M. Matsuo, K. Koga, and S. Hotta, Japanese Patent, 74 61,070 (1974).
84: 30978b — C. E. Olsen, *Acta Chem. Scand., Ser. B*, **29,** 953 (1975).
84: P91661q — P. S. Littlewood and A. V. Mercer, German Patent, 2,522,139 (1975).
84: 135554p — J. A. C. Allison, H. S. El Khadem, and C. A. Wilson, *J. Heterocycl. Chem.*, **12,** 1275 (1975).
84: 164334x — J. Hocker, H. Giesecke, and R. Merten, *Angew. Chem.*, **88,** 151 (1976).
84: P164786w — R. T. Buckler, H. E. Hartzler, and S. Hayao, U.S. Publ. Pat. Appl. B, 488,111 (1976).
84: P181615g — F. Fleck, A. V. Mercer, and R. Paver, German Patent, 2,535,069 (1976).
85: P123926u — R. T. Buckler, H. E. Hartzler, S. Hayao, and G. Nichols, U.S. Patent, 3,948,932 (1976).

CHAPTER 4

1,2,3-Triazolecarboxylic Acids and Their Functional Derivatives

4.1. 1*H*-1,2,3-TRIAZOLE-1- AND -5-CARBOXYLIC ACIDS AND THEIR FUNCTIONAL DERIVATIVES

L'abbé and his collaborators have made important contributions to our understanding of the synthesis of these compounds through their studies of ethyl azidoformate (**4.1-1**) with α-keto phosphorus ylides. Their first report,[1] which claimed good yields of the 1-(ethoxycarbonyl) product (**4.1-2**), was soon followed by evidence for the 2-(ethoxycarbonyl) structure (**4.1-3**) (Eq. 1).[2] It has been shown that **4.1-2** is an intermediate that can be isolated with low temperature, proper solvent, and short reaction time.[3] Huisgen and Blaschke[4] have studied the regiospecificity of the addition of **4.1-1** to phenylacetylene (Eq. 2). Their presumed 1,4-product (**4.1-4**) is actually the 2,4-isomer.[2] The synthesis of 1-cyano-1,2,3-triazoles (**4.1-5**) with cyanogen azide gives good yields of an equilibrium mixture in which the open-chain α-diazo-*N*-cyanoimine is greatly favored (Eq. 3).[5]

The preparation of 1,2,3-triazolecarboxylic acids without nitrogen substituents should produce readily interconvertible tautomers (**4.1-6** and

$$Ph_3P{=}CHCOR + EtO_2CN_3 \longrightarrow \left[\text{Ph}_3\text{P}^+\text{O}^- \text{ adduct (H, R; N–N–N ring; } CO_2Et)\right] \longrightarrow \textbf{4.1-2} \longrightarrow \textbf{4.1-3} \qquad (1)$$

4.1-1   **4.1-2**   **4.1-3**

R = Me   Ph   4-NO_2Ph
% **4.1-3** = 65   46   88 (with excess azide[3])

$$PhC{\equiv}CH + EtO_2CN_3 \longrightarrow$$ 1-(CO$_2$Et)-4-Ph-triazole (47 pts.) + **4.1-4** (**4.1-3** Ph) 1-(CO$_2$Et)-5-Ph-triazole (53 pts.) (2)

$$RC{\equiv}CR' + N_3CN \longrightarrow$$ R, R' = H, Me, *n*-Bu; **4.1-5** (1-NC-4,5-R-triazole) $\rightleftharpoons$ $RC(R)(CN_2)$=N–NC (3)

4.1-7) with intramolecular hydrogen bonding favoring **4.1-7** (Eq. 4). Both 4- and 5-substituted products have been reported for these compounds and are described in this chapter. The structure suggested by the authors is given.

4.1-6 (HO_2C, N–H) $\rightleftharpoons$ **4.1-7** (HO–C=O···H–N) (4)

A large number of unsaturated substrates have been shown to react with hydrazoic acid or sodium azide, usually in excellent yield (Eqs. 5 to 9)[6–10]. In the last example Miller and his collaborators reported no yields for 4-chloro-, 2- and 3-nitrophenyl substituents, but another article shows the

$$HC{\equiv}CCO_2H + HN_3 \longrightarrow$$ HO_2C-triazole N–H, 71% (5)

$$PhC{\equiv}CCHO + HN_3 \longrightarrow$$ Ph, CHO triazole N–H, 90% $\xrightarrow[Na^{+-}OH]{AgNO_3}$ Ph, CO_2H triazole N–H (6)

$$Ph_3P^+CH{=}CHCO_2Me + Na^{+-}N_3 \longrightarrow$$ CO_2Me triazole N–H, 79% (7)

$$\text{R(H)C=C(CO}_2\text{Et)(NO}_2) + \text{Na}^{+}\,{}^{-}\text{N}_3 \xrightarrow{\text{HCONMe}_2} \text{4-R-5-CO}_2\text{Et-1}H\text{-1,2,3-triazole}\ (45\text{–}60\%) \quad (8)$$

R = Me, Et, *n*-Pr, *i*-Pr, Ph

$$\text{ArC}\equiv\text{CCO}_2\text{Me} + \text{N}_3^- \xrightarrow{\text{HCONMe}_2} \textbf{4.1-8} \quad (9)$$

Ar =	Ph	4-MeOPh	2-ClPh	4-NO_2Ph	4-ClPh[11]
% **4.1-8** =	91	76	86	69	92

4-chlorophenyl substrate reacting in excellent yield.[11] The azide ion adds in excellent yield to vinyl sulfones bearing an α-electron-withdrawing substituent (Eq. 10).[12] The preparation of 5-cyano-1,2,3-triazoles has been accomplished in moderate yield (Eqs. 11,12).[13,14]

$$\text{ArCH=C(SO}_2\text{Ph)Y} + \text{Na}^{+}\,{}^{-}\text{N}_3 \xrightarrow{\text{HCONMe}_2} \text{4-Ar-5-Y-1}H\text{-1,2,3-triazole} \quad (10)$$

Ar = Ph, 4-MeOPh, 4-NO_2Ph, 2-furyl, 3-pyridyl, 9-anthryl, etc.
Y = CN, $CONH_2$, CO_2Me

$$\text{HC}\equiv\text{CCN} + \text{Al(N}_3)_3 \xrightarrow{\text{THF}} \text{5-cyano-1}H\text{-1,2,3-triazole}\ (32\%) + \text{5-(1}H\text{-tetrazol-5-yl)-1}H\text{-1,2,3-triazole}\ (47\%) \quad (11)$$

$$\text{PhCH=C(CN)}_2 + \text{Na}^{+}\,{}^{-}\text{N}_3 \xrightarrow{\text{Me}_2\text{SO}} \text{4-Ph-5-CN-1}H\text{-1,2,3-triazole}\ (40\%) \quad (12)$$

The indirect formation of 1,2,3-triazolecarboxylic acid derivatives by trimethylsilylazide addition and subsequent hydrolysis (Eqs. 13,14) proceeds in excellent yield.[15,16] The methyl ester (**4.1-9**) was also converted to the carbonyl azide (80%) and hydrazide (93%).[14]

$$RC{\equiv}CCO_2SiMe_3 + Me_3SiN_3 \longrightarrow \text{4-R-5-}(CO_2SiMe_3)\text{-2-}(SiMe_3)\text{-2}H\text{-1,2,3-triazole} \quad (13)$$

R = H or Me 97%

$$MeC{\equiv}CO_2Me + Me_3SiN_3 \longrightarrow \text{4-}(MeO_2C)\text{-5-Me-1}H\text{-1,2,3-triazole (4.1-9)} \quad (14)$$

4.1-9
85%

The introduction of a nonenolizable substituent on nitrogen allows the separation of 5-carboxylic acids and their derivatives. Such compounds have been prepared in a variety of ways, including the oxidation of alkyl groups (Eq. 15)[17,18] or formyl groups (Eq. 16).[17] Sheehan and Robinson carried out a similar reaction sequence (Eq. 17) using a diethylacetal.[7]

$$\text{5-R-1-Ph-1,2,3-triazole} \xrightarrow{KMnO_4} \text{1-Ph-1,2,3-triazole-5-}CO_2H \quad (15)$$

R = Me, Et, $CH_2{=}CH$, $CH_2CH_2NMe_2$ (65%)

$$PhC{\equiv}CCHO + PhN_3 \longrightarrow \text{4-Ph-5-CHO-1-Ph-triazole (4.1-10)} + \text{4-OHC-5-Ph-1-Ph-triazole} \quad (16)$$

4.1-10 22%; 40% (90% total crude)

$$\text{4.1-10} \xrightarrow{AgNO_3,\ Na^{+}\ {}^{-}OH} \text{4-Ph-1-Ph-1,2,3-triazole-5-}CO_2H$$

90%

L'abbé and his collaborators have reported the isomer distribution found when vinyl azides or their iodo precursors are added to methyl propynoic acid (Eq. 18).[19] The ratio of **4.1-11** to **4.1-12** is generally 8:1 except with R = —CH(Ph)CH_2I where 9:1 was found.

$$HC\equiv CCH(OEt)_2 + PhN_3 \longrightarrow \text{4-}CH(OEt)_2\text{-1-phenyl-1,2,3-triazole} \xrightarrow{H^+}$$

OHC-(1-Ph-1,2,3-triazol-4-yl) 52% + CHO-(1-Ph-1,2,3-triazol-5-yl) 23% (17)

$\downarrow$ $AgNO_3$, $Na^{+-}OH$

1-Ph-1,2,3-triazole-5-CO_2H 90%

$$MeO_2CC\equiv CH + RN_3 \longrightarrow$$ MeO$_2$C-(1-R-1,2,3-triazol-4-yl) **4.1-11** + CO_2Me-(1-R-1,2,3-triazol-5-yl) **4.1-12** (18)

R = PhC=CHMe, MeC=CHMe, $CH_2CH(I)Bu\text{-}t$, etc.

An interesting substitution reaction has been reported by Smith's laboratory for the synthesis of 5-substituted 1,2,3-triazoles.[20,21] The reaction (Eqs. 19,20) is apparently very sensitive to small amounts of moisture.

4-Ph-5-Cl-1-Ph-1,2,3-triazole + $Na^{+-}CN$ $\xrightarrow{Me_2SO\text{, anhydrous}}$ 4-Ph-5-CN-1-Ph-1,2,3-triazole 75% (19)

4-Ph-5-Cl-1-Ph-1,2,3-triazole + $Na^{+-}CN$ $\xrightarrow{Me_2SO\ H_2O}$ 4-Ph-5-CO_2NH_2-1-Ph-1,2,3-triazole 58% (20)

REFERENCES

1. **70:** 77878v
2. **74:** 76375z
3. **76:** 25192w
4. **63:** 16337h
5. **67:** 116854u
6. **55:** 17626i
7. **45:** 9037f
8. **70:** 96727m
9. **80:** 146082b
10. **80:** 108452q
11. **80:** 133389g
12. **80:** 3439n
13. **71:** 13067t
14. **75:** 140767d
15. **68:** 13061z
16. **85:** 123875b
17. **44:** 1102c
18. **61:** 3096g
19. **72:** 132635g
20. **68:** 78204t
21. **80:** 145179b

4.2 1,2,3-TRIAZOLE-4-CARBOXYLIC ACIDS AND THEIR FUNCTIONAL DERIVATIVES

The synthesis of 4-carboxyl-1,2,3-triazoles and derivatives has been an active field of study, and the variety of methods developed is impressive. One of the earliest reactions is the oxidation, usually with silver oxide, of the formyl group (Eq. 21).[22] Sheehan and Robinson have reported[23] the first example of phenyl azide addition in which both isomeric products were isolated (Eq. 22), and they demonstrated the structures of **4.2-1** and **4.2-2** by oxidation. Similar results were obtained using the diethylacetal of propynal followed by hydrolysis.

$$HC{\equiv}CCHO + PhN_3 \longrightarrow \text{1-phenyl-4-formyl-1,2,3-triazole (75\%)} \xrightarrow{Ag_2O} \text{1-phenyl-1,2,3-triazole-4-carboxylic acid (high)} \qquad (21)$$

$$PhC{\equiv}CCHO + PhN_3 \longrightarrow \underset{\textbf{4.2-1},\ 40\%}{\text{4-formyl-1,5-diphenyl-1,2,3-triazole}} + \underset{\textbf{4.2-2},\ 22\%}{\text{5-formyl-1,4-diphenyl-1,2,3-triazole}} \xrightarrow[Na^+\ ^-OH]{AgNO_3}$$

$$\underset{97\%}{\text{1,5-diphenyl-1,2,3-triazole-4-carboxylic acid}} + \underset{90\%}{\text{1,4-diphenyl-1,2,3-triazole-5-carboxylic acid}} \qquad (22)$$

The analogous chemistry of 2-*N*-substituted 1,2,3-triazoles has been reported by Riebsomer and his collaborators[24,25] starting with 2-phenyl-4-D-glucosotriazole (**4.2-3**) and potassium permanganate (Eq. 23). The subsequent conversions to acid chloride, esters, and amides were made in good to excellent yield. The intermediate aldehyde (**4.2-4**) was prepared quantitatively and transformed to the 4-nitrile (Eq. 24).[25]

HOCH$_2$(CHOH)$_3$-(2-phenyl-1,2,3-triazol-4-yl) **4.2-3** —$KMnO_4$→ 2-Ph-1,2,3-triazole-4-CO$_2$H (HO$_2$C) ⟶ various acid derivatives (23)

4.2-3 —$Na^+\,{}^-IO_4$→ OHC-(2-Ph-1,2,3-triazol-4-yl) **4.2-4** —$HONH_2$, H^+→

H(C=NOH)-(2-Ph-1,2,3-triazol-4-yl) —$SOCl_2$, C_6H_6→ NC-(2-Ph-1,2,3-triazol-4-yl) 70% (24)

Klingsberg has cyclized the nitrogen derivatives of α,β-diketobutaneanilides (**4.2-5**) to the 1,2,3-triazolecarboxylic acids (Eq. 25) in good yield and has converted the products to amides in excellent yields.[26]

MeC(=NOH)C(=NNHAr)CONHPh **4.2-5** —(1) $K^+\,{}^-OH$, (2) CO_2→ 4-Me-5-CO$_2$H-2-Ar-1,2,3-triazole —(1) PCl_5, (2) PhNHMe→ 4-Me-5-C(O)N(Ph)Me-2-Ar-1,2,3-triazole (25)

Ar	acid	amide
Ph	58%	95%
4-MePh	80%	95%

Recent studies on the reaction of 1,2,3-triazoles with glycosyl halides have shown promise in the preparation of glycosyltriazoles (Eq. 26).[27] The use of 5-diazouracils (e.g., **4.2-6**) has been shown to be an important approach (Eq. 27) with many examples and excellent yields.[28,29] El Khadem's laboratory[30] has provided several examples of 2-aryl- derivatives from the osatriazoles (Eq. 28). The extent of interest in this field is illustrated by the publication of an undergraduate laboratory experiment based on the conversion of sucrose to 2-phenyl-1,2,3-triazole-4-carboxylic acid.[31]

(26)

MeCN, 100°

4.2-6

(27)

D-gluco, $KMnO_4$, CO_2H, Ar

4.2-7

(28)

Ar	= 4-NO_2Ph	3-Me-4-NO_2Ph	3-Cl-4-NO_2Ph	4-Cl-3-NO_2Ph	4-AcNHPh
% **4.2-7** =	76	40	50	30	66

The addition of azides to acetylenic acids or esters has been used with good results (Eqs. 29 to 32).[32–36] The ease with which these acids decarboxylate is indicated by the presence of **4.2-8** in Equation 29. This property has been useful in many structure proof problems (e.g., Reference 2). An interesting variant of this approach involves the Iotsich complexes (e.g., **4.2-9**) (Eq. 33).[37]

$PhC{\equiv}CCO_2H + PhCH_2N_3 \longrightarrow$

4.2-8

62% (total)

(29)

$HC{\equiv}CCO_2H + RN_3 \xrightarrow[\text{heat}]{C_6H_6}$

57%, 70%

R = Me,

(30)

$$R\overset{NO_2}{\underset{NO_2}{C}}CH_2N_3 + HC{\equiv}CCO_2H \xrightarrow{CHCl_3} \text{(triazole: } HO_2C,\ N{=}N{-}N,\ CH_2\overset{R}{C}(NO_2)_2) \quad (31)$$

R = F, Me

90% (both)

$$HC{\equiv}CCO_2Me + \text{(}N_3,\ Cl,\ Ph{-}C{=}O\text{ benzene)} \longrightarrow \text{(}MeO_2C\text{ triazole, } O{=}C(Ph),\ Cl\text{)} \quad (32)$$

82% (crude)

$$PhC{\equiv}CMgBr + PhN_3 \longrightarrow \text{(BrMg, Ph, N, N, N, Ph)} \xrightarrow{CO_2} \text{(}HO_2C\text{, Ph, N, N, N, Ph)} \quad (33)$$

4.2-9

54.5%

The reaction of azides with active methylene compounds is an especially attractive route to the acid derivatives. An early report involving aryliodo-azides (Eq. 34) showed excellent yields.[38] The methyltropolonyl azide (**4.2-11**) reacts smoothly with several active methylene substrates (Eq. 35),[39] most of which give amino products that are to be discussed in Chapter X. Sodamide has also shown good results as a catalyst for these reactions (Eq. 36).[40]

$$ArN_3 + MeCOCH_2CO_2Et \xrightarrow[EtOH]{Na} \text{(}HO_2C\text{, Me, N, N, N, Ar)} \quad (34)$$

4.2-10

Ar	=3-IPh	4-IPh	3,4-I_2Ph
% **4.2 10** =	88	94	99

$$\text{(O, MeO tropone)}{-}N_3 + 2MeCOCH_2CO_2Et \longrightarrow$$

4.2-11

$$\text{(}EtO_2C\text{, Me, N, N, N, tropone with O and OMe)} \quad (35)$$

90%

$$PhCOCH_2CO_2Et + PhCH_2N_3 \xrightarrow{Na^{+-}NH_2} \text{1-benzyl-5-phenyl-1,2,3-triazole-4-carboxylic acid } (HO_2C, Ph, CH_2Ph) \quad 61\% \tag{36}$$

Once again L'abbé's laboratory has made important contributions in the study of these syntheses (Eq. 37).[41] This productive group has also demonstrated the ratio of isomeric products in acetylene additions (Eq. 38)[42] and has prepared a large number of 1-aryl-1,2,3-triazoles in good to excellent yield. These investigators have defined the regiospecificity of their phosphorus ylide method by comparison with triazoles prepared by the Dimroth method via **4.2-12** (Eq. 39).[43] Recently the L'abbé group has carried out an addition involving azides containing α,β-unsaturated carbonyl substituents (**4.2-13**) and has shed light on subsequent Michael and cyclization chemistry (Eq. 40).[44]

$$RCOCH_2CO_2Et + N_3CH{=}CH\text{Bu-}t \longrightarrow \text{1-(CH=CHBu-}t\text{)-5-R-1,2,3-triazole-4-}CO_2H \quad 60\text{–}80\% \tag{37}$$

R = Me, Ph

$$MeO_2CC{\equiv}CH + RN_3 \longrightarrow \text{1-R-4-}CO_2Me\text{-1,2,3-triazole (about 80 pts.)} + \text{1-R-5-}CO_2Me\text{-1,2,3-triazole (about 20 pts.)} \tag{38}$$

R = PhC=CHMe, MeC=CHMe, CH_2CHBu-*t* (with I on CH), etc.

$$RCOCH_2CO_2Et + ArN_3 \xrightarrow[EtOH]{NaOEt} \text{1-Ar-5-R-1,2,3-triazole-4-}CO_2H \ (\mathbf{4.2\text{-}12}) \xrightarrow{\Delta} \text{1-Ar-5-R-1,2,3-triazole} \tag{39}$$

R =	Me	Ph	4-NO_2Ph	4-NO_2Ph	Ph	4-NO_2Ph	4-NO_2Ph
Ar =	4-MeOPh	4-BrPh	4-NO_2Ph	Ph	4-MeOPh	4-BrPh	4-MeOPh
% **4.2-12** =	62	60	100	76	52	93	62

MeCOCH$_2$CO$_2$Et + (PhCO)(N$_3$)C=C(H)(Ar) **4.2-13** $\xrightarrow{Et_3N}$ 4-EtO$_2$C-5-Me-1-[C(C(=O)Ph)=CH(Ar)]-1,2,3-triazole (40)

Pocar and his collaborators[45] have obtained excellent yields by extension of their enamine method (Eq. 41).

(R)(H)C=C(Y)(R′) + ArN$_3$ ⟶ 4-R-5-R′-1-Ar-1,2,3-triazole (41)

R = CO$_2$Et, PhNHCO R′ = Me, Me$_2$CHCH$_2$ Ar = 4-NO$_2$Ph, 2,4-(NO$_2$)$_2$Ph
Y = 4-morpholinyl, NMe$_2$

Two reactions of diazo compounds indicate promise for future development (Eq. 42)[46] and (Eq. 43).[47] In the latter case product **4.2-14** was also prepared by azide-acetylene addition, and the 66% total yield was shown to contain 70% 1,5-diphenyl and 30% 1,4-diphenyl.[47]

EtO$_2$CC(N$_2$)CHO + PhNH$_2$ $\xrightarrow[EtOH]{HOAc}$ 4-EtO$_2$C-1-Ph-1,2,3-triazole, 74% (42)

(Ph)(PhNH)CHC(N$_2$)CO$_2$Me + KMnO$_4$ ⟶ 4-MeO$_2$C-5-Ph-1-Ph-1,2,3-triazole **4.2-14**, 53% (43)

The addition of tosyl azide to an enamine (Eq. 44) represents a potentially interesting approach.[48] Finally, it may be possible to improve the low yields obtained in the photochemical reaction of arylnitriles with ethyl diazoacetate (Eq. 45).[49]

EtO$_2$CCH=C(Me)(NHPh) + 4-MeC$_6$H$_4$SO$_2$N$_3$ $\xrightarrow[H_2O]{Na^{+}\,^{-}OH}$ 4-EtO$_2$C-5-Me-1-Ph-1,2,3-triazole, 68% (44)

$$ArCN + N_2CHCO_2Et \xrightarrow[Cu]{h\nu} \text{4-}EtO_2C\text{-2-Ar-1,2,3-triazole} \quad (45)$$

Ar = Ph 18%
2,6-Me_2Ph 10%

REFERENCES

22. **37:** 119^2
23. **45:** 9037f
24. **43:** 2619b
25. **46:** 6123e
26. **78:** 16102f
27. **77:** 126981d
28. **79:** 5526s
29. **84:** 122189y
30. **58:** 11454f
31. **72:** 38736u
32. **46:** 8651i
33. **55:** 17626i
34. **67:** 53900h
35. **74:** 99215m
36. **84:** 74239r
37. **67:** 100071a
38. **57:** 7255b
39. **69:** 19088f
40. **51:** 13854i
41. **72:** 121447w
42. **72:** 132635g
43. **74:** 141988t
44. **83:** 193178j
45. **58:** 12560d
46. **67:** 43759e
47. **76:** 59538d
48. **73:** 14769b
49. **79:** 104890n

4.3. 1,2,3-TRIAZOLE-4,5-DICARBOXYLIC ACIDS AND THEIR FUNCTIONAL DERIVATIVES

In the early development of *v*-triazole chemistry a number of reactions leading to the 4,5-dicarboxylic acids and their derivatives were discovered and should be explored further. Fries[50] oxidized benzotriazoles (Eqs. 46,47) and prepared a broad range of acid derivatives. Some especially promising compounds in this series have been prepared from what was then called a hydrazoic acid polymer (**4.3-1**) by reaction with nitrous acid (Eqs. 48,49).[51,52] Although the yields were not reported, the chemistry is intriguing. The oxidation of a 1,2,3-triazole methyl group and conversion to acid

$$\text{5-}H_2N\text{-2-phenylbenzotriazole} + KMnO_4 \longrightarrow \text{2-Ph-1,2,3-triazole-4,5-}(CO_2H)_2 \quad (46)$$

50%

$$\text{5-}H_2N\text{-1-phenylbenzotriazole} + KMnO_4 \longrightarrow \text{1-Ph-1,2,3-triazole-4,5-}(CO_2H)_2 \quad (47)$$

derivatives has been accomplished in good yield (Eq. 50).[53] The 4,5-diformyl- and diacid compounds are available in high yield from the tetraethyl bisacetal of 1,4-but-2-ynediol (**4.3-2**) through hydrolysis and oxidation (Eq. 51).[54]

HN=C(CN)–H₂NCHCN **4.3-1** + $\xrightarrow{HNO_2}$ 4,5-dicyano-1H-1,2,3-triazole (NC, CN; N, N, N, H) $\xrightarrow{H_2O}$ 1H-1,2,3-triazole-4,5-dicarboxylic acid (HO_2C, CO_2H; N, N, N, H) (48)

4.3-1 $\xrightarrow{\text{Ac derivative, } HNO_2}$ NC, $CONH_2$ triazole (N, N, N, H) + NC, CN triazole (N, N, N, H) $\xrightarrow{H_2O}$ HO_2C, CO_2H triazole (49)

HO_2C, Me triazole (N, N, N, Ph) $\xrightarrow{KMnO_4}$ HO_2C, CO_2H triazole (N, N, N, Ph) (50)

$(EtO)_2CHC{\equiv}CCH(OEt)_2 + RN_3 \longrightarrow$ $(EtO)_2CH$, $CH(OEt)_2$ triazole (N, N, N, R) $\xrightarrow{H^+}$
4.3-2 Ph 82%

OHC, CHO triazole (N, N, N, R) $\xrightarrow{KMnO_4}$ HO_2C, CO_2H triazole (N, N, N, R) (51)
Ph 92% Ph high

Probably no acetylene has been more popular for demonstrating 1,3-dipolar azide addition than dimethyl acetylenedicarboxylate (**4.3-3**). In the 1920s both methyl azidoacetate (Eq. 52)[55] and 2-azido-1,4-dimethylbenzene (Eq. 52)[56] were shown to add in excellent yield. In both instances various acid derivatives were also prepared. Later additional aryl azides were examined with comparable results. Of special interest are the additions of 2-nitrophenylazide[58] and 2-azido-3,4,5,6-tetrachloropyridine,[59] which in some instances have been reported to add to acetylenes with difficulty.

$$MeO_2CC{\equiv}CCO_2Me + RN_3 \longrightarrow \textbf{4.3-4} \quad (52)$$

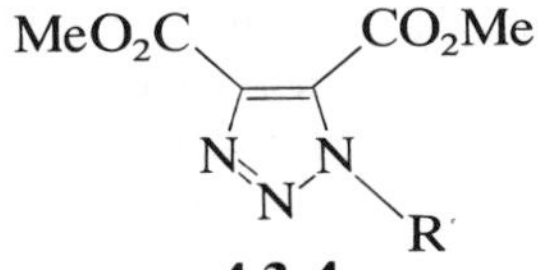

4.3-3

R	= CH_2CO_2Me	2,5-Me_2Ph	4-NO_2Ph[57]	4-MeOPh[57]	2-NO_2Ph[58]
% **4.3-4** =	100	82	87	97	good

Eq.52 R = (tetrachloropyridyl) 85%[59]; (2-benzoyl-4-chlorophenyl) 67%[60]; (tetrazolopyridazinyl) 16%[61]; (3-ethoxycarbonyl-4-methylpyrrolyl) 95%[62]; (tetrazolopyridopyridazinyl) 52%[61]

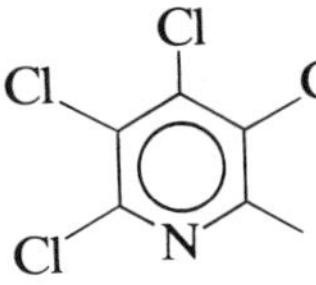

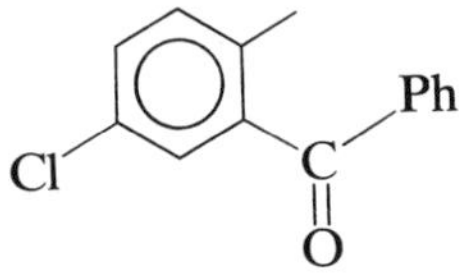

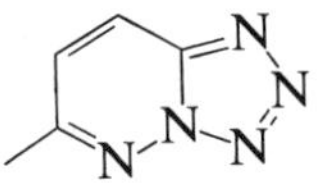

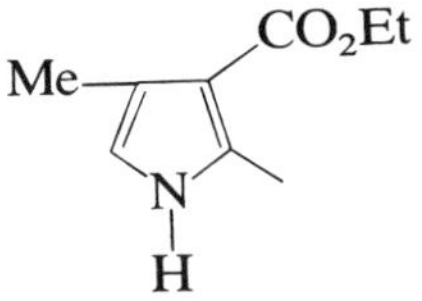

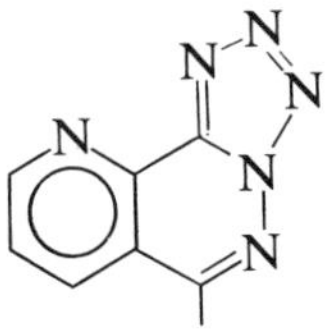

Certain complex heterocyclic azides show wide differences in their ability to add to **4.3-3**.[61,62]

A large number of sugars have been converted to azides that add smoothly to **4.3-3** and may be converted to other derivatives in high yield.[63]

Tropylium azide (**4.3-5**) is especially interesting in that the adducts can often be hydrolyzed quantitatively to the 1,2,3-triazole without a nitrogen substituent (Eq. 53).[64] The compound **4.3-6** has also been prepared in fair

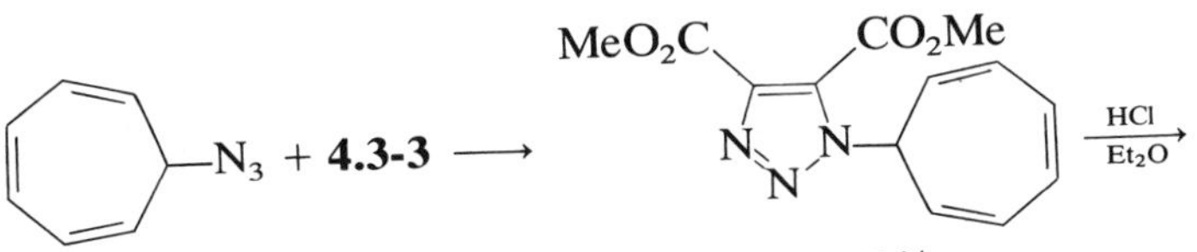

$$\textbf{4.3-5}\text{-}N_3 + \textbf{4.3-3} \longrightarrow \text{adduct (73\%)} \xrightarrow[Et_2O]{HCl} \textbf{4.3-6} \quad (53)$$

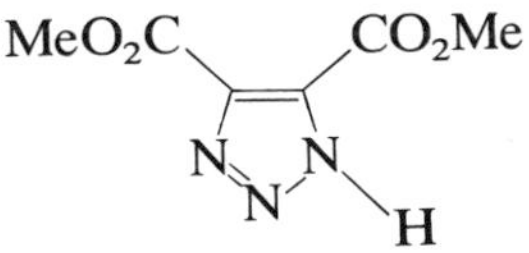

yield by the direct addition of the azide ion (Eq. 54)[65,66] or trimethylsilylazide and the hydrolysis of the trimethylsilyl group (Eq. 55).[67] Later studies suggest that **4.3-7** is probably the 2-(trimethylsilyl) isomer.[68]

$$\textbf{4.3-3} + Na^{+}\,{}^{-}N_3 \xrightarrow{HCONMe_2} \textbf{4.3-6}\ 60\% \quad (54)$$

$$\textbf{4.3-3} + Me_3SiN_3 \longrightarrow \textbf{4.3-7}\ 85\% \xrightarrow{H_2O} \textbf{4.3-6}\ 100\% \quad (55)$$

(**4.3-7**: 4,5-bis(CO_2Me)-1,2,3-triazole with N–$SiMe_3$)

The two reports of the addition of organometallic azides to **4.3-3** show promise for future development (Eqs. 56,57).[69,70]

4.3-3 (Me or Et)

$$+ (n\text{-Bu})_3SnN_3 \longrightarrow \text{4,5-bis}(CO_2Et)\text{-triazole–}Sn(Bu\text{-}n)_3\ \ 80\% \quad (56)$$

$$+ Ph_3PbN_3 \longrightarrow \text{4,5-bis}(CO_2Me)\text{-triazole–}PbPh_3\ \ \text{good} \quad (57)$$

An internal salt involving a rather complex relationship of functional groups (**4.3-8**) adds well to **4.3-3** (Eq. 58).[71] The product (**4.3-9**) hydrolyzes to produce good yields of aldehydes and **4.3-6**.

$$RCH(O^-)N(SiMe_3)N_2^+ \ (\textbf{4.3-8}) + \textbf{4.3-3} \longrightarrow \textbf{4.3-9} \xrightarrow{MeOH} \textbf{4.3-6} + RCHO + Me_3SiOMe \quad (58)$$

(**4.3-9**: 4,5-bis(CO_2Me)-1,2,3-triazole with N–$CHR(OSiMe_3)$; **4.3-6**: 4,5-bis(CO_2Me)-1,2,3-triazole with N–H)

R = *i*-Pr, *t*-Bu
% **4.3-9** = 81, 77

Finally, the decomposition of certain tricyclic azide adducts (**4.3-10**) provides good-to-excellent yields of dimethyl 1-phenyl-1,2,3-dicarboxylate (**4.3-11**) (Eq. 59).[72,73]

$$\text{bicyclic diester (X)} + PhN_3 \longrightarrow \mathbf{4.3\text{-}10} \longrightarrow \mathbf{4.3\text{-}11} \quad (59)$$

X	= O	NCO_2Et
% **4.3-11**	= 100	64

The parent compound (**4.3-6**) can be alkylated to produce an approximately equimolar mixture of 1- and 2-substitution product (Eq. 60).[74] Tanaka and Miller have studied the effect of base and solvent on this isomer distribution.[74]

$$\mathbf{4.3\text{-}6} + ClCH_2CO_2Et \longrightarrow \mathbf{4.3\text{-}12} + \mathbf{4.3\text{-}13} \quad (60)$$

Base	Solvent	% **4.3-12**	% **4.3-13**	% total
$Na^{+-}OMe$	EtOH	53	47	
	$HCONMe_2$	44	56	56
1,8-bis(Me_2N)-naphthalene	EtOH	62	38	
	$HCONMe_2$	45	55	35

REFERENCES

50. **21:** 2690[4]
51. **22:** 423[5]
52. **31:** 7849[3]
53. **37:** 5404[9]
54. **38:** 1743[1]
55. **24:** 3215[1]
56. **22:** 3411[2]
57. **64:** 5076e
58. **67:** 64311t
59. **83:** 192957a
60. **84:** 74239r
61. **83:** 97213w
62. **73:** 98718v
63. **81:** 163240q
64. **62:** 13138g
65. **73:** 25364s
66. **80:** 108452q
67. **60:** 5537a
68. **68:** 13061z
69. **60:** 14532c
70. **67:** 82249h
71. **83:** 10269x
72. **82:** 16641p
73. **85:** 46469t
74. **80:** 108451p

TABLE 4. 1,2,3-TRIAZOLECARBOXYLIC ACIDS AND THEIR FUNCTIONAL DERIVATIVES

Compound	Reference
4.1. 1*H*-1,2,3-Triazole-1- and -5-Carboxylic Acids and Their Functional Derivatives	
-4-acetic acid, 5-carboxy-2-phenyl-	**80:** 37050u
-4-acetic acid, 5-(methoxycarbonyl)-2-phenyl-, methyl ester	**80:** 37050u
-4-acetic acid, 5-[(1-methylethoxy)carbonyl]-2-phenyl-, isopropyl ester	**80:** 37050u
-1-acrylic acid, 5-carboxy-, dimethyl ester	**73:** 25364s
-1-carbonitrile	**67:** 116854u
-5-carbonitrile, 4-[4-(acetamido)phenyl]-	**78:** P72161m
-1-carbonitrile, 4-butyl-	**67:** P64403z
-1-carbonitrile, 5-butyl-	**67:** P64403z
-1-carbonitrile, 4,5-dimethyl-	**67:** 116854u
-5-carbonitrile, 1,4-diphenyl-	**68:** 78204t, **80:** 145179b
-1-carbonitrile, 4-methyl-	**67:** 116854u
-1-carbonitrile, 5-methyl-	**67:** 116854u
-1-carboxamide, *N*-butyl-	**57:** 4649h
-5-carboxamide, 1,4-diphenyl-	**80:** 145179b
-5-carboxamide, 1-β-D-galactopyranosyl-	**74:** 112402s
-5-carboxamide, 1-β-D-glucopyranosyl-	**74:** 112402s
-5-carboxamide, 1-β-D-ribofuranosyl-	**78:** 84715h
-1-carboxanilide	**57:** 4649i
-1-carboximidamide, *N*′-(4-chlorophenyl)-*N*,*N*-dimethyl-	**82:** P72999y
-1-carboximidamide, *N*′-(3,4-dichlorophenyl)-*N*,*N*-dimethyl-	**82:** P72999y
-1-carboximidamide, *N*,*N*-dimethyl-*N*′-phenyl-	**82:** P72999y
-1-carboxylic acid, ethyl ester	**76:** 99184s
-5-carboxylic acid, 1-(2-acetamido-2-deoxy-β-D-glucopyranosyl)-, methyl ester, 3′,4′,6′-triacetate ester	**74:** 112402s
-5-carboxylic acid, 1-benzyl-(?)	**55:** 23505a
-5-carboxylic acid, 1-benzyl-, 2-nitroethyl ester (?)	**55:** 23505a
-5-carboxylic acid, 1-benzyl-4-[2,3,5-*tris*-*O*-benzyl-D-ribofuranosyl]-, methyl ester	**83:** 147684y
-5-carboxylic acid, 1-(4-bromophenyl)-	**69:** 10402w
-5-carboxylic acid, 1-(4-chlorophenyl)-	**53:** 16120e, **69:** 10402w
-5-carboxylic acid, 4-(4-chlorophenyl)-1-phenyl-, ethyl ester	**81:** 3850k
-5-carboxylic acid, 4-(4-chlorophenyl)-1-phenyl, hydrazide	**81:** 3850k
-5-carboxylic acid, 1-(2,4-dichlorophenyl)-	**53:** 16120e
-5-carboxylic acid, 1-(2,5-dichlorophenyl)-	**53:** 16120e
-5-carboxylic acid, 1-(1,3-dioxo-2-phenyl-2-indanyl)-	**69:** P96741r
-5-carboxylic acid, 1-(3,3-dimethyl-1-butenyl)-, methyl ester, (E)-	**72:** 132635g
-5-carboxylic acid, 1,4-diphenyl-	**45:** 9037i, **80:** 145179b

TABLE 4 (*Continued*)

Compound	Reference
4.1. 1*H*-1,2,3-Triazole-1- and -5-Carboxylic Acids and Their Functional Derivatives (*Continued*)	
-5-carboxylic acid, 1,4-diphenyl-, ethyl ester	**81:** 3850k, **83:** 147900r
-5-carboxylic acid, 1-ethenyl-, butyl ester	**81:** 37519z, **83:** 147900r
-5-carboxylic acid, 1-ethenyl-, isobutyl ester	**83:** 147900r
-5-carboxylic acid, 1-ethenyl-, isopentyl ester	**83:** 147900r
-5-carboxylic acid, 1-ethenyl-, isopropyl ester	**83:** 147900r
-5-carboxylic acid, 1-ethenyl-, pentyl ester	**83:** 147900r
-5-carboxylic acid, 1-ethenyl-, propyl ester	**83:** 147900r
-5-carboxylic acid, 1-ethyl-, butyl ester	**81:** 37519z
-5-carboxylic acid, 1-β-D-galactopyranosyl-, methyl ester, 2′,3′,4′,6′-tetraacetate ester	**74:** 112402s
-5-carboxylic acid, 1-β-D-glucopyranosyl-, methyl ester, 2′,3′,4′,6′-tetraacetate ester	**74:** 112402s
-5-carboxylic acid, 4-(hydroxymethyl)-1-phenyl-	**49:** 3948i
-5-carboxylic acid, 1-(2-iodo-3,3-dimethylbutyl)-, methyl ester	**72:** 132635g
-5-carboxylic acid, 1-[α-(iodomethyl)benzyl]-, methyl ester	**72:** 132635g
-5-carboxylic acid, 1-(4-iodophenyl)-	**69:** 10402w
-1-carboxylic acid, 5-methyl-, ethyl ester	**70:** 77878v
-5-carboxylic acid, 4-(4-methylphenyl)-1-phenyl-, ethyl ester	**81:** 3850k
-1-carboxylic acid, 5-(4-nitrophenyl)-, ethyl ester	**70:** 77878v, **74:** 76375z, **76:** 25192w, **79:** 31992k
-2-carboxylic acid, 4-(4-nitrophenyl)-, ethyl ester	**76:** 25192w, **79:** 31992k
-5-carboxylic acid, 1-(4-nitrophenyl)-	**58:** 12560h, **67:** 100071a, **69:** 10402w
-5-carboxylic acid, 1-(4-nitrophenyl)-, methyl ester	**58:** 12561a
-1-carboxylic acid, 4-phenyl-, ethyl ester	**61:** 5626h, **74:** 76375z
-1-carboxylic acid, 5-phenyl-, ethyl ester	**74:** 76375z, **76:** 25192w, **79:** 31992k
-2-carboxylic acid, 4-phenyl-, ethyl ester	**74:** 76375z, **76:** 25192w, **79:** 31992k
-5-carboxylic acid, 1-phenyl-	**44:** 1102c, **48:** 2685a, **53:** 17108g, **61:** 3097a, **67:** 100071a, **69:** 10402w
-5-carboxylic acid, 1-phenyl-, methyl ester	**49:** 3948f, **64:** 5076e
-5-carboxylic acid, 1-(1-phenylethenyl)-, methyl ester, (E)-	**72:** 132635g
-5-carboxylic acid, 1-(1-phenylpropenyl)-, methyl ester, (E)-	**72:** 132635g
-5-carboxylic acid, 1-(2,3,5-tri-*O*-benzoyl-β-D-ribofuranosyl)-, methyl ester	**74:** 112402s, **78:** 84715h
2-[4-(ethoxycarbonyl)phenyl]-4-methyl-5-[(phenylamino)-carbonyl]-	**77:** P7313c
-4-prop-2-enoic acid, 5-cyano-, (Z)-	**84:** 135548q

TABLE 4 (*Continued*)

Compound	Reference
4.2. 1,2,3-Triazole-4-Carboxylic acids and Their Functional Derivatives	
-5-acetic acid, 4-carboxyl-1-phenyl-	**53:** 17108g, **61:** 3096g
-5-acetic acid, 4-carboxy-1-phenyl-, dimethyl ester	**61:** 3096g
-5-acetic acid, 4-(ethoxycarbonyl)-1-imidazo[1,2-*b*]-pyridazin-6-yl-, ethyl ester	**82:** 16745a
-5-acetic acid, 4-(ethoxycarbonyl)-1-(4-oxo-4*H*-pyrido-[1,2-*a*]pyrimidin-2-yl)-, ethyl ester	**83:** 9958q
-5-acetic acid, 4-(ethoxycarbonyl)-1-tetrazolo[1,5-*b*]-pyridazin-6-yl-, ethyl ester	**82:** 16745a
-5-acetic acid, 4-(ethoxycarbonyl)-1-(1,2,4-triazolo-[4,3-*b*]pyridazin-6-yl)-, ethyl ester	**82:** 16745a
4-[1-[4-[3-butylmethylamino)propyl]azacyclohexyl]-carbonyl]-2-phenyl-, 1:2 adduct with maleic acid	**68:** P59441b
-4-carbonitrile	**69:** 77179x, **71:** 13067t
-4-carbonitrile, 5-(3-aminophenyl)-	**78:** P72161m
-4-carbonitrile, 5-(9-anthracenyl)-	**80:** 3439n
-4-carbonitrile, 1-(2-benzoylpropenyl)-5-phenyl-	**67:** 32420m
-4-carbonitrile, 2-(4-bromophenyl)-5-phenyl-	**63:** P11574e
-4-carbonitrile, 5-(4-chlorophenyl)-	**80:** 3439n
-4-carbonitrile, 2-(4-chlorophenyl)-5-(4-methoxyphenyl)-	**79:** P18720a
-4-carbonitrile, 1-(5′-deoxy-5′-thymidinyl)-	**81:** 163240q
-4-carbonitrile, 5-(1,2-dihydro-5-acenaphthylenyl)-	**80:** 3439n
-4-carbonitrile, 5-[4-(dimethylamino)phenyl]-, monohydrochloride	**80:** 3439n
-4-carbonitrile, 2,5-diphenyl-	**63:** P11574e
-4-carbonitrile, 5-(2-furanyl)-	**80:** 3439n
-4-carbonitrile, 5-(3-hydroxyphenyl)-	**78:** P72161m
-4-carbonitrile, 5-(4-hydroxyphenyl)-	**80:** 3439n
-4-carbonitrile, 1-(4-methoxyphenyl)-	**64:** 11198g
-4-carbonitrile, 2-(3-methoxyphenyl)-	**76:** 114090d
-4-carbonitrile, 5-(4-methoxyphenyl)-	**80:** 3439n
-4-carbonitrile, 5-methyl-1-phenyl-	**62:** 545f
-4-carbonitrile, 5-(1-naphthalenyl)-	**78:** P72161m
-4-carbonitrile, 5-(3-nitrophenyl)-	**78:** P72161m
-4-carbonitrile, 5-(4-nitrophenyl)-	**80:** 3439n
-4-carbonitrile, 1-phenyl-	**24:** 592[9]
-4-carbonitrile, 2-phenyl-	**46:** 6123d, **76:** 114090d
-4-carbonitrile, 5-phenyl-	**61:** 14665a, **67:** 32420m, **75:** 140767d, **77:** 126517g, **80:** 3439n
-4-carbonitrile, 5-propyl-	**78:** P72161m
-4-carbonitrile, 5-(2-thienyl)-	**80:** 3439n
-4-carbonitrile, 1-(2,3,5-tri-*O*-acetyl-β-D-ribofuranosyl)-	**78:** 84715h
-4-carbonitrile, 2-(2,3,5-tri-*O*-acetyl-β-D-ribofuranosyl)-	**78:** 84715h

TABLE 4 (*Continued*)

Compound	Reference
4.2. 1,2,3-Triazole-4-Carboxylic Acids and Their Functional Derivatives (*Continued*)	
-4-carbonitrile, 5-(3,4,5-trimethoxyphenyl)-	**80:** 3439n
-4-carbonyl azide	**51:** 14697h, **72:** 31710x
-4-carbonyl azide, 5-(4-chlorophenyl)-1-phenyl-	**81:** 3850k
-4-carbonyl azide, 1,5-diphenyl-	**81:** 3850k
-4-carbonyl azide, 5-methyl-	**85:** 123875b
-4-carbonyl azide, 5-methyl-1-phenyl-	**80:** 47912r
-4-carbonyl azide, 5-(4-methylphenyl)-1-phenyl-	**81:** 3850k
-4-carbonyl chloride, 2-[4-(chlorocarbonyl)phenyl]-	**83:** P61732a
-4-carbonyl chloride, 2-[4-[(4,5-dihydro-3-methyl-5-oxo-1-phenyl-1*H*-pyrazol-4-yl)azo]phenyl]-5-methyl-	**80:** P122396f
-4-carbonyl chloride, 1,5-diphenyl-	**81:** 3850k
-4-carbonyl chloride, 2,5-diphenyl-	**84:** P166274b
-4-carbonyl chloride, 2-[4-[(2-hydroxy-1-naphthalenyl)azo]phenyl]-5-methyl-	**80:** P122396f
-4-carbonyl chloride, 2-(3-methoxyphenyl)-	**76:** 114090d
-4-carbonyl chloride, 1-methyl-	**82:** P43446k
-4-carbonyl chloride, 5-methyl-2-[4-[[2-oxo-1-[(phenylamino)carbonyl]propyl]azo]phenyl]-	**80:** P122396f
-4-carbonyl chloride, 5-methyl-1-phenyl-	**47:** 3087e, **52:** 17246b
-4-carbonyl chloride, 5-methyl-2-phenyl-	**76:** 114090d
-4-carbonyl chloride, 2-phenyl-	**43:** 2619d, **50:** 4924a, **76:** 114090d
-4-carbothioamide, 1-β-D-ribofuranosyl-	**78:** 84715h
-4-carbothioamide, 2-β-D-ribofuranosyl-	**78:** 84715h
-4-carbothioamide, 1-(2,3,5-tri-*O*-acetyl-β-D-ribofuranosyl)-	**78:** 84715h
-4-carbothioamide, 2-(2,3,5-tri-*O*-acetyl-β-D-ribofuranosyl)-	**78:** 84715h
-4-carbothioic acid, 1-methyl-	**82:** P43446k
-4-carboxamide	**51:** 14697g, **84:** 122189y
-4-carboxamide, 1-(2-acetamido-2-deoxy-β-D-glucopyranosyl)-	**74:** 112402s
-4-carboxamide, 2-[4-(aminocarbonyl)phenyl]-5-methyl-	**79:** P18722c
-4-carboxamide, 1-(4-aminophenyl)-*N*,*N*-diethyl-5-methyl-	**55:** 2926g
-4-carboxamide, 2-(4-bromophenyl)-	**61:** 5738c
-4-carboxamide, 2-(4-chlorophenyl)-	**61:** 5738c
-4-carboxamide, 2-(4-chlorophenyl)-5-methyl-	**79:** P18722c
-4-carboxamide, 2-[4-(2-cyano-2-phenylethenyl)-3-methoxyphenyl]-5-phenyl-	**81:** P51152f
-4-carboxamide, 1-(2-deoxy-β-D-*erythro*-pentofuranosyl)-	**84:** 122189y
-4-carboxamide, 1-(5′-deoxy-5′-thymidinyl)-	**81:** 163240q
-4-carboxamide, 1-(5′-deoxy-5′-thymidinyl)-, 3′-acetate ester	**81:** 163240q
-4-carboxamide, 5-(3,4-dichlorophenyl)-	**78:** P111330n

TABLE 4 (*Continued*)

Compound	Reference
4.2. 1,2,3-Triazole-4-Carboxylic Acids and Their Functional Derivatives (*Continued*)	
-4-carboxamide, *N,N*-diethyl-5-methyl-1-phenyl-	**36:** 835[7]
-4-carboxamide, *N,N*-diisobutyl-5-methyl-1-phenyl-	**36:** 835[7]
-4-carboxamide, *N*,1-dimethyl-	**84:** 122189y
-4-carboxamide, *N,N*-[4-[4-(dimethylamino)butyl]aza-cyclohexyl]-2-phenyl-, monohydrochloride	**68:** P59441b
-4-carboxamide, 5-[4-(dimethylamino)phenyl]-	**80:** 3439n
-4-carboxamide, *N,N*-[4-[3-(dimethylamino)propyl]aza cyclohexyl]-2-(2,4-dinitrophenyl)-	**68:** P59441b
-4-carboxamide, *N*,5-dimethyl-*N*,2-diphenyl-	**78:** 16102f
-4-carboxamide, *N*,5-dimethyl-2-(4-methylphenyl)-*N*-phenyl-	**78:** 16102f
-4-carboxamide, *N*,2-diphenyl-5-methyl-	**78:** 16102f
-4-carboxamide, *N*-ethyl-5-methyl-1-phenyl-	**80:** 47912r
-4-carboxamide, 2-(4-fluorophenyl)-	**61:** 5738c
-4-carboxamide, 5-(2-furanyl)-	**80:** 3439n
-4-carboxamide, 1-β-D-galactopyranosyl-	**74:** 112402s
-4-carboxamide, 1-β-D-glucopyranosyl-	**74:** 112402s
-4-carboxamide, 5-(3-hydroxy-4-methoxyphenyl)-	**80:** 3439n
-4-carboxamide, 5-(4-hydroxy-3-methoxyphenyl)-	**78:** P111330n
-4-carboxamide, 5-(4-hydroxyphenyl)-	**80:** 3439n
-4-carboxamide, 2-(2-hydroxy-5-sulfophenyl)-5-methyl-, monosodium salt	**76:** P99674b
-4-carboxamide, *N*-(methoxycarbonyl)-	**84:** 122189y
-4-carboxamide, 2-(4-methoxyphenyl)-5-methyl-	**79:** P18722c
-4-carboxamide, *N*-(6-methoxy-8-quinolyl)-2-phenyl-	**46:** 6123h
-4-carboxamide, *N*-methyl-	**84:** 122189y
-4-carboxamide, 1-methyl-	**77:** 126981d, **84:** 122189y
-4-carboxamide, 5-methyl-*N*,2-diphenyl-, 3-oxide	**80:** 95840n
-4-carboxamide, 5-methyl-2-phenyl-	**76:** 114090d
-4-carboxamide, 2-(4-nitrophenyl)-	**61:** 5738c
-4-carboxamide, 5-(4-nitrophenyl)-	**80:** 3439n
-4-carboxamide, 1-phenyl-	**63:** P7019b
-4-carboxamide, 2-phenyl-	**43:** 2619d
-4-carboxamide, 5-phenyl-	**61:** 14665a, **77:** 126517g, **80:** 3439n
-4-carboxamide, 1-β-D-ribofuranosyl-	**74:** 112402s, **77:** 126981d, **78:** 84715h, **79:** 5526s, **84:** 122189y
-4-carboxamide, 2-β-D-ribofuranosyl-	**78:** 84715h
-4-carboxamide, 5-β-D-ribofuranosyl-	**84:** 44571a
-4-carboxamide, 5-(2,3,5-tri-*O*-acetyl-β-D-ribofuranosyl)-	**84:** 44571a
-4-carboxamide, 5-(3,4,5-trimethoxyphenyl)-	**80:** 3439n
-4-carboxamide, *N,N*,5-trimethyl-1-phenyl-	**36:** 835[7]
-4-carboxamidine	**63:** 9942g
-4-carboxanilide, 5-benzyl-1-phenyl-	**33:** 4249[9]
-4-carboxanilide, 1-(4-chlorophenyl)-5-isobutyl-	**58:** 12561a
-4-carboxanilide, 1-(4-chlorophenyl)-5-methyl-	**58:** 12560f

TABLE 4 (*Continued*)

Compound	Reference
4.2. 1,2,3-Triazole-4-Carboxylic Acids and Their Functional Derivatives (*Continued*)	
-4-carboxanilide, 4′-[4-[3-(dimethylamino)propyl]-1-piperazinyl]-2-phenyl-, (1:2) adduct with maleic acid	**68:** P59616n
-4-carboxanilide, 1-(2,4-dinitrophenyl)-5-methyl-	**61:** 5633b
-4-carboxanilide, 5-isobutyl-1-(2-nitrophenyl)-	**58:** 12561a
-4-carboxanilide, 5-isobutyl-1-(4-nitrophenyl)-	**58:** 12560h
-4-carboxanilide, 5-methyl-	**61:** 5633b
-4-carboxanilide, 5-methyl-1-(2-nitrophenyl)-	**58:** 12560f
-4-carboxanilide, 5-methyl-1-(4-nitrophenyl)-	**58:** 12560f
-4-carboxanilide, 5-methyl-1-phenyl-	**37:** 5405[3]
-4-carboxanilide, 2-phenyl-	**43:** 2619d
-4-carboxanilide, 2′,4′,6′-triiodo-1-(4-iodophenyl)-5-methyl-	**57:** 7255g
-4-carboximidic acid, 5-cyano-, isopropyl ester	**85:** P63074c
-4-carboxylic acid	**36:** 2863[3], **49:** 3949f, **49:** 13227f, **50:** 9392f, **55:** 17626i, **61:** 14665b, **77:** 126517g, **84:** 122189y
-4-carboxylic acid, butyl ester	**81:** 37519z, **82:** 140021a, **85:** 78055w
-4-carboxylic acid, ethyl ester	**79:** 92148m, **82:** 140021a, **85:** 78055w
-4-carboxylic acid, hexyl ester	**82:** 140021a, **85:** 78055w
-4-carboxylic acid, hydrazide	**55:** P571c, **72:** P67724u
-4-carboxylic acid, isobutyl ester	**82:** 140021a, **85:** 78055w
-4-carboxylic acid, isopentyl ester	**82:** 140021a, **85:** 78055w
-4-carboxylic acid, isopropyl ester	**82:** 140021a, **83:** 97148d, **85:** 78055w
-4-carboxylic acid, 2-isopropyl hydrazide	**84:** 144561r
-4-carboxylic acid, methyl ester	**64:** 5076f, **70:** 96727m, **73:** 25364s, **82:** 140021a, **84:** 122189y, **85:** 78055w
-4-carboxylic acid, pentyl ester	**82:** 140021a, **85:** 78055w
-4-carboxylic acid, propyl ester	**82:** 140021a, **85:** 78055w
-4-carboxylic acid, propyl ester, sodium salt	**85:** 47115m
-4-carboxylic acid, 1-(2-acetamido-2-deoxy-β-D-glucopyranosyl)-, 3′,4′,6′-triacetate ester	**74:** 112402s
-4-carboxylic acid, 1-(2-acetamido-2-deoxy-β-D-glucopyranosyl)-, methyl ester, 3′,4′,6′-triacetate ester	**74:** 112402s
-4-carboxylic acid, 2-(4-acetamidophenyl)-	**58:** 11454f
-4-carboxylic acid, 1-(3-acetyl-2-oxo-2*H*-cyclohepta[*b*]furan-5-yl)-5-methyl-, ethyl ester	**76:** 140726k
-4-carboxylic acid, 2-(2-amino-4-arsonophenyl)-5-methyl-	**40:** 7191[9]
-4-carboxylic acid, 1-(2-amino-6-cycloheptimidazolyl)-5-methyl-, ethyl ester	**76:** 140726k
-4-carboxylic acid, 1-(2-amino-1,3-dicarboxy-6-azulenyl)-5-methyl-, triethyl ester	**76:** 140726k

TABLE 4 (*Continued*)

Compound	Reference
4.2. 1,2,3-Triazole-4-Carboxylic Acids and Their Functional Derivatives (*Continued*)	
-4-carboxylic acid, 1-(2-amino-1,3-dicyano-6-azulenyl)-5-methyl-, ethyl ester	**76:** 140726k
-4-carboxylic acid, 2-(4-amino-3-methylphenyl)-	**58:** 11454g
-4-carboxylic acid, 2-(4-aminophenyl)-	**57:** 7353g, **58:** 11454g
-4-carboxylic acid, 2-(4-arsono-2-nitrophenyl)-5-methyl-	**40:** 7191[7]
-4-carboxylic acid, 2-(4-arsono-2-nitrophenyl)-5-methyl-, ethyl ester	**40:** 7191[7]
-4-carboxylic acid, 2-(4-arsonophenyl)-	**40:** 7191[8]
-4-carboxylic acid, 2-(4-arsonophenyl)-5-(2-carboxyphenyl)-	**24:** 1378[4]
-4-carboxylic acid, 2-(4-arsonophenyl)-5-methyl-	**40:** 7191[7]
-4-carboxylic acid, 2-(4-arsonophenyl)-5-methyl-, ethyl ester	**40:** 7191[7]
-4(or 5)-carboxylic acid, 1-(2-benzothiazolyl)-	**79:** P78815v
-4-carboxylic acid, 1-(2-benzothiazolyl)-5-methyl-, ethyl ester	**79:** P78815v
-4-carboxylic acid, 1-[1-benzoyl-2-(3-nitrophenyl)-4-oxo-4-phenylbutyl]-5-phenyl-, ethyl ester	**83:** 193178j
-4-carboxylic acid, 1-(1-benzoyl-4-oxo-2,4-diphenylbutyl)-5-phenyl-, ethyl ester	**83:** 193178j
-4-carboxylic acid, 1-benzyl-	**73:** 98912d
-4-carboxylic acid, 1-benzyl-5-methyl-	**83:** 193178j
-4-carboxylic acid, 1-benzyl-5-phenyl-	**46:** 8651r, **51:** 13854i
-4-carboxylic acid, 5-benzyl-1-phenyl-	**33:** 4249[9], **37:** 5405[4]
-4-carboxylic acid, 1-benzyl-5-(2,3,5-tris-*O*-benzyl-D-ribofuranosyl)-, methyl ester	**83:** 147684y
-4-carboxylic acid, 1,5-bis(4-nitrophenyl)-	**74:** 141988t
-4-carboxylic acid, 2-(4-bromo-3-carboxyphenyl)-	**55:** 3563d
-4-carboxylic acid, 2-(3-bromo-4-chlorophenyl)-	**55:** 27071g
-4-carboxylic acid, 2-(4-bromo-2-chlorophenyl)-	**63:** 1850h
-4-carboxylic acid, 2-(4-bromo-3-chlorophenyl)-	**55:** 27071g
-4-carboxylic acid, 1-[5-bromo-1-(2-cyanoethyl)-1,6-dihydro-6-oxo-4-pyridazinyl]-5-methyl-, ethyl ester	**76:** 153691v
-4-carboxylic acid, 1-[5-bromo-1,6-dihydro-1-(2-hydroxyethyl)-6-oxo-4-pyridazinyl]-5-methyl-, ethyl ester	**76:** 153691v
-4-carboxylic acid, 1-(5-bromo-1,6-dihydro-1-methyl-6-oxo-4-pyridazinyl)-5-methyl-, ethyl ester	**76:** 153691v
-4-carboxylic acid, 1-(5-bromo-1,6-dihydro-6-oxo-1-phenyl-4-pyridazinyl)-5-methyl-, ethyl ester	**73:** 77172x, **76:** 153691v
-4-carboxylic acid, 2-(4-bromo-3-fluorophenyl)-	**59:** 14095f
-4-carboxylic acid, 2-(4-bromo-2-methylphenyl)-	**54:** 1504g
-4-carboxylic acid, 2-(3-bromophenyl)-	**54:** 1504h
-4-carboxylic acid, 2-(4-bromophenyl)-	**53:** 344i
-4-carboxylic acid, 2-(4-bromophenyl)-, ethyl ester	**61:** 5738c
-4-carboxylic acid, 2-(4-bromophenyl)-, methyl ester	**61:** 5738c
-4-carboxylic acid, 1-(4-bromophenyl)-5-methyl-	**69:** 10402w

TABLE 4 (*Continued*)

Compound	Reference
4.2. 1,2,3-Triazole-4-Carboxylic Acids and Their Functional Derivatives (*Continued*)	
-4-carboxylic acid, 1-(4-bromophenyl)-5-(4-nitrophenyl)-	**74:** 141988t
-4-carboxylic acid, 1-(4-bromophenyl)-5-phenyl-	**74:** 141988t
-4-carboxylic acid, 2-(4-butylphenyl)-	**76:** 114090d
-4-carboxylic acid, 2-(3-carboxy-4-chlorophenyl)-	**55:** 27071g
-4-carboxylic acid, 2-(5-carboxy-2-hydroxyphenyl)-5-(2-carboxyphenyl)-, triethyl ester	**71:** P124445j
-4-carboxylic acid, 2-(5-carboxy-2-hydroxyphenyl)-5-(2-carboxyphenyl)-, trimethyl ester	**71:** P124445j
-4-carboxylic acid, 2-(4-carboxy-3-methylphenyl)-	**60:** 643h
-4-carboxylic acid, 2-(3-carboxyphenyl)-	**54:** 1504g, **55:** 3563d
-4-carboxylic acid, 2-(4-carboxyphenyl)-	**53:** 344i, **55:** 3563d
-4-carboxylic acid, 5-(2-carboxyphenyl)-2-(4-chloro-2-hydroxyphenyl)-, dimethyl ester	**71:** P124445j
-4-carboxylic acid, 5-(2-carboxyphenyl)-2-(4-hydroxy-3-biphenylyl)-, dimethyl ester	**71:** P124445j
-4-carboxylic acid, 5-(2-carboxyphenyl)-2-(2-hydroxyphenyl)-	**71:** P124445j
-4-carboxylic acid, 5-(2-carboxyphenyl)-2-(2-hydroxyphenyl)-, bis(2-ethoxyethyl) ester	**71:** P124445j
-4-carboxylic acid, 5-(2-carboxyphenyl)-2-(2-hydroxyphenyl)-, dimethyl ester	**71:** P124445j
-4-carboxylic acid, 5-(2-carboxyphenyl)-2-(2-hydroxyphenyl)-, dioctadecyl ester	**71:** P124445j
-4-carboxylic acid, 5-(2-carboxyphenyl)-2-(2-hydroxyphenyl)-, dioctyl ester	**71:** P124445j
-4-carboxylic acid, 5-(2-carboxyphenyl)-2-(2-methoxyphenyl)-	**71:** P124445j
-4-carboxylic acid, 1-(4-carboxyphenyl)-5-methyl-	**69:** 19088f, **79:** 5300p
-4-carboxylic acid, 2-(4-carboxyphenyl)-5-methyl-	**77:** P7313c
-4-carboxylic acid, 4-(2-carboxyphenyl)-2-(2-nitrophenyl)-	**23:** 4217[1]
-4-carboxylic acid, 4-(2-carboxyphenyl)-2-(3-nitrophenyl)-	**23:** 4217[1]
-4-carboxylic acid, 4-(2-carboxyphenyl)-2-(4-nitrophenyl)-	**23:** 4217[1]
-4-carboxylic acid, 1-(2-carboxy-α-phenylphenacyl)-, disodium salt	**73:** 56037s
-4-carboxylic acid, 2-(4-carboxyphenyl)-5-phenyl-	**77:** P7313c
-4-carboxylic acid, 5-(2-carboxyphenyl)-1-phenyl-	**21:** 742[4]
-4-carboxylic acid, 5-(2-carboxyphenyl)-2-phenyl-	**36:** 2862[7]
-4-carboxylic acid, 5-(2-carboxy-3-pyridyl)-2-phenyl-	**21:** 2129[9]
-4-carboxylic acid, 5-chloro-1-(3-chlorophenyl)-, ethyl ester	**74:** P22847m
-4-carboxylic acid, 5-chloro-1-(4-chlorophenyl)-, ethyl ester	**74:** P22847m
-4-carboxylic acid, 1-(5-chloro-1,6-dihydro-6-oxo-1-phenyl-4-pyridazinyl)-5-methyl-, ethyl ester	**78:** 136200h

TABLE 4 (*Continued*)

Compound	Reference
4.2. 1,2,3-Triazole-4-Carboxylic Acids and Their Functional Derivatives (*Continued*)	
-4-carboxylic acid, 5-chloro-1-(3-fluorophenyl)-, ethyl ester	**74:** P22847m
-4-carboxylic acid, 5-chloro-1-(4-fluorophenyl)-, ethyl ester	**74:** P22847m
-4-carboxylic acid, 2-(5-chloro-2-hydroxyphenyl)-5-[2-(ethoxycarbonyl)phenyl]-, ethyl ester	**77:** P115464m
-4-carboxylic acid, 2-(5-chloro-2-hydroxyphenyl)-5-methyl-	**76:** P99674b
-4-carboxylic acid, 2-(5-chloro-2-hydroxyphenyl)-5-methyl-, ethyl ester	**76:** P99674b
-4-carboxylic acid, 2-(3-chloro-4-nitrophenyl)-	**58:** 11454f
-4-carboxylic acid, 2-(4-chloro-3-nitrophenyl)-	**58:** 11454f
-4-carboxylic acid, 1-(2-chloro-4-methylphenyl)-5-methyl-	**31:** 3888[1]
-4-carboxylic acid, 2-(3-chlorophenyl)-	**55:** 27071g
-4-carboxylic acid, 2-(3-chlorophenyl)-, ethyl ester	**61:** 5738c
-4-carboxylic acid, 2-(4-chlorophenyl)-	**55:** 27071g
-4-carboxylic acid, 2-(4-chlorophenyl)-, ethyl ester	**61:** 5738c
-4-carboxylic acid, 2-(4-chlorophenyl)-, methyl ester	**61:** 5738c
-4-carboxylic acid, 5-(2-chlorophenyl)-, ethyl ester	**80:** 108452q
-4-carboxylic acid, 5-(4-chlorophenyl)-, ethyl ester	**80:** 108452q
-4-carboxylic acid, 5-(4-chlorophenyl)-, methyl ester	**80:** 133389g
-4-carboxylic acid, 1-[2-(4-chlorophenyl)-1,3-dioxo-2-indanyl]-	**73:** 56037s
-4-carboxylic acid, 1-(2-chlorophenyl)-5-methyl-	**53:** 16120e
-4-carboxylic acid, 1-(4-chlorophenyl)-5-methyl-	**69:** 10402w
-4-carboxylic acid, 1-(4-chlorophenyl)-5-methyl-, ethyl ester	**58:** 12560e
-4-carboxylic acid, 2-(2-chlorophenyl)-5-methyl-, methyl ester	**61:** 10670h, **63:** 4268h
-4-carboxylic acid, 2-(3-chlorophenyl)-5-methyl-	**63:** 4268h
-4-carboxylic acid, 2-(3-chlorophenyl)-5-methyl-, ethyl ester	**63:** 4268h
-4-carboxylic acid, 2-(3-chlorophenyl)-5-methyl-, methyl ester	**63:** 4268h
-4-carboxylic acid, 5-(4-chlorophenyl)-1-phenyl-, ethyl ester	**81:** 3850k
-4-carboxylic acid, 5-(4-chlorophenyl)-1-phenyl-, hydrazide	**81:** 3850k
-4-carboxylic acid, 5-(2-chlorophenyl)-2-phenyl-	**83:** P206291t
-4-carboxylic acid, 5-(4-chlorophenyl)-2-phenyl-	**83:** P206291t
-4-carboxylic acid, 1-(5*H*-cyclohepta[*b*]quinoxalin-8-yl)-5-methyl-, ethyl ester	**76:** 140726k
-4-carboxylic acid, 1-(5′-deoxy-5′-thymidinyl)-	**81:** 163240q
-4-carboxylic acid, 1-(5′-deoxy-5′-thymidinyl)-, ethyl ester	**81:** 163240q
-4-carboxylic acid, 2-(2,4-dibromophenyl)-	**63:** 1850h

TABLE 4 (*Continued*)

Compound	Reference
4.2. 1,2,3-Triazole-4-Carboxylic Acids and Their Functional Derivatives (*Continued*)	
-4-carboxylic acid, 2-(3,4-dibromophenyl)-	**55:** 3563e
-4-carboxylic acid, 2-(2,5-dichlorophenyl)-	**76:** 114090d
-4-carboxylic acid, 2-(3,4-dichlorophenyl)-	**55:** 27071g
-4-carboxylic acid, 1-(2,4-dichlorophenyl)-5-methyl-	**53:** 18946h
-4-carboxylic acid, 1-(2,4-dichlorophenyl)-5-methyl-, ethyl ester	**53:** 18946h
-4-carboxylic acid, 1-(2,5-dichlorophenyl)-5-methyl-	**53:** 18946h
-4-carboxylic acid, 1-[4-(diethylamino)-5-oxo-1,3,6-cycloheptatrien-1-yl]-5-methyl-, ethyl ester, monopicrate	**76:** 140726k
-4-carboxylic acid, 1-(3,4-diiodophenyl)-5-methyl-	**57:** 7255b, **59:** 6408d
-4-carboxylic acid, 5-[4-(dimethylamino)phenyl]-, methyl ester	**80:** 3439n
-4-carboxylic acid, 1-(3,3-dimethyl-1-butenyl)-, methyl ester, (E)-	**72:** 132635g
-4-carboxylic acid, 1-(3,3-dimethyl-1-butenyl)-5-methyl-, (E)-	**72:** 121447w
-4-carboxylic acid, 1-(3,3-dimethyl-1-butenyl)-5-phenyl-, (E)-	**72:** 121447w
-4-carboxylic acid, 2-(2,5-dimethylphenyl)-	**60:** 643h
-4-carboxylic acid, 1-(2,6-dimethylphenyl)-, ethyl ester	**79:** 104890n
-4-carboxylic acid, 1-(4,6-dimethyl-2-pyrimidinyl)-, methyl ester	**71:** 81296s
-4-carboxylic acid, 2-(2,4-dinitrophenyl)-	**72:** 2972d
-4-carboxylic acid, 1-(2,4-dinitrophenyl)-5-methyl-, ethyl ester	**61:** 5633a
-4(or 5)-carboxylic acid, 1-(2,2-dinitropropyl)- (?)	**74:** 99215m
-4-carboxylic acid, 1-(1,3-dioxo-2-phenyl-2-indanyl)-	**67:** 53900h, **73:** 56037s
-4-carboxylic acid, 1,5-diphenyl-	**44:** 1102c, **64:** 9713h, **67:** 100071a, **76:** 113139q, **81:** 3850k
-4-carboxylic acid, 1,5-diphenyl-, ethyl ester	**81:** 3850k
-4-carboxylic acid, 1,5-diphenyl-, hydrazide	**81:** 3850k
-4-carboxylic acid, 1,5-diphenyl-, methyl ester	**64:** 9713h, **67:** 100071a, **76:** 5938d
-4-carboxylic acid, 2,5-diphenyl-	**25:** 1827[4], **76:** 114090d
-4-carboxylic acid, 2,5-diphenyl-, ethyl ester	**77:** P7313c
-4-carboxylic acid, 1-[5-(ethoxycarbonyl)-2-oxo-4,6-diphenyl-3-cyclohexen-1-yl]-5-phenyl-, ethyl ester, (1α, 5α, 6β)-	**83:** 193178j
-4-carboxylic acid, 1-[5-(ethoxycarbonyl)-4-oxo-2,6-diphenyl-2-cyclohexen-1-yl]-5-methyl-, ethyl ester, (1α, 5α, 6β)-	**83:** 193178j
-4-carboxylic acid, 1-[5-(ethoxycarbonyl)-4-oxo-6-(3-nitrophenyl)-2-phenyl-2-cyclohexen-1-yl]-5-methyl-, ethyl ester, (1α, 5α, 6β)-	**83:** 193178j

TABLE 4 (*Continued*)

Compound	Reference
4.2. 1,2,3-Triazole-4-Carboxylic Acids and Their Functional Derivatives (*Continued*)	
-4-carboxylic acid, 1-[3-(ethoxycarbonyl)-4-hydroxy-4-methyl-6-oxo-2-phenylcyclohexyl]-5-methyl-, ethyl ester, (1α, 2β, 3α, 4α)-	**83:** 193178j
-4-carboxylic acid, 1-[5-(ethoxycarbonyl)-2-hydroxy-6-(3-nitrophenyl)-4-oxo-2-phenylcyclohexyl]-5-methyl-, ethyl ester, (1α, 2α, 5α, 6β)-	**83:** 193178j
-4-carboxylic acid, 1-[3-(ethoxycarbonyl)-4-hydroxy-6-oxo-2,4-diphenylcyclohexyl]-5-phenyl-, ethyl ester, (1α, 2β, 3α, 4α)-	**83:** 193178j
-4-carboxylic acid, 1-[5-(ethoxycarbonyl)-2-hydroxy-4-oxo-2,6-diphenylcyclohexyl]-5-methyl-, ethyl ester, (1α, 2α, 5α, 6β)-	**83:** 193178j
-4-carboxylic acid, 1-[5-(ethoxycarbonyl)-4-methyl-2-oxo-6-phenyl-3-cyclohexen-1-yl]-5-methyl-, ethyl ester, (1α, 5α, 6β)-	**83:** 193178j
-4-carboxylic acid, 1-(4-ethoxy-5-oxo-1,3,6-cycloheptatrien-1-yl)-5-phenyl-, ethyl ester	**79:** 5300p
-4-carboxylic acid, 2-(3-ethoxy-3-oxo-1-phenyl-1-propenyl)-5-phenyl-, ethyl ester	**80:** 108451p
-4-carboxylic acid, 5-ethyl-, ethyl ester	**80:** 146082b
-4-carboxylic acid, 1-[4-(ethylamino)-5-oxo-1,3,6-cycloheptatrien-1-yl]-5-methyl-, ethyl ester	**76:** 140726k
-4-carboxylic acid, 1-[2-[2-(3-ethyl-4-oxo-2-thioxo-5-thiazolidinylidene)ethylidene]-3-ethyl-5-benzothiazolidinyl]-5-methyl-	**74:** 48048b
-4(or 5)-carboxylic acid, 1-(2-fluoro-2,2-dinitroethyl)- (?)	**74:** 99215m
-4-carboxylic acid, 2-(2-fluorophenyl)-	**80:** 48283y
-4-carboxylic acid, 2-(3-fluorophenyl)-	**59:** 14095f
-4-carboxylic acid, 2-(3-fluorophenyl)-, ethyl ester	**61:** 5738c
-4-carboxylic acid, 2-(4-fluorophenyl)-	**59:** 14095f
-4-carboxylic acid, 2-(4-fluorophenyl)-, ethyl ester	**61:** 5738c
-4-carboxylic acid, 5-(2-fruanyl)-, methyl ester	**80:** 3439n
-4-carboxylic acid, 1-β-D-galactopyranosyl-, methyl ester, 2′,3′,4′,6′-tetraacetate ester	**74:** 112402s
-4-carboxylic acid, 1-β-D-glucopyranosyl-, methyl ester, 2′,3′,4′,6′-tetraacetate ester	**74:** 112402s, **77:** 102129f
-4-carboxylic acid, 1-α-D-glucopyranosyl-, methyl ester, 3′,4′,6′-triacetate ester	**74:** 112402s
-4-carboxylic acid, 5-heptadecyl-2-(2-hydroxyphenyl)-	**76:** 114090d
-4-carboxylic acid, 5-heptadecyl-2-(2-methoxy-5-methylphenyl)-, ethyl ester	**77:** P7313c
-4-carboxylic acid, 1-(4-hydrazino-5-oxo-1,3,6-cycloheptatrien-1-yl)-5-methyl-, ethyl ester	**76:** 140726k
-4-carboxylic acid, 2-(2-hydroxy-4,5-dimethylphenyl)-5-methyl-	**76:** P99674b
-4-carboxylic acid, 2-(2-hydroxy-4,5-dimethylphenyl)-5-methyl-, ethyl ester	**76:** P99674b

TABLE 4 (*Continued*)

Compound	Reference
4.2. 1,2,3-Triazole-4-Carboxylic Acids and Their Functional Derivatives (*Continued*)	
-4-carboxylic acid, 5-(3-hydroxy-4-methoxyphenyl)-, methyl ester	**80:** 3439n
-4-carboxylic acid, 2-(2-hydroxy-4-methoxyphenyl)-5-methyl-	**76:** P99674b
-4-carboxylic acid, 2-[2-hydroxy-4-(methylsulfonyl)-phenyl]-5-methyl-	**77:** P7313c
-4-carboxylic acid, 1-(4-hydroxy-5-oxo-1,3,6-cycloheptatrien-1-yl)-5-methyl-	**76:** 140726k
-4-carboxylic acid, 1-(4-hydroxy-5-oxo-1,3,6-cycloheptatrien-1-yl)-5-phenyl-	**76:** 140726k
-4-carboxylic acid, 2-(2-hydroxyphenyl)-	**76:** 114090d
-4-carboxylic acid, 5-(4-hydroxyphenyl)-, methyl ester	**80:** 3439n
-4-carboxylic acid, 2-(2-hydroxyphenyl)-5-methyl-	**76:** 114090d
-4-carboxylic acid, 2-(2-hydroxyphenyl)-5-methyl-, ethyl ester	**71:** P124445j
-4-carboxylic acid, 2-(2-hydroxyphenyl)-5-phenyl-, ethyl ester	**71:** P124445j
-4-carboxylic acid, 2-(2-hydroxy-5-sulfophenyl)-5-methyl-, ethyl ester, monosodium salt	**76:** P99674b
-4-carboxylic acid, 5-hydroxy-1-tetrazolo[5,1-*a*]-phthalazin-6-yl-, ethyl ester, sodium salt	**81:** 37518y
-4-carboxylic acid, 1-(2-iodo-3,3-dimethylbutyl)-, methyl ester	**72:** 132635g
-4-carboxylic acid, 1-[α-(iodomethyl)benzyl]-, methyl ester	**72:** 132635g
-4-carboxylic acid, 2-(3-iodophenyl)-	**55:** 27071g
-4-carboxylic acid, 2-(4-iodophenyl)-	**55:** 27071g
-4-carboxylic acid, 1-(2-iodophenyl)-5-methyl-	**57:** 7255c(?), **59:** 6408d(?), **69:** 10402w
-4-carboxylic acid, 1-(3-iodophenyl)-5-methyl-	**57:** 7255b, **59:** 6408d
-4-carboxylic acid, 1-(4-iodophenyl)-5-methyl-	**57:** 7255b, **59:** 6408d, **69:** 10402w
-4-carboxylic acid, 1-imidazo[1,2-*b*]pyridazin-6-yl-5-methyl-, ethyl ester	**82:** 16745a
-4-carboxylic acid, 1-imidazo[1,2-*b*]pyridazin-6-yl-5-phenyl-, ethyl ester	**82:** 16745a
-4-carboxylic acid, 5-isobutyl-1-(4-nitrophenyl)-	**58:** 12560h
-4-carboxylic acid, 5-isopropyl-, ethyl ester	**80:** 146082b
-4-carboxylic acid, 1-[4-(methoxycarbonyl)phenyl]-5-methyl-, ethyl ester	**69:** 19088f, **79:** 5300p
-4-carboxylic acid, 2-(5-methoxy-3-methyl-4-isoazolyl)-5-methyl-, methyl ester	**81:** 135286r
-4-carboxylic acid, 2-(2-methoxy-5-methylphenyl)-5-phenyl-, ethyl ester	**77:** P7313c
-4-carboxylic acid, 1-(4-methoxy-5-oxo-1,3,6-cyclo-heptatrien-1-yl)-5-methyl-, ethyl ester	**69:** 19088f, **79:** 5300p
-4-carboxylic acid, 1-(4-methoxy-5-oxo-1,3,6-cyclo-heptatrien-1-yl)-5-methyl-, methyl ester	**79:** 5300p

TABLE 4 (*Continued*)

Compound	Reference
4.2. 1,2,3-Triazole-4-Carboxylic Acids and Their Functional Derivatives (*Continued*)	
-4-carboxylic acid, 1-(4-methoxyphenyl)-	**64:** 11198h
-4-carboxylic acid, 2-(2-methoxyphenyl)-	**55:** 3562i
-4-carboxylic acid, 2-(3-methoxyphlenyl)-	**55:** 3562i, **76:** 114090d
-4-carboxylic acid, 2-(4-methoxyphenyl)-	**55:** 3562i
-4-carboxylic acid, 5-(4-methoxyphenyl)-, ethyl ester	**80:** 108452q
-4-carboxylic acid, 1-[2-(4-methoxyphenyl)-1,3-dioxo-2-indanyl]-	**73:** 56037s
-4-carboxylic acid, 1-(4-methoxyphenyl)-5-methyl-	**74:** 141988t
-4-carboxylic acid, 2-(2-methoxyphenyl)-5-methyl-	**71:** P124445j
-4-carboxylic acid, 2-(2-methoxyphenyl)-5-methyl-, ethyl ester	**77:** P7313c
-4-carboxylic acid, 1-(4-methoxyphenyl)-5-(4-nitrophenyl)-	**74:** 141988t
-4-carboxylic acid, 1-(4-methoxyphenyl)-5-phenyl-	**62:** 16248g, **74:** 141988t
-4-carboxylic acid, 5-(4-methoxyphenyl)-1-phenyl-, ethyl ester	**81:** 3850k
-4-carboxylic acid, 5-(4-methoxyphenyl)-1-phenyl-, hydrazide	**81:** 3850k
-4-carboxylic acid, 1-methyl-	**42:** 2599h, **50:** 9392f, **55:** 17627a, **84:** 122189y
-4-carboxylic acid, 1-methyl-, ethyl ester	**77:** 126981d
-4-carboxylic acid, 1-methyl-, methyl ester	**84:** 122189y
-4-carboxylic acid, 5-methyl-	**26:** 455[3], **36:** 772[1], **50:** 9392g, **57:** 5911h, **58:** 13944a
-4-carboxylic acid, 5-methyl-, ethyl ester	**58:** 13944a, **61:** 5633b
-4-carboxylic acid, 1-(1-methyl-1*H*-benzimidazol-2-yl)-, methyl ester	**79:** 66246x
-4-carboxylic acid, 2-(1-methyl-1*H*-benzimidazol-2-yl)-, methyl ester	**79:** 66246x
-4-carboxylic acid, 5-methyl-1-(2-methyl-6-benzothiazolyl)-	**56:** P4291g
-4-carboxylic acid, 5-methyl-1-(2-methylphenyl)-	**53:** 18946h
-4-carboxylic acid, 5-methyl-2-(2-methylphenyl)-	**79:** P18720a
-4-carboxylic acid, 5-methyl-2-(4-methylphenyl)-	**78:** 16102f
-4-carboxylic acid, 5-methyl-1-(2-nitrophenyl)-	**58:** 12560f
-4-carboxylic acid, 5-methyl-1-(4-nitrophenyl)-	**58:** 12560e, **62:** 16247h, **67:** 82171b, **69:** 10402w
-4-carboxylic acid, 5-methyl-1-(4-nitrophenyl)-, ethyl ester	**58:** 12560d, **67:** 82171b
-4-carboxylic acid, 5-methyl-1-(4-nitrophenyl)-, methyl ester	**67:** 82171b, **80:** 95655f
-4-carboxylic acid, 5-methyl-2-(4-nitrophenyl)-	**79:** P18720a
-4-carboxylic acid, 5-methyl-2-(4-nitrophenyl)-, ethyl ester	**79:** P18720a
-4-carboxylic acid, 2-(3-methyl-4-nitrophenyl)-	**58:** 11454f
-4-carboxylic acid, 5-methyl-1-(4-oxo-4*H*-pyrido-[1,2-*a*]pyrimidin-2-yl)-, ethyl ester	**83:** 9958q

TABLE 4 (*Continued*)

Compound	Reference
4.2. 1,2,3-Triazole-4-Carboxylic Acids and Their Functional Derivatives (*Continued*)	
-4-carboxylic acid, 5-methyl-1-phenyl-	**37:** 5404[9], **44:** 1102c, **52:** 17246b, **59:** 6408d, **69:** 10402w
-4-carboxylic acid, 5-methyl-1-phenyl-, 2-benzylhydrazide	**60:** 9268c
-4-carboxylic acid, 5-methyl-1-phenyl-, benzylidenehydrazide	**60:** 9268d
-4-carboxylic acid, 5-methyl-1-phenyl-, ethyl ester	**37:** 5404[9], **73:** 14769b
-4-carboxylic acid, 5-methyl-1-phenyl-, hydrazide	**60:** 9268c
-4-carboxylic acid, 5-methyl-1-phenyl-, O-free radical	**52:** 17246a
-4-carboxylic acid, 5-methyl-2-phenyl-	**35:** 3638[4], **42:** 2970e, **76:** 114090d, **78:** 16102f
-4-carboxylic acid, 5-methyl-2-phenyl-, ethyl ester	**84:** P181615g
-4-carboxylic acid, 5-methyl-2-phenyl-, methyl ester	**65:** 16960d
-4-carboxylic acid, 5-methyl-2-phenyl-, 2-phenylhydrazide	**78:** 16102f
-4-carboxylic acid, 2-(4-methylphenyl)-	**54:** 1504g
-4-carboxylic acid, 5-methyl-1-(1-phenylethenyl)-	**72:** 121447w
-4-carboxylic acid, 5-(4-methylphenyl)-1-phenyl-, ethyl ester	**81:** 3850k
-4-carboxylic acid, 5-(4-methylphenyl)-1-phenyl-, hydrazide	**81:** 3850k
-4-carboxylic acid, 5-methyl-1-(4-quinolyl)-, ethyl ester	**58:** 4545d
-4-carboxylic acid, 5-methyl-1-(4-quinolyl)-, oxide	**58:** 4545d
-4-carboxylic acid, 5-methyl-1-styryl-, (E)-	**72:** 121447w
-4-carboxylic acid, 5-methyl-1-tetrazolo[5,1-*a*]-phthalazin-6-yl-, ethyl ester	**82:** 16745a
-4-carboxylic acid, 5-methyl-1-tetrazolo[1,5-*b*]-pyridazin-6-yl-, ethyl ester	**82:** 16745a
-4-carboxylic acid, 5-methyl-1-(1,2,4-triazolo[4,3-*b*]-pyridazin-6-yl)-, ethyl ester	**82:** 16745a
-4-carboxylic acid, 5-methyl-2-(trimethylsilyl)-, trimethylsilyl ester	**68:** 13061z
-4-carboxylic acid, 1-(4-nitrophenyl)-	**47:** 8738c, **58:** 12561c, **64:** 3523g
-4-carboxylic acid, 1-(4-nitrophenyl)-, methyl ester	**58:** 12561c
-4-carboxylic acid, 2-(4-nitrophenyl)-	**48:** 5805c, **58:** 11454f
-4-carboxylic acid, 2-(4-nitrophenyl)-, ethyl ester	**61:** 5738c
-4-carboxylic acid, 2-(4-nitrophenyl)-, methyl ester	**61:** 5738c
-4-carboxylic acid, 5-(2-nitrophenyl)-, ethyl ester	**77:** 164604r, **80:** 108452q
-4-carboxylic acid, 5-(3-nitrophenyl)-, ethyl ester	**77:** 164604r, **80:** 108452q
-4-carboxylic acid, 5-(4-nitrophenyl)-, ethyl ester	**77:** 164604r, **80:** 108452q
-4-carboxylic acid, 5-(4-nitrophenyl)-, methyl ester	**80:** 3439n
-4-carboxylic acid, 5-(3-nitrophenyl)-2-(3-oxo-1-phenyl-1-propenyl)-, ethyl ester, (E)-	**80:** 108451p
-4-carboxylic acid, 5-(3-nitrophenyl)-2-(3-oxo-1-phenyl-1-propenyl)-, ethyl ester, (Z)-	**80:** 108451p

TABLE 4 (*Continued*)

Compound	Reference
4.2. 1,2,3-Triazole-4-Carboxylic Acids and Their Functional Derivatives (*Continued*)	
-4-carboxylic acid, 5-(4-nitrophenyl)-2-(3-oxo-1-phenyl-1-propenyl)-, ethyl ester, (E)-	**80:** 108451p
-4-carboxylic acid, 5-(4-nitrophenyl)-2-(3-oxo-1-phenyl-1-propenyl)-, ethyl ester, (Z)-	**80:** 108451p
-4-carboxylic acid, 1-(4-nitrophenyl)-5-phenyl-	**62:** 16248a
-4-carboxylic acid, 5-(4-nitrophenyl)-1-phenyl-	**74:** 141988t
-4-carboxylic acid, 2-(3-oxo-1-phenyl-1-propenyl)-5-phenyl-, ethyl ester, (E)-	**80:** 108451p
-4-carboxylic acid, 2-(3-oxo-1-phenyl-1-propenyl)-5-phenyl-, ethyl ester, (Z)-	**80:** 108451p
-4-carboxylic acid, 1-(4-oxo-4*H*-pyrido[1,2-*a*]-pyrimidin-2-yl)-5-phenyl-, ethyl ester	**83:** 9958q
-4-carboxylic acid, 1-phenyl-	**24:** 592^9, **37:** 119^2, **48:** 2685a, **50:** 4924f, **64:** 3523g, **64:** 5076h, **75:** 88891y
-4-carboxylic acid, 1-phenyl-, *tert*-butyl ester	**63:** P7019b
-4-carboxylic acid, 1-phenyl-, ethyl ester	**64:** 11198b, **67:** 43759e, **79:** 104890n
-4-carboxylic acid, 1-phenyl-, methyl ester	**44:** 1102d, **64:** 5076e, **64:** 11198b, **75:** 88891y
-4-carboxylic acid, 2-phenyl-	**43:** 2619a, **53:** 344i, **58:** 11454g, **72:** 38736u, **76:** 114090d
-4-carboxylic acid, 2-phenyl-, ethyl ester	**43:** 2619a, **61:** 5738c
-4-carboxylic acid, 2-phenyl-, hydrazide	**50:** 362h
-4-carboxylic acid, 5-phenyl-	**61:** 14665b, **62:** 6441b, **77:** 126517g
-4-carboxylic acid, 5-phenyl-, ethyl ester	**61:** 14664h, **77:** 126517g, **80:** 108452q, **80:** 146082b
-4-carboxylic acid, 5-phenyl-, methyl ester	**80:** 3439n
-4-carboxylic acid, 1-(1-phenylethenyl)-, methyl ester	**72:** 132635g
-4-carboxylic acid, 5-phenyl-1-(1-phenylethenyl)-	**72:** 121447w, **82:** 156182q
-4-carboxylic acid, 1-(1-phenylpropenyl)-, methyl ester, (E)-	**72:** 132635g
-4-carboxylic acid, 5-phenyl-1-tetrazolo[5,1-*a*]-phthalazin-6-yl-, ethyl ester	**82:** 16745a
-4-carboxylic acid, 5-phenyl-1-tetrazolo[1,5-*b*]-pyridazin-6-yl-, ethyl ester	**82:** 16745a
-4-carboxylic acid, 5-phenyl-1-(1,2,4-triazolo-[4,3-*b*]pyridazin-6-yl)-, ethyl ester	**82:** 16745a
-4-carboxylic acid, 5-propyl-, ethyl ester	**80:** 146082b
-4-carboxylic acid, 1-(2-pyridyl)-, methyl ester	**71:** 81296s
-4-carboxylic acid, 1-β-D-ribofuranosyl-, methyl ester	**84:** 122189y
-4-carboxylic acid, 1-β-D-ribofuranosyl-, methyl ester, 2′,3′,5′-tribenzoate ester	**74:** 112402s, **78:** 84715h
-4-carboxylic acid, 5-β-D-ribofuranosyl-, methyl ester, 2′,3′,5′-tribenzoate ester	**74:** 112402s, **78:** 84715h

TABLE 4 (*Continued*)

Compound	Reference
4.2. 1,2,3-Triazole-4-Carboxylic Acids and Their Functional Derivatives (*Continued*)	
-4-carboxylic acid, 5-β-D-ribofuranosyl-, methyl ester	**84:** 44571a
-4-carboxylic acid, 1-styryl-, methyl ester, (E)-	**72:** 132635g
-4-carboxylic acid, 1-(2,3,5-tri-*O*-benzoyl-β-D-ribofuranosyl)-, ethyl ester	**77:** 126981d
-4-carboxylic acid, 1-(2,3,5-tri-*O*-benzoyl-β-D-ribofuranosyl)-, methyl ester	**84:** 122189y
-4-carboxylic acid, 2-(2,3,5-tri-*O*-benzoyl-β-D-ribofuranosyl)-, methyl ester	**78:** 84715h
-4-carboxylic acid, 5-(3,4,5-trimethoxyphenyl)-, methyl ester	**80:** 3439n
-4-carboxylic acid, 2-(trimethylsilyl)-, trimethylsilyl ester	**68:** 13061z
4-[1-[4-[3-(dimethylamino)propyl]azacyclohexyl]-carbonyl]-2-phenyl-	**68:** P59441b
4-[1-[4-[3-(dimethylamino)propyl]-1,4-diazacyclo-hexyl]carbonyl]-2-(2,4-dinitrophenyl)-	**68:** P59611g
4.3. 1,2,3-Triazole-4,5-Dicarboxylic Acids and Their Functional Derivatives	
-1-acetamide, 4,5-dicarbamoyl-	**24:** 3215[1]
-1-acetic acid, 4,5-dicarboxy-	**24:** 3215[1]
-1-acetic acid, 4,5-dicarboxy-α-methyl-	**24:** 3215[2]
-4-carboxamide, 5-cyano-	**84:** P17361a
-4-carboxamide, 5-cyano-, ethyl ester	**84:** P175141h
-4-carboximidic acid, 5-cyano-, ethyl ester	**51:** 4364f
-4-carboxylic acid, 5-carbamoyl-	**51:** 14697h
-4-carboxylic acid, 5-cyano-, ethyl ester	**51:** 4364f
-4,5-dicarbonitrile	**22:** 4475[8], **22:** 423[5], **31:** 6237[4], **31:** 7849[3]
-4,5-dicarbonitrile, adduct with benzenamine (1:1)	**83:** P43334t
-4,5-dicarbonitrile, adduct with morpholine (1:1)	**83:** P43334t
-4,5-dicarbonitrile, adduct with piperidine (1:1)	**83:** P43334t
-4,5-dicarbonitrile, adduct with pyridine (1:1)	**83:** P43334t
-4,5-dicarbonitrile, 1-(aminomethyl)-	**83:** P43333s
-4,5-dicarbonitrile, 1-(hydroxymethyl)-	**83:** P43333s
-4,5-dicarbonitrile, 1-methyl-	**82:** 73727v
-4,5-dicarbonitrile, 2-methyl-	**82:** 73727v
-4,5-dicarbonitrile, 1-(4-morpholinylmethyl)-	**83:** P43333s
-4,5-dicarbonitrile, 1-(1-piperidinylmethyl)-	**83:** P43333s
-4,5-dicarbonitrile, 1-(1-piperazinylmethyl)-	**83:** P43333s
-4,5-dicarbonyl azide, 2-phenyl-	**21:** 2690[4]
-4,5-dicarbonyl dichloride	**82:** 140021a
-4,5-dicarbonyl dichloride, 2-[4-(chloroformyl)phenyl-	**21:** 2690[5]
-4,5-dicarbonyl dichloride, 1-methyl-	**82:** 73727v
-4,5-dicarbonyl dichloride, 2-methyl-	**82:** 73727v
-4,5-dicarbonyl dichloride, 1-phenyl-	**37:** 5404[9]
-4,5-dicarbonyl dichloride, 2-phenyl-	**21:** 2690[4]
-4,5-dicarboxamide (?)	**52:** 16359e
-4,5-dicarboxamide, 1-benzyl-	**24:** 3232[1]

TABLE 4 (*Continued*)

Compound	Reference
4.3. 1,2,3-Triazole-4,5-Dicarboxylic Acids and Their Functional Derivatives (*Continued*)	
-4,5-dicarboxamide, 2-(4-carbamylphenyl)-	**21:** 2690[5]
-4,5-dicarboxamide, 1-(5′-deoxy-5′-thymidinyl)-	**81:** 163240q
-4,5-dicarboxamide, 1-(5′-deoxy-5′-thymidinyl)- *N,N*′-dibenzoyl	**81:** 163240q
-4,5-dicarboxamide, 1-(5′-deoxy-5′-thymidinyl)- *N,N*′-di-(2-hydroxyethyl)	**81:** 163240q
-4,5-dicarboxamide, 1-(5′-deoxy-5′-thymidinyl)- *N,N*′-dipropyl-	**81:** 163240q
-4,5-dicarboxamide, 1-β-D-glucopyranosyl-	**52:** 16359d
-4,5-dicarboxamide, 2-phenyl-(?)	**21:** 2269[1], **21:** 2690[4]
-4,5-dicarboxamide, 1-β-D-ribofuranosyl-	**53:** 6245c
-4,5-dicarboxamide, 1-β-D-xylopyranosyl-	**52:** 16359d
-4,5-dicarboxamide, 2-(4-nitrophenyl)-	**21:** 2690[4]
-4,5-dicarboxanilide, 1-phenyl-	**37:** 5404[9]
-4,5-dicarboxanilide, 2-phenyl-	**21:** 2690[4]
-4,5-dicarboxanilide, 2-(4-phenylcarbamylphenyl)-	**21:** 2690[5]
-4,5-dicarboxylato-N^1-, iron, dicarbonyl($*^5$-2,4-cyclopentadien-1-yl)(dimethyl),	**82:** 16926k
-1,4-dicarboxylic acid, isopropyl ester (?)	**61:** 5626h, **63:** 16337h
-1,5-dicarboxylic acid, isopropyl ester (?)	**63:** 16337h,
-4,5-dicarboxylic acid	**31:** 7849[3], **50:** 15517e, **78:** 111025s, **82:** 140021a, **83:** 97148d
-4,5-dicarboxylic acid, dibutyl ester	**82:** 140021a
-4,5-dicarboxylic acid, diethyl ester	**82:** 140021a
-4,5-dicarboxylic acid, diisobutyl ester	**82:** 140021a
-4,5-dicarboxylic acid, diisopentyl ester	**82:** 140021a
-4,5-dicarboxylic acid, diisopropyl ester	**82:** 140021a
-4,5-dicarboxylic acid, dimethyl ester	**60:** 5537a, **62:** 13138h, **73:** 25364s, **80:** 108452q, **82:** 140021a, **83:** 10269x
-4,5-dicarboxylic acid, dipentyl ester	**82:** 140021a
-4,5-dicarboxylic acid, dipropyl ester	**82:** 140021a
-4,5-dicarboxylic acid, hydrazide	**55:** P571c
-4,5-dicarboxylic acid, monoisopropyl ester	**83:** 97148d, **85:** 78055w
-4,5-dicarboxylic acid, monopotassium salt	**82:** 49831d, **82:** 49832e
-4,5-dicarboxylic acid, monosilver(1+) salt	**80:** 14221m
-4,5-dicarboxylic acid, 1-(4-amino-5-bromo-6-phenyl-2-pyrimidinyl)-, dimethyl ester	**72:** P55493e
-4,5-dicarboxylic acid, 1-(4-amino-5-butyl-6-phenyl-2-pyrimidinyl)-, dimethyl ester	**72:** P55493e
-4,5-dicarboxylic acid, 1-(5-amino-1,6-dihydro-6-oxo-1-phenyl-4-pyridazinyl)-, dimethyl ester	**78:** 72036z
-4,5-dicarboxylic acid, 1-[4-amino-5-(2-ethoxyethyl)-6-phenyl-2-pyrimidinyl]-, dimethyl ester	**72:** P55493e
-4,5-dicarboxylic acid, 1-(4-amino-5-methyl-6-phenyl-2-pyrimidinyl)-, dimethyl ester	**72:** P55493e
-4,5-dicarboxylic acid, 1-[2-(aminophenylmethyl)-4-chlorophenyl]-, diethyl ester	**82:** P4322x

TABLE 4 (*Continued*)

Compound	Reference
4.3. 1,2,3-Triazole-4,5-Dicarboxylic Acids and Their Functional Derivatives (*Continued*)	
-4,5-dicarboxylic acid, 2-(4-arsonophenyl)-	**40:** 7191[8]
-4,5-dicarboxylic acid, 1-(4-azido-5-bromo-6-phenyl-2-pyrimidinyl]-, dimethyl ester	**72:** P55493e
-4,5-dicarboxylic acid, 1-[4-azido-5-(2-butynyl)-6-phenyl-2-pyrimidinyl]-, dimethyl ester	**72:** P55493e
-4,5-dicarboxylic acid, 1-[4-azido-5-(2-ethoxyethyl)-6-phenyl-2-pyrimidinyl]-, dimethyl ester	**72:** P55493e
-4,5-dicarboxylic acid, 1-(4-azido-5-methyl-6-phenyl-2-pyrimidinyl)-, dimethyl ester	**72:** P55493e
-4,5-dicarboxylic acid, 1-[4-azido-6-phenyl-5-(2-propynl)-2-pyrimidinyl]-, dimethyl ester	**72:** P55493e
-4,5-dicarboxylic acid, 1-(2-benzothiazolyl)-, dimethyl ester	**79:** P78815v
-4,5-dicarboxylic acid, 1-benzyl-	**24:** 3232[1], **50:** 15517d
-4,5-dicarboxylic acid, 1-benzyl-, bis-(2-nitroethyl) ester	**55:** 23505a
-4,5-dicarboxylic acid, 1-[2,3-bis(methoxycarbonyl)-1,4-dimethyl-2-cyclobuten-1-yl]-, dimethyl ester	**79:** 92123z
-4,5-dicarboxylic acid, 1-[9,10-bis(methoxycarbonyl)-4-iodotricyclo[4.2.2.0^{2,5}]dec-7-en-3-yl]-, dimethyl ester, (1α, 2α, 3α, 4α, 5α, 6α, 9R*, 10S*)-	**82:** 170175w
-4,5-dicarboxylic acid, 1-[2,3-bis(methoxycarbonyl)-1-methyl-1,3-pentadienyl]-, dimethyl ester, (E,E)-	**79:** 92123z
-4,5-dicarboxylic acid, 1-[2,3-bis(methoxycarbonyl)-1-methyl-1,3-pentadienyl]-, dimethyl ester, (Z,Z)-	**79:** 92123z
-4,5-dicarboxylic acid, 1-[9,10-bis(methoxycarbonyl)-tricyclo[4.2.2.0^{2,5}]dec-7-en-3-yl]-, dimethyl ester	**84:** 4547r
-4,5-dicarboxylic acid, 1-[9,10-bis(methoxycarbonyl)-tricyclo[4.2.2.0^{2,5}]dec-7-en-3-yl-, dimethyl ester (1α, 2α, 3α, 4α, 5α, 6α, 9R*, 10S*)-	**84:** 4328v, **84:** 4547r
-4,5-dicarboxylic acid, 1-[9,10-bis(methoxycarbonyl)-tricyclo[4.2.2.0^{2,5}]dec-7-en-3-yl]-, dimethyl ester, (1α, 2α, 3α, 4α, 5α, 6α, 9S*, 10R*)-	**84:** 4547r
-4,5-dicarboxylic acid, 1-butyl-	**50:** 4129c
-4,5-dicarboxylic acid, 1-[5-(butylamino)-1,6-dihydro-6-oxo-1-phenyl-4-pyridazinyl]-, dimethyl ester	**78:** 72036z
-1,4(or 1,5)-dicarboxylic acid, 5(or 4)-carbamoyl-, dimethyl ester	**24:** 3215[3]
-4,5-dicarboxylic acid, 1-carbamoyl-, dimethyl ester	**24:** 3215[3]
-4,5-dicarboxylic acid, 1-[α-(carboxyiodomethyl)-benzyl]-, trimethyl ester, *erythro-*	**72:** 132635g
-4,5-dicarboxylic acid, 2-(4-carboxyphenyl)-	**21:** 2690[5]
-4,5-dicarboxylic acid, 2-(4-carboxyphenyl)-, triethyl ester	**21:** 2690[5]
-4,5-dicarboxylic acid, 1-(5-chloro-1,6-dihydro-1-methyl-6-oxo-4-pyridazinyl)-, diethyl ester	**78:** 72035y
-4,5-dicarboxylic acid, 1-(5-chloro-1,6-dihydro-1-methyl-6-oxo-4-pyridazinyl)-, dimethyl ester	**78:** 72035y

TABLE 4 (*Continued*)

Compound	Reference
4.3. 1,2,3-Triazole-4,5-Dicarboxylic Acids and Their Functional Derivatives (*Continued*)	
-4,5-dicarboxylic acid, 1-(5-chloro-1,6-dihydro-6-oxo-1-phenyl-4-pyridazinyl)-, dimethyl ester	**74:** 125600e, **78:** 72036z
-4,5-dicarboxylic acid, 1-(5-chloro-1,6-dihydro-6-oxo-4-pyridazinyl)-, diethyl ester	**78:** 72035y
-4,5-dicarboxylic acid, 1-(5-chloro-1,6-dihydro-6-oxo-4-pyridazinyl)-, dimethyl ester	**78:** 72035y
-4,5-dicarboxylic acid, 1-[4-chloro-2-[(hydroxyimino)-phenylmethyl]phenyl]-, dimethyl ester, (E)-	**84:** 74239r
-4,5-dicarboxylic acid, 1-[4-chloro-2-[(hydroxyimino)-phenylmethyl]phenyl]-, dimethyl ester, (Z)-	**84:** 74239r
-4,5-dicarboxylic acid, 1-[5-(α-cyanostyryl)-2-methylphenyl]-, dimethyl ester	**53:** 13101e
-4,5-dicarboxylic acid, 1-(2,4,6-cycloheptatrien-1-yl)-, dimethyl ester	**62:** 13138g
-4,5-dicarboxylic acid, 1-(1,3-cyclooctadien-2-yl)-, dimethyl ester	**72:** 132635g
-4,5-dicarboxylic acid, 1-(1,3,3a,4,4a,5,6,6a,7,7a-decahydro-6-iodo-1,3-dioxo-4,7-ethenocyclo-but[*f*]isobenzofuran-5-yl)-, dimethyl ester, (3aα, 4β, 4aα, 5α, 6α, 6aα, 7β, 7aα)-	**84:** 4547r
-4,5-dicarboxylic acid, 1-(1,3,3a,4,4a,5,6,6a,7,7a-decahydro-6-iodo-1,3-dioxo-4,7-ethenocyclobut-[*f*]isobenzofuran-5-yl)-, dimethyl ester, (3aα, 4β, 4aα, 5β, 6α, 6aα, 7β, 7aα)-	**81:** 105136z
-4,5-dicarboxylic acid, 1-(5′-deoxy-5′-thymidinyl)-, dimethyl ester	**81:** 163240q
-4,5-dicarboxylic acid, 1-(5′-deoxy-5′-thymidinyl)-, dihydrazide	**81:** 163240q
-4,5-dicarboxylic acid, 1-(5*H*-dibenzo[*a,d*]cyclo-hepten-5-yl)-, dimethyl ester	**74:** P53342x
-4,5-dicarboxylic acid, 1-(7,8-dibromo-2-iodo-bicyclo[4.2.0]oct-4-en-3-yl)-, dimethyl ester	**78:** 83889f
-4,5-dicarboxylic acid, 1-[2-[(1,2-dicarboxy-ethenyl)amino]phenyl]-, tetramethyl ester	**67:** 90739h
-4,5-dicarboxylic acid, 1-[(3,5-dicarboxy-4-methyl-pyrrol-2-yl)methyl]-, diisopropyl ester	**73:** 98718v
-4,5-dicarboxylic acid, 1-[4-[2-[(dichloroacetyl)-amino]-1,3-dihydroxypropyl]phenyl]-, dimethyl ester, R-(R*,R*)-	**84:** 4609n
-4,5-dicarboxylic acid, 1-(1,6-dihydro-5-hydroxy-6-oxo-1-phenyl-4-pyridazinyl)-, dimethyl ester	**78:** 72036z
-4,5-dicarboxylic acid, 1-[1,6-dihydro-5-(4-morphol-inyl)-6-oxo-1-phenyl-4-pyridazinyl]-, dimethyl ester	**78:** 72036z
-4,5-dicarboxylic acid, 1-(2,5-dimethylphenyl)-	**22:** 3411[2]
-4,5-dicarboxylic acid, 1-(4,6-dimethyl-2-pyrimi-dinyl)-, dimethyl ester	**71:** 81296s
-4,5-dicarboxylic acid, 1-[2,2-dimethyl-1-[(tri-methylsilyl)oxy]propyl]-, dimethyl ester	**83:** 10269x

TABLE 4 (*Continued*)

Compound	Reference
4.3. 1,2,3-Triazole-4,5-Dicarboxylic Acids and Their Functional Derivatives (*Continued*)	
-4,5-dicarboxylic acid, 1-[(diphenylphosphinyl)-methyl]-, dimethyl ester	**74:** 112127f
-4,5-dicarboxylic acid, 1-(5,6-diphenylpyrazinyl)-, dimethyl ester	**74:** 75924r
-4,5-dicarboxylic acid, 1-(2-ethoxy-2-oxoethyl)-, dimethyl ester	**80:** 108451p
-4,5-dicarboxylic acid, 2-(2-ethoxy-2-oxoethyl)-, dimethyl ester	**80:** 108451p
-4,5-dicarboxylic acid, 2-(3-ethoxy-3-oxo-1-propenyl)-, dimethyl ester, (E)-	**80:** 108451p
-4,5-dicarboxylic acid, 2-(3-ethoxy-3-oxo-1-propenyl)-, dimethyl ester, (Z)	**80:** 108451p
-4,5-dicarboxylic acid, 1-ethyl-	**53:** 17902b
-4,5-dicarboxylic acid, 1-(ferrocenylmethyl)-, dimethyl ester	**73:** 25632c
-4,5-dicarboxylic acid, 1-β-D-glucopyranosyl-, dimethyl ester, tetraacetate ester	**52:** 16359d
-4,5-dicarboxylic acid, 1-(5-hydrazino-1,6-dihydro-6-oxo-1-phenyl-4-pyridazinyl)-, dimethyl ester	**78:** 72036z
-4,5-dicarboxylic acid, 2-(2-hydroxyphenyl)-, diethyl ester	**77:** P115464m
-4,5-dicarboxylic acid, 2-(2-hydroxyphenyl)-, dimethyl ester	**71:** P124445j
-4,5-dicarboxylic acid, 1-(2-hydroxy-2-phenylethyl)-, dimethyl ester	**83:** 10269x
-4,5-dicarboxylic acid, 1-[4-iodo-9,10-bis(methoxy carbonyl)tricyclo[4.2.2.0^{2,5}]dec-7-en-3-yl]-, dimethyl ester, (1α, 2α, 3α, 4α, 5α, 6α, 9R*, 10S*)-	**84:** 4328v, **84:** 4547r
-4,5-dicarboxylic acid, 1-(8-iodo-4-cycloocten-1-yl)-, (1α, 4Z, 8β)-	**76:** 126021v
-4,5-dicarboxylic acid, 1-(2-iodocyclooctyl)-, dimethyl ester	**84:** 73724h
-4,5-dicarboxylic acid, 1-(4-iodocyclooctyl)-, dimethyl ester	**84:** 73724h
-4,5-dicarboxylic acid, 1-(2-iodo-3,3-dimethyl-butyl)-, dimethyl ester	**72:** 132635g
-4,5-dicarboxylic acid, 1-[α-(1-iodethyl)benzyl]-, dimethyl ester, *erythro*-	**72:** 132635g
-4,5-dicarboxylic acid, 1-[α-(α-iodophenacyl)-benzyl]-, dimethyl ester, *erythro*-	**72:** 132635g
-4,5-dicarboxylic acid, 1-isobutyl-	**53:** 17901e
-4,5-dicarboxylic acid, 2-[3-methoxy-1-(methoxy-carbonyl)-3-oxo-1-propenyl]-, dimethyl ester	**80:** 108451p
-4,5-dicarboxylic acid, 1-(4-methoxyphenyl)-, dimethyl ester	**64:** 5076h
-4,5-dicarboxylic acid, 2-(2-methoxyphenyl)-	**71:** P124445j
-4,5-dicarboxylic acid, 1-methyl-	**49:** 5454b, **50:** 4129b, **82:** 73727v
-4,5-dicarboxylic acid, 1-methyl-, dimethyl ester	**51:** 14697f

TABLE 4 (*Continued*)

Compound	Reference
4.3. 1,2,3-Triazole-4,5-Dicarboxylic Acids and Their Functional Derivatives (*Continued*)	
-4,5-dicarboxylic acid, 1-methyl-, monopotassium salt	**82:** 73727v
-4,5-dicarboxylic acid, 2-methyl-, monopotassium salt	**82:** 73727v
-4,5-dicarboxylic acid, 1-(1-methyl-1*H*-benzimidazol-2-yl)-, dimethyl ester	**79:** 66246x
-4,5-dicarboxylic acid, 2-(1-methyl-1*H*-benzimidazol-2-yl)-, dimethyl ester	**79:** 66246x
-4,5-dicarboxylic acid, 1-(1-methylpropenyl)-, dimethyl ester, (E)-	**72:** 132635g, **79:** 92123z
-4,5-dicarboxylic acid, 1-(1-methylpropenyl)-, dimethyl ester, (Z)-	**72:** 132635g, **79:** 92123z
-4,5-dicarboxylic acid, 1-[2-methyl-1-[(trimethylsilyl)oxo]propyl]-, dimethyl ester	**83:** 10269x
-4,5-dicarboxylic acid, 1-(2-nitrophenyl)-, dimethyl ester	**67:** 64311t
-4,5-dicarboxylic acid, 2-(4-nitrophenyl)-	**21:** 2690[4]
-4,5-dicarboxylic acid, 1-(4-nitrophenyl)-, dimethyl ester	**21:** 2690[4], **64:** 5076h
-4,5-dicarboxylic acid, 2-(3-oxo-1-phenyl-1-propenyl)-, dimethyl ester, (E)-	**80:** 108451p
-4,5-dicarboxylic acid, 2-(3-oxo-1-phenyl-1-propenyl)-, dimethyl ester, (Z)-	**80:** 108451p
-4,5-dicarboxylic acid, 1-phenyl-	**38:** 1743[1], **48:** 2685b, **56:** 9990h
-4,5-dicarboxylic acid, 1-phenyl-, dihydrazide	**21:** 2690[5], **65:** 15378f
-4,5-dicarboxylic acid, 1-phenyl-, dimethyl ester	**44:** 4430g, **49:** 3948h, **82:** 16641p, **83:** 114301t, **85:** 46469t
-4,5-dicarboxylic acid, 2-phenyl-	**42:** 2970f, **55:** 13413h
-4,5-dicarboxylic acid, 1-(1-phenylpropenyl)-, dimethyl ester, (E)-	**72:** 132635g
-4,5-dicarboxylic acid, 1-(1-phenylpropenyl)-, dimethyl ester, (Z)-	**72:** 132635g
-4,5-dicarboxylic acid, 1-[2-phenyl-2-[(trimethylsilyl)oxy]ethyl]-, dimethyl ester	**83:** 10269x
-4,5-dicarboxylic acid, 1-pyrido[2,3-*d*]tetrazolo-[1,5-*b*]pyridazin-6-yl-, dimethyl ester	**83:** 97213w
-4,5-dicarboxylic acid, 1-(2-pyridyl)-, dimethyl ester	**71:** 81296s
-4,5-dicarboxylic acid, 1-β-D-ribofuranosyl-, dimethyl ester, tribenzoate ester	**53:** 6245c
-4,5-dicarboxylic acid, 1-(3,4,5,6-tetrachloro-2-pyridinyl)-, dimethyl ester	**83:** 192957a
-4,5-dicarboxylic acid, 1-tetrazolo[1,5-*b*]pyridazin-6-yl-, dimethyl ester	**74:** 75924r, **83:** 97213w
-4,5-dicarboxylic acid, 1-(2,3,4-tri-*O*-acetyl-α-D-lyxopyranosyl)-, diethyl ester	**81:** 49979n
-4,5-dicarboxylic acid, 1-(2,3,4-tri-*O*-acetyl-β-D-ribopyranosyl)-, diethyl ester	**81:** 49979n

TABLE 4 (*Continued*)

Compound	Reference
4.3. 1,2,3-Triazole-4,5-Dicarboxylic Acids and Their Functional Derivatives (*Continued*)	
-4,5-dicarboxylic acid, 1-(tributylstannyl)-, diethyl ester	**60:** 14532c
-4,5-dicarboxylic acid, 1-(trimethylsilyl), dimethyl ester	**60:** 5537a
-4,5-dicarboxylic acid, 1-(triphenylplumbyl)-, dimethyl ester	**67:** 82249h
-4,5-dicarboxylic acid, 1-β-D-xylopyranosyl-, dimethyl ester	**52:** 16359d
-1-malonic acid, 4,5-dicarboxy-α-phenyl-, diethyl dimethyl ester	**53:** 11390b
-1-propanoic acid, 4,5-dicarbamoyl-, methyl ester	**24:** 3215[2]
-1-propanoic acid, 4,5-dicarboxy-	**24:** 3215[2]
-1,4,5-tricarboxamide	**24:** 3215[3]
-1,4,5-tricarboxylic acid, 1-ethyl-4,5-dimethyl ester	**61:** 5626h, **63:** 16337h

REFERENCES

21: 742[4] G. Charrier, *Atti Accad. Lincei*, **4,** 312 (1926).

21: 2129[9] A. Beretta, *Gazz. Chim. Ital.*, **57,** 179 (1927).

21: 2269[1] A. Beretta, *Gazz. Chim. Ital.*, **57,** 173 (1927).

21: 2690[4–5] K. Fries, *Justus Liebigs Ann. Chem.*, **454,** 164 (1927).

22: 423[5] J. A. Fialkoff, *Bull. Soc. Chim. Fr.*, **41,** 1209 (1927).

22: 3411[2] A. Bertho and F. Holder, *J. Prakt. Chem.*, **119,** 189 (1928).

22: 4475[8] E. Gryszkiewicz-Trochimowski, *Roczniki Chem.*, **8,** 165 (1928).

23: 4217[1] M. Gallatti and A. Ercoli, *Gazz. Chim. Ital.*, **59,** 207 (1929).

24: 592[9] H. Wieland, W. Frank, and Z. Kitasato, *Justus Liebigs Ann. Chem.*, **475,** 42 (1929).

24: 1378[4] G. Charrier and A. Neri, *Gazz. Chim. Ital.*, **59,** 742 (1929).

24: 3215[1–3] T. Curtius and W. Klavehn, *J. Prakt. Chem.*, **125,** 498 (1930).

24: 3232[1] T. Curtius and K. Raschig, *J. Prakt. Chem.*, **125,** 466 (1930).

25: 1827[4] M. Gallotti, G. Barro, and L. Salto, *Gazz. Chem. Ital.*, **60,** 866 (1930).

26: 455[3] A. Quilico and M. Freri, *Gazz. Chim. Ital.*, **61,** 484 (1931).

31: 3888[1] R. Justoni, *Atti V. Congr. Nazl. Chim. Pura Applicata, Rome*, **I,** 370 (1935).

31: 6237[4] J. A. Bilton, R. P. Linstead, and J. M. Wright, *J. Chem. Soc.*, 922 (1937).

31: 7849[3] L. E. Hinkel, G. O. Richards, and O. Thomas, *J. Chem. Soc.*, 1432 (1937).

33: 4249[9] W. Borsche and H. Hahn, *Justus Liebigs Ann. Chem.*, **537,** 219 (1939).

35: 3638[4] T. Ajello and S. Cusmano, *Gazz. Chem. Ital.*, **70,** 770 (1940).

36: 772[1] A. Quilico and C. Musante, *Gazz. Chim. Ital.*, **71,** 327 (1941).

36: 835[7] R. W. Cunningham, E. J. Fellows, and A. E. Livingston, *J. Pharmacol.*, **73,** 312 (1941).

36: 2863[2–7] E. Ghigi and T. Pozzo-Balbi, *Gazz. Chim. Ital.*, **71,** 228 (1941).

37: 119[2]	R. Hüttel, *Chem. Ber.*, **74B,** 1680 (1941).
37: 5404–5[9–4]	W. Borsche, H. Hahn, and M. Wagner-Roemmich, *Justus Liebigs Ann. Chem.*, **554,** 15 (1943).
38: 1743[1]	K. Henkel and F. Weygand, *Chem. Ber.*, **76B,** 812 (1943).
40: 7191[7–9]	R. F. Coles and C. S. Hamilton, *J. Amer. Chem. Soc.*, **68,** 1799 (1946).
42: 2599h	R. Hüttel and A. Gebhardt, *Justus Liebigs Ann. Chem.*, **558,** 34 (1947).
42: 2970e,f	T. Ajello and B. Tornetta, *Gazz. Chem. Ital.*, **77,** 332 (1947).
43: 2619a–d	J. L. Riebsomer and G. Sumrell, *J. Org. Chem.*, **13,** 807 (1948).
44: 1102c,d	Ramart-Lucas, J. Hoch, and Grumez, *Bull. Soc. Chim. Fr.*, 447 (1949).
44: 4430g	K. Alder and W. Trimborn, *Justus Liebigs Ann. Chem.*, **566,** 58 (1950).
45: 9037f	J. C. Sheehan and C. A. Robinson, *J. Amer. Chem. Soc.*, **73,** 1207 (1951).
46: 6123d–h	J. L. Riebsomer and D. A. Stauffer, *J. Org. Chem.*, **16,** 1643 (1951).
46: 8651i	Fr. Moulin, *Helv. Chim. Acta*, **35,** 167 (1952).
47: 3087e	W. Cooper, *J. Chem. Soc.*, 2408 (1952).
47: 8738c	G. Caronna and S. Palazzo, *Gazz. Chim. Ital.*, **82,** 292 (1952).
48: 2685a	E. Mugnaini and P. Grünanger, *Atti Accad. Naz. Lincei., Rend., Cl. Sci. Fis. Mat. Nat.*, **14,** 95 (1953).
48: 5850c	C. T. Bishop, *Science*, **117,** 715 (1953).
49: 3948f–i	E. Mugnaini and P. Grünanger, *Atti Accad. Naz. Lincei., Rend., Cl. Sci. Fis. Mat. Nat.*, **14,** 275 (1953).
49: 3949f	L. W. Hartzel and F. R. Benson, *J. Amer. Chem. Soc.*, **76,** 667 (1954).
49: 5454b	G. W. E. Plaut, *J. Amer. Chem. Soc.*, **76,** 5801 (1954).
49: 13227f	R. H. Wiley, N. R. Smith, D. M. Johnson, and J. Moffat, *J. Amer. Chem. Soc.*, **76,** 4933 (1954).
50: 362h	L. Toldy, T. Nógrádi, L. Vargha, G. Ivánovics, and I. Koczka, *Acta Chim. Acad. Sci. Hung.*, **4,** 303 (1954).
50: 4129b,c	R. H. Wiley, N. R. Smith, D. M. Johnson, and J. Moffat, *J. Amer. Chem. Soc.*, **77,** 3412 (1955).
50: 4924a	M. L. Stein and L. D'Antoni, *Il Farmaco (Pavia) Ed. Sci.*, **10,** 235 (1955).
50: 9392f,g	R. Hüttel and G. Welzel, *Justus Liebigs Ann. Chem.*, **593,** 207 (1955).
50: 15517d,e	R. H. Wiley, K. F. Hussung, and J. Moffat, *J. Org. Chem.*, **21,** 190 (1956).
51: 4364f	H. Bredereck and G. Schmötzer, *Justus Liebigs Ann. Chem.*, **600,** 95 (1956).
51: 13854i	R. Gompper, *Chem. Ber.*, **90,** 382 (1957).
51: 14697f–h	S. Yamada, T. Mizoguchi, and A. Avata, *Yokugaku Zasshi*, **77,** 452 (1957).
52: 16359d,e	J. Baddiley, J. G. Buchanan, and G. O. Osborne, *J. Chem. Soc.*, 1651 (1958).
52: 17246a,b	M. C. Ford and D. Mackay, *J. Chem. Soc.*, 1290 (1958).
53: 344i	H. El Khadem and Z. M. El-Shafei, *J. Chem. Soc.*, 3117 (1958).
53: 6245c	J. Baddiley, J. G. Buchanan, and G. O. Osborne, *J. Chem. Soc.*, 3606 (1958).
53: 11390b	O. Červinka, *Collect. Czech. Chem. Commun.*, **24,** 1146 (1959).
53: 13101e	K. Hohenlohe-Oehringen, *Monatsh. Chem.*, **89,** 635 (1958).
53: 16120e	D. Dal Monte and P. Veggetti, *Boll. Sci. Fac. Chim. Ind. Bologna*, **16,** 1 (1958).
53: 17108g	S. Corsano and R. Inverardi, *Ric. Sci.*, **29,** 74 (1959).
53: 17901–2e–b	P. A. S. Smith, J. M. Clegg, and J. Lakritz, *J. Org. Chem.*, **23,** 1595 (1958).
53: 18946h	D. Dal Monte, A. M. Casoni, and M. Fernando, *Gazz. Chim. Ital.*, **88,** 1035 (1958).

54: 1504g,h	H. El Khadem and Z. M. El-Shafei, *J. Chem. Soc.*, 1655 (1959).
55: P571c	S. Yamada and T. Mizoguchi, Japanese Patent, 1819 (1960).
55: 2926g	R. Aston and H. Cullumbine, *Arch. Intern. Pharmacodynamie*, **126,** 219 (1960).
55: 3562–3i–e	H. El Khadem, Z. M. El-Shafei, and Y. S. Mohammed, *J. Chem. Soc.*, 3993 (1960).
55: 13413h	L. Anderson and J. N. Aronson, *J. Org. Chem.*, **24,** 1812 (1959).
55: 17626–7i–a	C. Pedersen, *Acta Chem. Scand.*, **13,** 888 (1959).
55: 23505a	E. Profft and W. Georgi, *Justus Liebigs Ann. Chem.*, **643,** 136 (1961).
55: 27071g	H. El Khadem, Z. M. El-Shafei, and M. H. Meshreki, *J. Chem. Soc.*, 2957 (1961).
56: P4291g	Yu. B. Vilenskii, B. M. Ivanov, V. Ya. Pochinok, L. F. Avramenko, and A. M. Shemet, Russian Patent, 138,487 (1960).
56: 9990h	G. Wittig and A. Krebs, *Chem. Ber.*, **94,** 3260 (1961).
57: 4649h,i	H. A. Staab and W. Benz, *Justus Liebigs Ann. Chem.*, **648,** 72 (1961).
57: 5911h	A. J. Boulton and A. R. Katritzky, *J. Chem. Soc.*, 2083 (1962).
57: 7255b–g	H. N. Bojarska-Dahlig, *Rec. Trav. Chim.*, **80,** 1348 (1961).
57: 7353g	H. El Khadem and M. H. Meshreki, *Nature*, **194,** 373 (1962).
58: 4545d	S. Kamiya, *Chem. Pharm. Bull.*, **10,** 471 (1962).
58: 11454f,g	B. B. Bishay, H. El Khadem, Z. M. El-Shafei, and M. H. Meshreki, *J. Chem. Soc.*, 3154 (1962).
58: 12560–1d–c	R. Fusco, G. Bianchetti, D. Pocar, and R. Ugo, *Gazz. Chim. Ital.*, **92,** 1040 (1962).
58: 13944a	R. Fusco, G. Bianchetti, D. Pocar, and R. Ugo, *Chem. Ber.*, **96,** 802 (1963).
59: 6408d	H. Bojarska-Dahlig and P. Nantka-Namirski, *Congr. Sci. Farm., Conf. Comun.*, **21,** *Pisa*, 203 (1961).
59: 14095f	H. El Khadem, A. M. Kolkaila, and M. H. Meshreki, *J. Chem. Soc.*, 3531 (1963).
60: 643h	B. B. Bishay, H. El-Khadem and Z. M. El-Shafei, *J. Chem. Soc.*, 4980 (1963).
60: 5537a	L. Birkofer, A. Ritter, and P. Richter, *Chem. Ber.*, **96,** 2750 (1963).
60: 9268c,d	H. Bojarska-Dahlig, *Acta Polon. Pharm.*, **19,** 273 (1962).
60: 14532c	J. G. A. Luijten and G. J. M. van der Kerk, *Rec. Trav. Chim.*, **83,** 295 (1964).
61: 3096–7g–a	D. Pocar, G. Bianchetti and P. D. Croce, *Chem. Ber.*, **97,** 1225 (1964).
61: 5626h	R. Huisgen and H. Blaschke, *Tetrahedron Lett.*, 1409 (1964).
61: 5633a,b	G. Bianchetti, D. Pocar, and P. D. Croce, *Gazz. Chim. Ital.*, **94,** 340 (1964).
61: 5738c	H. El Khadem, M. H. Meshreki, and G. H. Labib, *J. Chem. Soc.*, 2306 (1964).
61: 10670h	L. A. Summers and D. J. Shields, *Chem. Ind.*, 1264 (1964).
61: 14664–5h–b	A. N. Nesmeyanov and M. I. Rybinskaya, *Dokl. Akad. Nauk SSSR*, **158,** 408 (1964).
62: 545f	W. Ried and H. Mengler, *Justus Liebigs Ann. Chem.*, **678,** 95 (1964).
62: 6441b	H. Reimlinger, F. Billiau, and M. Peiren, *Chem. Ber.*, **97,** 3493 (1964).
62: 13138g,h	J. J. Looker, *J. Org. Chem.*, **30,** 638 (1965).
62: 16247–8g–a	R. Huisgen, L. Möbius, and G. Szeimies, *Chem. Ber.*, **98,** 1138 (1965).
63: 1850h	H. El Khadem, Z. M. El-Shafei, and M. H. Meshreki, *J. Chem. Soc.*, 1524 (1965).
63: 4268h	L. A. Summers, P. F. H. Freeman, and D. J. Shields, *J. Chem. Soc.*, 3312 (1965).
63: P7019b	CIBA Ltd., French Patent, 1,389,779 (1965).
63: 9942g	Y. F. Shealy and C. A. O'Dell, *J. Org. Chem.*, **30,** 2488 (1965).

63: P11574e	B. Hirsch, East German Patent, 36,137 (1965).
63: 16337h	R. Huisgen and H. Blaschke, *Chem. Ber.*, **98,** 2985 (1965).
64: 5323g	W. König, M. Coenen, W. Lorenz, F. Bahr, and A. Bassl, *J. Prakt. Chem.*, **30,** 96 (1965).
64: 5076e–h	R. Huisgen, R. Knorr, L. Möbius, and G. Szeimies, *Chem. Ber.*, **98,** 4014 (1965).
64: 9713h	G. S. Akimova, V. N. Chistokletov, and A. A. Petrov. *Zh. Org. Khim.*, **1,** 2077 (1965).
64: 11198b–h	R. Huisgen, G. Szeimies, and L. Möbius, *Chem. Ber.*, **99,** 475 (1966).
65: 15378f	L. Erichomovitch and F. L. Chubb, *Can. J. Chem.*, **44,** 2095 (1966).
65: 16960d	L. A. Summers, *Experientia*, **22,** 499 (1966).
67: 32420m	S. Maiorana, *Ann. Chim.* (*Rome*), **56,** 1531 (1966).
67: 43759e	F. M. Stojanovič and Z. Arnold, *Collect. Czech, Chem. Commun.*, **32,** 2155 (1967).
67: 53900h	E. Gudriniece, N. R. Burvele, and A. Ieviņš, *Dokl. Akad. Nauk SSSR*, **171,** 869 (1966).
67: 64311t	J. C. Kauer and R. A. Carboni, *J. Amer. Chem. Soc.*, **89,** 2633 (1967).
67: P64403z	F. D. Marsh, U.S. Patent, 3,322,782 (1967).
67: 82171b	G. Bianchetti, D. Pocar, P. D. Croce, and R. Stradi, *Gazz. Chim. Ital.*, **97,** 304 (1967).
67: 82249h	H. Gorth and M. C. Henry, *J. Organometal. Chem.*, **9,** 117 (1967).
67: 90739h	S. K. Khetan and M. V. George, *Can. J. Chem.*, **45,** 1993 (1967).
67: 100071a	G. S. Akimova, V. N. Chistokletov, and A. A. Petrov. *Zh. Org. Khim.*, **3,** 968 (1967).
67: 116854u	M. E. Hermes and F. D. Marsh, *J. Amer. Chem. Soc.*, **89,** 4760 (1967).
68: 13061z	L. Birkofer and P. Wegner, *Chem. Ber.*, **100,** 3485 (1967).
68: P59441b	A. S. Tomcufcik, P. F. Fabio, and A. M. Hoffman, British Patent, 1,077,173 (1967).
68: P59611g	A. S. Tomcufcik and A. M. Hoffman, U.S. Patent, 3,331,830 (1967).
68: P59616n	A. S. Tomcufcik, P. F. Fabio, and A. M. Hoffman, U.S. Patent, 3,331,845 (1967).
68: 78204t	P. A. S. Smith and J. G. Wirth, *J. Org. Chem.*, **33,** 1145 (1968).
69: 10402w	H. El Khadem, H. A. R. Mansour, and M. H. Meshreki, *J. Chem. Soc., C,* 1329 (1968).
69: 19088f	T. Nozoe, H. Horino, and T. Toda, *Tetrahedron Lett.*, 5349 (1967).
69: 77179x	N. S. Zefirov and N. K. Chapovskaya, *Zh. Org. Khim.*, **4,** 1300 (1968).
69: P96741r	E. Gudriniece and N. Bruvele, Russian Patent, 216,738 (1968).
70: 77878v	G. L'abbé and H. J. Bestmann, *Tetrahedron Lett.*, 63 (1969).
70: 96727m	M. Rasberger and E. Zbiral, *Monatsh. Chem.*, **100,** 64 (1969).
71: 13067t	C. Arnold, Jr. and D. N. Thatcher, *J. Org. Chem.*, **34,** 1141 (1969).
71: 81296s	R. Huisgen, K. Von Fraunberg, and H. J. Sturm, *Tetrahedron Lett.*, 2589 (1969).
71: P124445j	Geigy, A.-G., French Patent, 1,559,131 (1969).
72: 2972d	D. J. Brecknell, R. M. Carman, H. C. Deeth, and J. J. Kibby, *Aust. J. Chem.*, **22,** 1915 (1969).
72: 31710x	N. S. Zefirov and N. K. Chapovskaya, *Vest. Mosk. Univ., Khim.*, **24,** 113 (1969).
72: 38736u	R. M. Carman and R. F. Evans, *J. Chem. Educ.*, **46,** 847 (1969).
72: P55493e	H. A. Wagner, U.S. Patent, 3,479,357 (1969).
72: P67724u	M. Minagawa and K. Nakagawa, German Patent, 1,927,447 (1970).
72: 121447w	G. L'abbé and A. Hassner, *J. Heterocycl. Chem.*, **7,** 361 (1970).
72: 132635g	G. L'abbé, J. E. Galle, and A. Hassner, *Tetrahedron Lett.*, 303 (1970).
73: 14769b	M. Regitz and G. Himbert, *Justus Liebigs Ann. Chem.*, **734,** 70 (1970).
73: 25364s	F. P. Woerner and H. Reimlinger, *Chem. Ber.*, **103,** 1908 (1970).

73: 25632c D. E. Bublitz, *J. Organometal. Chem.*, **23,** 225 (1970).

73: 56037s N. Bruvele and E. Gudriniece, *Latv. PSR Zinat. Akad. Vestis, Kim. Ser.*, 198 (1970).

73: 77172x E. Gudriniece and V. V. Solov'eva, *Latv. PSR Zinat. Akad. Vestis, Kim. Ser.*, 378 (1970).

73: 98718v A. Treibs and K. Jacob, *Justus Liebigs Ann. Chem.*, **737,** 176 (1970).

73: 98912d N. Bruvele and E. Gudriniece, *Latv. PSR Zinat. Akad. Vestis, Kim. Ser.*, 368 (1970).

74: P22847m I. Pinter, A. Messmer, F. Oerdogh, and L. Pallos, German Patent, 2,012,943 (1970).

74: 48048b L. F. Avramenko, Yu. B. Vilenskii, B. M. Ivanov, N. V. Kudryavskays, I. A. Ol'shevskaya, V. Ya. Pochinok, L. I. Skripnik, L. M. Fedorova, and I. P. Fedorova, *Usp. Nauch. Fotogr.* **14,** 12 (1970).

74: P53342x J. J. Looker, U.S. Patent, 3,533,786 (1970).

74: 75924r T. Sasaki, K. Kanematsu and M. Murata, *J. Org. Chem.*, **36,** 446 (1971).

74: 76375z P. Ykman, G. L'abbé, and G. Smets, *Tetrahedron*, **27,** 5623 (1971).

74: 99215m H. G. Adolph, *J. Org. Chem.*, **36,** 806 (1971).

74: 112127f D. Seyferth and P. Hilbert, *Org. Prep. Proced. Int.*, **3,** 51 (1971).

74: 112402s G. Alonso, M. T. García-López, G. García-Muñoz, R. Madroñero, and M. Rico, *J. Heterocycl. Chem.*, **7,** 1269 (1970).

74: 125600e E. Gudriniece and V. Urbans, *Latv. PSR Zinat. Akad. Vestis, Kim. Ser.*, 82 (1971).

74: 141988t P. Ykman, G. L'abbé, and G. Smets, *Tetrahedron*, **27,** 845 (1971).

75: 88891y H. El Khadem, D. Horton, and M. H. Meshreki, *Carbohyd. Res.*, **16,** 409 (1971).

75: 140767d N. S. Zefirov, N. K. Chapovskaya, and V. V. Kolesnikov, *Chem. Commun.*, 1001 (1971).

76: 25192w P. Ykman, G. L'abbé, and G. Smets, *Tetrahedron*, **27,** 5623 (1971).

76: 59538d F. Texier and R. Carrié, *Bull. Soc. Chim. Fr.*, 3642 (1971).

76: 99184s T. Sasaki and Y. Suzuki, *Yuki Gosei Kagaku Kyokai Shi*, **29,** 880 (1971).

76: P99674b A. F. Strobel and M. L. Whitehouse, German Patent, 2,129,864 (1971).

76: 113139q R. Jacquier, C. Pellier, C. Petrus, and F. Petrus, *Bull. Soc. Chim. Fr.*, 4078 (1971).

76: P114090d J. Rody and H. Lind, German Patent, 2,111,538 (1971).

76: 126021v T. Sasaki, K. Kanematsu, and Y. Yukimoto, *J. Org. Chem.*, **37,** 890 (1972).

76: 140726k H. Horino and T. Toda, *Bull. Chem. Soc. Jap.*, **45,** 584 (1972).

76: 153691v E. Gudriniece and V. V. Solov'eva, *Latv. PSR Zinat. Akad. Vestis, Kim. Ser.*, 81 (1972).

77: P7313c H. Lind, German Patent, 2,133,012 (1972).

77: 102129f M. T. García-López, G. García-Muñoz, and R. Madroñero, *J. Heterocycl. Chem.*, **9,** 717 (1972).

77: P115464m R. Kirchmayr, H. J. Heller, and J. Rody, Swiss Patent, 522,432 (1972).

77: 126517g A. N. Nesmeyanov and M. I. Rybinskays, *Issled. Obl. Org. Khim. Izbr. Tr. Nesmeyanov, A.N.*, 291 (1971).

77: 126981d O. Makabe, S. Fukatsu, and S. Umezawa, *Bull. Chem. Soc. Jap.*, **45,** 2577 (1972).

77: 164604r Y. Tanaka and S. I. Miller, *J. Org. Chem.*, **37,** 3370 (1972).

78: 16102f E. Klingsberg, *Synthesis*, 475 (1972).

78: 72035y V. Urbans, E. Gudriniece, and L. A. Khoroshanskays, *Latv. PSR Zinat. Akad. Vestis, Kim. Ser.*, 712 (1972).

78: 72036z	V. Urbans, E. Gudriniece, and L. A. Khoroshanskays, *Latv. PSR Zinat. Akad. Vestis, Kim. Ser.*, 715 (1972).
78: P72161m	D. Guenther, German Patent, 2,126,839 (1972).
78: 83889f	T. Sasaki, K. Kanematsu, and Y. Yukimoto, *J. Chem. Soc., Perkin Trans.* **1,** 375 (1973).
78: 84715h	F. A. Lehmkuhl, J. T. Witkowski, and R. K. Robins, *J. Heterocycl. Chem.*, **9,** 1195 (1972).
78: 111025s	D. J. Anderson, T. L. Gilchrist, G. E. Gymer, and C. W. Rees, *J. Chem. Soc., Perkin Trans.* **1,** 550 (1973).
78: P111330n	G. Beck and D. Guenther, German Patent, 2,138,522 (1973).
78: 136200h	V. V. Solov'eva and E. Gudriniece, *Khim. Geterotsikl. Soedin.*, 256 (1973).
79: 5300p	H. Horino and T. Toda, *Bull. Chem. Soc. Jap.*, **46,** 1212 (1973).
79: 5526s	T. C. Thurber and L. B. Townsend, *J. Amer. Chem. Soc.*, **95,** 3081 (1973).
79: P18720a	B. Hirsch, H. J. Heckemann, and K. P. Thi, East German Patent, 95,006 (1973).
79: P18722c	B. Hirsch, H. J. Heckemann, and K. P. Thi, East German Patent, 94,383 (1972).
79: 31992k	G. L'abbé, *Verh. Kon. Acad. Wetensch., Lett. Schome Kunsten Belg., Kl. Wetensch*, **34,** 49 pp (1972).
79: 66246x	Y. Shiokawa and S. Ohki, *Chem. Pharm. Bull.*, **21,** 981 (1973).
79: P78815v	E. Aufderhaar and A. Meyer, German Patent, 2,263,878 (1973).
79: 92123z	A. N. Thakore, J. Buchshriber, and A. C. Oehlschlager, *Can. J. Chem.*, **51,** 2406 (1973).
79: 92148m	E. R. Wagner, *J. Org. Chem.*, **38,** 2976 (1973).
79: 104890n	M. Muramatsu, N. Obata, and T. Takizawa, *Tetrahedron Lett.*, 2133 (1973).
80: 3439n	G. Beck and D. Günther, *Chem. Ber.*, **106,** 2758 (1973).
80: 14221m	F. Capitan, F. Salinas, and E. J. Alonso, *Bol. Soc. Quim. Peru*, **39,** 1 (1973).
80: 37050u	V. M. Belikov, T. F. Savel'eva, and E. N. Safonova, *Izv. Akad. Nauk SSSR, Ser. Khim.*, 2060 (1973).
80: 47912r	C. E. Olsen, *Acta Chem. Scand.*, **27,** 1987 (1973).
80: 48283y	H. S. El Khadem and D. L. Swartz, *Carbohydr. Res.*, **30,** 400 (1973).
80: 95655f	P. D. Croce, *Ann. Chim. (Rome)*, **63,** 29 (1973).
80: 96840n	A. M. Talati and B. V. Shah, *Indian J. Chem.*, **11,** 1077 (1973).
80: 108451p	Y. Tanaka and S. I. Miller, *Tetrahedron*, **29,** 3285 (1973).
80: 108452q	Y. Tanaka, S. R. Velen, and S. I. Miller, *Tetrahedron*, **29,** 3271 (1973).
80: P122396f	B. Hirsch, H. J. Heckemann, and K. P. Thi, East German Patent, 99,998 (1973).
80: 133389g	E. M. Burgess and J. P. Sanchez, *J. Org. Chem.*, **39,** 940 (1974).
80: 145179b	P. A. S. Smith and E. M. Bruckmann, *J. Org. Chem.*, **39,** 1047 (1974).
80: 146082b	C. Shin, Y. Yonezawa, and J. Yoshimura, *Tetrahedron Lett.*, 7 (1974).
81: 3850k	F. G. Badder, M. N. Basyouni, F. A. Fouli, and W. I. Awad, *J. Indian Chem. Soc.*, **50,** 589 (1973).
81: 37518y	A. Karlina, E. Gudriniece, and L. K. Bronfman, *Latv. PSR Zinat. Akad. Vestis, Kim. Ser.*, 202 (1974).
81: 37519z	B. I. Mikhant'ev, G. V. Shatalov, and V. D. Galkin, *Monomery Vysokomol. Soedin.*, 87 (1973).
81: 49979n	H. Paulsen, Z. Györgydeák, and M. Friedmann, *Chem. Ber.*, **107,** 1590 (1974).

81: P51152f E. Asaumi and E. Kobayashi, Japanese Patent, 73 37,325 (1973).

81: 105136z T. Sasaki, K. Kanematsu, and A. Kondo, *J. Org. Chem.*, **39,** 2246 (1974).

81: 135286r M. Croci, G. Fronza, and P. Vita-Finzi, *Gazz. Chim. Ital.*, **103,** 1045 (1973).

81: 163240q J. J. Baker, P. Mellish, C. Riddle, A. R. Somerville, and J. R. Tittensor, *J. Med. Chem.*, **17,** 764 (1974).

82: P4322x A. Gagneux, R. Heckendorn, and R. Meier, German Patent, 2,363,515 (1974).

82: 16641p D. N. Reinhoudt and C. G. Kouwenhoven, *Tetrahedron Lett.*, 2163 (1974).

82: 16745a A. Gorup, M. Kovačič, B. Kranje Škraba, B. Mihelčič, S. Simonič, B. Stanovnik, and M. Tišler, *Tetrahedron*, **30,** 2251 (1974).

82: 16926k A. Rosan and M. Rosenblum, *J. Organomet. Chem.*, **80,** 103 (1974).

82: P43446k J. M. Essery and L. C. Cheney, U.S. Patent, 3,780,032 (1974).

82: 49831d D. Rahner, G. Fischer, and K. W. Junge, *J. Signalaufzeichnungsmaterialien*, **2,** 39 (1974).

82: 49832e D. Rahner, G. Fischer, and K. W. Junge, *J. Signalaufzeichnungsmaterialien*, **1,** 367 (1973).

82: P72999y E. Regel, L. Eue, and R. R. Schmidt, German Patent, 2,321,330 (1974).

82: 73727v R. N. Macdonald, A. Cairncross, J. B. Sieja, and W. H. Sharkey, *J. Polym. Sci. Polym. Chem. Ed.*, **12,** 663 (1974).

82: 140021a B. I. Mikhat'ev, G. V. Shatalov, V. D. Galkin, B. Kh. Kimsanov, and K. Khuseinov, *Dokl. Akad. Nauk Tadzh. SSR*, **17,** 38 (1974).

82: 156182q T. L. Gilchrist, G. E. Gymer, and C. W. Rees, *J. Chem. Soc., Perkin Trans.*, **1,** 1 (1975).

82: 170175w G. Mehta, P. K. Dutta, and P. N. Pandey, *Tetrahedron Lett.*, 445 (1975).

83: 9958q M. Kovačič, S. Polanc, B. Stanovnik, and M. Tišler, *J. Heterocycl. Chem.*, **11,** 949 (1974).

83: 10269x L. Birkofer and W. Kaiser, *Justus Liebigs Ann. Chem.*, 266 (1975).

83: P43333s T. Takagi, K. Aida, H. Segawa, and H. Nazai, Japanese Patent, 74 127,974 (1974).

83: P43334t T. Takagi, K. Aida, H. Segawa, and H. Nagai, Japanese Patent, 74 127,975 (1974).

83: P61732a A. Dorlars and O. Neuner, German Patent, 2,258,276 (1974).

83: 97148d B. I. Mikhant'ev, G. V. Shatalov, and V. D. Galkin, *Monomery Vysokomol. Soedin.*, 6 (1974).

83: 97213w B. Mihelčič, S. Simonič, B. Stanovnik, and M. Tišler, *Croat. Chem. Acta*, **46,** 275 (1974).

83: 114301t F. Texier and J. Bourgois, *J. Heterocycl. Chem.*, **12,** 505 (1975).

83: 147900r B. I. Mikhant'ev, G. V. Shatalov, V. D. Galkin, V. S. Voishchev, and O. V. Voishcheva, *Vysokomol. Soedin., Ser. B*, **17,** 467 (1975).

83: 147684y J. G. Buchanan, A. R. Edgar, M. J. Power, and G. C. Williams, *J. Chem. Soc., Chem. Commun.*, 501 (1975).

83: 192957a R. A. Abramovitch, S. R. Challand, and Y. Yamada, *J. Org. Chem.*, **40,** 1541 (1975).

83: 193178j G. L'abbé, G. Mathys, and S. Toppet, *J. Org. Chem.*, **40,** 1549 (1975).

83: P206291t H. Arie, A. Saika, M. Yamanaka, I. Saito, Y. Inai, T. Sato, K. Ema, and S. Nomoto, Japanese Patent, 75 96,593 (1975).

84: 4328v G. Mehta and P. N. Pandey, *J. Org. Chem.*, **40,** 3631 (1975).

84: 4547r	T. Sasaki, K. Kanematsu, and A. Kondo, *Tetrahedron*, **31,** 2215 (1975).
84: 4609n	P. E. Nielsen, V. Leick, and O. Buckhardt, *Acta Chem. Scand., Ser. B*, **B29,** 662 (1975).
84: P17361a	T. Toshiaki, K. Aida, and H. Segawa, Japanese Patent, 73 104,158 (1973).
84: 44571a	F. G. De Las Heras, S. Y. K. Tam, R. S. Klein, and J. J. Fox, *J. Org. Chem.*, **41,** 84 (1976).
84: 73724h	R. C. Hayward and G. H. Whitman, *J. Chem. Soc., Perkin Trans.*, **1,** 2267 (1975).
84: 74239r	D. L. Coffen, R. I. Fryer, D. A. Katonak, and F. Wong, *J. Org. Chem.*, **40,** 894 (1975).
84: 122189y	T. C. Thurber and L. B. Townsend, *J. Org. Chem.*, **41,** 1041 (1976).
84: 135548q	V. Cerè, D. Dal Monte, S. Pollicino, and E. Sandri, *Gazz. Chim. Ital.*, **105,** 723 (1975).
84: 144561r	C. L. Johnson, *J. Med. Chem.*, **19,** 600 (1976).
84: P166274b	G. Kormany, G. Kabas, H. Schlaepfer, and A. E. Siegrist, German Patent, 2,535,615 (1976).
84: P175141h	N. Nitanai, N. Komoto, A. Ohkubo, O. Morikawa, Y. Hirohisa, T. Toyama, and H. Tachibana, Japanese Patent, 76 01,641 (1976).
84: P181615g	F. Fleck, A. V. Mercer, and R. Paver, German Patent, 2,535,069 (1976).
85: P6010e	J. T. Witkowski, R. K. Robins, and F. A. Lehmkuhl, U.S. Patent, 3,948,885 (1976).
85: 46469t	D. N. Reinhoudt and C. G. Kouwenhoven, *Recl. Trav. Chim. Pays-Bas*, **95,** 67 (1976).
85: 47115m	K. Khuseinov, G. V. Shatalov and B. I. Mikhantev, I *Vyssh. Uchebn. Zaved. Khim. Khim. Tekhnol*, **19,** 655 (1976).
85: P63074c	A. Ohkubo, K. Nitanai, T. Toyama, O. Morikawa, H. Tachibana, and H. Yanagida, Japanese Patent, 76 48,435 (1976).
85: 78055w	B. I. Mikhant'ev, G. V. Shatalov, and V. D. Galkin, *Deposited Publ.* VINITI 5825–73, 10 pp (1973).
85: 123875b	R. S. Klein, F. G. De Las Heras, S. Y. K. Tam, I. Wempen, and J. J. Fox, *J. Heterocycl. Chem.*, **13,** 589 (1976).

CHAPTER 5

Formyl- and Acyl-1,2,3-Triazoles

The preparation of *N*-acyl-1,2,3-triazoles by reaction of *N*-trimethylsilyl derivatives with acetyl chloride has been reported by Birkofer and Wegner (Eq. 1).[1] In a later paper[2] they showed that the 1-*N*-product (**5.1-1**) is obtained at low temperature (15°), and the 2-*N*-product (**5.1-2**) at high temperature (120°). The yields are generally high (*ca.* 90%), and the isomer ratio for an unsymmetrical case was determined (Eq. 2). Zbiral and his collaborators have added acetyl azide to α-keto phosphorus ylides and have obtained generally good yields of *N*-acetyl-1,2,3-triazoles (Eq. 3).[3] The isomer ratio for unsymmetrical products (**5.1-3**) was not determined.

The analogous direct reaction with benzoyl chloride has been reported to fail,[4] but there are several practical alternative routes. L'abbé and his

R R' N N N SiMe$_3$ + AcCl ⟶ R R' N N N Ac (1)

5.1-1

R = H, Me, *n*-Pr, *n*-Bu, HC≡C(CH$_2$)$_2$, Ph   R' = Me, Ph

Me N N N SiMe$_3$ + AcCl $\xrightarrow{15°}$ Me N N N Ac + Me N N N Ac

75%   25%

(2)

$\xrightarrow{120°}$ Me N N N Ac

5.1-2

$$Ph_3P{=}C(OCR^1)(R^2) + MeCON_3 \xrightarrow{CH_2Cl_2} \textbf{5.1-3} \quad (3)$$

R^1	= n-Pr	i-Pr	n-Pr	n-Pr	c-C_4H_7
R^2	= n-Pr	Me	MeCH=CH	Me(CH=CH)$_2$	n-Pr
% **5.1-3**	= 72	81	40	27	73

collaborators have studied a large array of aroyl azides and triphenylphosphonium enolates (**5.1-4**) (Eq. 4).[5] For the most part, yields are 50% or better, and reaction times are reasonable. The 1-N-products (**5.1-5**) are obtained in excellent yield with short reaction time. Harvey has described a similar very high yield reaction.[6] In one case Tanaka and Miller[7] found high substitution yield using the internal salt **5.1-6** (Eq. 5). The subsequent basic hydrolysis of the triphenylphosphorus group was not carried out, but it should proceed smoothly on the basis of several similar examples.

$$\textbf{5.1-4} + ArCON_3 \xrightarrow[\text{or } C_6H_6]{CH_2Cl_2} \textbf{5.1-5} \longrightarrow \quad (4)$$

R = Me, Ph, 4-NO_2Ph
Ar = Ph, 3- or 4-NO_2Ph, 4-ClPh, 4-MeOPh

Ph; 3,4-Cl_2Ph 97%[6]

$$\textbf{5.1-6} + PhCOCl \longrightarrow \quad (5)$$

The addition of azides to acetylene mono-aldehydes has been used often and with good results (e.g., Eqs. 6 to 8).[8–11] The preparation of 4,5-bisaldehydes is best accomplished through the tetraethyl bisacetal (**5.1-7**)

$$HC{\equiv}CCHO + RN_3 \longrightarrow \quad (6)$$

R	= H	Me	Ph	tropyl[9]
%	= 78	67	75	67

$$PhC{\equiv}CCHO \xrightarrow{HN_3} \text{4-formyl-5-phenyl-1H-1,2,3-triazole (OHC, Ph; N–H)}\ 90\% \quad (7)$$

$$PhC{\equiv}CCHO \xrightarrow{PhN_3} \text{(OHC, Ph; N–Ph)}\ 53\% + \text{(Ph, CHO; N–Ph)}\ 22\%\ (90\%\ \text{total crude}) \quad (8)$$

and subsequent hydrolysis (Eq. 9)[12]. The oxidation of racemic *myo*-inosose-2 osatriazole (**5.1-8**) produces the 2-phenyl bisaldehyde (**5.1-9**), but the yield of the cyclization step could not be improved (Eq. 10).[13] The analogous oxidations of osatriazoles has proven to be an excellent synthesis of monaldehydes (Eq. 11).[14–16] In El Khadem's laboratory yields of 80 to 95% are generally obtained using periodic acid.[16] A method involving substitution and reduction has been developed by Smith and his collaborators (Eq. 12).[17] A novel method giving a very high yield based on diazomalonaldehyde (**5.1-10**) has been published (Eq. 13).[18] Although its authors call the preparation of **5.1-10** simple, it appears to be a compound that is difficult to handle.

$$(EtO)_2CHC{\equiv}CCH(OEt)_2 + RN_3 \longrightarrow \text{(EtO)}_2\text{CH, CH(OEt)}_2\text{; N–R triazole} \xrightarrow{H^+} \text{OHC, CHO; N–R triazole} \quad (9)$$

R	%	%
Ph	95	92
$PhCH_2$	90	90
$n\text{-}C_6H_{13}$	90	90
$n\text{-}C_{12}H_{25}$	75	98

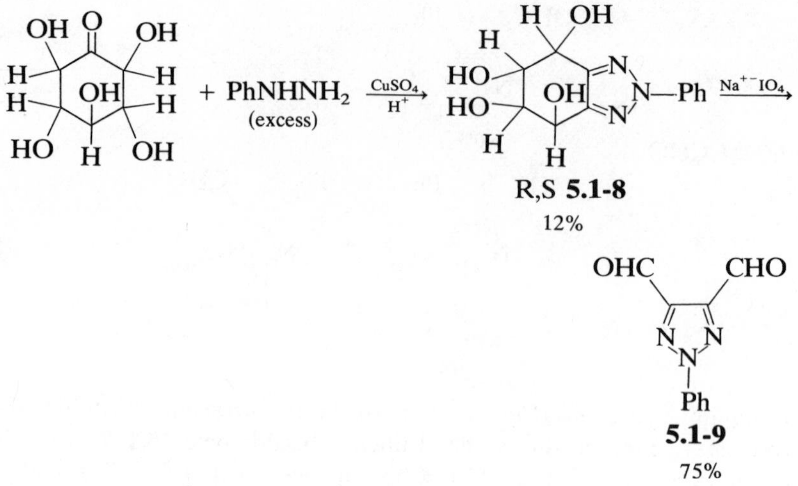

+ PhNHNH$_2$ (excess) $\xrightarrow[H^+]{CuSO_4}$ R,S **5.1-8** 12% $\xrightarrow{Na^{+-}IO_4}$ **5.1-9** 75% (10)

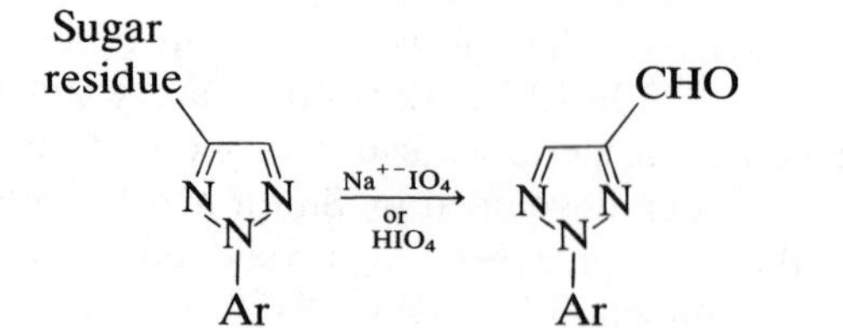

$\xrightarrow[\text{or } HIO_4]{Na^{+-}IO_4}$ (11)

Ar = Ph, RPh, XPh, NO_2Ph, MeOPh, AcNHPh, etc.

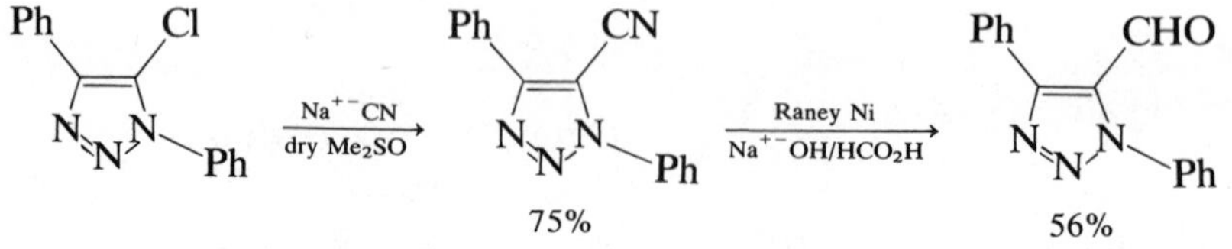

$\xrightarrow[\text{dry } Me_2SO]{Na^{+-}CN}$ 75% $\xrightarrow[Na^{+-}OH/HCO_2H]{\text{Raney Ni}}$ 56% (12)

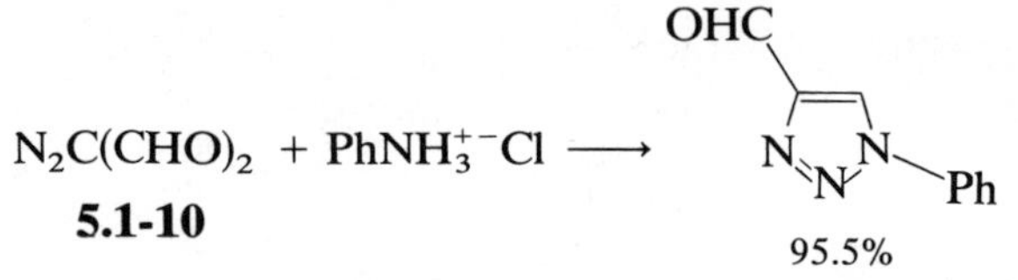

$$N_2C(CHO)_2 + PhNH_3^{+-}Cl \longrightarrow \quad (13)$$

5.1-10 95.5%

Many examples of 4- or 5-acetyl-1,2,3-triazoles have been prepared and a number of methods developed. For example, the addition of phenyl azide to acetylphenylacetylene (Eq. 14) gives an excellent yield, and both isomeric products are obtained.[19,20] The reaction of sodium azide with β-acetylstyrylsulfones proceeds in good yield (Eq. 15).[21] The diacetyl phenylhydrazone **5.1-12** is smoothly converted to a 1,2,3-triazole under novel conditions that deserve further exploration (Eq. 16).[22] The heterocyclic azide **5.1-13** reacts with 2,4-pentanedione under basic conditions (Eq. 17).[23] The treatment of the amides **5.1-14** with a Grignard reagent produces

$$PhC{\equiv}CCOR + PhN_3 \longrightarrow \text{4-RCO-5-Ph-1-Ph-1,2,3-triazole} + \text{4-Ph-5-COR-1-Ph-1,2,3-triazole} \quad (14)$$

R = Me 67% 23%
Ph 86%[20]

$$ArCH{=}C(SO_2Ph)(Ac) + Na^{+}\,^{-}N_3 \xrightarrow{HCONMe_2} \text{4,5-diacetyl-1-phenyl-1,2,3-triazole } (\mathbf{5.1\text{-}11}) \quad (15)$$

Ar	= Ph	4-MeOPh	3-HO-4-MeOPh
% **5.1-11**	= 64	70	50

$$Ac_2C{=}NNHPh\ (\mathbf{5.1\text{-}12}) \xrightarrow[EtOH]{Cu(OAc)_2} \xrightarrow[\text{sealed tube}]{NH_3} \text{4-Ac-5-Me-2-Ph-1,2,3-triazole} \quad (16)$$

84%

$$\mathbf{5.1\text{-}13} + AcCH_2Ac \xrightarrow[EtOH]{Et_3N} \text{4-Ac-5-Me-1,2,3-triazole (1-pyridopyrimidinonyl)} \quad (17)$$

85%

excellent yields of the corresponding 1,2,3-triazole (Eq. 18).[24] The 5-(2-propenyl)-1,2,3-triazole, formed from an Iotsich complex (**5.1-15**), can be oxidized in fair yield to the acetyl-1,2,3-triazole (Eq. 19).[25] Olsen has carried out a related sequence with substantially higher yields (Eq. 20).[26] Other 4-acyl-1,2,3-triazoles have been prepared in fair yield by the reaction of phenylazide with α-ketovinyl chlorides (Eq. 21).[27]

$$\mathbf{5.1\text{-}14}\ (\text{4-Me-5-C(O)N(Ph)Me-2-Ar-1,2,3-triazole}) + MeMgCl \longrightarrow \text{4-Me-5-Ac-2-Ar-1,2,3-triazole} \quad (18)$$

Ar = Ph	90%
4-MePh	95%

$CH_2{=}C(Me)C{\equiv}CMgBr$ (**5.1-15**) + PhN_3 ⟶ 1-phenyl-5-(1-methylvinyl)-1,2,3-triazole (67%) $\xrightarrow{KMnO_4}$ 5-acetyl-1-phenyl-1,2,3-triazole (30%) (19)

$Me_2C{=}CHCOMe + PhN_3 \xrightarrow{K^{+}\,{}^{-}OBu\text{-}t}$ 4-(1-methylvinyl)-5-methyl-1-phenyl-1,2,3-triazole (91%) $\xrightarrow[50\%\ HOAc]{KMnO_4}$ 4-acetyl-5-methyl-1-phenyl-1,2,3-triazole (58%) (20)

$RCOCH{=}CHCl + PhN_3 \longrightarrow$ 4-RCO-1-phenyl-1,2,3-triazole (**5.1-16**) (21)

R	= Me	*n*-Pr	*i*-Bu	$ClCH_2$
% **5.1-16**	= 38	29	21	61

A similar situation obtains with the benzoyl- and aroyl-substituted 1,2,3-triazoles. Pocar and his collaborators have used their enamine method in both acetyl and benzoyl cases with good results (Eq. 22).[28,29] An interesting sidelight of this study was the smooth cleavage of a 1-(2,4-dinitrophenyl) group with hydrazine (Eq. 23).[29] Looker has shown a similar acid cleavage with a 1-cycloheptatrienyl group (Eq. 24).[10] Under these conditions the 4-formyl-1,2,3-triazole analog cited earlier in this chapter did not cleave.[10]

$RHC{=}C(R')Y + ArN_3 \longrightarrow$ 4-R-5-R'-1-Ar-1,2,3-triazole (22)

Ar = 4-ClPh, 2,4-$(NO_2)_2$Ph Y = NMe_2, NMePh, 4-morpholinyl
R = Ac, PhCO R' = H, Me, Ph, Me_2CHCH_2

$\xrightarrow{H_2NNH_2}$ (23)

$-N_3$ + PhCOC≡CCOPh ⟶ 58%

$\xrightarrow[Et_2O]{HCl}$ 100% (24)

Another example of this potentially important method of cleavage under mild conditions has been reported (Eq. 25).[30] Two early reactions that may deserve further study involve the tricarbonyl derivative **5.1-17** cyclization and the heterocyclic (**5.1-18**) rearrangement, both shown to proceed in good yield (Eqs. 26, 27).[31]

The addition of azide ions to sulfonic acid salts produces good yields of 4-aroyl-1,2,3-triazoles (**5.1-19**) (Eqs. 28, 29).[32,33] Zbiral and his collaborators

PhCOC≡CCOPh + (ArN_3) ⟶ 86% $\xrightarrow{base}$ 86% (25)

5.1-17 $\xrightarrow{Ac_2O}$ 80% (26)

5.1-18 $\xrightarrow[\text{Cu or sand}]{\text{heat}}$ 70% (27)

$$ArCOCH{=}CHSO_3^{-}{}^{+}M + Na^{+}{}^{-}N_3 \xrightarrow[(2)\ HCl]{(1)\ H_2O} \textbf{5.1-19} \quad (28)$$

Ar	= Ph	4-MeOPh	4-ClPh	4-BrPh
M^+	= K	K	NMe_3	NMe_3
% **5.1-19**	= 71	96	78	66

$$ArCOCH(SO_3^{-}{}^{+}Na)_2 + Na^{+}{}^{-}N_3 \xrightarrow{HCONMe_2} \textbf{5.1-19} \quad (29)$$

Ar	= Ph	4-MeOPh	4-ClPh	4-BrPh
% **5.1-19**	= 70	71	53	65

have studied a large number of reactions involving the addition of azide ions to α-keto triphenylphosphonium compounds (Eqs. 30, 31).[34,35] Although the yields vary from poor to excellent, these methods are of genuine importance in the synthesis of a broad range of acyl-1,2,3-triazoles.

The heterocyclic azide mentioned earlier (**5.1-13**) also reacts with dibenzoylmethane, and the isomeric products have been separated (Eq. 32).[23]

$$Ph_3P^{+}CH{=}CHCOR + Na^{+}{}^{-}N_3 \longrightarrow \longrightarrow \textbf{5.1-20} \quad (30)$$

R	= Me	Et	*n*-Pr	*i*-Pr	Ph
% **5.1-20**	= 81	70	74	95	30

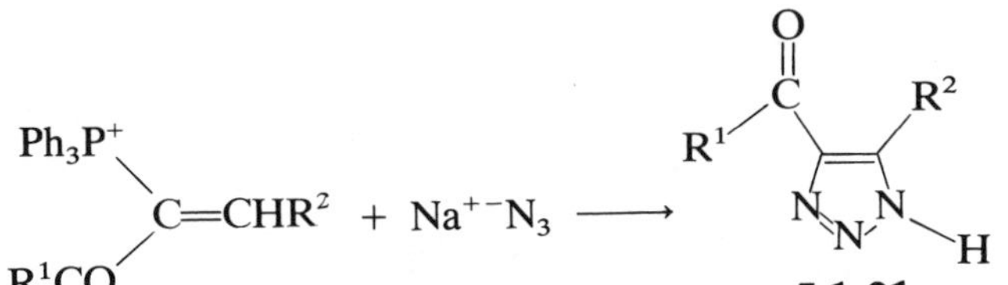

$Ph_3P^+C(COR^1){=}CHR^2 + Na^{+}\,{}^{-}N_3 \longrightarrow$ **5.1-21** (31)

R^1	= *i*-Pr	*t*-Bu	*c*-C_6H_{11}	*c*-C_3H_5	*i*-Pr	*c*-C_6H_{11}
R^2	= Et	Et	Et	Et	H	H
% **5.1-21**	= 30	41	15	50	25	28

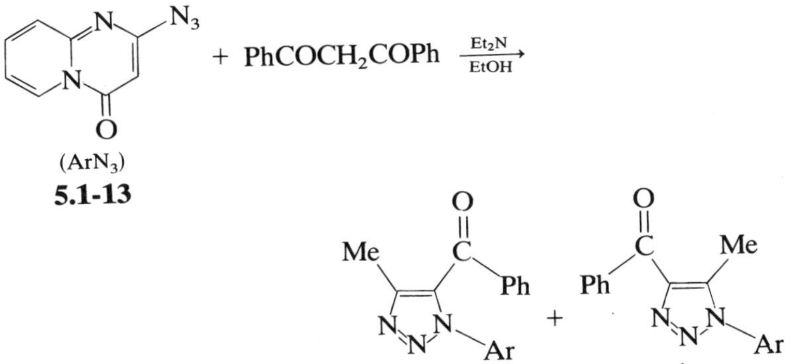

(ArN_3) **5.1-13** $+ PhCOCH_2COPh \xrightarrow[EtOH]{Et_3N}$ (32)

2 pts. + 1 pt.

(56% total)

Essentially quantitative yields of 4-benzoyl-1,2,3-triazoles have been obtained from two reactions that certainly deserve further study (Eqs. 33,34).[36,37] Several examples illustrating high yields of 4-acyl- and 4-aroyl-1,2,3-triazoles are based on the triazolecarboxylic acid derivatives (e.g., Eqs. 36 to 37).[38]

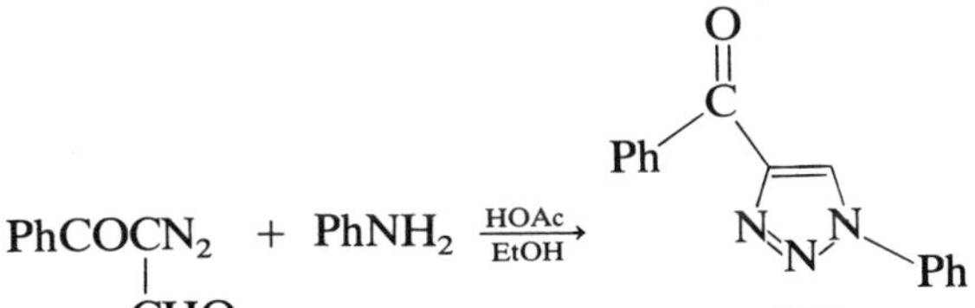

$PhCOC(N_2)CHO + PhNH_2 \xrightarrow[EtOH]{HOAc}$ 97% (33)

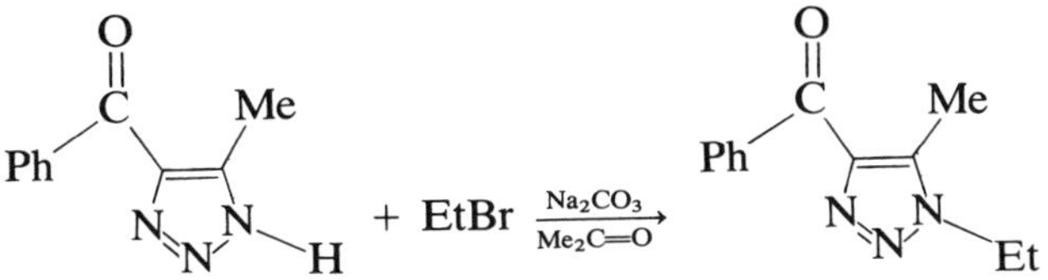

$+ EtBr \xrightarrow[Me_2C{=}O]{Na_2CO_3}$ 100% (34)

Y = Cl, $C_6H_6/AlCl_3$ → 4-benzoyl-1-phenyl-1,2,3-triazole (PhC(=O)–, N–Ph) (35)

Y = OEt, HOAc/Na → $EtO_2CCH_2C(=O)$–triazole–N–Ph, H_2O → Ac–triazole–N–Ph (36)

Y = OMe, PhCOMe/Na → $PhC(=O)CH_2C(=O)$–triazole–N–Ph (37)

REFERENCES

1. **65:** 15414g
2. **68:** 13061z
3. **71:** 112865h
4. **56:** 1388h
5. **76:** 25192w
6. **64:** 19597a
7. **79:** 66254y
8. **37:** 119[2]
9. **42:** 2599g
10. **62:** 13138g
11. **45:** 9037f
12. **38:** 1742[9]
13. **55:** 13413f
14. **43:** 2619b
15. **59:** 14095d
16. **60:** 643h
17. **68:** 78204t
18. **79:** 136425f
19. **76:** 59538d
20. **64:** 5077a
21. **80:** 3439n
22. **68:** 2862k
23. **83:** 9958q
24. **78:** 16102f
25. **67:** 100071a
26. **80:** 47912r
27. **50:** 4924d
28. **58:** 12560f
29. **61:** 5633b
30. **64:** 14184h
31. **22:** 2945[9]
32. **64:** 17581f
33. **64:** 17579a
34. **70:** 96727m
35. **71:** 91581v
36. **67:** 43759e
37. **82:** 171481e
38. **37:** 5404[9]

TABLE 5. ACYL-1,2,3-TRIAZOLES

Compound	Reference
-4-acetaldehyde, 2-(4-bromophenyl)-5-formyl-	**76:** 4098e
1-(4-acetamido-2,3-dihydro-3-oxo-2-phenyl-4-pyridazinyl)-4-acetyl-5-methyl-	**78:** 16120k
1-acetyl-	**54:** 4547d, **68:** 13061z
2-acetyl-	**65:** 15414g, **68:** 13061z
4-acetyl-	**65:** 5454h, **70:** 96727m
4-acetyl-1-(4-amino-3-oxo-2-phenyl-2*H*-pyridazin-5-yl)-5-methyl-	**78:** 16120k
4-acetyl-1-(4-azido-3-oxo-2-phenyl-2*H*-pyridazin-5-yl)-5-methyl-	**78:** 16120k
4-acetyl-1-[4-(benzylamino)-3-oxo-2-phenyl-2*H*-pyridazin-5-yl]-5-methyl-	**78:** 16120k
4-acetyl-1-[4-bromo-2-(2-cyanoethyl)-3-oxo-3*H*-pyridazin-5-yl]-5-methyl-	**76:** 153691v
4-acetyl-1-[4-bromo-2-(2-hydroxyethyl)-3-oxo-2*H*-pyridazin-5-yl]-5-methyl-	**76:** 153691v
4-acetyl-1-(4-bromo-1-methyl-3-oxo-2*H*-pyridazin-5-yl)-5-methyl-	**76:** 153691v
4-acetyl-1-(4-bromo-3-oxo-2-phenyl-2*H*-pyridazin-5-yl)-5-methyl-	**73:** 77172x, **76:** 153691v
4-acetyl-1-(4-bromophenyl)-5-methyl-	**75:** 109587w
4-acetyl-2-(4-bromophenyl)-5-methyl-	**63:** P11574f
1-acetyl-4-butyl-	**68:** 13061z
1-acetyl-5-butyl-	**68:** 13061z
2-acetyl-4-butyl-	**68:** 13061z
4-acetyl-1-[4-(butylamino) -3-oxo-2-phenyl-2*H*-pyridazin-5-yl]-5-methyl-	**78:** 16120k
4-acetyl-1-(4-chloro-3-oxo-2-phenyl-2*H*-pyridazin-5-yl)-5-methyl-	**78:** 136200h
4-acetyl-2-(2-chlorophenyl)-5-methyl-, oxime	**82:** P126614q
4-acetyl-1-(4-chlorophenyl)-5-methyl-	**75:** 109587w
4-acetyl-2-(4-chlorophenyl)-5-methyl-	**79:** P18724e
4-acetyl-2-(4-cyanophenyl)-5-methyl-	**84:** P181615g
1-acetyl-5-cyclobutyl-4-propyl-	**71:** 112865h
4-acetyl-1-[4-(cyclohexylamino)-3-oxo-2-phenyl-2*H*-pyridazin-5-yl]-5-methyl-	**78:** 16120k
4-acetyl-1-(3,7-dibromo-2-hydroxy-1-oxo-2,4,6-cycloheptatrien-5-yl)-5-methyl-	**76:** 140726k
4-acetyl-1-[4-(diethylamino)-3-oxo-2-phenyl-2*H*-pyridazin-5-yl]-5-methyl-	**78:** 16120k
1-acetyl-4,5-dimethyl-	**54:** 4547f, **65:** 15414g **68:** 13061z
2-acetyl-4,5-dimethyl-	**65:** 15414g, **68:** 13061z
4-acetyl-1-(3,5-dimethyl-4-isoxazolyl)-5-methyl-	**57:** 5911g
4-acetyl-1-(3,5-dimethyl-4-isoxazolyl)-5-methyl-, (2,4-dinitrophenyl) hydrazone	**57:** 5911g
4-acetyl-1,5-diphenyl-	**76:** 59538d
5-acetyl-1,4-diphenyl-	**76:** 59538d
1-acetyl-4,5-dipropyl-	**71:** 112865h

TABLE 5 (*Continued*)

Compound	Reference
4-acetyl-1-(3-ethyl-2-methyl-2-phenylbenzimidazolio-5-yl)-4-methyl-, iodide	**74:** 48048b
4-acetyl-1-[4-[(2-hydroxyethyl)amino]-3-oxo-2-phenyl-2*H*-pyridazin-5-yl]-5-methyl-	**78:** 16120k
4-acetyl-5-(3-hydroxy-4-methoxyphenyl)-	**80:** 3439n
4-acetyl-1-(4-hydroxy-3-oxo-2-phenyl-2*H*-pyridazin-5-yl)-5-methyl-	**78:** 16120k
4-acetyl-1-[4-[(3-hydroxyphenyl)amino]-3-oxo-2-phenyl-2*H*-pyridazin-5-yl]-5-methyl-	**78:** 16120k
4-acetyl-2-(4-methoxyphenyl)-5-methyl-	**84:** P181615g
1-acetyl-4-methyl-	**54:** 4547f, **68:** 13061z
1-acetyl-5-methyl-	**68:** 13061z, **74:** 76375z
2-acetyl-4-methyl-	**68:** 13061z, **74:** 76375z
4-acetyl-1-imidazo[1,2-*b*]pyridazin-6-yl-5-methyl-	**82:** 16745a
4-acetyl-1-imidazo[1,2-*b*]pyridazin-6-yl-5-methyl-, hydrazone	**82:** 16745a
4-acetyl-1-imidazo[1,2-*b*]pyridazin-6-yl-5-methyl-, phenylhydrazone	**82:** 16745a
1-acetyl-5-isopropyl-4-methyl-	**71:** 112865h
4-acetyl-1-(2-methoxy-1-oxo-2,4,6-cycloheptatrien-5-yl)-5-methyl-	**69:** P67394z
4-acetyl-5-(4-methoxyphenyl)-	**80:** 3439n
4-acetyl-1-[4-[(4-methoxyphenyl)amino]-3-oxo-2-phenyl-2*H*-pyridazin-5-yl]-5-methyl-	**78:** 16120k
4-acetyl-2-(2-methoxyphenyl)-5-methyl-	**77:** P7313c, **79:** P18724e
4-acetyl-5-methyl-	**36:** 771[9], **75:** 48991z
4-acetyl-5-methyl-1-[4-[(1-methylethyl)amino]-3-oxo-2-phenyl-2*H*-pyridazin-5-yl]-	**78:** 16120k
4-acetyl-5-methyl-2-(4-methylphenyl)-	**78:** 16102f
4-acetyl-5-methyl-2-(4-methylphenyl)-, phenylhydrazone	**78:** 16102f
4-acetyl-5-methyl-1-(2-methyl-1-phenyl-5-benzimidazolyl)-	**74:** 48048b
4-acetyl-5-methyl-1-[4-(4-morpholinyl)-3-oxo-2-phenyl-2*H*-pyridazin-5-yl]-	**78:** 16120k
4-acetyl-5-methyl-1-[3-oxo-2-phenyl-4-(1-piperidinyl)-2*H*-pyridazin-5-yl]-	**78:** 16120k
4-acetyl-5-methyl-1-(4-oxo-4*H*-pyrido[1,2-*a*]pyrimidin-2-yl)-	**83:** 9958q
4-acetyl-5-methyl-1-phenyl-	**37:** 5405[3], **80:** 47912r
4-acetyl-5-methyl-2-phenyl-	**68:** 2862k
4-acetyl-5-methyl-2-phenyl-, (2,4-dinitrophenyl) hydrazone	**37:** 5405[3], **65:** 5454h
4-acetyl-5-methyl-2-phenyl-, (4-nitrophenyl)hydrazone	**78:** 16102f
4-acetyl-5-methyl-2-phenyl-, oxime	**68:** 2862k
4-acetyl-5-methyl-1-[4-(phenylamino)-3-oxo-2-phenyl-2*H*-pyridazin-5-yl]-	**78:** 16120k
4-acetyl-5-methyl-1-tetrazolo[5,1-*a*]phthalazin-6-yl-	**80:** 82860u
4-acetyl-5-methyl-1-tetrazolo[1,5-*b*]pyridazin-6-yl-	**82:** 16745a
4-acetyl-5-methyl-1-(1,2,4-triazolo[1,5-*b*]pyridazin-6-yl)-	**82:** 31304z

TABLE 5 (*Continued*)

Compound	Reference
4-acetyl-5-methyl-1-(1,2,4-triazolo[4,3-*b*]pyridazin-6-yl)-	**82:** 16745a
4-acetyl-5-methyl-1-(1,2,4-triazolo[4,3-*b*]pyridazin-6-yl)-, hydrazone	**82:** 16745a
N-acetyl-4 (or 5)-[2-(5-nitro-2-furanyl)ethenyl]-	**77:** 101470y
4-acetyl-1-(2-nitrophenyl)-5-phenyl-	**58:** 12560d
4-acetyl-1-(4-nitrophenyl)-5-phenyl-	**58:** 12560d
4-[4-(acetyloxy)-2-hydroxybenzoyl]-5-methyl-2-phenyl-	**76:** 114090d
1-acetyl-4-(1,3-pentadienyl)-5-propyl-	**71:** 112865h
1-acetyl-4-phenyl-	**63:** 16338e, **68:** 13061z
2-acetyl-4-phenyl-	**55:** 8394b, **68:** 13061z
4-acetyl-1-phenyl-	**37:** 5405[1]
4-acetyl-1-phenyl-, (2,4-dinitrophenyl)hydrazone	**37:** 5405[1]
4-acetyl-5-phenyl-	**80:** 3439n
5-acetyl-1-phenyl-	**67:** 100071a
4(or 5)-acetyl-1-phenyl-5(or 4)-(trimethylsilyl)-	**60:** 5538a
1-acetyl-4-propenyl-5-propyl-	**71:** 112865h
1-acetyl-4-propyl-	**68:** 31061z
1-acetyl-5-propyl-	**68:** 13061z
2-acetyl-4-propyl-	**68:** 13061z
4-acetyl-1-(4-oxo-4*H*-pyrido[1,2-*a*]pyrimidin-2-yl)-5-phenyl-	**83:** 9958q
4-[*N*-(4-aminophenyl)formimidoyl]-2-phenyl-	**64:** 8276b
1-benzoyl-	**51:** 14698b
4-benzoyl-	**61:** 14664h, **67:** 32420m **77:** 126519j
4-benzoylacetyl-1-phenyl-	**37:** 5405[2]
4-benzoyl-1-(4-bromophenyl)-5-methyl-	**75:** 109587w
4-benzoyl-1-(4-bromo-3-oxo-2-phenyl-2*H*-pyridazin-5-yl)-5-phenyl-	**76:** 153691v
4-benzoyl-1-(4-chlorophenyl)-5-methyl-	**75:** 109587w
4-benzoyl-1-(4-chloro-3-oxo-2-phenyl-2*H*-pyridazin-5-yl-methyl-	**78:** 136200h
4-benzoyl-1-(4-chloro-3-oxo-2-phenyl-2*H*-pyridazin-5-yl)-5-phenyl-	**78:** 136200h
4-benzoyl-2-(4-chlorophenyl)-5-methyl-	**63:** P11574f
4-benzoyl-1-(2,6-dimethylphenyl)-	**79:** 104890n
1-benzoyl-4,5-diphenyl-	**56:** 1388h
4-benzoyl-1,5-diphenyl-	**64:** 5077a
5-benzoyl-1,4-diphenyl-	**80:** 145179b
5-benzoyl-1,4-diphenyl-, hydrazone	**80:** 145179b
4-benzoyl-1,5-diphenyl-	**81:** 3850k
1-benzoyl-4,5-diphenyl-	**75:** 62695n
4-benzoyl-1-(2-ethoxy-1-oxo-2,4,6-cycloheptatrien-5-yl)-5-phenyl-	**79:** 5300p
4-benzoyl-1-ethyl-5-methyl-	**82:** 171481e
4-benzoyl-1-(7-hydroxy-1-oxo-2,4,6-cycloheptatrien-4-yl)-5-phenyl-	**76:** 140726k
4-benzoyl-1-imidazo[1,2-*b*]pyridazin-6-yl-5-methyl-	**82:** 16745a
1-benzoyl-5-methyl-	**76:** 25192w

TABLE 5 (*Continued*)

Compound	Reference
2-benzoyl-4-methyl-	**76:** 25192w, **79:** 31992k
4-benzoyl-5-methyl-	**61:** 5633b, **75:** 48991z **82:** 171481e
4-benzoyl-5-methyl-1-(2-nitrophenyl)-	**58:** 12560c
4-benzoyl-5-methyl-1-(4-nitrophenyl)-	**58:** 12560c
4-benzoyl-5-methyl-1-(4-oxo-4*H*-pyrido[1,2-*a*]pyrimidin-2-yl)-	**83:** 9958q
4-benzoyl-5-methyl-1-phenyl-	**37:** 5405^{3}
4-benzoyl-5-methyl-1-phenyl-, (2,4-dinitrophenyl) hydrazone	**37:** 5405^{3}
4-benzoyl-5-methyl-2-phenyl-	**22:** 2945^{9}, **22:** 2946^{2}
4-benzoyl-5-methyl-2-phenyl-, (4-nitrophenyl) hydrazone	**22:** 2946^{2}
4-benzoyl-5-methyl-1-tetrazolo[5,1-*a*]phthalazin-6-yl-	**80:** 82860u
4-benzoyl-5-methyl-1-tetrazolo[1,5-*b*]pyridazin-6-yl-	**82:** 16745a
4-benzoyl-5-methyl-1-(1,2,4-triazolo[4,3-*b*]pyridazin-6-yl)-	**82:** 16745a
4-benzoyl-1-phenyl-	**37:** 5404^{9}, **67:** 43759e, **79:** 104890n
4-benzoyl-1-phenyl-, (2,4-dinitrophenyl)hydrazone	**37:** 5404^{9}
1-benzoyl-5-phenyl-	**76:** 25192w
2-benzoyl-4-phenyl-	**76:** 25192w, **79:** 31992k
4-benzoyl-5-phenyl-1-tetrazolo[5,1-*a*]phthalazin-6-yl-	**80:** 82860u
2-benzoyl-2-phenyl-4-(triphenylphosphonio)-, chloride	**79:** 66254y
4,5-bis(1-oxo-3-phenyl-2-propynyl)-1-phenyl-	**82:** 140029j, **83:** 192890y
4-(bromoacetyl)-2-(4-cyanophenyl)-5-methyl-	**84:** P181615g
4-(bromoacetyl)-2-(4-methoxyphenyl)-5-methyl-	**84:** P181615g
4-(bromoacetyl)-5-methyl-2-phenyl-	**84:** P181615g
4-(4-bromobenzoyl)-	**77:** 126519j
4-[*N*-(3-bromo-4-chlorophenyl)formimidoyl]-2-phenyl-	**64:** 8276b
4-[*N*-(4-bromophenyl)formimidoyl]-2-phenyl-	**64:** 8276b
2-(4-bromophenyl)-5-methyl-4-[3-(1-piperidinyl)-1-oxopropyl]-	**79:** P18725f
-1-butanoic acid, α(or β)-dodecenyl-α-oxo-	**78:** P113645t
-1-butanoic acid, α(or β)-isooctadecyl-α-oxo-	**78:** P113645t
2-(4-butylphenyl)-4-(2,4-dihydroxybenzoyl)-	**76:** 114090d
2-(4-butylphenyl)-4-[4-(hexyloxy)-2-hydroxybenzoyl]-	**76:** 114090d
2-(4-butylphenyl)-4-(2-hydroxy-4-methoxybenzoyl)-	**76:** 114090d
2-(4-butylphenyl)-4-[2-hydroxy-4-(3-methylbutoxy) benzoyl]-	**76:** 114090d
2-(4-butylphenyl)-4-[2-hydroxy-4-[(2-methyl-2-propenyl)oxy]benzoyl]-	**76:** 114090d
2-(4-butylphenyl)-4-[2-hydroxy-4-(octadecyloxy) benzoyl]-	**78:** P137370a
2-(4-butylphenyl)-4-[2-hydroxy-4-(octyloxy)benzoyl]-	**76:** 114090d
2-(4-butylphenyl)-4-[2-hydroxy-4-(1-oxopropoxy) benzoyl]-	**76:** 114090d
4-[4-[4-(*tert*-butylphenyl)methoxy]benzoyl]-2-phenyl-	**78:** P137370a
4-[4-[4-(*tert*-butylphenyl)methoxy]-2-hydroxybenzoyl] -2-phenyl-	**76:** 114090d

TABLE 5 (*Continued*)

Compound	Reference
-4-carboxaldehyde,	**37:** 119[2], **43:** 6620f, **80:** 108452q
-4-carboxaldehyde, thiosemicarbazone	**65:** 7121e, **81:** 33139c
-4-carboxaldehyde, 2-[4-(2-benzoxazolyl)phenyl]-5-methyl-	**82:** P17894k
-4-carboxaldehyde, 1-benzyl-	**50:** 4129c
-4-carboxaldehyde, 1-benzyl-, (2,4-dinitrophenyl) hydrazone	**50:** 4129c
-4-carboxaldehyde, 2-(4-bromo-3-chlorophenyl)-	**60:** 643h
-4-carboxaldehyde, 2-methylphenyl-	**59:** 14095e
-4-carboxaldehyde, 2-(4-bromo-3-methylphenyl)-	**59:** 14095e
-4-carboxaldehyde, 2-(3-bromophenyl)-	**59:** 14095e
-4-carboxaldehyde, 2-(4-bromophenyl)-	**59:** 14095e, **66:** 37841p
-4-carboxaldehyde, 2-(4-bromophenyl)-, phenylhydrazone	**66:** 37841p
-4-carboxaldehyde, 2-(4-bromophenyl)-5-methyl-	**66:** 37841p
-4-carboxaldehyde, 2-(4-bromophenyl)-5-methyl-, phenylhydrazone	**66:** 37841p
-4-carboxaldehyde, 2-(4-carboxyphenyl)-	**79:** P66367n
-4-carboxaldehyde, 2-(4-chloro-3-methylphenyl)-	**59:** 14095e
-4-carboxaldehyde, 2-(4-chlorophenyl)-5-methyl-	**84:** P166274b
-4-carboxaldehyde, 2-(4-chloro-3-nitrophenyl)-	**59:** 14095e
-4-carboxaldehyde, 2-(3-chlorophenyl)-	**59:** 10495d
-4-carboxaldehyde, 2-(4-chlorophenyl)-	**59:** 14095e
-4-carboxaldehyde, 2-(4-chlorophenyl)-, 2-(chlorophenyl) anil	**84:** P166274b
-4-carboxaldehyde, 2-(4-chlorophenyl)-5-methyl-, 4-(chlorophenyl) anil	**84:** P166274b
-4-carboxaldehyde, 2-(4-chlorophenyl)-5-phenyl-	**84:** P166274b
-4-carboxaldehyde, 2-(4-chlorophenyl)-5-phenyl-, 4-(chlorophenyl) anil	**84:** P166274b
-4-carboxaldehyde, 2-(3-cyano-4-methylphenyl)-	**84:** P166272z
-4-carboxaldehyde, 2-(4-cyanophenyl)-	**84:** P166272z
-4-carboxaldehyde, 1-(2,4,6-cycloheptatrien-1-yl)-	**62:** 13138g
-4-carboxaldehyde, 2-(3,4-dibromophenyl)-	**59:** 14095e
-4-carboxaldehyde, 2-(3,4-dichlorophenyl)-	**63:** 1850n
-4-carboxaldehyde, 2-(2,5-dimethylphenyl)-	**63:** 1850h
-4-carboxaldehyde, 2-(3,4-dimethylphenyl)-	**60:** 643h
-4-carboxaldehyde, 1,5-diphenyl-	**45:** 9037h
-4-carboxaldehyde, 1,5-diphenyl-, hydrazone	**68:** 78204t
-4-carboxaldehyde, 1,5-diphenyl-, (4-methylphenyl-sulfo)hydrazone	**68:** 78204t
-4-carboxaldehyde, 1,5-diphenyl-, oxime	**45:** 9037h
-4-carboxaldehyde, 1,5-diphenyl-, thiosemicarbazone	**47:** 7477i
-4-carboxaldehyde, 2,5-diphenyl-	**84:** P166274b
-4-carboxaldehyde, 2,5-diphenyl-, 4-(chlorophenyl) anil	**84:** P166274b
-5-carboxaldehyde, 1,4-diphenyl-	**45:** 9037h
-5-carboxaldehyde, 1,4-diphenyl-, hydrazone	**68:** 78204t, **80:** 145179b
-5-carboxaldehyde, 1,4-diphenyl-, (4-methylphenyl-sulfo)hydrazone	**68:** 78204t
-5-carboxaldehyde, 1,4-diphenyl-, oxime	**45:** 9037h

TABLE 5 (*Continued*)

Compound	Reference
-5-carboxaldehyde, 1,4-diphenyl-, thiosemicarbazone	**47:** 7477i
-4-carboxaldehyde, 2-(4-ethylphenyl)-	**60:** 643h
-4-carboxaldehyde, 2-(2-fluorophenyl)-	**80:** 48283y
-4-carboxaldehyde, 2-(3-fluorophenyl)-	**59:** 14095d
-4-carboxaldehyde, 2-(4-fluorophenyl)-	**59:** 14095d
-4-carboxaldehyde, 2-(4-formylphenyl)-	**83:** P61732a
-4-carboxaldehyde, 2-(3-methoxyphenyl)-	**60:** 643h
-4-carboxaldehyde, 2-(4-methoxyphenyl)-	**84:** P166274b
-4-carboxaldehyde, 2-(4-methoxyphenyl)-, 2-(chlorophenyl) anil	**84:** P166274b
-4-carboxaldehyde, 2-(4-methoxyphenyl)-, oxime	**84:** P166274b
-4-carboxaldehyde, 1-methyl-	**42:** 2599g, **67:** 16500g
-4-carboxaldehyde, 2-[3,4-(methylenedioxy)phenyl]-	**63:** 1850h
-4-carboxaldehyde, 2-(4-methyl-3-nitrophenyl)-	**60:** 643h
-4-carboxaldehyde, 5-methyl-2-(4-nitrophenyl)-, phenylhydrazone	**66:** 37841p
-4-carboxaldehyde, 2-(3-methylphenyl)-	**60:** 643h
-4-carboxaldehyde, 2-(4-methylphenyl)-	**60:** 643h, **63:** 11540f
-4-carboxaldehyde, 5-methyl-2-phenyl-	**66:** 37841p
-4-carboxaldehyde, 5-methyl-2-phenyl-, 4-(chlorophenyl) anil	**84:** P166274b
-4-carboxaldehyde, 5-methyl-2-phenyl-, phenylhydrazone	**66:** 37841p
-4-carboxaldehyde, 5-methyl-2-(4-sulfophenyl)-, sodium salt	**81:** P51142c
-4-carboxaldehyde, 2-(4-nitrophenyl)-	**48:** 5805c
-4-carboxaldehyde, 2-(4-nitrophenyl)-, (2,4-dinitrophenyl)hydrazone	**64:** 8276b
-4-carboxaldehyde, 2-(4-nitrophenyl)-, phenylhydrazone	**64:** 8276b
-4-carboxaldehyde, 2-(3-oxo-1-phenyl-1-propenyl)-5-phenyl-, (E)-	**80:** 108451p
-4-carboxaldehyde, 2-(3-oxo-1-phenyl-1-propenyl)-5-phenyl, (Z)-	**80:** 108451p
-4-carboxaldehyde, 1-phenyl-	**37:** 119[2], **45:** 9038b, **49:** 6241h, **79:** 136425f **84:** 180505j
-4-carboxaldehyde, 1-phenyl-, (2,4-dinitrophenyl) hydrazone	**75:** 88891y
-4-carboxaldehyde, 2-phenyl-	**38:** 3622[4], **41:** 3766a, **41:** 4459f, **41:** 5860c, **48:** 2608a, **48:** 5805b, **64:** 8276b, **67:** 32382a
-4-carboxaldehyde, 2-phenyl-, 2-azine with 2,4-dioxo-5-thiazolidineactic acid	**70:** 76696x, **70:** 96686x
-4-carboxaldehyde, 2-phenyl-, hydrazone	**64:** 8276b
-4-carboxaldehyde, 2-phenyl-, (methylphenyl)hydrazone	**57:** 3546f, **64:** 8276b
-4-carboxaldehyde, 2-phenyl-, (4-nitrophenyl)hydrazone	**49:** 13901d, **64:** 8276b
-4-carboxaldehyde, 2-phenyl-, oxime	**64:** 8276b
-4-carboxaldehyde, 2-phenyl-, phenyl anil	**78:** P85943t

TABLE 5 (*Continued*)

Compound	Reference
-4-carboxaldehyde, 2-phenyl-, phenylhydrazone	**49:** 12451g, **57:** 3546g **64:** 8276b
-4-carboxaldehyde, 2-phenyl-, thiosemicarbazone	**50:** 362f
-4-carboxaldehyde, 2-phenyl-, (4,5,6,7-tetrahydro-2-benzothiazolyl) hydrazone	**70:** 76696x, **70:** 96685x
-4-carboxaldehyde, 2-phenyl-, thiosemicarbazone	**70:** 76696x
-4-carboxaldehyde, 5-phenyl-	**45:** 9037f, **80:** 108451p, **80:** 108452q
-4-carboxaldehyde, 5-phenyl-, thiosemicarbazone	**47:** 7477i
-5-carboxaldehyde, 1-phenyl-	**45:** 9038b, **84:** 180505j
-4-carboxaldehyde, 2-(4-sulfamoylphenyl)-	**79:** P66367n
-4-carboxaldehyde, 2-(3-sulfophenyl)-	**80:** P16460n
-4-carboxaldehyde, 2-(4-sulfophenyl)-, sodium salt	**79:** P106146y
4-(4-chlorobenzoyl)-	**64:** 17579a, **64:** 17581f, **77:** 126519j
4-(4-chlorobenzoyl)-1,5-diphenyl-	**81:** 3850k
5-(4-chlorobenzoyl)-1,4-diphenyl-	**81:** 3850k
1-(4-chlorobenzoyl)-5-methyl-	**76:** 25192w, **79:** 31992k
2-(4-chlorobenzoyl)-4-methyl-	**76:** 25192w, **79:** 31992k
1-(4-chlorobenzoyl)-5-(4-nitrophenyl)-	**76:** 25192w
2-(4-chlorobenzoyl)-4-(4-nitrophenyl)-	**76:** 25192w, **79:** 31992k
1-(4-chlorobenzoyl)-5-phenyl-	**76:** 25192w
2-(4-chlorobenzoyl)-4-phenyl-	**76:** 25192w, **79:** 31992k
4-(2-chloro-*N*-fluoroacetimidoyl)-	**74:** 111499s
4-[*N*-(2-chlorophenyl)formimidoyl]-2-phenyl-	**64:** 8276h
4-[*N*-(3-chlorophenyl)formimidoyl]-2-phenyl-	**64:** 8276h
2-(4-chlorophenyl)-5-methyl-4-[3-(4-morpholinyl)-1-oxopropyl]-	**79:** P18724e, **79:** P18725f
2-(4-chlorophenyl)-5-methyl-4-[3-(4-morpholinyl)-1-oxopropyl]-, hydrochloride	**79:** P18724e
2-(3-chlorophenyl)-5-methyl-4-[1-oxo-3-(1-piperidinyl)propyl]-	**79:** P126505t
2-(x-chloro-y-sulfophenyl)-4-formyl-, sodium salt (?)	**79:** P106146y
2-(3-cyano-4-methylphenyl)-4-formyl-	**79:** 78518a
2-(4-cyanophenyl)-4-formyl-	**79:** P66367n
4-cyclohexyloxomethyl-	**71:** 91581v
4-cyclohexyloxomethyl-5-ethyl-	**71:** 91581v
4-cyclopropyloxomethyl-4-ethyl-	**71:** 91581v
4-[4-(decyloxy)-2-hydroxybenzoyl]-2-phenyl-	**76:** 114090d
4-[1-[2-(1,2-dicarbadodecaboran(12)-1-yl)]ethanoyl]-1-phenyl-	**71:** 39048n
4,5-dibenzoyl-	**62:** 13138h, **64:** 14184h, **80:** 82858z
4,5-dibenzoyl-1-(2,4,6-cycloheptatrien-1-yl)-	**62:** 13138h
4,5-dibenzoyl-1-(5*H*-dibenzo[*a,d*]cyclohepten-5-yl)-	**74:** P53342x
4,5-dibenzoyl-1-(10,11-dihydro-5*H*-dibenzo[*a,d*]cyclohepten-5-yl)-	**77:** P95371y
-4,5-dicarboxaldehyde, 1-benzyl-	**38:** 1743[1]
-4,5-dicarboxaldehyde, 1-benzyl-, bis(diethyl acetal)	**38:** 1743[1]
-4,5-dicarboxaldehyde, 2-(4-bromophenyl)-	**76:** 4098e

TABLE 5 (*Continued*)

Compound	Reference
-4,5-dicarboxaldehyde, 2-(4-bromophenyl)-, bis(phenylhydrazone)	**76:** 4098e
-4,5-dicarboxaldehyde, 1-dodecyl-	**38:** 1743[2]
-4,5-dicarboxaldehyde, 1-dodecyl-, bis(diethyl acetal)	**38:** 1743[2]
-4,5-dicarboxaldehyde, 1-hexyl-	**38:** 1743[2]
-4,5-dicarboxaldehyde, 1-hexyl-, bis(diethyl acetal)	**38:** 1743[2]
-4,5-dicarboxaldehyde, 1-phenyl-	**38:** 1742[9], **80:** 108454s, **82:** 140029j
-4,5-dicarboxaldehyde, 1-phenyl- bis(diethyl acetal)	**38:** 1742[9]
-4,5-dicarboxaldehyde, 2-phenyl-	**55:** 13413f
-4,5-dicarboxaldehyde, 2-phenyl-, bis(2,4-dinitrophenyl)hydrazone	**55:** 13413f
1-(3,4-dichlorobenzoyl)-5-methyl-	**76:** 25192w
1-(3,4-dichlorobenzoyl)-5-phenyl-	**64:** 19597a, **76:** 25192w, **79:** 31992k
2-(3,4-dichlorobenzoyl)-4-methyl-	**76:** 25192w, **79:** 31992k
2-(3,4-dichlorobenzoyl)-4-phenyl-	**76:** 25192w, **79:** 31992k
2-(2,4-dichlorophenyl)-4-(2,4-dihydroxybenzoyl)-	**76:** 114090d
2-(2,5-dichlorophenyl)-4-(2-hydroxy-4-methoxybenzoyl)-	**76:** 114090d
2-(2,5-dichlorophenyl)-4-[2-hydroxy-4-(octadecyloxy) benzoyl]-	**76:** 114090d
4-[3-(diethylamino)-1-oxo-3-phenyl-2-propenyl]-5-phenyl-	**80:** 108452q
4-(2,4-dihydroxybenzoyl)-2,5-diphenyl-	**76:** 114090d
4-(2,4-dihydroxybenzoyl)-2-(2-hydroxyphenyl)-	**76:** 114090d
4-(2,4-dihydroxybenzoyl)-2-(2-hydroxyphenyl)-5-methyl-	**76:** 114090d
4-(2,4-dihydroxybenzoyl)-5-methyl-2-phenyl-	**76:** 114090d
4-[3-(dimethylamino)-1-oxopropyl]-5-methyl-2-(4-methylphenyl)-, hydrochloride	**79:** P18724e
4-[3-(dimethylamino)-1-oxopropyl]-5-methyl-2-phenyl-	**79:** P126505t
4-[3-(dimethylamino)-1-oxopropyl]-5-methyl-2-phenyl-, monohydrochloride	**75:** 62010k
4,5-dimethyl-1-(3-nitrobenzoyl)-	**74:** 76375z, **76:** 25192w, **79:** 31992k
4,5-dimethyl-2-(3-nitrobenzoyl)-	**74:** 76375z, **76:** 25192w, **79:** 31992k
4-(2,2-dimethyl-1-oxopropyl)-5-ethyl-	**71:** 91581v
4-(2,2-dimethyl-1-oxopropyl)-1-phenyl-	**80:** 47912r
1,4-diphenyl-5-(4-methoxybenzoyl)-	**81:** 3850k
1,5-diphenyl-4-(4-methoxybenzoyl)-	**81:** 3850k
2,5-diphenyl-4-(4-ethoxy-2-hydroxybenzoyl)-	**76:** 114090d
2,5-diphenyl-4-(2-hydroxy-4-methoxybenzoyl)-	**76:** 114090d
2,5-diphenyl-4-[2-hydroxy-4-(octadecyloxy)benzoyl]-	**76:** 114090d
4-[4-(dodecyloxy)-2-hydroxybenzoyl]-2-(3-methoxyphenyl)-	**76:** 114090d
4-[4-(dodecyloxy)-2-hydroxybenzoyl]-5-methyl-2-phenyl-	**76:** 114090d
5-ethyl-4-(2-methyl-1-oxopropyl)-	**71:** 91581v
4-[*N*-(3-fluorophenyl)formimodoyl]-2-phenyl-	**64:** 8276b
4-formyl-5-methyl-2-(sulfophenyl)-, sodium salt (?)	**81:** P65249e

TABLE 5 (*Continued*)

Compound	Reference
4-formyl-2-(sulfophenyl)-(?)	**81:** P65249e
-4-glycolaldehyde, 2-phenyl-, (4-nitrophenyl) hydrazone	**49:** 13901d
-4-glyoxylanilide, 1-phenyl-, oxime	**24:** 592[9]
5-heptadecyl-2-(2-hydroxyphenyl)-5-(2-hydroxy-4-methoxybenzoyl)-	**76:** 114090d
4-[4-(hexadecyloxy)-2-hydroxybenzoyl]-2-(2-hydroxy-phenyl)-5-methyl-	**76:** 114090d
4-(2-hydroxy-4-methoxybenzoyl)-2-(2-hydroxyphenyl)-	**76:** 114090d
4-(2-hydroxy-4-methoxybenzoyl)-2-(2-hydroxyphenyl)-5-methyl-	**76:** 114090d
4-(2-hydroxy-4-methoxybenzoyl)-2-(3-methoxyphenyl)-	**76:** 114090d
4-(2-hydroxy-4-methoxybenzoyl)-5-methyl-2-phenyl-	**76:** 114090d
4-(2-hydroxy-4-methoxybenzoyl)-2-phenyl-	**76:** 114090d
4-[[2-hydroxy-4-(octadecanoyl)oxy]benzoyl]-2-(2-hydroxyphenyl)-	**76:** 114090d
4-[2-hydroxy-4-(octadecyloxy)benzoyl]-5-methyl-2-phenyl-	**76:** 114090d
4-[2-hydroxy-4-(octyloxy)benzoyl]-5-methyl-2-phenyl-	**76:** 114090d
4-[2-hydroxy-4-(octyloxy)benzoyl]-2-phenyl-	**76:** 114090d
2-(2-hydroxyphenyl)-4-[2-hydroxy-4-(tetradecyloxy) benzoyl]-	**76:** 114090d
4-(2-hydroxy-4-propoxybenzoyl)-2-phenyl-	**76:** 114090d
4-[*N*-(4-iodophenyl)formimodyl]-2-phenyl-	**64:** 8276b
4-(4-methoxybenzoyl)-	**64:** 17579a, **64:** 17581f, **77:** 126519i
1-(4-methoxybenzoyl)-5-methyl-	**74:** 76375z, **76:** 25192w, **79:** 31992k
2-(4-methoxybenzoyl)-4-methyl-	**74:** 76375z, **76:** 25192w, **79:** 31992k
2-(4-methoxybenzoyl)-4-(4-nitrophenyl)-	**74:** 76375z
1-(4-methoxybenzoyl)-5-phenyl-	**74:** 76375z, **76:** 25192w, **79:** 31992k
2-(4-methoxybenzoyl)-4-phenyl-	**74:** 76375z, **76:** 25192w, **79:** 31992k
4-[*N*-(2-methoxyphenyl)formimidoyl]-2-phenyl-	**64:** 8276b
4-[*N*-(4-methoxyphenyl)formimidoyl]-2-phenyl-	**64:** 8276b
4-[3-[[2-(4-methoxyphenyl)-1-methylethyl]methylamino]-1-oxopropyl]-5-methyl-2-phenyl-, monohydrochloride	**75:** 62010k
2-(2-methoxyphenyl)-5-methyl-4-[1-oxo-3-(1-piperi-dinyl)propyl]-, hydrochloride	**79:** P18724e
2-(4-methoxyphenyl)-5-methyl-4-[1-oxo-3-(1-piperi-dinyl)propyl]-, monohydrochloride	**79:** P126505t
4-[1-[2-(2-methyl-1,2-dicarbadodecaboran(12)-1-yl)] ethanoyl]-1-phenyl-	**71:** 39048n
5-methyl-4-[3-(4-morpholinyl)-1-oxopropyl]-2-phenyl-, monohydrochloride	**75:** 62010k
4-methyl-2-(3-nitrobenzoyl)-	**76:** 25192w, **79:** 31992k
4-methyl-2-(4-nitrobenzoyl)-	**74:** 76375z, **76:** 25192w, **79:** 31992k

TABLE 5 (*Continued*)

Compound	Reference
5-methyl-1-(3-nitrobenzoyl)-	**76:** 25192w
5-methyl-1-(4-nitrobenzoyl)-	**74:** 76375z, **76:** 25192w, **79:** 31992k
5-methyl-2-(4-nitrophenyl)-4-[1-oxo-3-(1-piperidinyl) propyl]-	**79:** P126505t
5-methyl-2-(4-nitrophenyl)-4-[1-oxo-3-(1-piperidinyl) propyl]-, monohydrochloride	**75:** 62010k
5-methyl-4-(1-oxo-2-butynyl)-	**80:** 108452q
5-methyl-4-[1-oxo-2-(1-piperidinyl)ethyl]-2-phenyl-	**79:** P18725f
5-methyl-4-[1-oxo-3-(1-piperidinyl)propyl]-2-phenyl-	**79:** P18724e, **79:** P126505t
1-(2-methyl-1-oxo-2-propenyl)-	**85:** 47115m
4-[*N*-(2-methylphenyl)formimidoyl]-2-phenyl-	**64:** 8276b
4-[*N*-(3-methylphenyl)formimidoyl]-2-phenyl-	**64:** 8276b
4-[*N*-(4-methylphenyl)formimidoyl]-2-phenyl-	**64:** 8276b
4-(*N*-naphthylformimidoyl)-2-phenyl-	**64:** 8276b
1-(3-nitrobenzoyl)-5-(4-nitrophenyl)-	**76:** 25192w
2-(3-nitrobenzoyl)-4-(4-nitrophenyl)-	**76:** 25192w, **79:** 31992k
1-(3-nitrobenzoyl)-5-phenyl-	**76:** 25192w
1-(4-nitrobenzoyl)-5-phenyl-	**76:** 25192w
2-(3-nitrobenzoyl)-4-phenyl-	**76:** 25192w, **79:** 31992k
2-(4-nitrobenzoyl)-4-phenyl-	**76:** 25192w, **79:** 31992k
4-[(5-nitro-2-furyl)oxomethyl]-1-phenyl-	**75:** 20354v
4-[*N*-(3-nitrophenyl)formimidoyl]-2-phenyl-	**64:** 8276b
4-[*N*-(4-nitrophenyl)formimidoyl]-2-phenyl-	**64:** 8276b
4-(1-oxobutyl)-	**77:** 126517g
1-(1-oxo-2-propenyl)-	**85:** 47115m
4-(1-oxopropyl)-	**70:** 96727m
4-(1-oxo-3-phenyl-3-phenylamino-2-propenyl)-5-phenyl-	**80:** 108452q
4-(1-oxo-3-phenyl-2-propenyl)-5-phenyl-	**80:** 108452q
4-(1-oxo-3-phenyl-2-propynyl)-5-phenyl-	**80:** 108452q
2-phenyl-4-(*N*-phenylformimidoyl)-	**64:** 8276b
-2-propanoic acid, 4-acetyl-5-phenyl-	**85:** P123926u
-4-propanoic acid, 5-methyl-β-oxo-1-phenyl, ethyl ester	**80:** 47912r

REFERENCES

22: 2945–6[9–2]	G. Wittig, F. Bangert, and H. Kleiner, *Chem. Ber.*, **61B,** 1140 (1928).
24: 529[9]	H. Wieland, W. Frank, and Z. Kitasato, *Justus Liebigs Ann. Chem.*, **475,** 42 (1929).
36: 771[9]	A. Quilico and C. Musante, *Gazz. Chim. Ital.*, **71,** 327 (1941).
37: 119[2]	R. Hüttel, *Chem. Ber.*, **74B,** 1680 (1941).
37: 5404–5[9–3]	W. Borsche, H. Hahn, and M. Wagner-Roemmich, *Justus Liebigs Ann. Chem.*, **554,** 15 (1943).
38: 1742–3[9–2]	K. Henkel and F. Weygand, *Chem. Ber.*, **76B,** 812 (1943).
38: 3622[4]	R. M. Hann and C. S. Hudson, *J. Amer. Chem. Soc.*, **66,** 735 (1944).
41: 3766a	P. P. Regna, *J. Amer. Chem. Soc.*, **69,** 246 (1947).
41: 4459f	W. T. Haskins, R. M. Hann, and C. S. Hudson, *J. Amer. Chem. Soc.*, **69,** 1050 (1947).

41: 5860c W. T. Haskins, R. M. Hann, and C. S. Hudson, *J. Amer. Chem. Soc.*, **69,** 1461 (1947).

42: 2599g R. Hüttel and A. Gebhardt, *Justus Liebigs Ann. Chem.*, **558,** 34 (1947).

43: 2619b J. L. Riebsomer and G. Sumrell, *J. Org. Chem.*, **13,** 807 (1948).

43: 6620f J. C. Sheehan and C. A. Robinson, *J. Amer. Chem. Soc.*, **71,** 1436 (1949).

45: 9037–8f–b J. C. Sheehan and C. A. Robinson, *J. Amer. Chem. Soc.*, **73,** 1207 (1951).

47: 7477i R. E. Hagenbach and H. Gysin, *Experientia*, **8,** 184 (1952).

48: 2608a E. Hardegger and E. Schreier, *Helv. Chim. Acta*, **35,** 623 (1952).

48: 5805c C. T. Bishop, *Science*, **117,** 715 (1953).

49: 6241h R. Hüttel, T. Schneiderhan, H. Hertwig, A. Leuchs, V. Reincke, and J. Miller, *Justus Liebigs Ann. Chem.*, **585,** 115 (1954).

49: 12451g H. J. Cothrell, D. L. Pain, and R. Slack, *J. Chem. Soc.*, 2968 (1954).

49: 13901d E. Schreier, G. Stöhr, and E. Hardegger, *Helv. Chim. Acta*, **37,** 574 (1954).

50: 362f L. Toldy, T. Nógrádi, L. Vargha, G. Ivánovics, and I. Koczka, *Acta Chim. Acad. Sci. Hung.*, **4,** 303 (1954).

50: 4129c R. H. Wiley, N. R. Smith, D. M. Johnson, and J. Moffat, *J. Amer. Chem. Soc.*, **77,** 3412 (1955).

50: 4924d N. K. Kochetkov, *Zhur. Obshcheĭ Khim.*, **25,** 1366 (1955).

51: 14698b S. Yamada, T. Mizoguchi, and A. Ayata, *Yakugaku Zasshi*, **77,** 452 (1957).

54: 4547d–f R. Hüttel and J. Kratzer, *Chem. Ber.*, **92,** 2014 (1959).

55: 8394b M. Ruccia, *Ann. Chim. (Rome)*, **50,** 1363 (1960).

55: 13413f L. Anderson and J. N. Aronson, *J. Org. Chem.*, **24,** 1812 (1959).

56: 1388h R. Huisgen and M. Seidel, *Chem. Ber.*, **94,** 2509 (1961).

57: 3546f,g G. Henseke, H. J. Binte, S. Schwerin, and G. Crawack, *Justus Liebigs Ann. Chem.*, **651,** 162 (1962).

57: 5911g A. J. Boulton and A. R. Katritzky, *J. Chem. Soc.*, 2083 (1962).

58: 12560f,g R. Fusco, G. Bianchetti, D. Pocar, and R. Ugo, *Gazz. Chim. Ital.*, **92,** 1040 (1962).

59: 14095d,e H. El Khadem, A. M. Kolkaila, and M. H. Meshreki, *J. Chem. Soc.*, 3531 (1963).

60: 643h B. B. Bishay, H. El Khadem, and Z. M. El-Shafei, *J. Chem. Soc.*, 4980 (1963).

60: 5538a L. Birkofer, A. Ritter, and P. Richter, *Chem. Ber.*, **96,** 2750 (1963).

61: 5633b G. Bianchetti, D. Pocar, and P. D. Croce, *Gazz. Chim. Ital.*, **94,** 340 (1964).

61: 14664h A. N. Nesmeyanov and M. I. Rybinskays, *Dokl. Akad. Nauk SSSR*, **158,** 408 (1964).

62: 13138g,h J. J. Looker, *J. Org. Chem.*, **30,** 638 (1965).

63: 1850h H. El Khadem, Z. M. El-Shafei, and M. H. Meshreki, *J. Chem. Soc.*, 1524 (1965).

63: 11540f H. J. Binte and G. Henseke, *Z. Chem.*, **5,** 268 (1965).

63: P11574f B. Hirsch, East German Patent, 36,137 (1965).

63: 16338e R. Huisgen and J. P. Anselme, *Chem. Ber.*, **98,** 2998 (1965).

64: 5077a R. Huisgen, R. Knorr, L. Möbius, and G. Szeimies, *Chem. Ber.*, **98,** 4014 (1965).

64: 8276b–h H. El Khadem, Z. M. El-Shafei, M. H. Meshreki, and M. A. Shaban, *J. Chem. Soc., C*, 91 (1966).

64: 14184h J. O. Fournier and J. B. Miller, *J. Heterocycl. Chem.*, **2,** 488 (1965).

64: 17579a A. N. Nesmeyanov and M. I. Rybinskaya, *Dokl. Akad. Nauk SSSR*, **167,** 109 (1966).

64: 17581f A. N. Nesmeyanov and M. I. Rybinskaya, *Dokl. Akad. Nauk SSSR*, **166,** 1362 (1966).

64: 19597a G. R. Harvey, *J. Org. Chem.*, **31,** 1587 (1966).

65: 5454h H. G. Garg, *Indian J. Chem.*, **4,** 200 (1966).

65: 7121e F. A. French and E. J. Blanz, Jr., *J. Med. Chem.*, **9,** 585 (1966).

65: 15414g L. Birkofer and P. Wegner, *Chem. Ber.*, **99,** 2512 (1966).
66: 37841p G. Henseke and I. Schmeisky, *J. Prakt. Chem.*, **33,** 256 (1966).
67: 16500g G. B. Barlin and T. J. Batterham, *J. Chem. Soc.*, *B*, 516 (1967).
67: 32382a Y. Usui and C. Matsumura, *Yakugaku Zasshi*, **87,** 43 (1967).
67: 32420m S. Maiorana, *Ann. Chim. (Rome)*, **56,** 1531 (1966).
67: 43759e F. M. Stojanovič and Z. Arnold, *Collect. Czech. Chem. Commun.*, **32,** 2155 (1967).
67: 100071a G. S. Akimova, V. N. Chistokletov, and A. A. Petrov, *Zh. Org. Khim.*, **3,** 968 (1967).
68: 2862k A. J. Boulton, A. R. Katritzky, and A. M. Hamid, *J. Chem. Soc.*, *C*, 2005 (1967).
68: 13061z L. Birkofer and P. Wegner, *Chem. Ber.*, **100,** 3485 (1967).
68: 78204t P. A. S. Smith and J. G. Wirth, *J. Org. Chem.*, **33,** 1145 (1968).
69: P67394z T. Nozoe, T. Mukai, T. Toda, and H. Horino, Japanese Patent, 68 00,709 (1968).
70: 76696x Y. Usui, *Ann. Rep. Takeda Res. Lab.*, **27,** 144 (1968).
70: 96685x Y. Usui, *Ann. Rep. Takeda Res. Lab.*, **27,** 112 (1968).
70: 96727m M. Rasberger and E. Zbiral, *Monatsh. Chem.*, **100,** 64 (1969).
71: 39048n L. I. Zakharkin and A. V. Grebennikov, *Zh. Obshch. Khim.*, **39,** 575 (1969).
71: 91581v E. Zbiral, M. Rasberger, and H. Hengstberger, *Justus Liebigs Ann. Chem.*, **725,** 22 (1969).
71: 112865h E. Zbiral and J. Stroh, *Monatsh. Chem.*, **100,** 1438 (1969).
73: 77172x E. Gudriniece and V. V. Solov'eva, *Latv. PSR Zinat. Akad. Vestis, Kim. Ser.*, 378 (1970).
74: 48048b L. F. Avramenko, Yu. B. Vilenskii, B. M. Ivanov, M. V. Kudryavskaya, I. A. Ol'shevskaya, V. Ya. Pochinok, L. I. Skripnik. L. N. Fedorova, and I. P. Fedorova, *Usp. Nauch. Fotogr.* **14,** 12 (1970).
74: P53342x J. J. Looker, U.S. Patent, 3,533,786 (1970).
74: 76375z P. Ykman, G. L'abbé, and G. Smets, *Tetrahedron Lett.*, 5225 (1970).
74: 111499s B. L. Dyatkin, K. N. Makarov, and I. L. Knunyants, *Tetrahedron*, **27,** 51 (1971).
75: 20354v H. Saikachi and H. Sugimoto, *Yakugaku Zasshi*, **91,** 64 (1971).
75: 48991z S. Maiorana and G. Pagani, *Chim. Ind.*, **53,** 470 (1971).
75: 62010*k* I. Gruebner, W. Klinger, and H. Ankermann, *Arch. Int. Pharmacodyn. Ther.*, **191,** 37 (1971).
75: 62695n F. Compernolle and M. Dekeirel, *Org. Mass Spectrom.*, **5,** 427 (1971).
75: 88891y H. El Khadem, D. Horton, and M. H. Meshreki, *Carbohyd. Res.*, **16,** 409 (1971).
75: 109587w J. Van Thielen, T. Van Thien, and F. C. DeSchryvet, *Tetrahedron Lett.*, 3031 (1971).
76: 25192w P. Ykman, G. L'abbé, and G. Smets, *Tetrahedron*, **27,** 5623 (1971).
76: 4098e A. J. Fatiadi, *Carbohyd. Res.*, **20,** 179 (1971).
76: 59538d F. Texier and R. Carrié, *Bull. Soc. Chim. Fr.*, 3642 (1971).
76: P114090d J. Rody and H. Lind, German Patent, 2,111,538 (1971).
76: 140726k H. Horino and T. Toda, *Bull. Chem. Soc. Jap.*, **45,** 584 (1972).
76: 153691v E. Gudriniece and V. V. Solov'eva, *Latv. PSR Zinat. Akad. Vestis. Kim. Ser.*, **81** (1972).
77: P7313c H. Lind, German Patent, 2,133,012 (1972).
77: P95371y R. C. Deselms, J. J. Looker, and C. V. Wilson, British Patent, 1,271,221 (1972).
77: 101470y I. Hirao and Y. Kato, *Bull. Chem. Soc. Jap.*, **45,** 2055 (1972).
77: 126517g A. N. Nesmeyanov and M. I. Rybinskaya, *Issled. Obl. Org. Khim. Izbr. Tr. Nesmeyanov, A.N.*, 291 (1971).

77: 126519j A. N. Nesmeyanov and M. I. Rybinskaya, *Issled. Obl. Org. Khim. Izbr. Tr. Nesmeyanov, A.N.*, 284 (1971).
78: 16102f E. Klingsberg, *Synthesis*, 475 (1972).
78: 16120k V. V. Solov'eva and E. Gudriniece, *Latv. PSR Zinat. Akad. Vestis, Khim. Ser.*, 572 (1972).
78: P113645t H. J. Andress, Jr. and A. B. Piotrowski, German Patent, 2,228,485 (1973).
78: 136200h V. V. Solov'eva and E. Gudriniece, *Khim. Geterotsikl. Soedin.*, 256 (1973).
78: P137370a J. Rody and H. Lind, Swiss Patent, 524,662 (1972).
79: 5300p H. Horino and T. Toda, *Bull. Chem. Soc. Jap.*, **46,** 1212 (1973).
79: P18724e B. Hirsch, D. Lohmann, and H. Froemmel, East German Patent, 94,195 (1972).
79: P18725f B. Hirsch, D. Lohmann, and H. Froemmel, East German Patent, 94,010 (1972).
79: 31992k G. L'abbé, *Verh. Kon. Acad. Wetensch. Lett. Schone Kunsten Belg., Kl. Wetensch.*, **34,** 49 pp (1972).
79: 66254y Y. Tanaka and S. I. Miller, *J. Org. Chem.*, **38,** 2708 (1973).
79: P66367n I. Okubo, M. Tsujimoto, R. Tsukahaara, and I. Nishizawa, Japanese Patent 73 11,098 (1973).
79: 104890n M. Muramatsu, N. Obata, and T. Takizawa, *Tetrahedron Lett.*, 2133 (1973).
79: P106146y K. Weber and H. Schlaepfer, German Patent, 2,262,633 (1973).
79: P126505t B. Hirsch, D. Lohmann, and H. Froemmel, East German Patent, 92,924 (1972).
79: 136425f Z. Arnold and J. Šauliová, *Collect. Czech. Chem. Commun.*, **38,** 2641 (1973).
80: 3439n G. Beck and D. Günther, *Chem. Ber.*, **106,** 2758 (1973).
80: P16460n P. Liechti and H. Schlaepfer, German Patent, 2,309,614 (1973).
80: 47912r C. E. Olsen, *Acta Chem. Scand.*, **27,** 1987 (1973).
80: 48283y H. S. El Khadem and D. L. Swartz, *Carbohydr. Res.*, **30,** 400 (1973).
80: 82858z T. L. Gilchrist, G. E. Gymer, and C. W. Rees, *J. Chem. Soc., Chem. Commun.*, 819 (1973).
80: 82860u A. Karklina and E. Gudriniece, *Latv. PSR Zinat. Akad. Vestis, Khim. Ser.*, 750 (1973).
80: 108451p Y. Tanaka and S. I. Miller, *Tetrahedron*, **29,** 3285 (1973).
80: 108452q Y. Tanaka, S. R. Velen, and S. I. Miller, *Tetrahedron*, **29,** 3271 (1973).
80: 108454s W. Winter and E. Müller, *Chem. Ber.*, **107,** 715 (1974).
80: 145179b P. A. S. Smith and E. M. Bruckmann, *J. Org. Chem.*, **39,** 1047 (1974).
81: 3850k F. G. Badder, M. N. Basyouni, F. A. Fouli, and W. I. Awad, *J. Indian Chem. Soc.*, **50,** 589 (1973).
81: 33139c F. A. French, E. J. Blanz, Jr., S. C. Shaddix, and R. W. Brockman, *J. Med. Chem.*, **17,** 172 (1974).
81: P51142c F. Fleck and H. Schmid, Swiss Patent, 544,103 (1973).
82: 16745a A. Gorup, M. Kovačič, B. Kranjc-Škraba, B. Mihelčič, S. Simonič, B. Stanovnik, and M. Tišler, *Tetrahedron*, **30,** 2251 (1974).
82: 31304z S. Polanc, B. Vercek, B. Sek, B. Stanovnik, and M. Tišler, *J. Org. Chem.*, **39,** 2143 (1974).
82: 140029j E. Müller and W. Winter, *Justus Liebigs Ann. Chem.*, 1876 (1974).
82: 171481e F. C. DeSchryver, T. V. Thien, S. Toppet, and G. Smets, *J. Polym. Sci., Polym. Chem. Ed.*, **13,** 227 (1975).
83: 9958q M. Kovačič, S. Polanc, B. Stanovnik, and M. Tišler, *J. Heterocycl. Chem.*, **11,** 949 (1974).
83: 192890y E. Müller, A. Scheller, W. Winter, F. Wagner, and H. Meier, *Chem.-Ztg.*, **99,** 155 (1975).
84: P45268u M. Minagawa and N. Kubota, Japanese Patent, 75 105,559 (1975).
84: P166272z H. Aebli, F. Fleck, and H. Schmid, U.S. Patent, 3,940,388 (1976).

84: P166274b G. Kormany, G. Kabas, H. Schlaepfer, and A. E. Siegrist, German Patent, 2,535,615 (1976).

84: 180505j D. Horton and A. Liav, *Carbohydr. Res.*, **47,** 81 (1976).

84: P181615g F. Fleck, A. V. Mercer, and R. Paver, German Patent, 2,535,069 (1976).

85: 47115m K. Khuseinov G. V. Shatalov, and B. I. Mikhantev, *Izv. Vyssh. Uchebn. Zaved. Khim. Khim. Tekhnol.*, **19,** 655 (1976).

85: P123926u R. T. Buckler, H. E. Hartzler, S. Hayao, and G. Nichols, U.S. Patent, 3,948,932 (1976).

CHAPTER 6

Amino- and Amido-1,2,3-Triazoles

v-Triazoles containing amino-, amido-, or sulfonamido-substituents have been studied extensively and a broad range of preparations attempted. Most of these approaches have dealt with isolated examples; a few of the more promising are illustrated here with the general thought that further study is appropriate (Eqs. 1 to 4).[1–6] An unsymmetrical substrate has been shown to provide some regiospecificity (Eq. 5).[5,7] Products like **6.1-1** are hydrolyzed under mild conditions to the amine.

The widely used addition of azides to acetylenes produces only fair yields of amino-1,2,3-triazoles in the limited number of examples reported (Eqs.

$$PhCH{=}C(NO_2)Br + PhN_3 \longrightarrow \text{4-}O_2N\text{-5-Ph-1-Ph-1,2,3-triazole} \xrightarrow[PtO_2]{H_2NNH_2H_2O} \underset{93\%}{\text{4-}H_2N\text{-5-Ph-1-Ph-1,2,3-triazole}} \quad (1)$$

$$\text{4-Ph-5-}N_3\text{-1-Ph-1,2,3-triazole} \xrightarrow[(2)\ Na_2S_2O_4]{(1)\ >50^\circ} \underset{91\%}{PhNHN{=}NCH(Ph)CN} \longrightarrow \text{4-Ph-5-}NH_2\text{-1-Ph-1,2,3-triazole} \quad (2)$$

$$O_2NCH{=}C(R)(\text{N-morpholino}) + ArN_3 \longrightarrow \underset{R = H,\ Ar = Ph,\ 61\%}{\text{4-}O_2N\text{-5-R-1-Ar-1,2,3-triazole}} \xrightarrow{[H]} \underset{\text{good}}{\text{4-}H_2N\text{-5-R-1-Ar-1,2,3-triazole}} \quad (3)$$

$$\text{RC(=NNHTs)–C(=NNHTs)R} \xrightarrow[\text{diglyme}]{\text{Na}^{+\,-}\text{OEt}} \text{4,5-R}_2\text{-1-NHTs-1,2,3-triazole } (\textbf{6.1-1}) \qquad (4)$$

R	= H	Me	Et	*n*-Pr	Ph
% **6.1-1**	= 76	79	87	84	81

$$\text{PhC(=NNHTs)–C(=NNHTs)Me} \xrightarrow[\text{HO(CH}_2)_2\text{OH}]{\text{K}^{+\,-}\text{OH}} \text{4-Ph-5-Me-1-NHTs-triazole (3pts.)} + \text{4-Me-5-Ph-1-NHTs-triazole (1 pt (72\% total))} \qquad (5)$$

6.1-1

6,7).[8,9] Product **6.1-2** can be hydrolyzed quantitatively to the 1*H*-1,2,3-triazole. Replacing the amino group with an ethoxy group produced unstable products in both cases. A recent study of additions to organometallic compounds shows promise in all but a few instances (Eq. 8).[10]

$$\text{PhC}{\equiv}\text{CNMe}_2 + \text{PhN}_3 \longrightarrow \text{4-Ph-5-NMe}_3\text{-1-Ph-1,2,3-triazole (59\%)} \qquad (6)$$

$$\text{Et}_2\text{NC}{\equiv}\text{CMe} + \text{R}_2\text{P(O)N}_3 \longrightarrow \text{4-Me-5-NEt}_2\text{-1-P(O)R}_2\text{-1,2,3-triazole (49\%)} \qquad (7)$$

R = Ph

6.1-2

$$\text{R}^1\text{R}^2\text{NC}{\equiv}\text{C}^{-\,+}\text{Li} + \text{R}^3_3\text{MX} \longrightarrow \text{R}^1\text{R}^2\text{NC}{\equiv}\text{CMR}^3_3 \xrightarrow{\text{4-NO}_2\text{PhN}_3} \text{4-R}^3_3\text{M-5-NR}^1\text{R}^2\text{-1-(4-NO}_2\text{C}_6\text{H}_4\text{)-1,2,3-triazole} \qquad (8)$$

R^1 = Me, Et, *i*-Pr, Ph R^2 = Ph R^3 = Me, Et, Ph

M = Si, Ge, Sn

50–75%

A great deal of effort has been expended by Hauptmann's laboratory in the study of the oxidation of α,β-bishydrazones (**6.1-3**) with metal oxides, but the yields are only fair (Eq. 9).[11–13] Comparable results have been obtained by El Khadem and his collaborators in the similar oxidation of osazone arylhydrazones.[14] Recently, Alexandrou has reported the important observation that bissemicarbazones react well with lead tetraacetate to produce ureido-1,2,3-triazoles (**6.1-4**), which can be hydrolyzed to the corresponding amines (Eq. 10).[15] Although the yields are still not very high, the method should receive further attention.

Ar(H)C(=NNH₂)–C(=NNH₂) **6.1-3** —HgO or MnO_2→ 4-Ar-1-amino-1,2,3-triazole, 25–60% (9)

Ar = Ph, 4-XPh, 4-RPh, 4-MeOPh, β-$C_{10}H_7$

R^1C(=NNHCONH₂)–C(=NNHCONH₂)R^2 —Pb(OAc)₄→ 4-R^1-5-R^2-1-(NHCONH₂)-1,2,3-triazole **6.1-4**, about 40% —HCl, heat→ 4-R^1-5-R^2-1-amino-1,2,3-triazole **6.1-5**, about 60%

R^1 = Me, Ph, 4-MePh, 4-BrPh
R^2 = H, Me, Ph (10)

The reaction of carbodiimides with diazo compounds has not provided exceptionally high yields of 1,2,3-triazoles, but the method may be improved as some experiments suggest (Eq. 11).[16] The 5-amino isomer (Eq. 12)[17] and 1,2,3-triazoles without a nitrogen substituent (Eq. 13)[18] have also been reported.

Me–C₆H₄–N=C=N–C₆H₄–Me + $(Me_3Sn)_2CN_2$ →

4-Me_3Sn-5-[N($SnMe_3$)(C₆H₄Me)]-1-(C₆H₄Me)-1,2,3-triazole, 96% —CH_2N_2→ 5-(NH–C₆H₄–Me)-1-(C₆H₄–Me)-1,2,3-triazole, 80% (11)

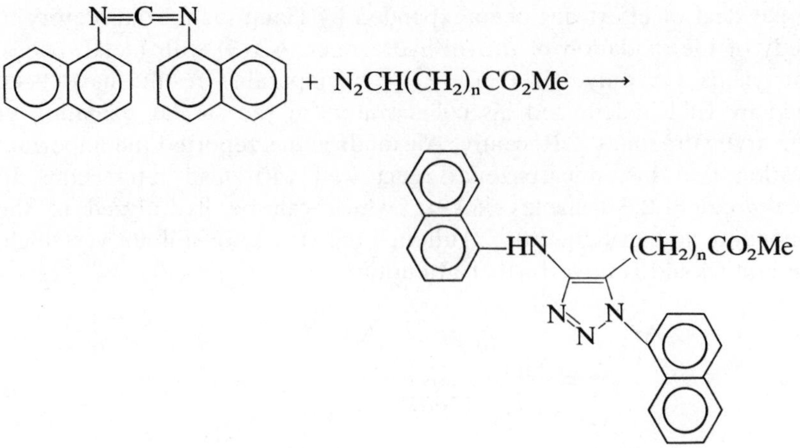

$$\text{(1-naphthyl)N=C=N(1-naphthyl)} + N_2CH(CH_2)_nCO_2Me \longrightarrow \text{1-(1-naphthyl)-4-(1-naphthyl)NH-5-}(CH_2)_nCO_2Me\text{-1,2,3-triazole} \quad (12)$$

n = 1 2 3 4
% = 40 43 27 45

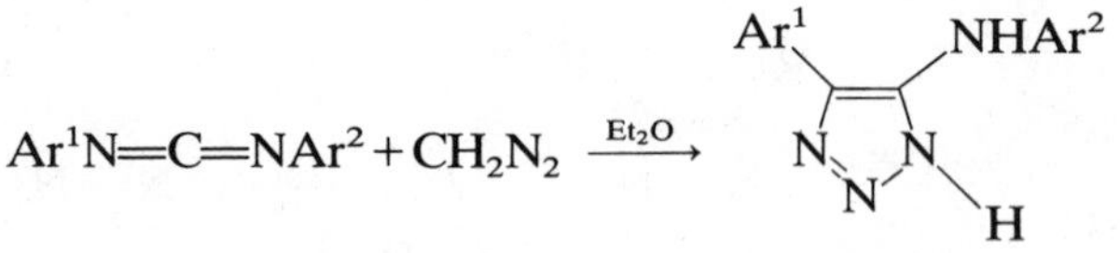

$$Ar^1N{=}C{=}NAr^2 + CH_2N_2 \xrightarrow{Et_2O} \textbf{6.1-6} \quad (13)$$

(6.1-6: 4-Ar^1-5-$NHAr^2$-1H-1,2,3-triazole)

Ar^1	$= Ar^2 =$ Ph	4-Me_2NPh	4-MeOPh	4-MePh	4-ClPh	4-BrPh	4-AcPh
% **6.16**	=41	34	36	53	17	39	16

Ar^1 = 4-MePh Ar^2 = Ph 17%

Perhaps the most intensively studied synthesis of amino-1,2,3-triazoles is the base-catalyzed condensation of nitriles with azides and the following rearrangement discovered by Dimroth (Eq. 14).[19] Lieber and Rao have made detailed studies of the relationship of cyclization to rearrangement and have shown the optimum conditions for the isolation of **6.1-7** and **6.1-8**.[20,21]

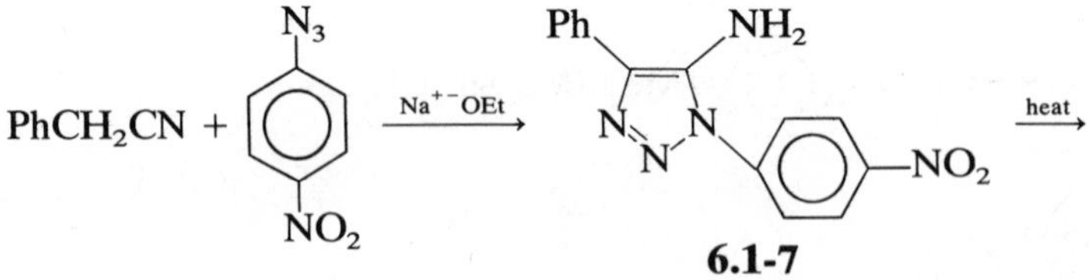

$$PhCH_2CN + 4\text{-}N_3C_6H_4NO_2 \xrightarrow{Na^{+}\,^{-}OEt} \textbf{6.1-7} \xrightarrow{heat} \textbf{6.1-8} \quad (14)$$

(6.1-7: 4-Ph-5-NH_2-1-(4-$NO_2C_6H_4$)-1,2,3-triazole; 6.1-8: 4-Ph-5-(4-$NO_2C_6H_4$NH)-1H-1,2,3-triazole)

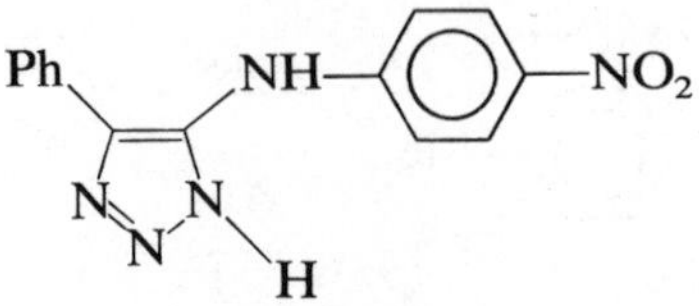

N-Substituent =	Ph	2-MePh	3- or 4-MePh	2-, 3- or 4-ClPh	3-NO_2Ph
% **6.1-7** =	99	59	91	95	51
% **6.1-8** =	92	98	90	80–96	100

N-Substituent =	4-NO_2Ph	4-MeOPh	4-BrPh	β-$C_{10}H_7$	PhCH_2	n-C_6H_{13}
% **6.1-7** =	82	82	71	89	78[a]	98
% **6.1-8** =	83	100	90	89	poor	—

[a] K^{+-}OBu-*t*/THF.

Several potentially important variants of this method have also been reported; for example, L'abbé and his collaborators have prepared 1-vinyl-1,2,3-triazoles without rearrangement (Eq. 15),[22] and Heep has investigated the use of phosphonates with some success (Eq. 16).[23]

$$PhCH_2CN + N_3CH_2CH(I)Bu\text{-}t \xrightarrow[\text{THF}]{K^{+-}OBu\text{-}t} \text{4-Ph-5-}NH_2\text{-1-}(CH_2CH(I)Bu\text{-}t)\text{-1,2,3-triazole (82.5\%)} \xrightarrow{K^{+-}OBu\text{-}t} \text{4-Ph-5-}NH_2\text{-1-}(CH{=}CHBu\text{-}t)\text{-1,2,3-triazole (91\%)} \quad (15)$$

$$(EtO)_2P(O)CH_2CN + ArN_3 \longrightarrow \text{4-}(EtO)_2P(O)\text{-5-}NH_2\text{-1-Ar-1,2,3-triazole} + \text{5-}H_2N\text{-4-}P(O)(OEt)_2\text{-3-Ar-isoxazole} \longrightarrow \text{4-}(EtO)_2P(O)\text{-5-NHAr-1H-1,2,3-triazole} \quad (16)$$

Ar =	%	%
Ph	76	—
4-ClPh	70	64
4-FPh	54	94
4-NO_2Ph	81	52

A single example of the application of perfluoro compounds suggests the desirability of further study (Eq. 17).[24]

The synthesis of 4,5-diamino-1,2,3-triazoles has been accomplished in several steps of modest yield (Eq. 18).[25]

$$F_2C{=}C(F)CF_3 + PhCH_2N_3 \longrightarrow \text{4,4,5-trifluoro-5-}CF_3\text{-1-}CH_2Ph\text{-triazoline} \xrightarrow{(Me_2N)_2C{=}C(NMe_2)_2}$$

$$\text{4-}Me_2N\text{-5-}CF_3\text{-1-}CH_2Ph\text{-1,2,3-triazole (53\%)} + \text{4-F-5-}CF_3\text{-1-}CH_2Ph\text{-1,2,3-triazole (43\%)} \quad (17)$$

$$\text{4-}EtO_2C\text{-5-}NH_2\text{-1-}CH_2Ph\text{-triazole} \xrightarrow[H_2O]{H_2NNH_2} \text{4-}H_2NNHC(O)\text{-5-}NH_2\text{-1-}CH_2Ph\text{-triazole} \xrightarrow{HNO_2}$$

$$\text{4-}N_3C(O)\text{-5-}NH_2\text{-1-}CH_2Ph\text{-triazole (74\%)} \xrightarrow[heat]{EtOH} \text{4-}EtOC(O)NH\text{-5-}NH_2\text{-1-}CH_2Ph\text{-triazole (52\%)} \xrightarrow[EtOH]{Na^{+-}OEt}$$

$$\text{4-}H_2N\text{-5-}NH_2\text{-1-}CH_2Ph\text{-triazole (55\%)} \quad (18)$$

Many amino-1,2,3-triazoles have been converted in good yield to amido- or sulfonamido- derivatives using standard (Schotten–Bauman) procedures, but a few direct and potentially useful methods have also been reported. For example, the oxidation of α,β-bisaroylhydrazones can produce either amido- or imido- products (Eqs. 19,20).[26] Although the yield of **6.1-9** is only fair and attempted extensions to other aroyl groups gave variable yields,[27] it was found that lead tetraacetate not only greatly improved the yield but gave regiospecificity in the cases examined.[28] The structure of these products and ways of improving the yields are under study by Alexandrou and his collaborators.

$$\begin{array}{l} \text{Me} \\ \quad \text{C=NNHCOPh} \\ \quad | \\ \quad \text{C=NNHCOPh} \\ \text{Me} \end{array} \xrightarrow{\text{HgO}}$$

6.1-9
40%

$$\xrightarrow{\text{HCl}} \text{4,5-dimethyl-2-(NHCOPh)-2H-1,2,3-triazole} \quad 60\% \qquad (19)$$

$$\xrightarrow{\text{heat}} \text{4,5-dimethyl-2-[N(COPh)}_2\text{]-2H-1,2,3-triazole} \quad 85\% \qquad (20)$$

The photochemical cyclization of the dianion of α,β-bistosylhydrazones can produce an excellent yield of the corresponding sulfonamido-1,2,3-triazole (Eq. 21), but the two examples reported show the great amount of further investigation required to make the approach generally useful.[29]

$$\begin{array}{l} \text{R} \\ \quad \text{C=NNTs}^- \\ \quad | \\ \quad \text{C=NNTs}^- \\ \text{R} \end{array} \xrightarrow{h\nu} \text{4,5-R}_2\text{-2-(NHTs)-2H-1,2,3-triazole} \qquad (21)$$

R = Me 81% (Corex)
Ph 25%

A recent report of the addition of hydrazoic acid to azosulfonimines shows great promise (Eq. 22).[30] The same group has also conducted a detailed study of the methylation of such sulfonamido compounds with diazomethane (Eq. 23).[31] The consistency of the data is remarkable. These compounds were also acetylated in high yield.

Finally, an interesting rearrangement of an oxadiazole has been reported, and the range of applicability of the reaction should be investigated (Eq. 24).[32]

$$RSO_2N{=}C(OR')(CHN_2) + HN_3 \longrightarrow \text{4-(NHSO}_2\text{R)-1H-1,2,3-triazole} \quad 80\text{–}90\% \qquad (22)$$

R = Me, Ph, 4-MePh, 4-BrPh, 4-NO_2Ph
R′ = Me, Et

(23)

Ar	%	%	%
Ph	15	54	30
4-MePh	16	60	22
4-BrPh	14	57	29
4-NO_2Ph	71		29

(24)

90%

REFERENCES

1. **66:** 10887k
2. **73:** 44674j
3. **70:** 37725m
4. **73:** 3612u
5. **47:** 10474h
6. **78:** 111219h
7. **76:** 72458x
8. **65:** 10524d
9. **73:** 25592q
10. **84:** 105706x
11. **68:** 59506b
12. **76:** 85756t
13. **83:** 95918f
14. **73:** 130944d
15. **85:** 177331q
16. **76:** 59718n
17. **67:** 100060w
18. **83:** 114297w
19. **22:** 423^7
20. **52:** 365a
21. **54:** 6752h
22. **72:** 121447w
23. **79:** 78884s
24. **64:** 12662f
25. **78:** 29681u
26. **60:** 1735g
27. **65:** 8814d
28. **77:** 75174h
29. **71:** 38087n
30. **85:** 143067t
31. **83:** 192266z
32. **74:** 141643b

TABLE 6. AMINO- AND AMIDO-1,2,3-TRIAZOLES

Compound	Reference
4-acetamido-5-[(acetylamino)methyl]-1-methyl-	**79:** 146462e
4-acetamido-5-[(acetylamino)methyl]-2-methyl-	**79:** 146462e
1-acetamido-4-(D-*arabino*-1,2,3,4-tetrahydroxybutyl)-, tetraacetate ester	**70:** 58163s
4-acetamido-1-benzyl-	**78:** 29681u
5-acetamido-1-benzyl-	**78:** 29681u
4-acetamido-2,5-diphenyl-	**55:** 4488f, **55:** 16522g
5-acetamido-1,4-diphenyl-	**74:** 125581z
4-acetamido-2-(2-hydroxyphenyl)-	**71:** P124445j
4-acetamidomethyl-1-benzyl-4-(α-ethoxyethylidene-amino)-	**88:** 121778c
4-acetamidomethyl-1-benzyl-4-ethoxymethyleneamino-	**84:** 121778c
4-acetamido-5-methyl-2-phenyl-	**38:** 4598^{3}
1-acetamido-4-phenyl-	**73:** 130944d
4-acetamido-1-phenyl-	**26:** 1287^{9}
4-acetamido-5-phenyl-	**55:** 4489b, **55:** 19911e, **75:** 20332m
4-[4-(acetamidophenyl)sulfonamido]-2-phenyl-	**78:** P4256w
-4-acetic acid, 1-(1-naphthyl)-5-(1-naphthylamino)-, methyl ester	**67:** 100060w
5-[(4-acetylphenyl)amino]-1-(4-acetylphenyl)-	**83:** 114297w
-1-amine	**78:** P49790g
-3-amine	**60:** 15069b
-4-amine	**51:** 2755i, **51:** 12107h, **51:** 14698a, **54:** 18555c, **74:** 22769n, **83:** 9910t **84:** 74210z
-4-amine, hydrochloride	**79:** 17821x
-5-amine, 4-(acetamidomethyl)-1-benzyl-	**73:** 77154t, **79:** 146462e
-4-amine, 5-(acetamidomethyl)-1-methyl-	**79:** 146462e
-5-amine, 4-(acetamidomethyl)-2-methyl-	**79:** 146462e
-4-amine, 5-(aminomethyl)-2-methyl-	**73:** 77154t, **79:** 146462e
-5-amine, 4-(aminomethyl)-1-methyl-	**73:** 77154t, **79:** 146462e
-4-amine, 2-(4-aminophenyl)-5-methyl-	**79:** P18722c, **80:** P16441g
-5-amine, 1-(4-aminophenyl)-4-phenyl-	**22:** 423^{7}, **52:** 3487h
-4-amine, *N*-(4-aminophenyl)-5-phenyl-	**22:** 423^{7}, **52:** 3487h
-5-amine, 4-(7-amino-3-phenyl-3*H*-*v*-triazolo[4,5-*d*] pyrimidin-5-yl)-*N*-phenyl-	**74:** 125581z
-5-amine, 4-(7-amino-3-phenyl-3*H*-*v*-triazolo[4,5-*d*] pyrimidin-5-yl)-1-phenyl-	**74:** 125581z
-4-amine, 1-benzenesulfonamido-	**81:** P27329y
-4-amine, 1-(1*H*-benzimidazol-2-yl)-*N*,*N*-diethyl-5-methyl-	**79:** P78815v
-5-amine, 1-(1*H*-benzimidazol-2-yl)-*N*,*N*-diethyl-4-methyl-	**79:** P78815v
-4-amine, 1-(2-benzothiazolyl)-*N*,*N*-diethyl-5-methyl-	**79:** P78815v
-5-amine, 1-(2-benzothiazolyl)-*N*,*N*-diethyl-4-methyl-	**79:** P78815v
-5-amine, 1-benzyl-	**51:** 2755h, **72:** 78952t, **78:** 29681u, **79:** 17821x

TABLE 6 (*Continued*)

Compound	Reference
-5-amine, 1-benzyl-4-(7-amino-3-benzyl-3*H*-*v*-triazolo [4,5-*d*]pyrimidin-5-yl)-	**72:** 78952t
-5-amine, 1-benzyl-4-[[(ethoxycarbonyl)amino]methyl]-	**79:** 146462e
-5-amine, 1-benzyl-4-[[(ethoxycarbonyl)methan-amido]methyl]-	**79:** 146462e
-5-amine, 1-benzyl-4-[[(ethylmercaptocarbonyl) amino]methyl]-	**79:** 146462e
-5-amine, 1-benzyl-4-formamidomethyl-	**73:** 77154t, **79:** 146462e
-5-amine, 1-benzyl-4-phenyl-	**52:** 365d, **54:** 6752i, **75:** 20332m, **84:** 16553c
-1-amine, 4-(1,1′-biphenyl-4-yl)-	**76:** 85756t
-5-amine, *N*,1-bis(4-bromophenyl)-	**83:** 114297w
-5-amine, *N*,1-bis(4-chlorophenyl)-	**83:** 114297w
-5-amine, *N*,1-bis[4-(dimethylamino)phenyl]-	**83:** 114297w
-5-amine, *N*,1-bis(4-methoxyphenyl)-	**83:** 114297w
-5-amine, *N*,1-bis(4-methylphenyl)-	**76:** 59718n, **83:** 114297w
-5-amine, *N*,1-bis(4-methyphenyl)-4-(trimethylsilyl)-	**76:** 59718n
-5-amine, *N*,1-bis(4-methylphenyl)-*N*,4-bis (trimethylstannyl)-	**76:** 59718n
-1-amine, 4-(3-bromophenyl)-	**76:** 85756t
-1-amine, 4-(4-bromophenyl)-	**68:** 59506b, **76:** 85756t
-4-amine, 2-(4-bromophenyl)-	**79:** P147417f
-1-amine, 4-(4-bromophenyl)-5-methyl-	**85:** 177331q
-4-amine, *N*-(4-bromophenyl)-5-phenyl-	**52:** 366b
-5-amine, 1-(4-bromophenyl)-4-phenyl-	**52:** 365d
-5-amine, 1-(1-butenyl)-4-phenyl-, (E)-	**73:** 120446d
-5-amine, 1-(1-butenyl)-4-phenyl-, (Z)-	**73:** 120446d
-1-amine, 4-[4-(*tert*-butyl)phenyl]-	**76:** 85756t
-1-amine, 4-(3-chlorophenyl)-	**76:** 85756t
-1-amine, 4-(4-chlorophenyl)-	**68:** 59506b, **76:** 85756t
-4-amine, 2-(2-chlorophenyl)-	**79:** P147417f
-4-amine, 2-(4-chlorophenyl)-	**79:** P147417f
-4-amine, 2-(4-chlorophenyl)-5-methyl-	**79:** P18722c
-1-amine, *N*-(chlorophenylmethylene)-4,5-dimethyl-, monohydrochloride	**76:** 140659r
-4-amine, *N*-(2-chlorophenyl)-5-phenyl-	**52:** 366b
-4-amine, *N*-(3-chlorophenyl)-5-phenyl-	**52:** 366b
-4-amine, *N*-(4-chlorophenyl)-5-phenyl-	**52:** 366b
-4-amine, 5-(4-chlorophenyl)-1-phenyl-	**81:** 3850k
-5-amine, 1-(2-chlorophenyl)-4-phenyl-	**52:** 365a
-5-amine, 1-(3-chlorophenyl)-4-phenyl-	**52:** 365a
-5-amine, 1-(4-chlorophenyl)-4-phenyl-	**52:** 365a
-1-amine, 4-(4-cyanophenyl)-	**83:** 95918f
-1-amine, 4-(4-cyclohexylphenyl)-	**76:** 85756t
-5-amine, 1-cyclohexyl-, monohydrochloride	**75:** 5801v
-5-amine, 1,4-dibenzyl-	**59:** 11494f
-1-amine, 4-(3,4-dichlorophenyl)-	**76:** 85756t
-5-amine, 4-(diethoxyphosphinyl)-1-(4-chlorophenyl)-	**79:** 78884s
-5-amine, 4-(diethoxyphosphinyl)-1-(4-fluorophenyl)-	**79:** 78884s
-5-amine, 4-(diethoxyphosphinyl)-1-(4-nitrophenyl)-	**79:** 78884s

TABLE 6 (*Continued*)

Compound	Reference
-5-amine, 4-(diethoxyphosphinyl)-1-phenyl-	**79:** 78884s
-5-amine, *N,N*-diethyl-1-(1-methyl-1*H*-benzimidazol-2-yl)-4-phenyl-	**79:** 66246x
-4-amine, *N,N*-diethyl-5-methyl-1-(4-methyl-6-isopropyl-2-pyrimidinyl)-	**79:** P78815v
-5-amine, *N,N*-diethyl-4-methyl-1-(4-methyl-6-isopropyl-2-pyrimidinyl)-	**79:** P78815v
-5-amine, *N,N*-diethyl-1-(4-nitrophenyl)-4-(triethylgermyl)-	**84:** 105706x
-5-amine, *N,N*-diethyl-1-(4-nitrophenyl)-4-(trimethylsilyl)-	**84:** 105706x
-5-amine, *N,N*-diethyl-1-(4-nitrophenyl)-4-(trimethylstannyl)-	**84:** 105706x
-5-amine, *N,N*-diethyl-1-(4-nitrophenyl)-4-(triphenylgermyl)-	**84:** 105706x
-5-amine, *N,N*-diethyl-1-(4-nitrophenyl)-4-(triphenylsilyl)-	**84:** 105706x
-5-amine, *N,N*-diethyl-1-(4-nitrophenyl)-4-(triphenylstannyl)-	**84:** 105706x
-5-amine, 4-(3,4-dihydro-6,7-dimethoxy-1-isoquinolinyl)-1-(phenylsulfonyl)-	**80:** 47806j
-1-amine, 4,5-dimethyl-	**78:** 111219h, **85:** 177331q
-5-amine, *N*,1-dimethyl, 3-oxide	**81:** 77865r
-5-amine, 1-(3,3-dimethyl-1-butenyl)-4-phenyl-, (E)-	**72:** 121447w
-5-amine, *N*,1-diphenyl-	**83:** 114297w
-1-amine, 4,5-diphenyl-	**78:** 111219h, **82:** 156182q, **85:** 177331q
-4-amine, 1,5-diphenyl-	**66:** 10887k, **81:** 3850k
-4-amine, 2,5-diphenyl-	**29:** 4739[3], **55:** 4488f
-4-amine, 5,5-diphenyl-(?)	**53:** 21902b
-5-amine, 1,4-diphenyl-	**44:** 1102d, **52:** 365c, **52:** 1045d, **61:** 654f, **73:** 44674j, **74:** 125581z
-5-amine, *N,N*-diphenyl-1-(4-nitrophenyl)-4-(triethylgermyl)-	**84:** 105706x
-5-amine, *N,N*-diphenyl-1-(4-nitrophenyl)-4-(trimethylsilyl)-	**84:** 105706x
-5-amine, *N,N*-diphenyl-1-(4-nitrophenyl)-4-(triphenylgermyl)-	**84:** 105706x
-5-amine, *N,N*-diphenyl-1-(4-nitrophenyl)-4-(triphenylsilyl)-	**84:** 105706x
-5-amine, *N,N*-diphenyl-1-(4-nitrophenyl)-4-(triphenylstannyl)-	**84:** 105706x
-4-amine, 5-[[(ethoxycarbonyl)amino]methyl]-1-methyl-	**79:** 146462e
-5-amine, 4-[[(ethoxycarbonyl)amino]methyl]-2-methyl-	**79:** 146462e
-5-amine, 4-[2-(ethoxycarbonyloxy)ethyl]-1-phenyl-	**76:** 34177x
-5-amine, 4-[2-(ethoxycarbonyloxy)-1-methylethyl]-1-phenyl-	**76:** 34177x
-5-amine, 1-(2-ethyl-1-butenyl)-4-phenyl-	**73:** 120446d
-1-amine, 4-(4-ethylphenyl)-	**76:** 85756t

TABLE 6 (*Continued*)

Compound	Reference
-5-amine, 1-ethyl-4-phenyl-	**52:** 365h
-5-amine, 1-(β-ethylstyryl)-4-phenyl-, (E)-	**73:** 120446d
-1-amine, 4-(4-fluorophenyl)-	**76:** 85756t
-5-amine, 1-hexyl-4-phenyl-	**54:** 6752h
-4-amine, 2-(2-hydroxyphenyl)-	**71:** P124445j
-5-amine, *N*-isopropyl-1-(4-nitrophenyl)-*N*-phenyl-4-(trimethylsilyl)-	**84:** 105706x
-1-amine, 4-(4-methoxyphenyl)-	**68:** 59506b, **76:** 85756t
-4-amine, 2-(4-methoxyphenyl)-	**79:** P147417f
-4-amine, 2-(4-methoxyphenyl)-5-methyl-	**79:** P18722c
-4-amine, *N*-(4-methoxyphenyl)-5-phenyl-	**52:** 366b
-5-amine, 1-(4-methoxyphenyl)-*N*-phenyl-	**83:** 114297w
-5-amine, 1-(4-methoxyphenyl)-4-phenyl-	**52:** 365d
-1-amine, 4-methyl-	**21:** 92[7]
-1-amine, 5-methyl-	**78:** P49790g
-4-amine, 1-methyl-	**55:** 17626h
-4-amine, 5-methyl-	**85:** 123875b
-4-amine, 5-methyl-, monohydrochloride	**85:** 123875b
-5-amine, 1-methyl-	**55:** 17626i, **72:** 21645r
-1-amine, 5-methyl-4-(4-methylphenyl)-	**85:** 177331q
-5-amine, *N*-methyl-1-(4-nitrophenyl)-*N*-phenyl-4-(triethylgermyl)-	**84:** 105706x
-5-amine, *N*-methyl-1-(4-nitrophenyl)-*N*-phenyl-4-(triphenylgermyl)-	**84:** 105706x
-5-amine, *N*-methyl-1-(4-nitrophenyl)-*N*-phenyl-4-(triphenylsilyl)-	**84:** 105706x
-5-amine, *N*-methyl-1-(4-nitrophenyl)-*N*-phenyl-4-(triphenylstannyl)-	**84:** 105706x
-1-amine, 4-(4-methylphenyl)-	**76:** 85756t, **83:** 95918f, **85:** 177331q
-1-amine, 4-methyl-5-phenyl-	**78:** 111219h, **82:** 156182q
-1-amine, 5-methyl-4-phenyl-	**78:** 111219h, **82:** 156182q, **85:** 177331q
-4-amine, 5-methyl-1-phenyl-	**38:** 4598[3], **70:** 37725m **80:** 47912r
-4-amine, 5-methyl-2-phenyl-	**58:** 11495f
-5-amine, 2-methyl-4-[(*N*-phenylhydrazinecarboxamido)methylene]-	**79:** 137053v
-5-amine, *N*-(2-methylphenyl)-4-phenyl-	**52:** 366b
-5-amine, *N*-(3-methylphenyl)-4-phenyl-	**52:** 366b
-5-amine, *N*-(4-methylphenyl)-4-phenyl-	**52:** 366b
-5-amine, 1-(2-methylphenyl)-4-phenyl-	**52:** 365a
-5-amine, 1-(3-methylphenyl)-4-phenyl-	**52:** 365a
-5-amine, 1-(4-methylphenyl)-4-phenyl-	**52:** 365a
-5-amine, 1-[(methlthio)methyl]-4-phenyl-	**72:** 43576w
-1-amine, 4-(2-naphthyl)-	**68:** 59506b, **76:** 85756t
-5-amine, 1-(2-naphthyl)-4-phenyl-	**52:** 365d
-1-amine, 4-(3-nitrophenyl)-	**83:** 95918f
-4-amine, *N*-(3-nitrophenyl)-5-phenyl-	**52:** 366b
-5-amine, *N*-(4-nitrophenyl)-4-phenyl-	**22:** 423[7], **52:** 366b

TABLE 6 (*Continued*)

Compound	Reference
-5-amine, 1-(3-nitrophenyl)-4-phenyl-	**52:** 365d
-5-amine, 1-(4-nitrophenyl)-4-phenyl-	**22:** 423[7], **52:** 365d
-5-amine, 1-(1-pentenyl)-4-phenyl, (E)-	**73:** 120446d
-5-amine, 1-(1-pentenyl)-4-phenyl-, (Z)-	**73:** 120446d
-1-amine, 4-phenyl-	**68:** 59506b, **76:** 85756t
-4-amine, 1-phenyl-	**26:** 1287[9]
-4-amine, 2-phenyl-	**54:** 6704c
-4-amine, 5-phenyl-	**55:** 4488h, **55:** 19910h, **73:** 130944d, **75:** 20332m, **79:** 5299v, **84:** 16553c
-5-amine, 1-phenyl-	**52:** 366a
-5-amine, 4-phenyl-1-(4-phenyl-1-butenyl)-, (E)-	**73:** 120446d
-5-amine, 4-phenyl-1-(4-phenyl-1-butenyl)-, (Z)-	**73:** 120446d
-5-amine, 4-phenyl-1-(3-phenylpropenyl)-, (E)-	**73:** 120446d
-5-amine, 4-phenyl-1-(3-phenylpropenyl)-, (Z)-	**73:** 120446d
-5-amine, 4-phenyl-1-styryl-, (E)-	**73:** 120446d
-5-amine, 4-phenyl-1-styryl-, (Z)-	**73:** 120446d
-5-amine, 4-phenyl-1-[2-(1*H*-1,2,4-triazol-3-yl)phenyl]-	**77:** 88421m
-4-amine, 1-*β*-D-ribofuranosyl-	**78:** 84715h
-4-amine, 2-*β*-D-ribofuranosyl-	**78:** 84715h
4-[(4-aminophenyl)sulfonamido]-2-phenyl-	**78:** P4256w
4-anilino-	**52:** 366d
5-anilino-4-[7-(diacetylamino-3-phenyl-3*H*-triazolo[4,5-*d*]-pyrimidin-5-yl]-	**74:** 125581z
1-anilino-4-methyl-	**67:** 32614c
5-anilino-1-methyl-	**55:** 17626i
4-anilino-5-phenyl-	**52:** 366b, **52:** 1045d
4-(D-*arabino*-1,2,3,4-tetrahydroxybutyl)-1-benzamido-	**70:** 58163s
4-(D-*arabino*-1,2,3,4-tetrahydroxybutyl)-1-(4-iodobenzamido)-	**69:** 36375q
4-(D-*arabino*-1,2,3,4-tetrahydroxybutyl)-1-(3-methylbenzamido)-	**69:** 36375q
4-(D-*arabino*-1,2,3,4-tetrahydroxybutyl)-1-(4-methylbenzamido)-	**69:** 36375q
4-benzamido-2-(4-bromophenyl)-5-phenyl-	**29:** 4739[5]
5-benzamido-1,4-dibenzyl-	**59:** 11494f
1-benzamido-4,5-dimethyl-	**60:** 1735g
4-benzamido-2,5-diphenyl-	**29:** 4739[3], **55:** 4488e, **55:** 17626g
1-benzamido-4,5-diphenyl-2-methyl-	**77:** 75174h
4-benzamido-2-(2-hydroxyphenyl)-	**71:** P124445j
1-benzamido-4-(4-methoxyphenyl)-	**79:** 126403h
1-benzamido-4-methyl-	**21:** 92[7], **84:** 135555q
1-benzamido-5-methyl-	**21:** 92[7]
4-benzamido-5-methyl-	**58:** 13944a
4-benzamido-5-(4-methylphenyl)-2-phenyl-	**29:** 4739[7]
1-benzamido-4-phenyl-	**73:** 130944d
4-benzamido-5-phenyl-	**55:** 4488g, **55:** 19911e
1-(*N*-benzoylbenzamido)-4-(4-methoxyphenyl)-	**79:** 126403h
1-(*N*-benzoylbenzamido)-4-phenyl-	**79:** 126403h

TABLE 6 (*Continued*)

Compound	Reference
4-(benzylamino)-	**72:** 78952t
4-(benzylamino)-5-phenyl-	**52:** 366b
1-benzyl-4,5-bis(acetamido)-	**78:** 29681u
1-benzyl-4-(dimethylamino)-5-(trifluoromethyl)-	**64:** 12662f
1-benzyl-5-[(ethoxycarbonyl)amino]-4-[[(ethoxycarbonyl)amino]methyl]-	**79:** 146462e
1-benzyl-4-[[2-(ethoxycarbonyl)-1-methylethenyl]amino]-	**78:** 29681u
1-benzyl-5-formamido-4-[(formylamino)methyl]-	**79:** 146462e
1-[(benzylidene)amino]-4-(4-bromophenyl)-	**68:** 59506b
4-[(benzylidene)amino]-2,5-diphenyl-	**29:** 4739[4]
4-[(benzylidene)amino]-4-methyl-	**21:** 92[7]
4-[(benzylidene)amino]-5-methyl-2-phenyl-	**38:** 4598[3]
1-bis(benzoylamino)-4-methyl-	**84:** 135555q
1-[bis(4-chlorobenzoyl)amino]-4,5-diphenyl-	**84:** 52562p
4-[2,2-bis(ethoxycarbonyl)ethenylamino]-5-phenyl-	**75:** 20332m
1,4-bis(2,4,6-trinitroanilino)-	**74:** 22769n
4-[4-bromo-*N*-methyl(phenylsulfonamido)]-1-methyl-	**83:** 192266z
4-[4-bromo-*N*-methyl(phenylsulfonamido)]-2-methyl-	**83:** 192266z
5-[4-bromo-*N*-methyl(phenylsulfonamido)]-1-methyl-	**83:** 192266z
4-(4-bromophenyl)-1-(diacetylamino)-	**68:** 59506b
4-(4-bromophenyl)-5-methyl-1-ureido-	**85:** 177331q
4-(4-bromophenyl)-1-(phenylureylene)-	**68:** 59506b
4-[(4-bromophenyl)sulfonamido]-	**83:** 192266z, **85:** 143067t
4-[(4-bromophenyl)sulfonamido]-, monoammonium salt	**85:** 143067t
5-[(4-bromophenyl)sulfonamido]-1-methyl-	**83:** 192266z
5-[(4-bromophenyl)sulfonamido]-1-methyl-, 1:1 adduct with methanamine	**85:** 143067t
-4-butanoic acid, 1-(1-naphthyl)-5-(1-naphthylamino)-, methyl ester	**67:** 100060w
1-butyl-5-[(4-bromophenyl)sulfonamido]-, 1:1 adduct with butanamine	**85:** 143067t
1-butyl-5-[(4-methylphenyl)sulfonamido]-	**85:** 143067t
-4-carbamic acid, benzyl ester	**51:** 14697i
-4-carbamic acid, ethyl ester	**51:** 14697i
-4-carbamic acid, 2-(2-hydroxyphenyl)-, methyl ester	**71:** P124445j
-4-carbamic acid, 2-(2-hydroxyphenyl)-, octyl ester	**71:** P124445j
-4-carbamic acid, 1-methyl-, ethyl ester	**55:** 17627a
4-(carbamimidoylamino)-5-(2,2-dicyanoethenyl)-	**79:** 146470f
1-(3-chlorobenzamido)-4-(D-*arabino*-1,2,3,4-tetrahydroxybutyl)-	**69:** 36375q
1-(4-chlorobenzamido)-4-(D-*arabino*-1,2,3,4-tetrahydroxybutyl)-	**69:** 36375q
1-(2-chlorobenzamido)-4,5-diphenyl-	**77:** 75174h, **85:** 191649f
1-(3-chlorobenzamido)-4-methyl-	**84:** 135555q
1-(4-chlorobenzamido)-4-methyl-	**84:** 135555q
4-[2-cyano-2-(ethoxycarbonyl)ethenyl]-5-ureido-	**79:** 146470f
5-(diacetylamino)-1,4-diphenyl-	**74:** 125581z
-4,5-diamine, 1-benzyl-	**78:** 29681u
-4,5-diamine, 1-benzyl-, 1:1 adduct with 2,4,6-trinitrophenol	**78:** 29681u

TABLE 6 (*Continued*)

Compound	Reference
-4,5-diamine, 1-benzyl-N^4-benzylidene-	**78:** 29681u
-4,5-diamine, N^4,N^{4*}-1,2-ethanediylidenebis(1-benzyl)-	**78:** 29681u
-4,5-diamine, 2-phenyl-	**81:** 105404k
1-(dibenzoylamino)-4,5-dimethyl-	**69:** 36042d
1-(dibenzoylamino)-4,5-diphenyl-	**68:** P91872d
1-(dibenzoylamino)-4-methyl-	**21:** $92^{7,8}$
1-(dibenzoylamino)-5-methyl-	**21:** $92^{7,8}$
4-(dibenzoylamino)-5-methyl-2-phenyl-	**38:** 4598^3
1-[(3,4-dibromo-3*H*-furan-5-yl-2-oxo)amino]-	**72:** P121548e
1-(3,5-di-*tert*-butyl-4-hydroxybenzamido)-	**84:** P45268u
-4,5-dicarbamic acid, 1-phenyl-, diethyl ester	**21:** 2690^4
1-(2,2-dichloroacetamido)-4,5-diphenyl-	**69:** 96327k
1-(2,4-dichlorobenzamido)-4,5-diphenyl-	**65:** 8814e
1-[(3,4-dichloro-3*H*-furan-5-yl-2-oxo)amino]-	**72:** P121548e
5-(2,2-dicyanoethenyl)-4-(methanimidoylamino)-	**79:** 146470f
5-(2,2-dicyanoethenyl)-4-thioureiod-, monopotassium salt	**79:** 146470f
5-(2,2-dicyanoethenyl)-4-ureido-	**79:** 146470f
4-(diethoxyphosphinyl)-5-[(4-chlorophenyl)amino]-	**79:** 78884s
4-(diethoxyphosphinyl)-5-(diethylamino)-	**73:** 25592q
4-(diethoxyphosphinyl)-5-[(4-fluorophenyl)amino]-	**79:** 78884s
4-(diethoxyphosphinyl)-5-[(4-nitrophenyl)amino]-	**79:** 78884s
5-(diethylamino)-1-(diphenylphosphinyl)-4-methyl-	**73:** 25592q
4-(diethylamino)-5-methyl-	**73:** 25592q
4,5-diethyl-1-[(4-methylphenyl)sulfonamido]-	**73:** 3612u
5-(dimethylamino)-1,4-diphenyl-	**65:** 10524d
N,2-dimethyl-4-[(4-methylphenyl)sulfonamido]-	**84:** 135549r
4,5-dimethyl-1-[(4-methylphenyl)sulfonamido]-	**47:** 10474h
4,5-dimethyl-1-[(4-methylphenyl)sulfonamido]-, ion (1^-)	**71:** 38087n
4,5-dimethyl-1-(phenylsulfonamido)-	**76:** 140659r
4-[*N*,4-dimethyl(phenylsulfonamido)]-1-methyl-	**83:** 192266z
4-[*N*,4-dimethyl(phenylsulfonamido)]-2-methyl-	**83:** 192266z
5-[*N*,4-dimethyl(phenylsulfonamido)]-1-methyl-	**83:** 192266z
4,5-dimethyl-1-ureido-	**85:** 177331q
1-(1,3-dioxo-1*H*-isoindol-2(2*H*)-yl)-4,5-dimethyl-	**76:** 72461t, **78:** 111219h
1-(1,3-dioxo-1*H*-isoindol-2(2*H*)-yl)-4,5-diphenyl-	**76:** 72461t, **78:** 111219h
1-(1,3-dioxo-1*H*-isoindol-2*H*-2-yl)-4-methyl-5-phenyl-	**76:** 72458x, **78:** 111219h
1-(1,3-dioxo-1*H*-isoindol-2*H*-2-yl)-5-methyl-4-phenyl-	**76:** 72458x, **78:** 111219h
4,5-diphenyl-2-[2-hydroxy-5-methyl-3-(phenylureylene)methyl]-	**82:** P17959k
4,5-diphenyl-1-[(4-methylphenyl)sulfonamido]-	**78:** 111219h
4,5-diphenyl-1-[(4-methylphenyl)sulfonamido]-, ion (1^-)	**71:** 38087n
4,5-diphenyl-1-(2-nitrobenzamido)-	**77:** 75174h, **85:** 191649f
4,5-diphenyl-1-(2,4,6-trimethylbenzamido)-	**65:** 8814d
4,5-diphenyl-1-(tritylamino)-	**58:** 4449a
4,5-diphenyl-1-ureido-	**85:** 177331q
4,5-dipropyl-1-[(4-methylphenyl)sulfonamido]-	**47:** 10474i
4-dodecanamido-2-(2-hydroxyphenyl)-	**71:** P124445j
-4-ethanol, 5-amino-α-methyl-1-phenyl-	**76:** 34177x

TABLE 6 (*Continued*)

Compound	Reference
-4-ethanol, 5-amino-1-phenyl-	**76:** 34177x
4-[(ethoxycarbonyl)amino]-5-[[(ethoxycarbonyl) amino]methyl]-1-methyl-, ethyl ester	**79:** 146462e
4-[(ethoxycarbonyl)amino]-5-methyl-	**85:** 123875b
4-[[2-(ethoxycarbonyl)-1-methylethenyl]amino]-5-phenyl-	**79:** 5299v
1-ethyl-5-[(4-bromophenyl)sulfonamido]-	**85:** 143067t
1-ethyl-5-[(4-methylphenyl)sulfonamido]-	**85:** 143067t
1-ethyl-5-(methylsulfonamido)-	**85:** 143067t
1-ethyl-5-[(4-nitrophenyl)sulfonamido]-	**85:** 143067t
1-ethyl-5-[(4-nitrophenyl)sulfonamido]-, 1:1 adduct with ethanamine	**85:** 143067t
1-ethyl-5-(phenylsulfonamido)-	**85:** 143067t
4-formamido-	**69:** 77238r
2-(2-hydroxyphenyl)-4-[(4-methylphenyl)sulfonamido]-	**71:** P124445j
2-(2-hydroxyphenyl)-4-(methylsulfonamido)-	**71:** P124445j
5-(iminophenylmethyl)-4-[(4-methylphenyl)sulfonamido]-	**84:** 135549r
-4-methanamine, 5-amino-1-benzyl-	**73:** 77154t, **79:** 146462e, **84:** 121778c
-4-methanamine, 5-amino-1-benzyl-, 2-hydroxy-1,2,3-propanetricarboxylate (1:1)	**79:** 146462e
-4-methanamine, 5-amino-1-benzyl-, monoacetate	**79:** 146462e, **84:** 121778c
-4-methanamine, 5-amino-1-benzyl-, monohydrochloride	**79:** 146462e, **84:** 121778c
-4-methanamine, 5-amino-1-benzyl-, phosphate (1:1)	**79:** 146462e
-4-methanamine, 5-amino-1-benzyl-, phosphate (2:1)	**79:** 146462e
-4-methanamine, 5-amino-2-methyl-	**84:** 121778c
-4-methanamine, 5-amino-2-methyl-, phosphate (1:1)	**79:** 146462e
-5-methanamine, 4-amino-1-methyl-	**73:** 77154t, **79:** 146462e, **84:** 121778c
-5-methanamine, 4-amino-1-methyl-, phosphate (1:1)	**79:** 146462e
-4-methanamine, 5-(dimethylamino)-*N*,*N*,α-trimethyl-1-(4-nitrophenyl)-, [4α(S^*), 5β]-($\pm$)-	**76:** 126879f
-4-methanol, 5-amino-1-benzyl-	**79:** 137053v
-4-methanol, 5-amino-1-benzyl-, conjugate monoacid	**79:** 137053v
-4-methanol, 5-amino-2-methyl-	**79:** 137053v
-4-methanol, 5-amino-2-methyl-, conjugate monoacid	**79:** 137053v
1-(4-methoxybenzamido)-4-methyl-	**84:** 135555q
4-[(methylamino)methyl]-1-[4-(methylamino)-3-oxo-2-phenyl-2*H*-pyridazin-5-yl]-	**72:** 55344g
5-(methylamino)-1-phenyl-	**51:** 10201c
1-methyl-5-[(4-bromophenyl)sulfonamido]-	**85:** 143067t
4-methyl-1-(3-methylbenzamido)-	**84:** 135555q
4-methyl-1-(4-methylbenzamido)-	**84:** 135555q
1-methyl-4-[*N*-methyl-(4-nitrophenyl)sulfonamido]-	**83:** 192266z
1-methyl-5-[*N*-methyl-(4-nitrophenyl)sulfonamido]-	**83:** 192266z
2-methyl-4-[*N*-methyl-(4-nitrophenyl)sulfonamido]-	**83:** 192266z
1-methyl-4-[*N*-methyl(phenylsulfonamido)]-	**83:** 192266z
1-methyl-5-[*N*-methyl(phenylsulfonamido)]-	**83:** 192266z
1-methyl-5-[(4-methylphenyl)sulfonamido]-	**83:** 192266z, **85:** 143067t
1-methyl-5-[(4-methylphenyl)sulfonamido]-, 1:1 adduct with methanamine	**85:** 143067t

TABLE 6 (*Continued*)

Compound	Reference
2-methyl-4-[*N*-methyl(phenylsulfonamido)]-	**83:** 192266z
4-methyl-1-[(4-methylphenyl)sulfonamido]-5-phenyl-	**47:** 10474i, **78:** 111219h
5-methyl-1-[(4-methylphenyl)sulfonamido]-4-phenyl-	**78:** 111219h
5-methyl-4-(4-methylphenyl)-1-ureido-	**85:** 177331q
1-methyl-5-(methylsulfonamido)-	**85:** 143067t
1-methyl-5-(methylsulfonamido)-, 1:1 adduct with methanamine	**85:** 143067t
1-methyl-5-[(4-nitrophenyl)sulfonamido]-	**83:** 192266z, **85:** 143067t
1-methyl-5-[(4-nitrophenyl)sulfonamido]-, 1:1 adduct with methanamine	**85:** 143067t
1-methyl-5-(phenylsulfonamido)-	**83:** 192266z, **85:** 143067t
1-[(4-methylphenyl)sulfonamido]-	**73:** 3612u
4-[(4-methylphenyl)sulfonamido]-	**83:** 192266z, **85:** 135549r
4-[(4-methylphenyl)sulfonamido]-, monoammonium salt	**85:** 143067t
5-methyl-4-phenyl-1-ureido-	**85:** 177331q
4-(4-methylphenyl)-1-ureido-	**85:** 177331q
4-(2-naphthylamino)-5-phenyl-	**52:** 366b
1-(1-naphthyl)-5-(1-naphthylamino)-	**26:** 455[9]
1-(2-naphthyl)-5-(2-naphthylamino)-	**26:** 455[9]
4-(3-nitrobenzamido)-2-(4-nitrophenyl)-5-phenyl-	**74:** 141643b
1-[(5-nitrofurfurylidene)amino]-	**58:** 9071g
4-[(4-nitrophenyl)sulfonamido]-	**83:** 192266z, **85:** 143067t
4-[(4-nitrophenyl)sulfonamido]-, monammonium salt	**85:** 143067t
4-[(3-oxobutanoyl)amino]-5-phenyl-	**79:** 5299v
-4-pentanoic acid, 1-(1-naphthyl)-5-(1-naphthylamino)-, methyl ester	**67:** 100060w
4-(phenylsulfonamido)-	**83:** 192266z, **85:** 143067t
4-(phenylsulfonamido)-, monoammonium salt	**85:** 143067t
-4-propanoic acid, 1-(1-naphthyl)-5-(1-naphthylamino)-, methyl ester	**67:** 100060w
-4-propenoic acid, 5-[(aminocarbonyl)amino]-2′-cyano-, ethyl ester, monopotassium salt	**79:** 146470f
-4-propenoic acid, 5-[(aminoiminomethyl)amino]-2′-cyano-, ethyl ester	**79:** 146470f
-4-propenoic acid, 5-[(aminothiooxomethyl)amino]-2′-cyano-, ethyl ester, monopotassium salt	**79:** 146470f
-4-propenoic acid, 2′-cyano-5-[(iminomethyl)amino]-, ethyl ester	**79:** 146470f
4-(2,4,6-trinitroanilino)-	**74:** 22769n

REFERENCES

21: 92[7,8]	R. Stollé, *Chem. Ber.*, **59B**, 1742 (1926).
21: 2690[4]	K. Fries, *Justus Liebigs Ann. Chem.*, **454,** 121 (1927).
22: 423[7]	O. Dimroth and W. Michalis, *Justus Liebigs Ann. Chem.*, **459,** 39 (1927).
26: 455[9]	A. Quilico and M. Freri, *Gazz. Chim. Ital.*, **61,** 484 (1931).
26: 1287[9]	H. Kleinfeller and G. Bönig, *J. Prakt. Chem.*, **132,** 175 (1931).

29: 4739[3–7]	E. Durio and P. Gramantieri, *Gazz. Chim. Ital.*, **65,** 102 (1935).
38: 4598[3]	A Quilico and C. Musdnte, *Gazz Chim. Ital.*, **72,** 399 (1942).
44: 1102d	Ramart-Lucas, J. Hoch, and Grumez, *Bull. Soc. Chim. Fr.*, 447 (1949).
47: 10474h,i	W. R. Bamford and T. S. Stevens, *J. Chem. Soc.*, 4735 (1952).
51: 2755h,i	J. R. E. Hoover and A. R. Day, *J. Amer. Chem. Soc.*, **78,** 5832 (1956).
51: 10201c	E. Lieber, C. N. R. Rao, and T. S. Chao, *Current Sci.*, **26,** 14 (1957).
51: 12107h	S. Yamada, T. Mizoguchi, and A. Avata, *Yakugaku Zasshi*, **77,** 441 (1957).
51: 14697–8i–a	S. Yamada, T. Mizoguchi, and A. Avata, *Yakugaku Zasshi*, **77,** 452 (1957).
52: 365a–h	E. Lieber, T-S. Chao, and C. N. R. Rao, *J. Org. Chem.*, **22,** 654 (1957).
52: 1045d	D. E. Bisgrove, J. F. Brown, and L. B. Clapp, *Org. Syntheses*, **37,** 23 (1957).
52: 3487h	E. Lieber, C. N. R. Rao, and T.-S. Chao, *J. Amer. Chem. Soc.*, **79,** 5962 (1957).
53: 21902b	K. Hohenlohe-Oehringen, *Monatsh. Chem.*, **89,** 557 (1958).
54: 6704c	Z. Muljiani, A. M. Kothare, and V. V. Nadkarny, *J. Univ. Bombay*, **27,** 21 (1959).
54: 6752h,i	E. Lieber, C. N. R. Rao, and T. V. Rajkumar, *J. Org. Chem.*, **24,** 134 (1959).
54: P18555c	S. Yamada and T. Mizoguchi, Japanese Patent, 10,468 (1959).
55: 4488–9e–b	M. Ruccia and D. Spinelli, *Gazz. Chim. Ital.*, **89,** 1654 (1959).
55: 16522g	M. Ruccia, G. Natale, and S. Cusmano, *Gazz. Chim. Ital.*, **90,** 831 (1960).
55: 17626–7g–a	C. Pedersen, *Acta Chem. Scand.*, **13,** 888 (1959).
55: 19910–1h–e	M. Ruccia and G. Natale, *Gazz. Chim. Ital.*, **90,** 1047 (1960).
58: 4449a	D. Y. Curtin, R. J. Crawford, and D. K. Wedegaertner, *J. Org. Chem.*, **27,** 4300 (1962).
58: 9071g	A. A. Ponomarev and M. D. Lipanova, *Zh. Obshch. Khim.*, **32,** 2974 (1962).
58: 11495f	M. Veselý, F. Mužík, and J. Poskočil, *Collect. Czech. Chem. Commun.*, **26,** 2530 (1961).
58: 13944a	R. Fusco, G. Bianchetti, D. Pocar, and R. Ugo, *Chem. Ber.*, **96,** 802 (1963).
59: 11494f	P. Yates and D. G. Farnum, *J. Amer. Chem. Soc.*, **85,** 2967 (1963).
60: 1735g	D. Y. Curtin and N. E. Alexandrou, *Tetrahedron*, **19,** 1697 (1963).
60: 15069b	G. J. Murtagh, *Australian J. Exptl. Agr. Animal Husbandry*, **3,** 173 (1963).
61: 654f	P. A. S. Smith, L. O. Krbechek, and W. Resemann, *J. Amer. Chem. Soc.*, **86,** 2025 (1964).
64: 12662f	W. Carpenter, A. Haymaker, and D. W. Moore, *J. Org. Chem.*, **31,** 789 (1966).
65: 8814d,e	N. E. Alexandrou, *Tetrahedron*, **22,** 1309 (1966).
65: 10524d	R. Fuks, R. Buijle, and H. G. Viehe, *Angew. Chem.*, **78,** 594 (1966).
66: 10887k	G. Rembarz, B. Kirchhoff, and G. Dongowski, *J. Prakt. Chem.*, **33,** 199 (1966).
67: 32614c	E. Schmitz and C. Hörig, *Chem. Ber.*, **100,** 2101 (1967).
67: 100060w	S. Hauptmann and K. Hirschberg, *J. Prakt. Chem.*, **36,** 82 (1967).
68: 59506b	S. Hauptmann, H. Wilde, and K. Moser, *Tetrahedron Lett.*, 3295 (1967).
68: P91872d	C. Yembrick, U.S. Patent, 3,364,030 (1968).
69: 36042d	N. E. Alexandrou and E. D. Micromastoras, *Tetrahedron Lett.*, 231 (1968).
69: 36375q	H. El Khadem, M. A. M. Nassr, and M. A. E. Shaban, *J. Chem. Soc., C,* 1465 (1968).

69: 77238r A. Albert, *J. Chem. Soc., C,* 2076 (1968).
69: 96327k A. Canonico, G. Casini, M. Felici, M. Ferappi, and P. Lucentini, *Farmaco, Ed. Sci.,* **23,** 383 (1968).
70: 37725m D. Pocar, S. Maiorana, and P. D. Croce, *Gazz. Chim. Ital.,* **98,** 949 (1968).
70: 58163s H. El Khadem and M. A. E. Shaban, *J. Chem. Soc., C,* 519 (1967).
71: 38087n P. K. Freeman and R. C. Johnson, *J. Org. Chem.,* **34,** 1746 (1969).
71: P124445j Geigy, A.-G., French Patent, 1,599,131 (1969).
72: 21645r A. Albert, *J. Chem. Soc., C,* 2379 (1969).
72: 43576w G. García-Muñoz, R. Madroñero, M. Rico, and M. C. Saldaña, *J. Heterocycl. Chem.,* **6,** 921 (1969).
72: 55344g A. Karklins and E. Gudriniece, *Latv. PSR Zinat. Akad. Vestis, Kim. Ser.,* 579 (1969).
72: 78952t A. Albert, *J. Chem. Soc., C,* 230 (1970).
72: 121447w G. L'abbé and A. Hassner, *J. Heterocycl. Chem.,* **7,** 361 (1970).
72: P121548e J. W. Baker and G. L. Bachman, U.S. Patent, 3,497,595 (1970).
73: 3612u K. Geibel and H. Mäder, *Chem. Ber.,* **103,** 1645 (1970).
73: 25592q K. D. Berlin, S. Rengaraju, T. E. Snider, and N. Mandava, *J. Org. Chem.,* **35,** 2027 (1970).
73: 44674j P. A. S. Smith, G. J. W. Breen, M. K. Hajek, and D. V. C. Awang, *J. Org. Chem.,* **35,** 2215 (1970).
73: 77154t A. Albert, *J. Chem. Soc., D,* 858 (1970).
73: 120446d K. Isomura, M. Okada, and H. Taniguchi, *Nippon Kagaku Zasshi,* **91,** 746 (1970).
73: 130944d H. El Khadem, M. A. E. Shaban, and M. A. M. Nassr, *J. Chem. Soc., C,* 2167 (1970).
74: 22769n P. N. Neuman, *J. Heterocycl. Chem.,* **7,** 1159 (1970).
74: 125581z D. R. Sutherland and G. Tennant, *J. Chem. Soc., C,* 706 (1971).
74: 141643b T. Sasaki, T. Yoshioka, and Y. Suzuki, *Bull. Chem. Soc. Jap.,* **44,** 185 (1971).
75: 5801v N. J. Cusack, G. Shaw, and G. J. Litchfield, *J. Chem. Soc., C,* 1501 (1971).
75: 20332m D. R. Sutherland, G. Tennant, *J. Chem. Soc., C,* 2156 (1971).
76: 34177x H. Wamhoff and P. Sohár, *Chem. Ber.,* **104,** 3510 (1971).
76: 59718n M. F. Lappert and J. S. Poland, *J. Chem. Soc., C,* 3910 (1971).
76: 72458x T. L. Gilchrist, G. E. Gymer, and C. W. Rees, *J. Chem. Soc., D,* 1519 (1971).
76: 85756t S. Hauptmann, H. Wilde, and K. Moser, *J. Prakt. Chem.,* **313,** 882 (1971).
76: 126879f D. Pocar, R. Stradi, and L. M. Rossi, *J. Chem. Soc., Perkin Trans.,* **1,** 619 (1972).
76: 140659r H. Bauer, A. J. Boulton, W. Fedeli, A. R. Katritzky, A. Majid-Hamid, F. Mazza, and A. Vaciago, *J. Chem. Soc., Perkin Trans.,* **2,** 662 (1972).
77: 75174h N. E. Alexandrou and E. D. Micromastoras, *J. Org. Chem.,* **37,** 2345 (1972).
77: 88421m R. A. Bowie and D. A. Thomason, *J. Chem. Soc., Perkin Trans.,* **1,** 1842 (1972).
78: P4256w S. Kawano and I. Yamazu, Japanese Patent, 72 40,050 (1972).
78: 29681u C. A. Lovelette and L. Long, Jr., *J. Org. Chem.,* **37,** 4124 (1972).
78: P49790g O. Fujimoto, German Patent, 2,149,196 (1972).
78: 84715h F. A. Lehmkuhl, J. T. Witkowski, and R. K. Robins, *J. Heterocycl. Chem.,* **9,** 1195 (1972).
78: 111219h T. L. Gilchrist, G. E. Gymer, and C. W. Rees, *J. Chem. Soc., Perkin Trans.,* **1,** 555 (1973).

79: 5299v D. R. Sutherland, G. Tennant, and R. J. S. Vevers, *J. Chem. Soc., Perkin Trans.*, **1,** 943 (1973).

79: 17821x H. Taguchi, *Tetrahedron Lett.*, 1137 (1973).

79: P18722c B. Hirsch, H. J. Heckemann, and N. V. Tai, East German Patent, 94,383 (1972).

79: 66246x Y. Shiokawa and S. Ohki, *Chem. Pharm. Bull.*, **21,** 981 (1973).

79: P78815v E. Aufderhaar and A. Meyer, German Patent, 2,263,878 (1973).

79: 78884s U. Heep, *Justus Liebigs Ann. Chem.*, 578 (1973).

79: 126403h K. S. Balanchandran and M. V. George, *Tetrahedron*, **29,** 2119 (1973).

79: 137053v A. Albert and H. Taguchi, *J. Chem. Soc., Perkin Trans.*, **1,** 1629 (1973).

79: 146462e A. Albert, *J. Chem. Soc., Perkin Trans.*, **1,** 1634 (1973).

79: 146470f A. Albert and W. Pendergast, *J. Chem. Soc., Perkin Trans.*, **1,** 1620 (1973).

79: P147417f H. J. Binte, T. Eulenberger, and H. Noack, East German Patent, 95,438 (1973).

80: P16441g B. Hirsch, H. J. Heckemann, and N. V. Tai, East German Patent, 96,078 (1973).

80: 47806j K. Harsányi, K. Takács, and E. Benedek, *Justus Liebigs Ann. Chem.*, 1606 (1973).

80: 47912r C. E. Olsen, *Acta Chem. Scand.*, **27,** 1987 (1973).

81: 3850k F. G. Badder, M. N. Basyouni, F. A. Fouli, and W. I. Awad, *J. Indian Chem. Soc.*, **50,** 589 (1973).

81: P27329y S. H. Roth, U.S. Patent, 3,783,017 (1974).

81: 77865r Y. Maki, M. Suzuki, K. Izuta, and S. Iwai, *Chem. Pharm. Bull.*, **22,** 1269 (1974).

81: 105404k A. Matsumoto, M. Yoshida, and O. Simamura, *Bull. Chem. Soc. Jap.*, **47,** 1493 (1974).

82: P17959k M. Matsuo, K. Koga, S. Hotta, and T. Akamatsu, Japanese Patent, 74 63,681 (1974).

82: 156182q T. L. Gilchrist, G. E. Gymer, and C. W. Rees, *J. Chem. Soc., Perkin Trans.*, **1,** 1 (1975).

83: 9910t J. Wilarrasa and R. Giranados, *J. Heterocycl. Chem.*, **11,** 867 (1974).

83: 95918f H. Wilde, S. Hauptmann, and E. Kleinpeter, *Z. Chem.*, **15,** 155 (1975).

83: 114297w A. Martvoň, Š. Stankovský, and J. Světlík, *Collect. Czech. Chem. Commun.*, **40,** 1199 (1975).

83: 192266z S. Auricchio, G. Vidari, and P. Vita-Finzi, *Gazz. Chim. Ital.*, **105,** 583 (1975).

84: 16553c C. Mayor and C. Wentrup, *J. Amer. Chem. Soc.*, **97,** 7467 (1975).

84: P45268u M. Minagawa and N. Kubota, Japanese Patent, 75 105,559 (1975).

84: 52562p S. C. Kokkou and P. J. Rentzeperis, *Acta Crystallog., Sect. B*, **B31,** 2788 (1975).

84: 74210z T. Novinson, P. Dea, and T. Okabe, *J. Org. Chem.*, **41,** 385 (1976).

84: 105706x G. Himbert, D. Frank, and M. Regitz, *Chem. Ber.*, **109,** 370 (1976).

84: 121778c A. Albert, *J. Chem. Soc., Perkin Trans.*, **1,** 291 (1976).

84: 135549r D. Stadler, W. Anschütz, M. Regitz, G. Keller, D. Van Assche, and J. P. Fleury, *Justus Liebigs Ann. Chem.*, 2159 (1975).

84: 135555q M. A. Shaban, M. A. Nassr, and M. A. Moustafa, *J. Heterocycl. Chem.*, **12,** 1295 (1975).

85: 123875b R. S. Klein, F. G. De Las Heras, S. Y. K. Tam, I. Wempen, and J. J. Fox, *J. Heterocycl. Chem.*, **13,** 589 (1976).

85: 177331q N. E. Alexandrou and S. Adamopoulos, *Synthesis*, 482 (1976).

85: 143067f S. Auricchio, G. Vidari, and P. Vita-Finzi, *Gazz. Chim. Ital.*, **106,** 1 (1976).

85: 191649f N. E. Alexandrou and E. D. Micromastoras, *Prakt. Panelleniou Chem. Synedriou, 4th*, **1,** 33 (1970, Pub. 1972).

CHAPTER 7

Azido-, Azo-, Diazo-, Triazeno- and Nitro-1,2,3-Triazoles

Starting with an amino-1,2,3-triazole, Smith and his collaborators have shown that the corresponding diazo- (**7.1-1**) and azido-1,2,3-triazole (**7.1-2**) can be prepared in excellent yield (Eq. 1).[1]

Ar′ NH$_2$ N N N Ar $\xrightarrow[\text{HCl}]{Na^+\ ^-NO_2}$ Ar′ $N_2^+Cl^-$ N N N Ar $\xrightarrow{Na^+\ ^-N_3}$ Ar′ N_3 N N N Ar (1)

7.1-1 **7.1-2**

Ar, Ar′ = Ph_2; Ph, 4-MePh (90% overall)

A variety of methods has been employed to prepare azo-derivatives of the 1,2,3-triazoles, but in general the yields have not been exceptional (Eqs. 2 to 5).[2–5]

O_2N Me Me O N + $ArN_2^+Cl^-$ ⟶ Me N=NAr N N N O Ph (2)

Ar = Ph, 4-MePh

PhN=N C=NNHPh C=NNHPh D-glucose residue $\xrightarrow{\text{HOAc}}$ PhN=N D-glucose residue N N N Ph 54% (3)

$Me_2C{=}O$ + PhN_3 $\xrightarrow{\text{KOBu-}t}$ $AcCH_2$ N=N Me N N N Ph 74% (4)

$$HSO_4^- \; {}^+N_2\text{-(1}H\text{-1,2,3-triazol-4-yl)} + \text{2-naphthol} \xrightarrow[H_2O]{Na^+ \, ^-OH} \text{4-(2-hydroxy-1-naphthyl-N=N-)-1}H\text{-1,2,3-triazole} \quad 100\% \qquad (5)$$

The synthesis of triazeno-1,2,3-triazoles has been accomplished using the Iotsich complexes and related organometallic materials (Eqs. 6 to 8).[6–8] The coupling

$$BrMgC{\equiv}CMgBr + 2\; \text{4-}BrC_6H_4N_3 \longrightarrow \text{4-}BrC_6H_4\text{-N=NHN-(1-(4-}BrC_6H_4\text{)-1,2,3-triazol-4-yl)} \qquad (6)$$

$$HC{\equiv}CMgBr + 2PhN_3 \longrightarrow PhN{=}NHN\text{-(1-Ph-1,2,3-triazol-4-yl)} \quad 30\% \qquad (7)$$

$$RC{\equiv}C^- \, {}^+Li + 2ArN_3 \longrightarrow ArN{=}NHN\text{-(5-R-1-Ar-1,2,3-triazol-4-yl)} \; \mathbf{7.1\text{-}3} \qquad (8)$$

R	= Ph	Ph	Ph	*n*-Pr	*n*-Pr	*n*-Pr
Ar	= Ph	4-NO_2Ph	4-BrPh	Ph	4-NO_2Ph	4-BrPh
% **7.1-3**	= 75	98	23	91	95	59

of the phenyldiazonium ion with an amino-1,2,3-triazole produces the 1-triazeno derivative (Eq. 9).[9]

$$\text{4-Me-5-}NH_2\text{-2-Ph-1,2,3-triazole} + PhN_2^+ \longrightarrow \text{4-Me-5-(N=NNHPh)-2-Ph-1,2,3-triazole} \quad 70\% \qquad (9)$$

The yields of nitro-1,2,3-triazoles are not usually high, but such products are useful intermediates in the preparation of many of the compounds previously discussed. For example, β-nitrostyrenes or enamines provide routes to the 4-amino compounds that are difficult to obtain by other methods (Eqs. 10 to 12).[10–13] When R = H, the yields are vastly superior. A still more promising route involves β-bromo-β-nitrostyrenes (Eq. 13).[13]

$$C_6H_5CH{=}CHNO_2 + PhN_3 \longrightarrow \text{4-}O_2N\text{-5-Ph-1-Ph-triazole} + \text{5-Ph-1-Ph-triazole} \quad (10)$$

$$(H)(O_2N)C{=}C(R)(N\text{-morpholino}) + ArN_3 \xrightarrow[\text{reflux}]{\text{EtOH}} \text{4-}O_2N\text{-5-R-1-Ar-triazole} \quad (11)$$

R = H, Me, Ph   Ar = Ph, 3- or 4-NO_2Ph, 3,5-$(NO_2)_2Ph$

$$O_2NCH{=}CH\text{—N(morpholino)} + TsN_3 \longrightarrow \text{4-}O_2N\text{-1-H-triazole} \quad (12)$$

60%

$$ArCH{=}C(NO_2)(Br) + Na^{+}\,{}^{-}N_3 \xrightarrow[\text{or EtOH}]{HCONMe_2,\ Me_2SO} \text{4-}O_2N\text{-5-Ar-1-H-triazole} \quad (13)$$

7.1-4

Ar =	Ph	4-MeOPh	4-BrPh	4-NO_2Ph	3-ClPh	3-NO_2Ph	α-$C_{10}H_7$
% **7.1-4** =	47	56	40	67	33	29	52

Both nucleophilic substitution (Eq. 14)[12,14] and direct nitration (Eq. 15)[12] have been employed with varying degrees of success. When 2,4-dinitrofluorobenzene is used in the substitution (Eq. 14), an equal amount of the *N*-2 isomer of **7.1-5** is obtained. The nitration yields (Eq. 15) are variable—for example, 83% with 2-(2,4,6-trinitrophenyl) and no reaction for 1-(2,4,6-trinitrophenyl)!

O_2N N N N H + N_3 Y NO_2 NO_2 ⟶ O_2N N N N NO_2 $-NO_2$ Y

Y = H 27%

NO_2 46%

+ O_2N N N N NO_2 NO_2 (14)

7.1-5
27%

N N N O_2N NO_2 NO_2 $\xrightarrow[85°]{HNO_3/H_2SO_4}$ NO_2 N N N O_2N NO_2 NO_2 $\xleftarrow[Na^+ \ ^-NO_3]{H_2SO_4 \cdot SO_3}$ N N N Ph (15)

REFERENCES

1. **61:** 654f
2. **38:** 4598[2]
3. **55:** 3562a
4. **80:** 47912r
5. **83:** 9910t
6. **26:** 1287[8]
7. **68:** 87243g
8. **68:** 105100q
9. **68:** 2857n
10. **66:** 55466z
11. **70:** 37725m
12. **74:** 99950x
13. **84:** 59325x
14. **74:** 53664d

TABLE 7. AZIDO-, AZO-, DIAZANEDIYL-, TRIAZENO-, AND NITRO-1,2,3 TRIAZOLES

Compound	Reference
4-[1-(aminocarbonyl)-1-cyanomethylene]diazanediyl-5-phenyl-	**78:** 72073j
2-(4-aminophenyl)-4-nitro-	**61:** P1873h
2-[4-(aminosulfonyl)phenyl]-4-[[4-(dimethylamino)-2-methoxyphenyl]azo]-	**79:** P147417f
5-azido-1,4-diphenyl-	**61:** 654f, **74:** 111399j
5-azido-1-(4-methylphenyl)-4-phenyl-	**61:** 645f
4-[(1-benzyl-1-cyanomethylene)diazanediyl]-5-phenyl-	**78:** 72073j
4[[1-benzyl-1-(ethoxycarbonyl)methylene]diazanediyl]-5-phenyl-	**78:** 72073j
4,4′-bis[(2-amino-8-hydroxy-6-sulfo-1-naphthalenyl)-azo]-5-methyl-2-phenyl-	**80:** P16441g
4,4′-bis[(3-carboxy-4-hydroxyphenyl)azo]-5-methyl-2-phenyl-	**80:** P16441g
4-[1-bis(ethoxycarbonyl)methylene]diazanediyl-5-phenyl-	**78:** 72073j
4-[[bis(2-hydroxyethyl)aminophenyl]azo]-5-methyl-2-phenyl-	**80:** P16441g
4,4′-bis[(2-hydroxy-1-naphthalenyl)azo]-5-methyl-2-phenyl-	**80:** P16441g
1-(4-bromophenyl)-4-[(4-bromophenyl)triazeno]-	**26:** 1287[8]
1-(4-bromophenyl)-4-[3-(4-bromophenyl)-2-triazeno]-5-phenyl-	**68:** 105100q
1-(4-bromophenyl)-4-[3-(4-bromophenyl)-2-triazeno]-5-propyl-	**68:** 105100q
2-(4-bromophenyl)-4-[[4-(dimethylamino)-3-methoxyphenyl]azo]-	**79:** P147417f
4-(4-bromophenyl)-5-nitro-	**84:** 59325x
5-*tert*-butyl-4-[(3,3-dimethyl-2-oxobutyl)azo]-1-phenyl-	**80:** 47912r
2-(5-chloro-2-methoxyphenyl)-4-nitro-	**61:** P1873g
2-(3-chloro-2-methylphenyl)-4-nitro-	**61:** P1873h
2-(4-chloro-2-methylphenyl)-4-nitro-	**61:** P1873h
2-(5-chloro-2-methylphenyl)-4-nitro-	**61:** P1873g
2-(2-chlorophenyl)-4-nitro-	**61:** P1873h
2-(3-chlorophenyl)-4-nitro-	**61:** P1873h
2-(4-chlorophenyl)-4-nitro-	**61:** P1873h
4-(3-chlorophenyl)-5-nitro-	**84:** 59325x
-4-diazonium (?)	**83:** 9910t
-4-diazonium, 5-phenyl-, chloride	**78:** 72073j
4-[(1,1-dibenzoylmethylene)diazanediyl]-5-phenyl-	**78:** 72073j
2-(2,5-dichlorophenyl)-4-nitro-	**58:** 6819f
2-(4-chlorophenyl)-4-[[4-(dimethylamino)-2-methoxyphenyl]azo]-	**79:** P147417f
2-(2-chlorophenyl)-4-[[4-(diethylamino)phenyl]azo]-	**79:** P147417f
4-[4-[[(2-cyanoethyl)ethylamino]phenyl]azo]-2-(4-methoxyphenyl)-	**79:** P147417f
4-[[4-(dimethylamino)-2-methoxyphenyl]azo]-2-(4-methoxyphenyl)-	**79:** P147417f
4-[[4-(dimethylamino)-2-methoxyphenyl]azo]-2-phenyl-	**79:** P147417f
2-(2,4-dinitrophenyl)-4-methyl-5-nitro-	**73:** 109746f

TABLE 7 (*Continued*)

Compound	Reference
1-(2,4-dinitrophenyl)-4-nitro-	**74:** 53664d
1-(3,4-dinitrophenyl)-4-nitro-	**74:** 99950x
1-(3,5-dinitrophenyl)-4-nitro-	**74:** 99950x
2-(2,4-dinitrophenyl)-4-nitro-	**58:** 6819f, **74:** 53664d
1,5-diphenyl-4-nitro-	**68:** 2857n
1,5-diphenyl-4-[(2-oxo-2-phenylethyl)azo]-	**80:** 47912r
2,4-diphenyl-5-(phenylazo)-	**77:** P103295u
1,5-diphenyl-4-(3-phenyl-2-triazeno)-	**68:** 105100q
4-[[1-(ethoxycarbonyl)-1-cyanomethylene]diazanediyl]-5-phenyl-	**78:** 72073j
2-[(2-hydroxy-4-methoxyphenyl)azo]-	**75:** 70942f
4-[(2-hydroxy-1-naphthalenyl)azo]-	**83:** 9910t
5-[(2-hydroxy-1-naphthalenyl)azo]-	**84:** 30969z
2-(2-hydroxyphenyl)-4-nitro-	**71:** P124445j
4-[(2-hydroxypropyl)azo]-5-methyl-1-phenyl-, acetate ester	**80:** 47912r
2-[(4-methoxy-2-phenolato)azo]-, cuprate(1−)	**75:** 70942f
2-(4-methoxyphenyl)-4-[(4-methoxyphenyl)azo]-5-methyl-	**77:** P103295u
2-(2-methoxyphenyl)-4-nitro-	**61:** P1873h
2-(4-methoxyphenyl)-4-nitro-	**61:** P1873h
4-(4-methoxyphenyl)-5-nitro-	**84:** 59325x
2-(2-methoxy-5-sulfophenyl)-4-[(2-methoxy-5-sulfophenyl)azo]-5-methyl-	**77:** P103295u
5-methyl-2-(4-methylphenyl)-4-[(4-methylphenyl)azo]-, 1-oxide	**38:** 4598[3]
4-methyl-5-nitro-	**73:** 109746f
5-methyl-4-nitro-1-(4-nitrophenyl)-	**70:** 37725m
5-methyl-4-nitro-1-phenyl-	**70:** 37725m
5-methyl-4-[(2-oxopropyl)azo]-1-phenyl-	**80:** 47912r
2-(2-methylphenyl)-4-nitro-	**61:** P1873h
2-(3-methylphenyl)-4-nitro-	**61:** P1873h
2-(4-methylphenyl)-4-nitro-	**61:** P1873h, **72:** P100716t
4-methyl-2-phenyl-5-(phenylazo)-	**55:** 3562a
5-methyl-2-phenyl-4-(phenylazo)-, 1-oxide	**38:** 4598[2]
5-methyl-1-phenyl-4-(3-phenyl-1-triazenyl)-	**80:** 47912r
2-(1-naphthyl)-4-nitro-	**61:** P1873h
2-(2-naphthyl)-4-nitro-	**61:** P1873g
4-(1-naphthyl)-5-nitro-	**84:** 59325x
4-nitro-	**66:** 55466z, **73:** 109746f, **74:** 99950x
4-nitro-, 1:1 adduct with morpholine	**70:** 37725m
4-nitro-1-(3-nitrophenyl)-	**74:** 99950x
4-nitro-1-(4-nitrophenyl)-	**66:** 55466z, **70:** 37725m, **74:** 99950x
4-nitro-2-(3-nitrophenyl)-	**61:** P1873h
4-nitro-2-(4-nitrophenyl)-	**58:** 6819f
4-nitro-5-(3-nitrophenyl)-	**84:** 59325x
4-nitro-5-(4-nitrophenyl)-	**84:** 59325x
4-nitro-1-phenyl-	**66:** 55466z, **70:** 37725m

TABLE 7 (*Continued*)

Compound	Reference
4-nitro-2-phenyl-	**61:** P1873h, **72:** P100716t, **79:** P137160c
4-nitro-5-phenyl-	**70:** 37725m, **84:** 59325x
1-(4-nitrophenyl)-4-[3-(4-nitrophenyl)-2-triazeno]-5-phenyl-	**68:** 105100q
1-(4-nitrophenyl)-4-[3-(4-nitrophenyl)-2-triazeno]-5-propyl-	**68:** 105100q
4-nitro-2-[5(or-3)-phenylpyrazol-3(or 5)-yl]-	**61:** P1873h
4-nitro-1-β-D-ribofuranosyl-	**78:** 84715h
4-nitro-2-β-D-ribofuranosyl-	**78:** 84715h
4-nitro-1-(2,3,5-tri-*O*-acetyl-β-D-ribofuranosyl)-	**78:** 84715h
4-nitro-2-(2,3,5-tri-*O*-acetyl-β-D-ribofuranosyl)-	**78:** 84715h
4-nitro-2-(1,2,4-triazol-3-yl)-	**71:** 112866j
4-nitro-1-(2,4,6-trinitrophenyl)-	**74:** 99950x
4-nitro-2-(2,4,6-trinitrophenyl)-	**74:** 99950x
1-phenyl-4-(phenyltriazeno)-(?)	**26:** 1287[9]
1-phenyl-4-(3-phenyl-2-triazeno)-	**68:** 87243g
1-phenyl-4-(3-phenyl-2-triazeno)-5-propyl-	**68:** 105100q

REFERENCES

26: 1287[8,9]	H. Kleinfeller and G. Bönig, *J. Prakt. Chem.*, **132,** 175 (1931).
38: 4598[2,3]	A. Quilico and C. Musdnte, *Gazz. Chim. Ital.*, **72,** 399 (1942).
55: 3562a	L. Mester and F. Weygand, *Bull. Soc. Chim. Fr.*, 350 (1960).
58: 6819f	B. M. Lynch and T-L. Chan, *Can. J. Chem.*, **41,** 274 (1963).
61: 645f	P. A. S. Smith, L. O. Krbechek, and W. Resemann, *J. Amer. Chem. Soc.*, **86,** 2025 (1964).
61: P1873g,h	R. Mohr and M. Zimmermann, German Patent, 1,168,437 (1964).
66: 55466z	S. Maiorana, D. Pocar, and P. D. Croce, *Tetrahedron Lett.*, 6043 (1966).
68: 2857n	P. D. Callaghan and M. S. Gibson, *Chem. Commun.*, 918 (1967).
68: 87243g	G. S. Akimova, V. N. Chistokletov, and A. A. Petrov, *Zh. Org. Khim.*, **3,** 2241 (1967).
68: 105100q	G. S. Akimova, V. N. Chistokletov, and A. A. Petrov, *Zh. Org. Khim.*, **4,** 389 (1968).
70: 37725m	D. Pocar, S. Maiorana, and P. D. Croce, *Gazz. Chim. Ital.*, **98,** 949 (1968).
71: 112866j	E. J. Browne, *Aust. J. Chem.*, **22,** 2251 (1969).
71: P124445j	Geigy, A.-G., French Patent, 1,559,131 (1969).
72: P100716t	Farbenfabriken Bayer A.-G., French Patent, 1,577,760 (1969).
73: 109746f	T. E. Eagles, M. A. Khan, and B. M. Lynch, *Org. Prep. Proced.*, **2,** 117 (1970).
74: 53664d	M. A. Khan and B. M. Lynch, *J. Heterocycl. Chem.*, **7,** 1237 (1970).
74: 99950x	P. N. Neuman, *J. Heterocycl. Chem.*, **8,** 51 (1971).
74: 111399j	J. J. Friar, *Diss. Abstr. Int. B*, **31,** 584 (1970).
75: 70942f	V. Chromy and L. Sommer, *Spisy Prirodoved. Fak. Univ. J. E. Purkyne Brne*, **517,** 407 (1970).

77: P103295u	M. Vesely and V. Kmonicek, Czechoslovakian Patent, 141,765 (1969).
78: 72073j	H. Mackie and G. Tennant, *Tetrahedron Lett.*, 4719 (1972).
78: 84715h	F. A. Lehmkuhl, J. T. Witkowski, and R. K. Robins, *J. Heterocycl. Chem.*, **9,** 1195 (1972).
79: P137160c	U. Claussen, H. Gold, and J. Schroeder, German Patent, 2,210,261 (1973).
79: P147417f	H. J. Binte, T. Eulenberger, and H. Noack, East German Patent, 95,438 (1973).
80: P16441g	B. Hirsch, H. J. Heckemann, and N. V. Tai, East German Patent, 96,708 (1973).
80: 47912r	C. E. Olsen, *Acta Chem. Scand.*, **27,** 1987 (1973).
83: 9910t	J. Vilarrasa and R. Granados, *J. Heterocycl. Chem.*, **11,** 867 (1974).
84: 30969z	G. Keller, J. P. Fleury, W. Anschütz, and M. Regitz, *Bull. Soc. Chim. Fr.*, 1219 (1975).
84: 59325x	G. Kh. Khisamutdinov, O. A. Bondarenko, and L. A. Kupriyanova, *Zh. Org. Khim.*, **11,** 2445 (1975).
85: P94660v	R. K. Robins and J. T. Witkowski, U.S. Patent, 3,968,103 (1976).

CHAPTER 8

O-Substituted Oxy- and *S*-Substituted Thio- or Sulfonyl-1,2,3-Triazoles

The synthesis of alkoxy-1,2,3-triazoles by azide addition to acetylenes appears to be highly regiospecific, but the yields vary widely with the nature of the azide (Eq. 1).[1] There are also some serious conflicting reports to be found. For examples, Huisgen and his collaborators have reported the addition of 4-nitrophenylazide (Eq. 1) to give an 83% yield.[2] Furthermore, Daeniker and Druey have found 4-ethoxy-1-phenyl-1,2,3-triazole (the isomer of **8.1-1**) from this same reaction (Eq. 2).[3] Clearly, these discrepancies must be cleared up before the full utility of the method can be properly assessed.

$$HC{\equiv}COEt + ArN_3 \longrightarrow \text{5-OEt-1-Ar-1,2,3-triazole} \quad (1)$$

8.1-1

Ar	= Ph	4-MePh	4-ClPh	4-BrPh	2-NO_2Ph	4-NO_2Ph	4-MeOPh
% **8.1-1**	= 87	51	49	59	6	12(83)[2]	35

$$\text{4-HO-1-Ph-1,2,3-triazole} + (EtO)_2SO_2 \longrightarrow \text{4-EtO-1-Ph-1,2,3-triazole} \longleftarrow HC{\equiv}COEt + PhN_3 \quad (2)$$

An alternative approach is the addition to bisenolethers (Eq. 3), but again the yields vary considerably.[4]

$$(EtO)_2C{=}CHR + PhN_3 \longrightarrow \text{4-R-5-OEt-1-Ph-1,2,3-triazole} \quad (3)$$

R	
H	78%
Me	63%
Ph	16%

An interesting cyclization has been reported by Zbiral and Stroh (Eq. 4).[5] L'abbé and his collaborators have used a similar approach, but they have found that in most cases the product is an open-chain triazene. Only one combination (R = Me; Ar = 4-NO_2Ph) produced a good yield (80%) of *v*-triazole (**8.1-2**)(Eq. 5).[6]

$$Ph_3P{=}C(CO_2Et)(Pr\text{-}n) + MeCON_3 \longrightarrow \text{1-acetyl-4-}n\text{-propyl-5-ethoxy-1,2,3-triazole (90\%)} \xrightarrow{OH^-} \text{4-}n\text{-propyl-5-ethoxy-1}H\text{-1,2,3-triazole (60\%)} \quad (4)$$

$$Ph_3P{=}C(CO_2Et)(R) + ArN_3 \longrightarrow \text{1-Ar-4-R-5-OEt-1,2,3-triazole } (\mathbf{8.1\text{-}2}) \quad (5)$$

R = Me, Ph

Scarpati and Graziano and their collaborators have shown that either base[7] or heat[8] can cause the rather exotic diazoacetal **8.1-3** to cyclize in excellent yield (Eq. 6). In spite of the starting material, the approach should be explored further.

$$\underset{\mathbf{8.1\text{-}3}}{N_2C(Ph)(NHCO_2Et)C(OMe)_2} \; \left\{ \begin{array}{l} \xrightarrow[\text{MeOH}]{K^{+}\,{}^{-}OH} \\ \xrightarrow[\text{HCONMe}_2 \text{ or } t\text{-BuOH}]{120^\circ} \end{array} \right. \text{4-Ph-5-OMe-1}H\text{-1,2,3-triazole (70–97\%)} \quad (6)$$

A large number of studies from the laboratories of Begtrup and Pedersen[9] show that the intricate chemistry of hydroxy-1,2,3-triazoles with diazomethane can, in certain instances, lead to good yields of *O*-methylation (e.g., Eqs. 7,8). These same scientists have shown that nucleophilic substitution will produce fair-to-good yields of methoxy-1,2,3-triazoles (Eq. 9).[10]

$$\text{4-Ph-5-OH-1}H\text{-1,2,3-triazole} + CH_2N_2 \longrightarrow \text{4-Ph-5-OMe-1}H\text{-1,2,3-triazole (60\%)} \quad (7)$$

R = H, Ph; + CH_2N_2 → 4-OMe-5-R-1-methyl-1,2,3-triazole: 72%, 64% (8)

R = Ph, $PhCH_2$; $Na^+ \ ^-OMe$ → 72%... 40%, 62% (9)

The peracid oxidation of a methylene group attached to the 1-nitrogen of a *v*-triazole leads to the interesting 1-alkoxy ethers (**8.1-4**) probably formed by a Meisenheimer rearrangement of the *N*-oxide (Eq. 10).[11] Once again the yields are highly variable.

CH_2R-triazole + $MeCO_3H$ → **8.1-4** (OCH_2R) (10)

R	= Ph	2,5-Me_2Ph	4-ClPh	Cl_5Ph	succinimido	phthalimido	benzotriazolyl
% **8.1-4**	= 45	30	25	15	80	80	20

4-Ph; 1-$PhCH_2$ 25%

A promising preparation of 4-aryloxy-1,2,3-triazoles (**8.1-6**) involves aryl cyano ethers and diazomethane (Eq. 11).[12]

$ArOCN + CH_2N_2$ → **8.1-5** (ArO, NH–COAr) $\xrightarrow{HCl}$ **8.1-6** (ArO, N–H) (11)

Ar	= Ph	4-MePh	4-MeOPh	4-ClPh
% **8.1-5**	= 83	79	81	67
% **8.1-6**	= 87	65	91	43

The same careful studies that led Begtrup and Pedersen to an understanding of alkylation in hydroxy-1,2,3-triazoles have produced important esterification methods.[9] An especially interesting early examples involves the preparation of the internal salt **8.1-7** and its reaction with benzoyl chloride (Eq. 12).[13] The direct reaction with benzoyl chloride often leads to both *N*- and *O*-benzoylation (Eq. 13), but the ester can be obtained in excellent yield by selective hydrolysis.[9] A new method developed in this work starts from the malonic ester anion (Eq. 14); although the yields are not great, the method is potentially important.[14]

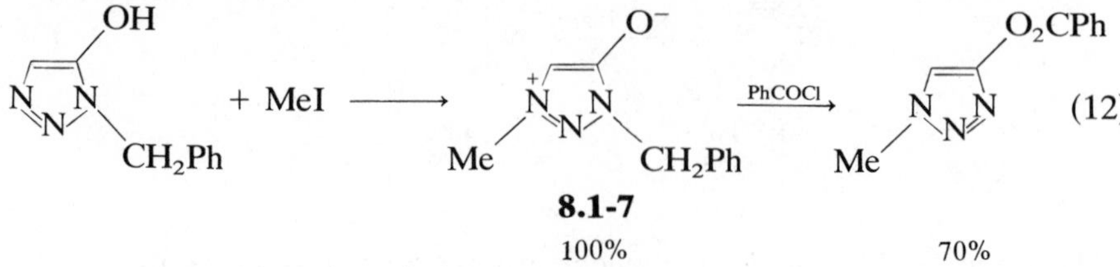

OH, N, N, N, CH_2Ph + MeI ⟶ O^-, +, N, N, N, Me, CH_2Ph —PhCOCl⟶ O_2CPh, N, N, N, Me (12)

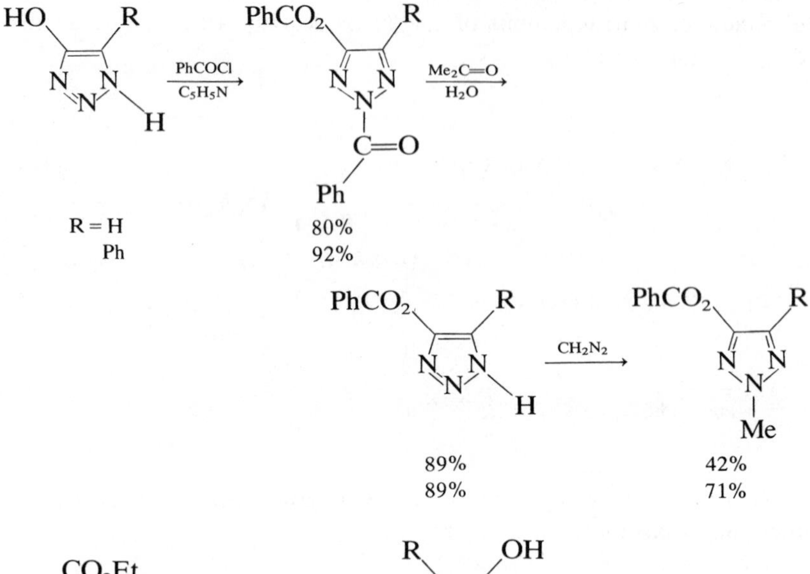

8.1-7
100% 70%

HO, R, N, N, N, H —PhCOCl / C_5H_5N⟶ $PhCO_2$, R, N, N, N, C=O, Ph —$Me_2C{=}O$ / H_2O⟶

R = H 80%
Ph 92%

$PhCO_2$, R, N, N, N, H —CH_2N_2⟶ $PhCO_2$, R, N, N, N, Me (13)

89% 42%
89% 71%

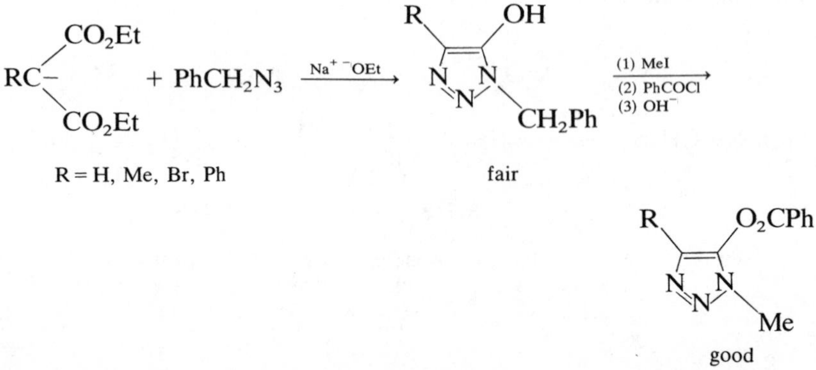

CO_2Et, RC$^-$, CO_2Et + $PhCH_2N_3$ —$Na^+\ ^-OEt$⟶ R, OH, N, N, N, CH_2Ph —(1) MeI (2) PhCOCl (3) OH^-⟶

R = H, Me, Br, Ph fair

R, O_2CPh, N, N, N, Me (14)

good

One of the most significant methods for the preparation of sulfides in the *v*-triazoles involves the rearrangement of 5-amino-1,2,3-thiadiazoles (**8.1-8**) under basic conditions (Eq. 15).[15] It was later suggested that the **8.1-8** structure is an intermediate in the reactions of isothiocyanates (**8.1-9**) with diazomethane (Eq. 16).[16] This reaction, with various **8.1-8** compounds, has been extended with promising results,[17] but the problem of product structure (compare **8.1-11** and **8.1-12** with **8.1-10** and Equation 15) remains to be examined (Eqs. 17, 18).

NHPh (thiadiazole) **8.1-8** $\xrightarrow{Na^+\,^-OH}$ S^-Na^+ (triazole, N-Ph) about 80% $\xrightarrow{MeI}$ SMe (triazole, N-Ph) 61% (15)

$RN{=}C{=}S + CH_2N_2$ **8.1-9** ⟶ [NHR (thiadiazole)] ⟶ SMe (triazole, N-R) **8.1-10** (16)

R	= MeC=CHCHMe$_2$(Z)	CH=CHEt(Z)	Me	Ph
% **8.1-10** =	90	87	40	71

NHR (thiadiazole) $\xrightarrow{Na^+\,^-OH}$ S^-Na^+ (triazole, N-R) $\xrightarrow{HCl}$ HS (triazole, N-R) **8.1-11** (17)

S^-Na^+ (triazole, N-R) $\xrightarrow{MeI}$ MeS (triazole, N-R) **8.1-12** (18)

R =	4-MePh	4-EtOPh	4-AcPh	4-BrPh
% **8.1-11** =	85	86	55	75
% **8.1-12** =	61	69	65	77

Begtrup has also made important contributions to the synthesis of sulfides. For example, in a reaction reminiscent of the ether studies exemplified in Equation 12, a rather regiospecific rearrangement has been observed (Eq. 19).[18] In a later and more detailed study[19] he has investigated the range and limitations of applicability of internal salts such as **8.1-13** (Eqs. 20,21). In this same study it was observed that the isomer of **8.1-13** gives a mixture of products but also some limited synthetic utility (Eq. 22).[19]

(19)

8.1-13 → 82% + 13%

(20)

C_6H_6, heat → 100%

(21)

MeI → 62% + 38%

(22)

MeI, $K^+ \ ^-OH$ → 43% + 20% + 18% + 10%

The direct methylation of 5-mercapto-1,2,3-triazole has been reported in excellent yield,[20] but subsequent *N*-methylation with diazomethane is somewhat less specific (Eq. 23).

(23)

MeI, $K^+ \ ^-OH$ → 86%; CH_2N_2 → about 50% + about 50%

In the *υ*-triazoles, as in most areas of contemporary organic chemistry, the tosyl derivatives have received a good deal of attention, and several excellent synthetic methods are available. An example is the cyclization of α,β-bistosylhydrazones (Eq. 24).[21] It is noteworthy that with R = R′ = Ph, diphenylacetylene is the only product isolated.

$$\text{RC(=NNHTs)–C(=NNHTs)R}' \xrightarrow[\text{HOC}_2\text{H}_4\text{OH}]{\text{K}^+\ ^-\text{OH}} \text{4-R-5-R}'\text{-1-Ts-1,2,3-triazole} \tag{24}$$

R = R′ = Me	79%
n-Pr	84%
R = Me; R′ = *n*-Pr	72%

Pocar and his collaborators have applied variations of their enamine synthesis with moderate success (Eqs. 25,26).[22,23]

$$\text{H(Ts)C=C(Me)(N-morpholino)} + \text{ArN}_3 \longrightarrow \text{4-Ts-5-Me-1-Ar-1,2,3-triazole} \tag{25}$$

Ar = 4-ClPh	42%
4-NO_2Ph	75%

$$\text{H(ArSO}_2\text{)C=C(H)(N-morpholino)} + \text{TsN}_3 \xrightarrow[\text{reflux}]{\text{EtOH}} \text{4-ArSO}_2\text{-1H-1,2,3-triazole} \tag{26}$$

Ar = Ph	50%
4-NO_2Ph	80%

Meek and Fowler have added azide ion to 1,2-ditosylethylene (**8.1-14**), which involves the intermediacy of **8.1-16** and they have found fair yields of the triazole under a variety of conditions (Eqs. 27 to 29).[24,25] A comparable yield of **8.1-15** was obtained by adding trimethylsilylazide to tosylacetylene.[25]

Harvey has shown that the addition of tosylazide to α-keptophosphorus ylides (**8.1-17**) can produce excellent yields of *N*-tosyl-5-substituted-1,2,3-triazoles (**8.1-18**) (Eq. 30).[26] The application of this method should be pursued vigorously.

$$\underset{\textbf{8.1-14}}{TsCH{=}CHTs} + Na^+ \, ^-N_3 \xrightarrow{HCONMe_2} \underset{\substack{\textbf{8.1-15}\\45\%}}{\text{4-Ts-1H-1,2,3-triazole}} \tag{27}$$

$$\underset{\textbf{8.1-16}}{TsCH{=}CHN_3} + Na^+ \, ^-N_3 + TsO^- \, ^+Na \xrightarrow{Me_2SO} \underset{57\%}{\textbf{8.1-15}} \tag{28}$$

$$TsCH{=}CHN_3 + TsO^- \, ^+Na \xrightarrow{Me_2SO} \underset{62\%}{\textbf{8.1-15}} \tag{29}$$

$$\underset{\textbf{8.1-17}}{Ph_3P{=}CHCOR} + TsN_3 \longrightarrow \underset{\textbf{8.1-18}}{\text{4-R-1-Ts-1,2,3-triazole}} \tag{30}$$

R	= Me	Ph	4-NO$_2$Ph	4-BrPh
% **8.1-18** =	98	98	87	80

REFERENCES

1. **55:** 11397b
2. **64:** 5077a
3. **58:** 7927f
4. **59:** 3913c
5. **71:** 112865h
6. **78:** 111442a
7. **75:** 88297e
8. **84:** 30609g
9. **71:** 124340w
10. **67:** 64305u
11. **73:** 3865d
12. **64:** 8171f
13. **67:** 100074d
14. **71:** 61296p
15. **58:** 2447c
16. **83:** 9919c
17. **85:** 94288e
18. **75:** 63704b
19. **77:** 114319d
20. **84:** 4862q
21. **47:** 10474f
22. **58:** 12561a
23. **66:** 55466z
24. **67:** 43749b
25. **68:** 77478y
26. **64:** 19597a

TABLE 8. *O*-SUBSTITUTED OXY- AND *S*-SUBSTITUTED THIO- OR SULFONYL-1,2,3-TRIAZOLES

Compound	Reference
-1-acetic acid, 5-ethoxy-	**67:** 54080c
-1-acetic acid, 5-ethoxy-, ethyl ester	**67:** 54080c
-1-acetic acid, 5-ethoxy-, methyl ester	**67:** 54080c
4-acetoxy-1-phenyl-	**58:** 7927f
5-acetoxy-1-phenyl-	**55:** 25923b
1-(4-acetylphenyl)-5-(methylthio)-	**85:** 94288e
1-(3-aminophenyl)-5-(ethylsulfonyl)-	**82:** 58044b
1[(1*H*-benzotriazol-1-yl)methoxy]-	**73:** 3865d
5-(benzoyloxy)-1-benzyl-	**67:** 100074d
4-(benzoyloxy)-1-methyl-	**67:** 100074d
5-[(benzoyloxy)methyl]-1-(1,3-dihydroxy-2-propyl)-4-(ethylsulfonyl)-, dibenzoate ester	**82:** 58044b
5-(benzoyloxy)-1-phenyl-	**55:** 11397d
4-(benzoylthio)-1-benzyl-	**75:** 63704b, **77:** 114317d
4-(benzoylthio)-1-methyl-	**77:** 114317d
5-(benzoylthio)-1-methyl-	**75:** 63704b, **77:** 114317d
1-benzyl-4-(benzylthio)-	**75:** 63704b, **77:** 114317d
1-benzyl-5-(benzylthio)-	**77:** 114317d
1-benzyl-4-methoxy-	**71:** 61296p
1-benzyl-5-methoxy-	**67:** 64305u
1-benzyl-5-methoxy-4-methyl-	**71:** 61296p
1-benzyl-5-methoxy-4-phenyl-	**71:** 61296p
1-benzyl-4-(methylthio)-	**77:** 114317d
1-benzyl-5-(methylthio)-	**77:** 114317d
1-(benzyloxy)-	**73:** 3865d
1-(benzyloxy)-4-phenyl-	**73:** 3865d
4-(benzylthio)-1-methyl-	**77:** 114317d
5-(benzylthio)-1-methyl-	**77:** 114317d
5-(benzylthio)-1-phenyl-	**77:** 114317d
-4,5-bis(oxyacetic acid), 1-(10,11-dihydro-5*H*-dibenzo-[*a,d*]cyclohepten-5-yl)-	**77:** P95371y
-4,5-bis(oxyacetic acid), 1-(5*H*-dibenzo[*a,d*]cyclo-hepten-5-yl)-	**77:** P95371y
2-(4-bromophenyl)-4-[(diethoxyphosphinethioylidyne)-oxy]-, *N*-oxide	**82:** P156328s
1-(4-bromophenyl)-5-ethoxy-	**55:** 11397d
1-(4-bromophenyl)-5-ethoxy-4-methyl-	**78:** 111442a
1-(4-bromophenyl)-5-ethoxy-4-phenyl-	**78:** 111442a
5-(4-bromophenyl)-1-[(4-methylphenyl)sulfonyl]-	**64:** 19597a, **79:** 31992k
1-(4-bromophenyl)-5-(methylthio)-	**85:** 94288e
1-[(4-bromophenyl)sulfonyl]-5-phenyl-	**64:** 11201e
1-(1-butenyl)-5-(methylthio)-, (Z)-	**83:** 9919c
1-[(4-chlorobenzyl)oxo]-	**73:** 3865d
1-(2-chloroethyl)-5-ethoxy-	**67:** 54080c
1-(4-chloro-3-nitrophenyl)-5-ethoxy-	**55:** 11397d, **64:** 8171f
2-(2-chlorophenyl)-4-[(diethoxyphosphinothioylidyne)-oxy]-, *N*-oxide	**82:** P156328s
2-(3-chlorophenyl)-4-[(diethoxyphosphinothioylidyne)-oxy]-, *N*-oxide	**82:** P156328s

TABLE 8 (*Continued*)

Compound	Reference
2-(4-chlorophenyl)-4-[(diethoxyphosphinothioylidyne)-oxy]-, *N*-oxide	**82:** P156328s
1-(4-chlorophenyl)-5-ethoxy-	**55:** 11397d
2-(3-chlorophenyl)-4-[(ethoxyethylphosphinothio-ylidyne)oxy]-, *N*-oxide	**82:** P156328s
2-(3-chlorophenyl)-4-[(ethoxypropylthiophosphin-ylidyne)oxy]-, *N*-oxide	**82:** P156328s
2-(4-chlorophenyl)-4-[(ethoxypropylthiophosphin-ylidyne)oxy]-, *N*-oxide	**82:** P156328s
1-(4-chlorophenyl)-5-methyl-4-[(2-methylphenyl)-sulfonyl]-	**58:** 12561b
2-(3,4-dichlorophenyl)-4-[(diethoxyphosphinothio-ylidyne)oxy]-, *N*-oxide	**82:** P156328s
4-[(diethoxyphosphinothioylidyne)oxy]-2-(4-fluoro-phenyl)-, *N*-oxide	**82:** P156328s
4-[(diethoxyphosphinothioylidyne)oxy]-2-(2-methoxy-phenyl)-, *N*-oxide	**82:** P156328s
4-[(diethoxyphosphinothioylidyne)oxy]-2-(4-methoxy-phenyl)-, *N*-oxide	**82:** P156328s
4-[(diethoxyphosphinothioylidyne)oxy]-2-(2-methyl-phenyl)-, *N*-oxide	**82:** P156328s
4-[(diethoxyphosphinothioylidyne)oxy]-2-(4-methyl-phenyl)-, *N*-oxide	**82:** P156328s
4-[(diethoxyphosphinothioylidyne)oxy]-2-phenyl-	**82:** P156328s
4-[(diethoxyphosphinothioylidyne)oxy]-2-phenyl-, *N*-oxide	**82:** P156328s
4-[(diethoxyphosphinothioylidyne)oxy]-2-[3-(trifluoro-methyl)phenyl]-, *N*-oxide	**82:** P156328s
5-[(*N*,*N*-diethylcarbamoyl)oxy]-4-methyl-1-phenyl-	**48:** P187h
5-[(*N*,*N*-diethylcarbamoyl)oxy]-1-phenyl-	**48:** P188a
1-(1,3-dihydroxy-2-propyl)-4-(ethylsulfonyl)-5-(hydroxymethyl)-	**82:** 58044b
4-[(dimethoxyphosphinothioylidyne)oxy]-2-phenyl-, *N*-oxide	**82:** P156328s
1-(1,3-dimethyl-1-butenyl)-5-(methylthio)-, (Z)-	**83:** 9919c
1-(2,4-dinitrophenyl)-4-ethyl-5-methoxy-	**77:** 151634a
-1-ethanol, 5-ethoxy-	**67:** 54080c
-1-ethanol, 5-ethoxy-, acetate ester	**67:** 54080c
4-ethoxy-	**67:** 54080c
5-ethoxy-1-(4-methyl-3-nitrophenyl)-	**55:** 11397d
5-ethoxy-1-(4-methylphenyl)-	**55:** 11397d
5-ethoxy-4-methyl-1-(4-nitrophenyl)-	**78:** 111442a
5-ethoxy-4-methyl-1-phenyl-	**59:** 3913c
5-ethoxy-1-(2-nitrophenyl)-	**55:** 11397e
5-ethoxy-1-(4-nitrophenyl)-	**55:** 11397c, **64:** 5077a
5-ethoxy-1-(4-nitrophenyl)-4-phenyl-	**78:** 111442a
4-ethoxy-1-phenyl-	**58:** 7927g
5-ethoxy-1-phenyl-	**55:** 11397b, **59:** 3913c
5-ethoxy-1-(3-phenyl-2-isoxazolin-5-yl)-	**67:** 54080c

TABLE 8 (*Continued*)

Compound	Reference
1-(4-ethoxyphenyl)-5-(methylthio)-	**85:** 94288e
4-ethoxy-5-propyl-	**71:** 112865h
4-[(ethoxypropylthiophosphinylidyne)oxy]-2-phenyl-	**82:** P156328s
4-[(ethoxypropylthiophosphinylidyne)oxy]-2-phenyl-, *N*-oxide	**82:** P156328s
5-ethoxy-1-vinyl-	**67:** 54080c
4-(ethylsulfonyl)-	**82:** 58044b
5-(ethylsulfonyl)-1-(3-nitrophenyl)-	**82:** 58044b
5-(ethylsulfonyl)-1-phenyl-	**82:** 58044b
5-(ethylthio)-1-phenyl-	**82:** 58044b
2-(2-hydroxyphenyl)-4-methyl-5-(methylsulfonyl)-	**71:** P124445j
2-(2-hydroxyphenyl)-4-methyl-5-(phenylsulfonyl)-	**71:** P124445j
2-(2-hydroxyphenyl)-4-(methylsulfonyl)-5-phenyl-	**71:** P124445j
2-(2-hydroxyphenyl)-4-phenyl-5-(phenylsulfonyl)-	**71:** P124445j
4-methoxy-1,5-dimethyl-	**71:** 61296p
5-methoxy-1,4-diphenyl-	**59:** 3913d, **67:** 64305u
4-methoxy-1-methyl-	**64:** 11200d
4-methoxy-2-methyl-	**64:** 11200d
5-methoxy-1-methyl-	**64:** 11200c
4-methoxy-1-methyl-5-phenyl-	**67:** 64305u
4-methoxy-2-methyl-5-phenyl-	**67:** 64305u
5-methoxy-1-methyl-4-phenyl-	**67:** 64305u
5-methoxy-4-methyl-1-phenyl-	**67:** 64305u
4-(4-methoxyphenoxy)-	**64:** 8171f, **64:** 11201d
4-methoxy-5-phenyl-	**75:** 88297e, **78:** 29677x, **84:** 30609g
5-methoxy-1-phenyl-	**67:** 64305u
2-(2-methoxyphenyl)-4-methyl-5-(methylsulfonyl)-	**77:** P7313c
5-(4-methoxyphenyl)-1-[(4-methylphenyl)sulfonyl]-	**64:** 11201e
2-(2-methoxyphenyl)-4-methyl-5-(phenylsulfonyl)-	**77:** P7313c
4-[(*N*-methyl-*O*-ethylphosphinamidothioylidyne)oxy]-2-phenyl-, *N*-oxide	**82:** P156328s
5-methyl-1-[(4-methylphenyl)sulfonyl]-	**64:** 19597a, **79:** 31992k
5-methyl-4-[(4-methylphenyl)sulfonyl]-1-(4-nitrophenyl)-	**58:** 12561a
4-methyl-1-[(4-methylphenyl)sulfonyl]-5-(10*H*-phenothiazin-10-yl)-	**80:** 37051v
1-methyl-4-(methylthio)-	**77:** 114317d, **84:** 4862q
1-methyl-5-(methylthio)-	**76:** 72454t, **77:** 114317d, **83:** 9919c, **84:** 4862q
2-methyl-4-(methylthio)-	**84:** 4862q
4-methyl-1-(1-naphthalenylsulfonyl)-5-(10*H*-phenothiazin-10-yl)-	**80:** 37051v
4-methyl-1-[(2-nitrophenyl)sulfonyl]-5-(10*H*-phenothiazin-10-yl)-	**80:** 37051v
4-methyl-1-[(3-nitrophenyl)sulfonyl]-5-(10*H*-phenothiazin-10-yl)-	**80:** 37051v
4-methyl-1-[(4-nitrophenyl)sulfonyl]-5-(10*H*-phenothiazin-10-yl)-	**80:** 37051v

TABLE 8. (*Continued*)

Compound	Reference
1-(4-methylphenyl)-5-(methylthio)-	**85:** 94288e
4-[(4-methylphenyl)oxy]-	**64:** 8171f
4-[(4-methylphenyl)sulfonyl]-	**67:** 43749b, **68:** 77478y
1-[(4-methylphenyl)sulfonyl]-5-(4-methylphenyl)-	**64:** 11201e
1-[(4-methylphenyl)sulfonyl]-5-(4-nitrophenyl)-	**64:** 19597a, **79:** 31992k
1-[(4-methylphenyl)sulfonyl]-5-phenyl-	**53:** 15065f, **79:** 31992k
5-(methylthio)-	**84:** 4862q
5-(methylthio)-1-phenyl-	**58:** 2447c, **76:** 72454t, **83:** 9919c
4-[(4-nitrophenyl)sulfonyl]-	**66:** 55466z
-4-ol, benzoate ester	**71:** 124340w
-4-ol, 1-benzyl-, benzoate ester	**71:** 61296p
-5-ol, 1-benzyl-, benzoate ester	**71:** 61296p
-4-ol, 2-(4-chlorophenyl)-, methanesulfonate ester	**82:** P156321j
-4-ol, 2-(4-chlorophenyl)-, methanesulfonate ester, 1-oxide	**82:** P156321j
-4-ol, 1,5-dimethyl-, benzoate ester	**71:** 61296p
-4-ol, 1-methyl-, benzoate ester	**71:** 61296p
-4-ol, 2-methyl-, benzoate ester	**71:** 124340w, **77:** 19583d
-4-ol, 1-methyl-5-phenyl-, benzoate ester	**71:** 61296p
-4-ol, 2-methyl-5-phenyl-, benzoate ester	**71:** 124340w
-4-ol, 2-phenyl-, benzoate ester	**77:** 19583d
-4-ol, 5-phenyl-, benzoate ester	**71:** 124340w
-4-ol, 2-phenyl-, methanesulfonate ester	**82:** P156321j
1-[(2,3,4,5,6-pentachlorobenzyl)oxy]-	**73:** 3865d
4-phenoxy-	**64:** 8171f
4-(phenylsulfonyl)-	**66:** 55466z
1-[(*N*-phthalimidoyl)methoxy]-	**73:** 3865d
1-[(*N*-succinimidoyl)methoxy]-	**73:** 3865d
-4-sulfonic acid, 2-(2-methoxyphenyl)-5-phenyl-, adduct with *N*-cyclohexylcyclohexanamine	**77:** P7313c

REFERENCES

47: 10474f — W. R. Bamford and T. S. Stevens, *J. Chem. Soc.*, 4735 (1952).

48: P187–8h,a — J. R. Geigy A.-G., British Patent, 681,376 (1952).

53: 15065f — J. H. Boyer, C. H. Mack, N. Goebel, and L. R. Morgan, Jr., *J. Org. Chem.*, **23,** 1051 (1958).

55: 11397b–e — P. Grünanger, P. V. Finzi, and E. Fabbri, *Gazz. Chim. Ital.*, **90,** 413 (1960).

55: 25923b — F. Korte and K. Störiko, *Chem. Ber.*, **94,** 1956 (1961).

58: 2447c — T. Kindt-Larsen and C. Pedersen, *Acta Chem. Scand.*, **16,** 1800 (1962).

58: 7927f,g — H. U. Daeniker and J. Druey, *Helv. Chim. Acta*, **45,** 2441 (1962).

58: 12561a,b — R. Fusco, G. Bianchetti, D. Pocar, and R. Ugo, *Gazz. Chim. Ital.*, **92,** 1040 (1962).

59: 3913c,d — R. Scarpati, D. Sica, and A. Lionetti, *Gazz. Chim. Ital.*, **93,** 90 (1963).

64: 5077a	R. Huisgen, R. Knorr, L. Möbius, and G. Szeimies, *Chem. Ber.*, **98,** 4014 (1965).
64: 8171f	D. Martin and A. Weise, *Chem. Ber.*, **99,** 317 (1966).
64: 11200c,d	M. Begtrup and C. Pedersen, *Acta Chem. Scand.*, **19,** 2022 (1965).
64: 11201d,e	P. V. Finzi, *Chim. Ind. (Milan)*, **47,** 1338 (1965).
64: 19597a	G. R. Harvey, *J. Org. Chem.*, **31,** 1587 (1966).
66: 55466z	S. Maiorana, D. Pocar, and P. D. Croce, *Tetrahedron Lett.*, 6043 (1966).
67: 43749b	J. S. Meek and J. S. Fowler, *J. Amer. Chem. Soc.*, **89,** 1967 (1967).
67: 54080c	P. V. Finzi and C. Scotti, *Atti Accad. Naz. Lincei, Rend., Cl. Sci. Fis., Mat. Nat.*, **41,** 204 (1966).
67: 64305u	M. Begtrup and C. Pedersen, *Acta Chem. Scand.*, **21,** 633 (1967).
67: 100074d	M. Begtrup, K. Hansen, and C. Pedersen, *Acta Chem. Scand.*, **21,** 1234 (1967).
68: 77478y	J. S. Meek and J. S. Fowler, *J. Org. Chem.*, **33,** 985 (1968).
71: 61296p	M. Begtrup and C. Pedersen, *Acta Chem. Scand.*, **23,** 1091 (1969).
71: 112865h	E. Zbiral and J. Stroh, *Monatsh. Chem.*, **100,** 1438 (1969).
71: 124340w	M. Begtrup, *Acta Chem. Scand.*, **23,** 2025 (1969).
71: P124445j	Geigy, A.-G., French Patent, 1,559,131 (1969).
73: 3865d	A. J. Hubert, *Bull. Soc. Chim. Belges*, **79,** 195 (1970).
75: 63704b	M. Begtrup, *Tetrahedron Lett.*, 1577 (1971).
75: 88297e	R. Scarpati and M. L. Graziano, *Tetrahedron Lett.*, 2085 (1971).
76: 72454t	M. Begtrup, *Acta Chem. Scand.*, **25,** 3500 (1971).
77: P7313c	H. Lind, German Patent, 2,133,012 (1972).
77: 19583d	M. Begtrup, *Acta Chem. Scand.*, **26,** 715 (1972).
77: P95371y	R. C. Deselms, J. J. Looker, and C. V. Wilson, British Patent, 1,271,221 (1972).
77: 114317d	M. Begtrup, *Acta Chem. Scand.*, **26,** 1243 (1972).
77: 151634a	G. Himbert and M. Regitz, *Chem. Ber.*, **105,** 2975 (1972).
78: 29677x	R. Scarpati and M. L. Graziano, *J. Heterocycl. Chem.*, **9,** 1087 (1972).
78: 111442a	P. Ykman, G. L'abbé and G. Smets, *Tetrahedron*, **29,** 195 (1973).
79: 31992k	G. L'abbé, *Verh. Kon. Acad. Wetensch., Lett. Schone Kunsten Belg., Kl. Wetensch.*, **34,** 49pp. (1972).
80: 37051v	G. Himbert and M. Regitz, *Justus Liebigs Ann. Chem.*, **9,** 1505 (1973).
82: 58044b	A. Farrington and L. Hough, *Carbohydr. Res.*, **38,** 107 (1974).
82: P156321j	B. Boehner, D. Dawes, W. Meyer, H. Kristinsson, and K. Ruefenacht, German Patent, 2,428,204 (1975).
83: 9919c	S. Hoff and A. P. Blok, *Recl. Trav. Chim. Pays-Bas*, **93,** 317 (1974).
84: 4862q	J. P. Daris and J. M. Muchowski, *J. Heterocycl. Chem.*, **12,** 761 (1975).
84: 30609g	M. L. Graziano, M. R. Iesce, R. Palombi, and R. Scarpati, *Ann. Chim.*, **64,** 843 (1974).
85: 94288e	M. Uher, A. Rybár, A. Martvoň, and J. Leško, *Collect. Czech. Chem. Commun.*, **41,** 1551 (1976).

CHAPTER 9

Halo-1,2,3-Triazoles

Only a single example of a fluoro-1,2,3-triazole has been reported in the literature (Eq. 1),[1] and certainly additional study of these compounds and the defluoronating reagent **9.1-1** is warranted.

$$CF_3CF{=}CF_2 + PhCH_2N_3 \longrightarrow \text{4,4,5-trifluoro-5-}CF_3\text{-1-}CH_2Ph\text{-triazoline (85\%)} \xrightarrow[\mathbf{9.1\text{-}1}]{(Me_2N)_2C{=}C(NMe_2)_2} \text{4-F-5-}CF_3\text{-1-}CH_2Ph\text{-1,2,3-triazole (43\%)} + \text{4-}Me_2N\text{-5-}CF_3\text{-1-}CH_2Ph\text{-1,2,3-triazole (53\%)} \quad (1)$$

Both direct and indirect syntheses of chloro-1,2,3-triazoles have been carried out. In the former approach, the reagent employed exerts an important influence on the product yield (Eqs. 2,3).[2] A large number of additional combinations were tried but without synthetically useful results. It has also been shown that cyanogen chloride reacts with trimethylsilyl-diazomethane to give chloro-1,2,3-triazoles (Eq. 4)[3] in fair yield.

$$\text{1-Me-1,2,3-triazole} \xrightarrow{Cl_2 \ (92\%) \ \text{or} \ Na^{+}\,{}^{-}OCl \ (65\%)} \text{5-Cl-1-Me-1,2,3-triazole} \quad (2)$$

A somewhat broader range of reactions has been explored for the synthesis of bromo-1,2,3-triazoles. Two reports of the direct bromination show the efficiency of such reactions (Eq. 5).[2,4] The reaction of **9.1-2** ($Y = H$) with two equivalents of sodium hypobromite also gives a very high yield of

60%

Me — $\xrightarrow[\text{CHCl}_3]{\text{Cl}_2}$ Me, Cl

N≈N–N–H $\xrightarrow{\text{Na}^{+-}\text{OCl}}$ $\text{N≈N–NH}_2^+\text{Cl}^-$

45%

(3)

$\text{ClCN} + \text{Me}_3\text{SiCHN}_2 \longrightarrow$ Cl; N–N–N; SiMe_3 + another 1:1 adduct

45%

(4)

$\xrightarrow[\text{H}_2\text{O}]{\text{EtOH}}$ Cl; N≈N–N–H

84%

4,5-dibromo-1,2,3-triazole, and the bromination of alkyl-1,2,3-triazoles produces fair product yields (Eq. 6,7).[2] It should be noted that 1-benzyl-1,2,3-triazole gives only a 33% yield of the 4-bromo product.[5]

Y; N≈N–N–H + Br_2 $\longrightarrow$ Br, Br; N≈N–N–H (5)

9.1-2

Y = H, CHO, CO_2H

about 100%

N≈N–N–Me + Br_2 $\longrightarrow$ Br; N≈N–N–Me (6)

68%

Me, Me; N≈N–N–H + Na^{+-}OBr $\longrightarrow$ Me, Me; N≈N–N–Br (7)

60%

The addition of azides to unsaturated bromo compounds has not produced exceptional yields (e.g., Eqs. 8,9).[6,7] In view of the general success of such methods these examples should be studied further.

$$BrC{\equiv}CC(OH)Me_2 + PhCH_2N_3 \longrightarrow \text{4-Br-5-}(CMe_2OH)\text{-1-}CH_2Ph\text{-1,2,3-triazole} \quad (8)$$

fair (trace of the isomer)

$$CH_2{=}C(Br)SO_2Cl + PhN_3 \longrightarrow \text{4-Br-1-Ph-1,2,3-triazole} \quad (9)$$

Diazotization of amino-1,2,3-triazoles could be an important route, but it has been reported in only one case that gave a moderate yield (Eq. 10).[8] The obvious alternative for this type of product i.e., methylation of bromo-1,2,3-triazoles has received only limited exploration with modest success (Eq. 11).[9] The use of dimethyl sulfate in this reaction gave 29% of each isomer.[10]

$$\text{4-}H_2N\text{-1-Me-1,2,3-triazole} + HBr \xrightarrow[\text{Cu-bronze}]{Na^{+-}NO_2} \text{4-Br-1-Me-1,2,3-triazole (41\%)} \quad (10)$$

$$\text{4,5-Br}_2\text{-1-H-1,2,3-triazole} + CH_2N_2 \longrightarrow \text{4,5-Br}_2\text{-1-Me-1,2,3-triazole (35\%)} + \text{4,5-Br}_2\text{-2-Me-1,2,3-triazole (46\%)} \quad (11)$$

The use of cyanogen bromide in Equation 4 gave somewhat better results—that is, 85% of the initial adduct and 61% overall.

The 1,2,3-triazoles can be iodinated directly and in quite good yield using several different reagents (Eqs. 12 to 14).[2,11]

$$\text{4-Me-1-H-1,2,3-triazole} + 2Na^{+-}OI \longrightarrow \text{4-Me-5-I-1-I-1,2,3-triazole (66\%)} \quad (12)$$

4,5-dimethyl-1H-1,2,3-triazole + I_2 (or $Na^{+-}OI$) ⟶ 4,5-dimethyl-1-iodo-1,2,3-triazole, 65% (50%) (13)

1H-1,2,3-triazole + Na ICl $\xrightarrow{EtOH}$ 1-iodo-1,2,3-triazole (60–80%) $\xrightarrow{\text{1,3,5-trimethyl-1,2,4-triazole}}$ 4,5-diiodo-1H-1,2,3-triazole (75%) (14)

REFERENCES

1. **64:** 12662g
2. **50:** 9392h
3. **78:** 111412r
4. **42:** 2600b
5. **74:** 141647f
6. **46:** 8651g
7. **50:** 3414r
8. **55:** 17626h
9. **72:** 43570q
10. **73:** 87865q
11. **73:** 66513h

TABLE 9. HALO-1,2,3-TRIAZOLES

Compound	Reference
1-benzyl-4-bromo-	**74:** 141647f
1-benzyl-5-chloro-	**67:** 64305u
1-benzyl-4-fluoro-5-(trifluoromethyl)-	**64:** 12662g
4-bromo-	**78:** 111412r
5-bromo-4-[(diethoxythiophosphinylidyne)oxy]-2-phenyl-	**82:** P156328s
1(or 2)-bromo-4,5-dimethyl-	**50:** 9393c
1(or 2)-bromo-4-methyl-	**50:** 9393f
4-bromo-1-methyl-	**50:** 9392i, **55:** 17626h
4-bromo-2-methyl-	**67:** 16500g
4-bromo-5-methyl-	**50:** 9393b
5-bromo-1-methyl-	**55:** 17626h
2-[4-(bromomethyl)-3-cyanophenyl]-4-chloro-5-(4-chlorophenyl)-	**83:** P81213k
4-bromo-1-phenyl-	**50:** 3414i
4-bromo-5-phenyl-	**70:** 69275u
4-chloro-	**78:** 111412r
4-chloro-2-(5-chloro-2-methoxyphenyl)-5-phenyl-	**74:** P100618d
4-chloro-2-(5-chloro-2-methoxyphenyl)-5-phenyl-, 3-oxide	**74:** P100618d
-4-chloro-5-(4-chlorophenyl)-2-[3-cyano-4-[(diethoxyphosphinyl)methyl]phenyl]-	**83:** P81213k
4-chloro-5-(4-chlorophenyl)-2-(3-cyano-4-methylphenyl)-	**83:** P81213k
5-chloro-4[(diethoxythiophosphinylidyne)oxy]-2-phenyl-	**82:** P156328s

TABLE 9 (*Continued*)

Compound	Reference
1(or 2)-chloro-4,5-dimethyl-	**50:** 9392i
4-chloro-1-methyl-	**50:** 9392i
4-chloro-5-methyl-	**50:** 9393a
4-chloro-1-phenyl-	**52:** 366a
5-chloro-1-phenyl-	**52:** 5866b
4,5-dibromo-	**42:** 2600b, **50:** 9393d, **67:** 64365p
4,5-dibromo-, 1:1 adduct with dimethylamine	**73:** 87865q
4,5-dibromo-, 1;1 adduct with triethylamine	**73:** 87865q
4,5-dibromo-2-(1,1-dimethyl-3-oxobutyl)-	**73:** 87865q
4,5-dibromo-2-(1,1-dimethyl-3-oxobutyl)-, semicarbazone	**73:** 87865q
1(or 2),5-dibromo-4-methyl-	**50:** 9393f
4,5-dibromo-1-methyl-	**72:** 43570q, **73:** 87865q
4,5-dibromo-2-methyl-	**50:** 9393e, **73:** 87865q
1(or 2),5-dichloro-4-methyl-	**50:** 9393e
4,5-diiodo-	**73:** 66513h
1(or 2),5-diiodo-4-methyl-	**50:** 9393b
4,5-diiodo-2-methyl-	**50:** 9393a
1(or 2)-iodo-	**50:** 9393h
4-iodo-	**50:** 9392h
1(or 2)-iodo-4,5-dimethyl-	**50:** 9393d
4-iodo-1-methyl-	**50:** 9392i
4-iodo-5-methyl-	**50:** 9393b
-4-methanol, 1-benzyl-5-bromo-α,α-dimethyl-	**46:** 8651g
-5-methanol, 1-benzyl-4-bromo-α,α-dimethyl-	**46:** 8651h
-4-methanol, 5-bromo-α,α-dimethyl-1-phenyl-	**46:** 8651h
1(or 2),4,5-tribromo-	**50:** 9393e

REFERENCES

42: 2600b — R. Hüttel and A. Gebhardt, *Justus Liebigs Ann. Chem.*, **558**, 34 (1947).

46: 8651g,h — Fr. Moulin, *Helv. Chim. Acta*, **35,** 167 (1952).

50: 3414i — C. S. Rondestvedt, Jr. and P. K. Chang, *J. Amer. Chem. Soc.*, **77,** 6532 (1955).

50: 9392–3h–i — R. Hüttel and G. Welzel, *Justus Liebigs Ann. Chem.*, **593,** 207 (1955).

52: 366a — E. Lieber, T. S. Chao, and C. N. R. Rao, *J. Org. Chem.*, **22,** 654 (1957).

53: 5866b — E. Lieber, C. N. R. Rao, T. S. Chao, and H. Rubinstein, *Can. J. Chem.*, **36,** 1441 (1958).

55: 17626h — C. Pedersen, *Acta Chem. Scand.*, **13,** 888 (1959).

64: 12662g — W. Carpenter, A. Haymaker, and D. W. Moore, *J. Org. Chem.*, **31,** 789 (1966).

67: 16500g — G. B. Barlin and T. J. Batterham, *J. Chem. Soc.*, *B*, 516 (1967).

67: 64305u — M. Begtrup and C. Pedersen, *Acta Chem. Scand.*, **21,** 633 (1967).

67: 64365p — P. Ferruti, D. Pocar, and G. Bianchetti, *Gazz. Chim. Ital.*, **97,** 109 (1967).

70: 68275u — J. H. Boyer and R. Selvarajan, *Tetrahedron Lett.*, 47 (1969).

72: 43570q — M. Begtrup and P. A. Kristensen, *Acta Chem. Scand.*, **23,** 2733 (1969).

73: 66513h	R. Miethchen, H. Albrecht, and E. Rachow, *Z. Chem.*, **10,** 220 (1970).
73: 87865q	J. E. Oliver and J. B. Stokes, *J. Heterocycl. Chem.*, **7,** 961 (1970).
74: P100618d	R. Kirchmayr, German Patent, 2,029,142 (1970).
74: 141647f	M. Begtrup, *Acta Chem. Scand.*, **25,** 249 (1971).
78: 111412r	J. M. Crossman, R. N. Haszeldine, and A. E. Tipping, *J. Chem. Soc., Dalton Trans.*, 483 (1973).
82: P156328s	B. Boehner, D. Dawes, H. Kristinsson, and W. Meyer, German Patent, 2,442,843 (1975).
83: P81213k	F. Fleck, H. Schmid, A. V. Mercer, and R. Paver, Swiss Patent, 561,709 (1975).

CHAPTER 10

1,2,3-Triazoles Containing More Than One Representative Function

10.1. AMINO-1,2,3-TRIAZOLECARBOXYLIC ACIDS AND THEIR FUNCTIONAL DERIVATIVES

The reaction of azides with active methylene compounds under basic conditions is the most widely used approach to the *v*-triazole amino acids (Eqs. 1,2).[1–3] This reaction, discovered by Dimroth,[4] is highly regiospecific and produces good to excellent yields for a variety of substituents (Eq. 3).[4]

$$NCCH_2CO_2Et + ArN_3 \xrightarrow{Na^{+}\,{}^{-}OEt} \textbf{10.1-1} \quad (1)$$

10.1-1 (4-CO_2Et, 5-NH_2, 1-Ar-1,2,3-triazole)

Ar	= Ph[2]	4-MePh	4-BrPh	4-MeOPh	4-NO_2Ph
% **10.1-1**	= 69	52	81	72	57

$$NCCH_2CONH_2 + PhCH_2N_3 \xrightarrow{\text{base}} \text{(4-}H_2NC(=O)\text{, 5-}NH_2\text{, 1-}CH_2Ph\text{-1,2,3-triazole)} \quad (2)$$

80%

$$NCCH_2R + R'N_3 \xrightarrow[MeOH]{Na^{+}\,{}^{-}OMe} \textbf{10.1-2} \quad (3)$$

10.1-2 (4-R, 5-NH_2, 1-R'-1,2,3-triazole)

R = CO_2NH_2, CN, CO_2Et

R′ = Me, Et, H_2NCOCH_2, Ph, 4-YPh (Y = Me, Cl, CO_2H, NO_2), 3-NO_2Ph

In this particular study a few cases of very low yield were found—for examples, $R = CONH_2$, $R' = Me$, Et or 2-MePh (7 to 13%).

Dimroth also reported a reversible rearrangement of the initially formed 1,2,3-triazole (**10.1-1** to **10.1-2**).[4] This process, which can significantly expand the utility of the addition, has been studied and applied by a number of groups (Eq. 4).[1 to 3,5,6] The addition of 4-nitrophenylazide in Equation 1 produces only the rearranged product (63% **10.1-1**). It was later shown that under the reaction conditions of Equation 3 most of the *N*-phenyl adduct rearranges.[2,5,6] Sutherland and Tennant also studied the factors that favor acetylation or rearrangement (Eqs. 5,6).[6]

10.1-1 or **10.1-2** ⟶ 4-R-5-NHR'-2H-1,2,3-triazole (N–H) (4)

5-amino-1-phenyl-1,2,3-triazole-4-carboxamide ($H_2NC(O)$, NH_2, N-Ph) —AcCl, H^+, 0°→ AcNHC(O)-, NHAc, N-Ph triazole 68% (5)

5-amino-1-phenyl-1,2,3-triazole-4-carboxamide —$(Ac)_2O$, reflux→ $H_2NC(O)$-, NHPh, N-Ac triazole 70% (6)

(prolonged heating: NHPh/N-Ac product → 68% product)

L'abbé and his collaborators have added azides to both active methylene compounds (Eq. 7)[7] and acetylenes (Eq. 8)[8] for the preparation of unusual examples of 1,2,3-triazoles. With ethyl azidoformate (**10.1-3**) the initially formed 1-ester (**10.1-4**) rearranges on workup to the 2-substituted product

H_2NCOCH_2CN + H(N$_3$)C=C(H)Bu-*t* —$Na^{+-}OMe$, MeOH→ $H_2NC(O)$-, NH_2, N–CH=CH–Bu-*t* triazole

62% (72% from iodide) (7)

(**10.1-5**) and small amounts of the third *N*-isomer (**10.1-6**) are also obtained (Eq. 9). This group has also prepared *N*-vinyl-1,2,3-triazoles by basic elimination of the corresponding alkyl iodide.[7]

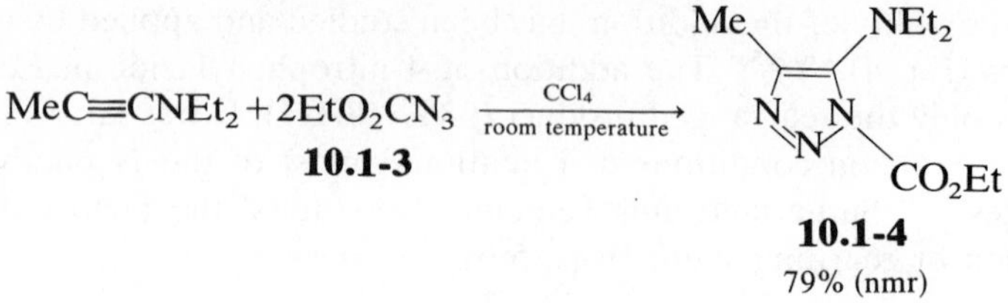

$$\mathrm{MeC{\equiv}CNEt_2 + 2EtO_2CN_3} \xrightarrow[\text{room temperature}]{\mathrm{CCl_4}} \text{10.1-4 (79\% (nmr))} \tag{8}$$

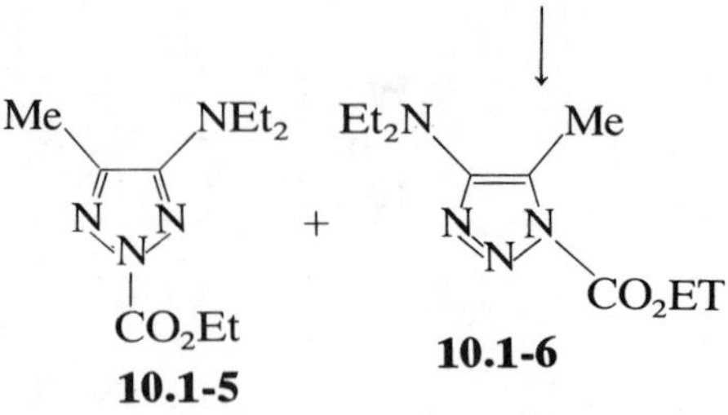

10.1-3 **10.1-4** 79% (nmr)

$$\text{10.1-4} \rightarrow \text{10.1-5} + \text{10.1-6} \tag{9}$$

10.1-5 **10.1-6**

A rather interesting and exotic example of active methylene–azide addition is found in the case of 5-azidotropolone (**10.1-7**) where the product varies with the ratio of the starting materials (Eqs. 10,11).[9, 10]

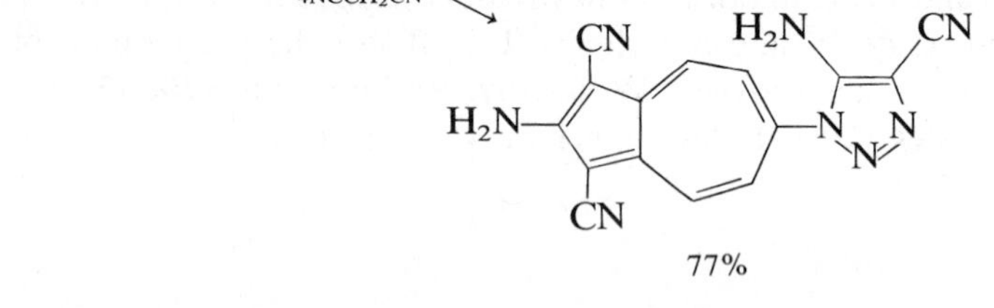

$$\text{10.1-7} \xrightarrow{\mathrm{3NCCH_2CN}} \text{product (50–90\%)} \tag{10}$$

$$\text{10.1-7} \xrightarrow{\mathrm{4NCCH_2CN}} \text{product (77\%)} \tag{11}$$

10.1-7 50–90% 77%

Albert has shown that, under appropriate conditions, a near quantitative yield of a 1,2,3-triazole anhydro dimer (**10.1-8**) may be obtained and then converted to the important derivative **10.1-9** (Eq. 12).[11] This compound (**10.1-9**) has played a role in the synthesis of various azapurines (e.g., **10.1-10**), which in turn are promising intermediates in *v*-triazole synthesis (Eqs. 13,14).[12–14] Albert has also made important contributions to our understanding of the optimal methods for preparing and isolating various

$PhCH_2N_3 + NCCH_2CONH_2 \xrightarrow{AcO_2CH}$

O
HCNH
H_2N
N
N—CH=N
N
C
O
O
C
NH_2
N
N
N
H

10.1-8

$\xrightarrow{(MeO)_2SO_2}$

O
C
H_2N
O
NHCH
N
N
N
Me

10.1-9

75%

(12)

SMe
N
N
N
N
N
N
R

10.1-10

R = H
Me
$PhCH_2$

$\xrightarrow{1\ M\ HCl}$

O
C
MeS
NH_2
N
N
N
R

100%
90%
85%

$\xrightarrow{NH_3\ aq.}$

O
C
H_2N
NH_2
N
N
N
R

100%

(13)

10.1-10 (R = Me) ⟶

[O
C
MeS
NH_2
N
N
N
Me]

⟶

$NH_2^+Cl^-$
C
MeS
NH_2
N
N
N
Me

60%

$\xrightarrow[pH\ 7]{heat}$

NC
NH_2
N
N
N
Me

(14)

methylated amino acid derivatives (Eqs. 15,16).[15] The mixture **10.1-11** proved difficult to separate and was first converted to the corresponding acids or esters.[15] A related synthetic method has been reported by Lovelette and Long, who claim improved procedures[16] but with the exception of the conversion of a hydrazide to an azide (Eq. 17) fail to provide yield data.

Two reactions of sulfonyl azides show promise in the preparation of *v*-triazole amino carboxamides (Eq. 18)[17] and nitriles (Eq. 19).[18]

$PhCH_2N_3 + NCCH_2CONH_2 \longrightarrow$

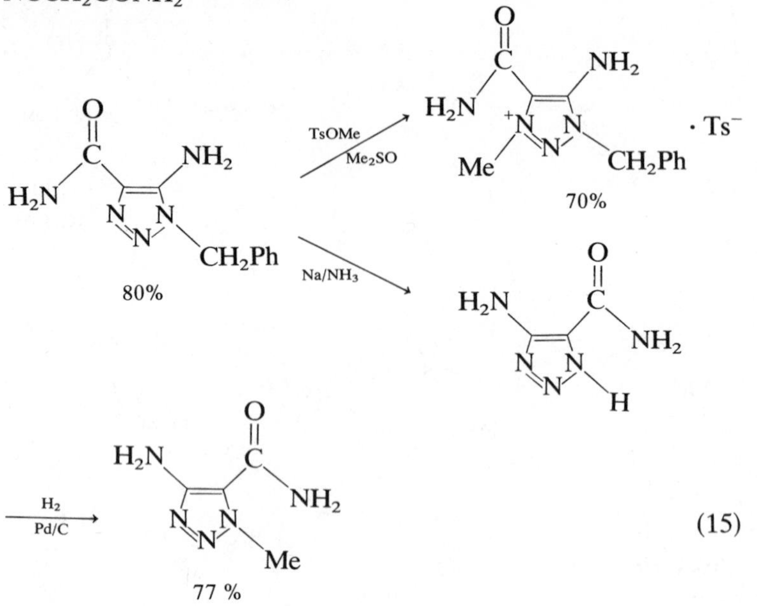

O
C
H_2N NH_2
N N
N
CH_2Ph
80%

TsOMe / Me_2SO →

O
C
H_2N NH_2
+N N
N
Me CH_2Ph
· Ts^-
70%

Na/NH_3 →

O
H_2N C
NH_2
N N
N
H

$\xrightarrow[Pd/C]{H_2}$

O
H_2N C
NH_2
N N
N
Me
77 %

(15)

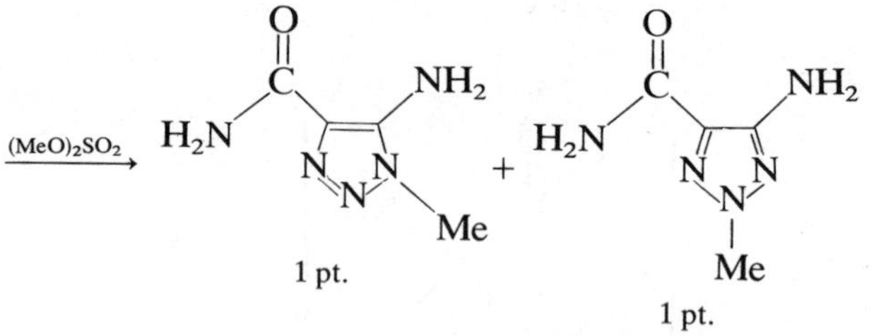

$\xrightarrow{(MeO)_2SO_2}$

O
C
H_2N NH_2
N N
N
Me
1 pt.

+

O
C
H_2N NH_2
N N
N
Me
1 pt.

10.1-11
60%

(16)

O
C
H_2NNH NH_2
N N
N
CH_2Ph

$\xrightarrow{HNO_2}$

O
C
N_3 NH_2
N N
N
CH_2Ph
74%

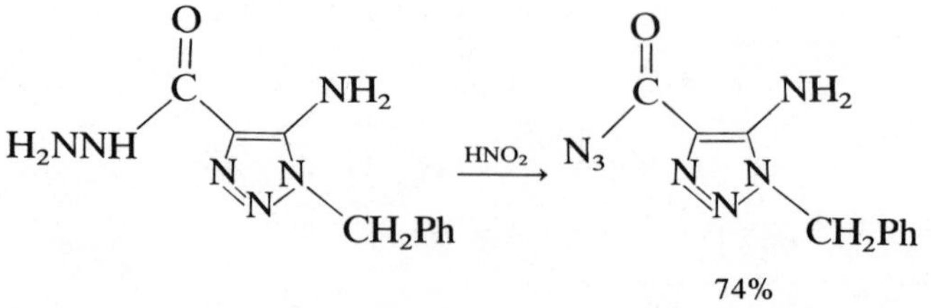

(17)

$$\text{HN}{=}\text{C}(\text{NH}_2)\text{CH}_2\text{CONH}_2\cdot\text{HCl} + \text{RSO}_2\text{N}_3 \xrightarrow{\text{Na}^{+}\,{}^{-}\text{OMe}} \text{5-amino-4-carbamoyl-1}H\text{-1,2,3-triazole (H}_2\text{NC(=O)-, NH}_2\text{, N-H)} \quad 70\text{–}90\% \qquad (18)$$

R = Me, Ph, 4-MePh, β-$C_{10}H_7$

$$(\text{NC})_2\text{CH}^- + \text{RSO}_2\text{N}_3 \longrightarrow \text{4-NC-5-NHSO}_2\text{R-1}H\text{-1,2,3-triazole} \quad \mathbf{10.1\text{-}12} \qquad (19)$$

R	= Me	Ph	4-MePh	4-MeOPh	4-NO_2Ph
% **10.1-12**	= 92	80	96	74	71

REFERENCES

1. **49:** 1710g
2. **52:** 366a
3. **68**: 68958c
4. **22:** 423[7]
5. **55:** 2661d
6. **74:** 125581z
7. **72:** 121447w
8. **78:** 43374c
9. **69:** 19088f
10. **79:** 5300p
11. **69:** 77238r
12. **72:** 21645r
13. **70:** 87690s
14. **71:** 13066s
15. **69:** 77238r
16. **78:** 29681u
17. **72:** 55346j
18. **80:** 95838t

10.2 MISCELLANEOUS 1,2,3-TRIAZOLES CONTAINING MORE THAN ONE REPRESENTATIVE FUNCTION

Of the various polyfunctional *v*-triazoles reported, an overwhelming number contain the sulfonamido group. One reason for this interest lies in the phenomenon of ring-open chain tautomerism (Eq. 20).[19–21] The position of the triazole-diazoamidine equilibrium (**10.2-1** and **10.2-2**) is heavily dependent on structure:

$$\text{R}^1\text{C}{\equiv}\text{CNEt}_2 + \text{R}^2\text{SO}_2\text{N}_3 \longrightarrow \underset{\mathbf{10.2\text{-}1}}{\text{4-R}^1\text{-5-NEt}_2\text{-1-SO}_2\text{R}^2\text{-1,2,3-triazole}} \rightleftharpoons \underset{\mathbf{10.2\text{-}2}}{\text{N}_2\text{C(R}^1\text{)C(NEt}_2){=}\text{NSO}_2\text{R}^2} \qquad (20)$$

R^1	R^2	% **10.2-1**
Me	Me	62
Me	Ph	67
Me	4-MePh	80
Me	4-MeOPh	92
Me	4-AcNHPh	78
Me	2,4,6-Me_3Ph	96
Ph	various Ar	0

Himbert and Regitz have shown that diphenylaminoacetylene is exceptional among β-H-ynamines in giving a good yield of 1,2,3-triazole (Eq. 21).[22] This group has also studied the reaction of alkoxyacetylenes with sulfonazides (Eq. 22).[23] With the exception of those listed, most combinations give only α-diazocarboximidates (**10.2-4**). The incorporation of a phosphorus group also failed, in general, to provide the expected 1,2,3-triazole (Eq. 23).[24] It was also shown that excellent yields of mono-addition can be obtained from certain diynamines, even with nitrophenylsulfonazides (Eq. 24).[25] A recent study by Himbert, Frank, and Regitz involved a triphenylmetal substituent and produced a fair yield of 1,2,3-triazoles in two exceptional instances (Eq. 25).[26]

$$Ph_2NC{\equiv}CH + 4\text{-}Me_2NC_6H_4SO_2N_3 \longrightarrow \text{4-}NPh_2\text{-1-}(4\text{-}Me_2NC_6H_4SO_2)\text{-1,2,3-triazole} \quad (74\%) \tag{21}$$

$$R^1OC{\equiv}CR^2 + RSO_2N_3 \longrightarrow \underset{\mathbf{10.2\text{-}3}}{\text{4-}R^2\text{-5-}OR^1\text{-1-}SO_2R\text{-1,2,3-triazole}} \rightleftharpoons \underset{\mathbf{10.2\text{-}4}}{N_2C(R^2)C(=NSO_2R)OR^1} \tag{22}$$

R =	4-Me$_2$NPh	2,4,6-Me$_3$Ph	4-Me$_2$NPh	4-Me$_2$NPh	4-Me$_2$NPh
R^1 =	Et	Et	Me	Me	Et
R^2 =	Me	Me	Et	Me	Et
% **10.2-3** =	75	42	60	66	69

$$R^1R^2NC{\equiv}CP(=X)Ph_2 + R^3SO_2N_3 \longrightarrow \text{4-}Ph_2P(=X)\text{-5-}NR^1R^2\text{-1-}SO_2R^3\text{-1,2,3-triazole} \rightleftharpoons \text{open-chain isomer} \tag{23}$$

R^1, R^2 = Me, Ph

X = S
R^3 = 4-MeOPh 60%
2,4,6-Me$_3$Ph 55%
X = NSO$_2$PhNMe$_2$-4
R^3 = PhCH$_2$ 85%

$$Me_2NC{\equiv}CC{\equiv}CNMe_2 + RSO_2N_3 \longrightarrow \text{4-}(Me_2NC{\equiv}C)\text{-5-}NMe_2\text{-1-}SO_2R\text{-1,2,3-triazole} \quad (75\text{–}100\%) \tag{24}$$

$$\begin{matrix} R^1 \\ \diagdown \\ NC{\equiv}CMPh_3 + RSO_2N_3 \longrightarrow \\ \diagup \\ R^2 \end{matrix} \quad \text{1-}SO_2R\text{-4-}NR^1R^2\text{-5-}Ph_3M\text{-1,2,3-triazole} \qquad (25)$$

$R^1, R^2 =$	M =	R =	
Ph_2	Si	4-MeOPh	66%
Ph, Me	Ge	$4\text{-}Me_2NPh$	61%

The sulfonamido group is found together with a number of other functional groups in isolated reports. For example, fair yields of 1,2,3-triazoles involving a carboxylic acid ester (Eq. 26),[27] good to excellent yields with a cyano group (Eq. 27),[28] and an old but unexplored reaction leading to an ether substituent (Eq. 28)[29] have been reported.

$$RC{\equiv}CCO_2R' + TsN_3 \longrightarrow \text{1-Ts-5-}R'O_2C\text{-4-R-1,2,3-triazole (10.2-5)} \qquad (26)$$

R'	= Me or Et		
R	= H	Ph	CO_2Me
% **10.2-5** =	37	49	63

$$(NC)_2CH^- + RSO_2N_3 \longrightarrow \text{1-H-5-NC-4-}NHSO_2R\text{-1,2,3-triazole (10.2-6)} \qquad (27)$$

R	= Me	Ph	4-MePh	4-MeOPh	$4\text{-}NO_2Ph$
% **10.2-6** =	92	80	96	74	71

$$(MeSO_2)_2C{=}NOMe + CH_2N_2 \longrightarrow \text{1-OMe-4,4-}(MeSO_2)_2\text{-triazoline (87\%)} \xrightarrow[O(CH_2CH_2)_2O]{\text{excess } CH_2N_2} \text{1-OMe-4-}MeSO_2\text{-1,2,3-triazole (53\%)} \qquad (28)$$

The reaction between tosylazide and the anion of malononitrile has been studied more fully, and the methylation of the product has been explored (Eq. 29).[30] The product (**10.2-7**) represents the major component of a mixture of dimethyl derivatives obtained in 85% yield.

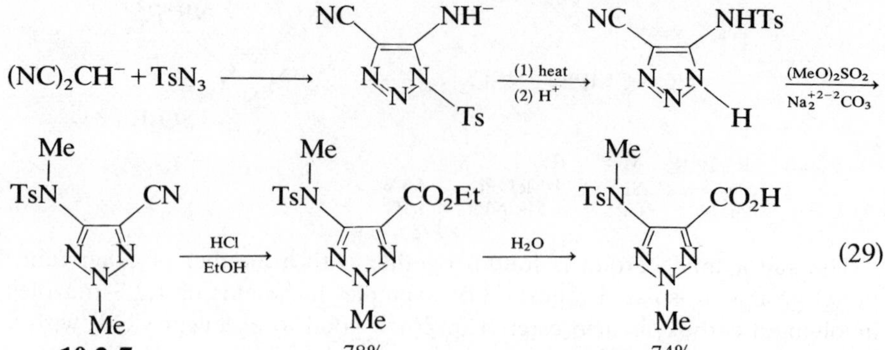

$$(NC)_2CH^- + TsN_3 \longrightarrow \text{[4-cyano-5-amido(NH}^-\text{)-1-tosyl-1,2,3-triazole]} \xrightarrow[(2)\ H^+]{(1)\ \text{heat}} \text{[4-cyano-5-(NHTs)-1H-1,2,3-triazole]} \xrightarrow[Na_2^{+2-2}CO_3]{(MeO)_2SO_2}$$

$$\underset{\mathbf{10.2\text{-}7}}{\text{[4-(TsNMe)-5-CN-2-Me-1,2,3-triazole]}} \xrightarrow[EtOH]{HCl} \underset{78\%}{\text{[4-(TsNMe)-5-}CO_2Et\text{-2-Me-triazole]}} \xrightarrow{H_2O} \underset{74\%}{\text{[4-(TsNMe)-5-}CO_2H\text{-2-Me-triazole]}} \quad (29)$$

Finally, the combination of acetamido and sulfonamido groups has been studied in some detail (Eq. 30).[31]

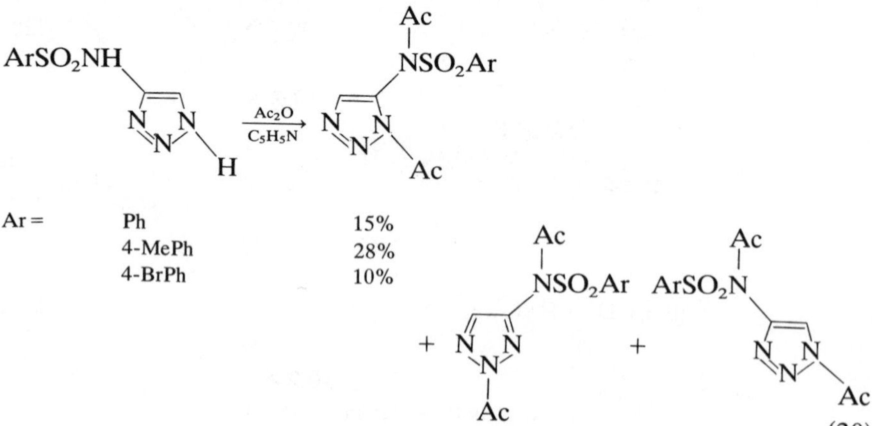

$$\text{[4-(}ArSO_2NH\text{)-1H-1,2,3-triazole]} \xrightarrow[C_5H_5N]{Ac_2O} \text{[4-(}N(Ac)SO_2Ar\text{)-1-Ac-triazole]} + \text{[4-(}N(Ac)SO_2Ar\text{)-2-Ac-triazole]} + \text{[4-(}ArSO_2N(Ac)\text{)-1-Ac-triazole]} \quad (30)$$

Ar =			
Ph	15%	60%	26%
4-MePh	28%	50%	22%
4-BrPh	10%	62%	29%

The presence of an acyl group characterizes the second most common disubstituted 1,2,3-triazole and in general may be obtained in excellent yield from, for example, aroyl azides and phosphorus ylides (Eq. 31).[32] A similar reaction has been used to prepare an acetyl-1,2,3-triazolecarboxylic acid

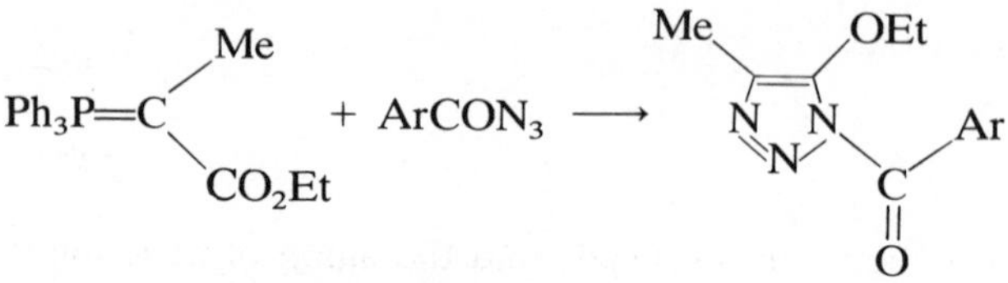

$$Ph_3P{=}C(Me)(CO_2Et) + ArCON_3 \longrightarrow \underset{\mathbf{10.2\text{-}8}}{\text{[4-Me-5-OEt-1-(ArC(=O))-1,2,3-triazole]}} \quad (31)$$

Ar =	Ph	$3\text{-}NO_2Ph$	$4\text{-}NO_2Ph$	4-MeOPh
% **10.2-8** =	63	73	77	70

ester (Eq. 32).[33] Although the structure of compound **10.2-9** was not determined, it is likely to be 2-acetyl in view of such evidence as the direct preparation of similar compounds (Eq. 33)[34] and the work of L'abbé (Eq. 34).[35]

$$Ph_3P{=}C(Pr\text{-}n)(CO_2Et) + MeCONO_3 \longrightarrow \text{4-}EtO_2C\text{-5-}(Pr\text{-}n)\text{-}N\text{-}Ac\text{-}1,2,3\text{-triazole} \quad (32)$$

10.2-9
90%

$$\text{4-HO-5-R-1}H\text{-1,2,3-triazole} + PhCOCl \xrightarrow[CH_2Cl_2]{C_5H_5N} \text{4-}PhCO_2\text{-5-R-2-}(PhC{=}O)\text{-2}H\text{-1,2,3-triazole} \quad (33)$$

R	Yield
H	80%
Ph	92%

$$PhC{\equiv}CNMe_2 + PhCON_3 \longrightarrow \text{4-Ph-5-}NMe_2\text{-2-}(PhC{=}O)\text{-2}H\text{-1,2,3-triazole} \quad (34)$$

The rearrangement of 1,2,3-thiadiazoles under basic conditions is a most attractive route to acyl- or carboxyl-1,2,3-triazole thiols (Eq. 35).[36,37]

$$\text{4-}RC(O)\text{-5-NHAr-1,2,3-thiadiazole} \xrightarrow[EtOH]{C_5H_{10}NH} \text{4-}RC(O)\text{-5-SH-1-Ar-1,2,3-triazole} \quad (35)$$

10.2-10

R^1	Ph	Ph	Me	OEt	OEt	OEt
Ar	Ph	4-NO_2Ph	Ph	Ph	4-NO_2Ph	4-MeOPh
% **10.2-10**	98	96	92	93	97	96

The reduction of 4-cyano-5-amino-1,2,3-triazoles, under carefully controled conditions, provides formyl derivatives which can be converted to potentially useful imines (Eq. 36).[38,39] Both processes take place in excellent yield.

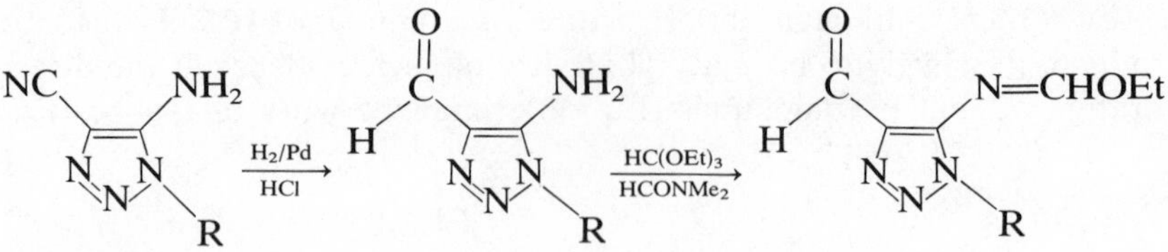

NC, NH_2 triazole (N-R) $\xrightarrow[HCl]{H_2/Pd}$ H–C(=O), NH_2 triazole (N-R) $\xrightarrow[HCONMe_2]{HC(OEt)_3}$ H–C(=O), N=CHOEt triazole (N-R) (36)

$R = Me, PhCH_2$

A combination of ether and ester (Eq. 37)[35] or iminoacid ester 1,2,3-triazoles (Eq. 38)[40] has been reported. In the first instance, L'abbé and his collaborators found that **10.2-11** and **10.2-12** are formed first, but can be converted to **10.2-13** during workup.

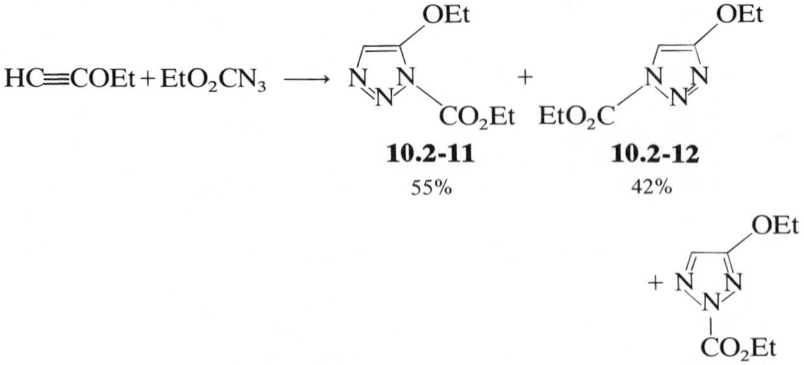

$HC{\equiv}COEt + EtO_2CN_3 \longrightarrow$ OEt, CO_2Et triazole **10.2-11** 55% + OEt, EtO_2C triazole **10.2-12** 42%

+ OEt, N-CO_2Et triazole **10.2-13** 4% (37)

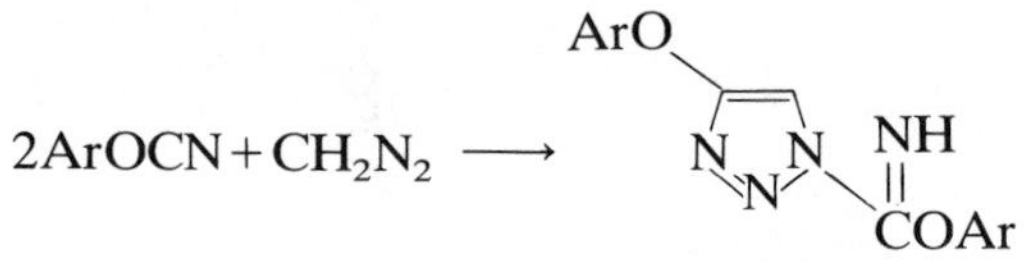

$2ArOCN + CH_2N_2 \longrightarrow$ ArO, NH=COAr triazole **10.2-14** (38)

Ar	= Ph	4-MePh	4-MeOPh	4-ClPh
% **10.2-14**	= 83	79	81	67

The diazotization of amino-1,2,3-triazoles and their subsequent conversion to azo- and triazino-substituents has produced potentially interesting compounds (Eqs. 39 to 41).[41,42]

Several methods have been developed for the introduction of halogen in the presence of other functional groups (Eqs. 42,43).[43,44]

O
C
H_2N NH_2
N N N H
$\xrightarrow[HOAc]{i\text{-}C_5H_{11}ONO}$
O C H_2N N_2^+ N N N
65%
ArH →
O C H_2N N=NAr N N N H
Ar = 4-Me_2NPh 84%
β-$C_{10}H_7$ 57%
(39)
R_2NH →
O C H_2N N=NNR_1 N N N H

10.2-15 (40)

R_2 =	Me_2	$(n\text{-Bu})_2$	C_4H_8N	H, n-Bu	Me, n-Bu	Me, i-Pr	Me, $PhCH_2$
% **10.2-15** =	94	86	92	87	80	69	53

H_2N NH_2 N N N Ph + PhNO $\xrightarrow[50\%\ Na^{+-}OH]{C_6H_6}$ H_2N N=NPh N N N Ph 72% $\xrightarrow[(2)\ Na^{+-}N_3]{(1)\ HNO_2}$

N_3 N=NPh N N N Ph 76% (41)

MeO_2C OH N N N Ph $\xrightarrow[anhydrous]{PCl_5}$

57% R = Me
RO_2C Cl N N N Ph
47% R = Et (42)

EtO_2C NH_2 N N N Ph $\xrightarrow[HCl]{i\text{-}C_5H_{11}ONO}$

Only two examples of 1,2,3-trizaoles bearing three different functional groups have been reported (Eq. 44).[45]

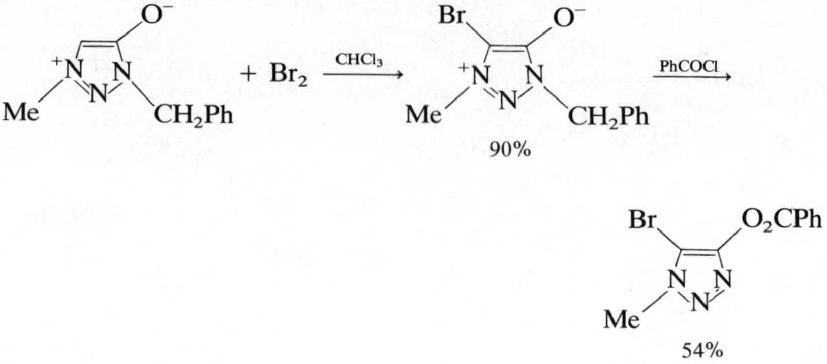

$$\text{1-Me-3-CH}_2\text{Ph-1,2,3-triazolium-4-olate} + Br_2 \xrightarrow{CHCl_3} \text{5-Br-1-Me-3-CH}_2\text{Ph-1,2,3-triazolium-4-olate (90\%)} \xrightarrow{PhCOCl} \text{5-Br-4-}O_2CPh\text{-1-Me-1,2,3-triazole (54\%)} \quad (43)$$

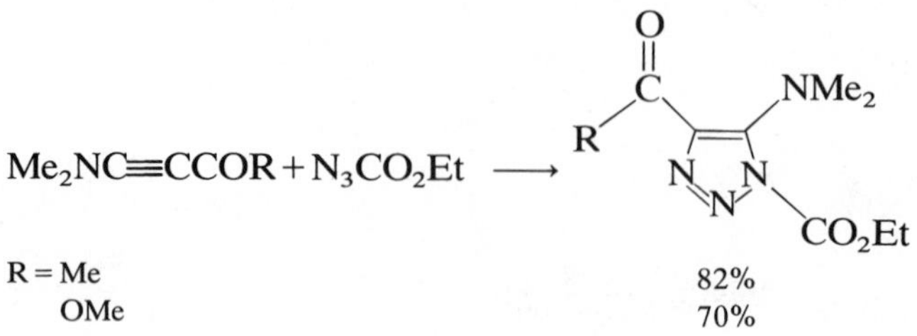

$$Me_2NC{\equiv}CCOR + N_3CO_2Et \longrightarrow \text{4-RCO-5-}NMe_2\text{-1-}CO_2Et\text{-1,2,3-triazole} \quad (44)$$

R = Me: 82%
R = OMe: 70%

REFERENCES

19. **73:** 77146s
20. **73:** 114792x
21. **73:** 119863f
22. **77:** 139556u
23. **77:** 151634a
24. **81:** 136244n
25. **80:** 37051v
26. **84:** 105706x
27. **64:** 5076f
28. **80:** 95838t
29. **45:** 6999g
30. **84:** 135549r
31. **83:** 192266z
32. **72:** 21176p
33. **71:** 112865h
34. **71:** 124340w
35. **78:** 43374c
36. **68:** 68936u
37. **70:** 77874r
38. **79:** 137053v
39. **80:** 37075f
40. **64:** 8171d
41. **65:** 12196h
42. **74:** 53665e
43. **53:** 18013d
44. **71:** 61296p
45. **72:** 31185e

TABLE 10. POLYFUNCTIONAL 1,2,3-TRIAZOLES

Compound	Reference
10.1. Amino-1,2,3-Triazolecarboxylic Acids and Their Functional Derivatives	
5-acetamido-1-benzyl-4-cyano-	**82:** 170802y
4-acetamido-5-cyano-1-methyl-	**80:** 82890d
-5-amine, 1-acetyl-4-(aminocarbonyl)-*N*-[[(2-ethoxycarbonyl)-1-methyl]ethenyl]-	**79:** 5299v
-5-amine, 1-[2-amino-1,3-bis(ethoxycarbonyl)azulen-6-yl]-4-(ethoxycarbonyl)-	**69:** 19088f, **79:** 5300p
-4-amine, 5-(aminocarbonyl)-*N*-[[(2-ethoxycarbonyl)-1-methyl]ethenyl]-	**75:** 52999v
-5-amine, 4-(aminocarbonyl)-1-[[(2-ethoxycarbonyl)-1-methyl]ethenyl]-	**79:** 5299v
-5-amine, 1-(2-amino-1,3-dicyanoazulen-5-yl)-4-cyano-	**69:** 19088f, **79:** 5300p
-5-amine, 1-(2-amino-1,3-dicyanoazulen-6-yl)-4-cyano-	**69:** 19088f, **79:** 5300p
1-benzyl-4-cyano-5-[(dimethylamino)methylideneamino]-	**72:** 78952t, **73:** 77154t
1-benzyl-4-cyano-5-formamido-	**80:** 95838t, **82:** 170802y,
-4-carbonitrile, 5-amino-	**51:** 2756b, **84:** 30969z, **84:** 135549r
-4-carbonitrile, 5-amino-, ion(1-)	**79:** 137053v
-4-carbonitrile, 5-amino-1-α-arabinofuranosyl-	**77:** 147448a
-4-carbonitrile, 5-amino-1-benzyl-	**51:** 2756a, **55:** 2662e, **72:** 78952t, **73:** 77154t **84:** 135549r
-4-carbonitrile, 5-amino-1-(4-chlorophenyl)-	**55:** 2662b
-4-carbonitrile, 5-amino-1-(4-hydroxy-5-oxo-1,3,6-cycloheptatrien-1-yl)-	**69:** 19088f, **79:** 5300p
-4-carbonitrile, 5-amino-1-(4-methoxy-5-oxo-1,3,6-cycloheptatrien-1-yl)-	**79:** 5300p
-5-carbonitrile, 4-amino-1-methyl-	**73:** 77154t, **79:** 146462e, **80:** 82890d, **82:** 170802y, **84:** 30969z
-5-carbonitrile, 4-amino-1-methyl-, conjugate monacid	**79:** 137053v
-4-carbonitrile, 5-amino-1-methyl-	**71:** 13066s, **72:** 21645r, **73:** 77154t, **79:** 146462e, **80:** 82890d, **82:** 170802y, **84:** 30969z
-4-carbonitrile, 5-amino-1-methyl-, phosphate (1:1)	**79:** 146462e
-4-carbonitrile, 5-amino-2-methyl-	**73:** 77154t, **79:** 137053v, **79:** 146462e, **80:** 37075f **80:** 82890d, **82:** 170802y
-4-carbonitrile, 5-amino-1-(4-nitrophenyl)-	**55:** 2662b
-4-carbonitrile, 5-amino-1-phenyl-	**55:** 2662b, **68:** 103631q, **74:** 125581z
-4-carbonitrile, 5-amino-2-(4-pyridinyl)-	**79:** 137053v
-4-carbonitrile, 5-amino-2-(4-pyridinyl)-, conjugate monoacid	**79:** 137053v
-4-carbonitrile, 5-amino-1-(2,3,5-tris-*O*-acetyl-α-D-arabinofuranosyl)-	**77:** 147448a

TABLE 10 (*Continued*)

Compound	Reference
10.1. Amino-1,2,3-Triazolecarboxylic Acids and Their Functional Derivatives (*Continued*)	
-4-carbonitrile, 5-amino-1-(2,3,5-tris-*O*-benzyl-α-D-arabinofuranosyl)-	**77:** 147448a
-4-carbonitrile, 4-amino-1-(2,3,5-tris-*O*-benzyl-β-D-arabinofuranosyl)-	**77:** 147448a
-4-carbonitrile, 1-benzyl-5-[(ethoxymethylene)amino]-	**80:** 82890d, **82:** 43350z
-4-carbonitrile, 1-benzyl-5-[(4-methylphenyl)-sulfonamido]-	**84:** 135549r
-5-carbonitrile, 4-[bis[(4-methylphenyl)-sulfonyl]amino]-	**84:** 30969z
-5-carbonitrile, 4-[bis[(4-methylphenyl)sulfonyl]-amino]-1-methyl-	**84:** 30969z
-4-carbonitrile, 1-methyl-5-[*N*-methyl-(4-methylphenyl)sulfonamido]-	**84:** 135549r
-5-carbonitrile, 1-methyl-4-[*N*-methyl-(4-methylphenyl)sulfonamido]-.	**84:** 135549r
-5-carbonitrile, 2-methyl-4-[*N*-methyl-(4-methylphenyl)sulfonamido]-	**84:** 135549r
-4-carbonitrile, 1-methyl-5-[(4-methylphenyl)-sulfonamido]-	**84:** 30969z, **84:** 135549r
-5-carbonitrile, 1-methyl-4-[(4-methylphenyl)-sulfonamido]-	**84:** 30969z, **84:** 135549r
-5-carbonitrile, 2-methyl-4-[(4-methylphenyl)-sulfonamido]-	**84:** 30969z, **84:** 135549r
-5-carbonitrile, 4-[(4-methylphenyl)sulfonamido]-	**84:** 30969z
-5-carbonitrile, 4-(methylsulfonamido)-, monohydrate	**85:** 102705f
-4-carbonitrile, 5-(4-nitroanilino)-	**55:** 2662c
-4-carbonyl azide, 5-amino-1-benzyl-	**78:** 29681u
-4-carbothioic acid, 5-amino-, S-methyl ester	**70:** 87690s, **72:** 21645r
-4-carbothioic acid, 5-amino-1-benzyl-, S-methyl ester	**72:** 21645r
-4-carbothioic acid, 5-amino-1-methyl-, S-methyl ester	**70:** 87690s, **72:** 21645r, **79:** 146462e
-4-carbothioic acid, 5-amino-2-methyl-, S-methyl ester	**70:** 87690s, **72:** 21645r
-5-carbothioic acid, 4-amino-1-methyl-, S-methyl ester	**70:** 87690s, **72:** 21645r, **76:** 126929x
-5-carbothioic acid, 4-[[(dimethylamino)methylene]-amino]-1-methyl-, S-methyl ester	**76:** 126929x
-4-carboxamide, 5-acetamido-	**52:** 1997d, **68:** 103631q, **75:** 20332m
-4-carboxamide, 5-acetamido-*N*-acetyl-	**52:** 1997c
-4-carboxamide, 5-acetamido-1-acetyl-	**75:** 20332m
-4-carboxamide, 5-acetamido-*N*-acetyl-1-phenyl-	**74:** 125581z
-4-carboxamide, 1-acetamido-5-amino-	**55:** 2662a
-4-carboxamide, 5-acetamido-2-(5-chloro-2-hydroxyphenyl)-	**74:** P125704s
-4-carboxamide, 5-acetamido-*N*,*N*-dimethyl-1-phenyl-	**74:** 125581z
-4-carboxamide, 5-acetamido-1-phenyl-	**55:** 2661e

TABLE 10 (*Continued*)

Compound	Reference
10.1. Amino-1,2,3-Triazolecarboxylic Acids and Their Functional Derivatives (*Continued*)	
-5-carboxamide, 4-(acetylamino)-1-(2,3,5-tri-*O*-benzoyl-β-D-ribofuranosyl)-	**77:** 126981d
-4-carboxamide, *N*-acetyl-5-anilino-	**74:** 125581z
-4-carboxamide, 1-acetyl-5-anilino-	**74:** 125581z
-4-carboxamide, 1-acetyl-5-anilino-*N*,*N*-dimethyl-	**74:** 125581z
-4-carboxamide, *N*-acetyl-5-(diacetylamino)-1-phenyl-	**74:** 125581z
-4-carboxamide, 5-amino-	**51:** 2755g, **51:** 3688g, **51:** 12075a, **52:** 1977c, **55:** 11422a, **65:** 12196h, **71:** 11523q, **72:** 55346j, **79:** 5299v, **79:** 137053v, **82:** 118340g
-4-carboxamide, 5-amino-, hydrochloride	**77:** 84962r
-4-carboxamide, 5-amino-1-α-D-arabinofuranosyl-	**77:** 147448a
-4-carboxamide, 5-amino-1-β-D-arabinofuranosyl-	**84:** P136010v
-4-carboxamide, 5-amino-1-benzyl-	**51:** 2755f, **51:** 14698d **68:** 103631q, **72:** 78952t, **80:** 82890d
-4-carboxamide, 5-amino-1-benzyl-*N*-methyl-	**72:** 21645r
-4-carboxamide, 5-amino-1-(benzylthio)-	**68:** 103631q
-4-carboxamide, 5-amino-*N*-[bis(methylamino)-methylene]-2-methyl-	**76:** 126928w
-4-carboxamide, 5-amino-1-(2-chlorophenyl)-	**55:** 2661h
-4-carboxamide, 5-amino-1-(3-chlorophenyl)-	**55:** 2661h
-4-carboxamide, 5-amino-1-(4-chlorophenyl)-	**55:** 2661h
-4-carboxamide, 5-amino-*N*,1-dimethyl-	**70:** 87690s
-4-carboxamide, 5-amino-*N*,2-dimethyl-	**72:** 21645r
-4-carboxamide, 5-amino-1-(3,3-dimethyl-1-butenyl)-, (E)-	**72:** 121447w
-4-carboxamide, 5-amino-*N*,*N*-dimethyl-1-phenyl-	**74:** 125581z
-4-carboxamide, 5-amino-1-ethyl-	**55:** 2662a
-4-carboxamide, 5-amino-*N*-methyl-	**70:** 87690s, **72:** 21645r
-4-carboxamide, 5-amino-1-methyl-	**52:** 16359f, **55:** 2661i, **69:** 77238r, **79:** 146462e, **84:** 30969z
-4-carboxamide, 5-amino-2-methyl-	**67:** 90731z, **69:** 77238r, **72:** 21645r, **84:** 30969z
-5-carboxamide, 4-amino-1-methyl-	**65:** 713g, **68:** 68958c, **84:** 30969z
-4-carboxamide, 5-amino-1-(2-methylphenyl)-	**55:** 2661f
-4-carboxamide, 5-amino-1-(3-methylphenyl)-	**55:** 2661f
-4-carboxamide, 5-amino-1-(4-methylphenyl)-	**55:** 2661f
-4-carboxamide, 5-amino-1-(3-nitrophenyl)-	**55:** 2661h
-4-carboxamide, 5-amino-1-(4-nitrophenyl)-	**55:** 2661h
-4-carboxamide, 5-amino-1-phenyl-	**51:** 14698c, **52:** 1997e, **55:** 2661d, **68:** 103631q, **74:** 125581z
-4-carboxamide, 5-amino-2-phenyl-	**51:** 2811d
-4-carboxamide, 5-amino-1-β-D-ribofuranosyl-	**77:** 160107h

TABLE 10 (*Continued*)

Compound	Reference
10.1. Amino-1,2,3-Triazolecarboxylic Acids and Their Functional Derivatives (*Continued*)	
-5-carboxamide, 4-amino-1-β-D-ribofuranosyl-	**77:** 126981d
-4-carboxamide, 5-amino-1-(phenylethenyl)-	**72:** 121447w
-4-carboxamide, 5-amino-1-(2,3,5-tris-*O*-acetyl-α-D-arabinofuranosyl)-	**77:** 147448a
-4-carboxamide, 5-amino-1-(1,2,3-tris-*O*-benzyl-α-D-arabinofuranosyl)-	**76:** 141226j, **77:** 147448a
-4-carboxamide, 5-amino-1-(2,3,5-tris-*O*-benzyl-β-D-arabinofuranosyl)-	**76:** 141226j, **77:** 147448a
-4-carboxamide, 5-anilino-	**52:** 1997f, **55:** 2661d
-4-carboxamide, 5-anilino-*N*,1-diacetyl-	**74:** 125581z
-4-carboxamide, 5-anilino-*N*,*N*-dimethyl-	**74:** 125581z
-4-carboxamide, 5-(benzylamino)-	**72:** 78952t
-4-carboxamide, 1-benzyl-5-(diformylamino)-	**70:** 47403u
-4-carboxamide, 2-benzyl-5-(formylamino)-	**76:** 126925t
-4-carboxamide, 1-benzyl-5-formamido-	**70:** 47403u
-4-carboxamide, 1-benzyl-5-(methylamino)-	**70:** 47403u
-4-carboxamide, 2-[*N*-(5-carbamoyl-1,2,3-triazol-4-yl)formimidoyl]-5-formamido-	**69:** 77238r
-2-carboxamide, 4-[[2-[[1-(dimethylamino)-carbonyl]1*H*-1,2,4-triazol-3-yl]amino]-2,4,6-cycloheptatrien-1-ylidene]amino]-*N*,*N*-dimethyl-	**77:** 113628u
-4-carboxamide, 5-[[(dimethylamino)methylene]-amino]-*N*-formyl-2-methyl-	**76:** 126929x
-5-carboxamide, 4-[[(dimethylamino)methylene]-amino]-*N*-formyl-1-methyl-	**76:** 126929x
-4-carboxamide, 5-[[(dimethylamino)methylene]-amino]-2-methyl-	**76:** 126929x
-5-carboxamide, 4-[[(dimethylamino)methylene]-amino]-1-methyl-	**76:** 126929x
-5-carboxamide, 4-[[(dimethylamino)methylene]-amino]-*N*,*N*,1-trimethyl-	**76:** 126929x
-4-carboxamide, *N*,1-dimethyl-5-(2-fluoroanilino)-	**69:** 96637m
-5-carboxamide, *N*,1-dimethyl-4-(2-fluoroanilino)-	**69:** 96637m
-4-carboxamide, *N*,1-dimethyl-5-(4-fluoroanilino)-	**69:** 96637m
-5-carboxamide, *N*,1-dimethyl-4-(4-fluoroanilino)-	**69:** 96637m
-4-carboxamide, 5-(4-fluoroanilino)-*N*-methyl-	**69:** 96637m
-4-carboxamide, 5-formamido-	**67:** 90731z, **68:** 103631q
-4-carboxamide, 5-formamido-1-methyl-	**69:** 77238r
-4-carboxamide, 5-formamido-2-methyl-	**67:** 90731z, **69:** 77238r
-4-carboxamide, 1-methyl-5-(methylamino)-	**69:** 77238r
-4-carboxamide, 2-methyl-5-(methylamino)-	**69:** 77238r
-4-carboxamide, 1-methyl-5-(*N*-methylformamido)-	**69:** 77238r
-4-carboxamide, 2-methyl-5-(*N*-methylformamido)-	**69:** 77238r
-4-carboxamide, 5-[(3-methylphenyl)amino]-	**55:** 2661f
-4-carboxamide, 5-[(4-methylphenyl)amino]-	**55:** 2661f
-4-carboxamide, 5-[[(4-methylphenyl)sulfonyl]amino]-	**84:** 135549r
-4-carboxamide, 1-phenyl-5-(2-phenylacetamido)-	**74:** 125581z

TABLE 10 (*Continued*)

Compound	Reference
10.1. Amino-1,2,3-Triazolecarboxylic Acids and Their Functional Derivatives (*Continued*)	
-4-carboxamide, 1-phenyl-5-(2-phenylacetamido)-*N*-(phenylacetyl)-	**74:** 125581z
-4-carboxamide, 1-phenyl-5-propanamido-*N*-propanoyl-	**74:** 125581z
-4-carboxamidine, 5-amino-	**55:** 11422a, **63:** 9943c, **68:** 103631q, **82:** 43350z
-4-carboxamidine, 5-amino-1-benzyl-	**68:** 103631q, **82:** 43350z
-4-carboxamidine, 5-amino-2-phenyl-	**51:** 2812c
-4-carboxamidoxime, 5-amino-	**55:** 1142li
-4-carboximidamide, 5-amino-, monohydrochloride	**82:** 43350z
-4-carboximidamide, 5-amino-*N*-(aminoiminoethyl)-1-benzyl-	**82:** 170802y
-4-carboximidamide, 5-amino-*N*-(aminoiminoethyl)-2-methyl-	**82:** 170802y
-4-carboximidamide, 5-amino-1-benzyl-, monohydrochloride	**82:** 43350z, **82:** 170802y
-4-carboximidamide, 5-amino-1-benzyl-*N*-butyl-	**82:** 43350z
-4-carboximidamide, 5-amino-1-benzyl-*N*-methyl-	**82:** 43350z
-4-carboximidamide, 5-amino-1-benzyl-*N*-methyl-, monohydrochloride	**82:** 43350z
-4-carboximidamide, 5-amino-*N*-butyl-, 1:1 adduct with formic acid	**82:** 43350z
-4-carboximidamide, 5-amino-*N*-butyl-2-methyl-	**82:** 43350z
-4-carboximidamide, 5-amino-*N*-butyl-2-methyl-, monohydrochloride	**82:** 43350z
-5-carboximidamide, 4-amino-*N*-butyl-1-methyl-	**82:** 43350z
-5-carboximidamide, 4-amino-*N*-butyl-1-methyl-, monohydrochloride	**82:** 43350z
-4-carboximidamide, 5-amino-*N*,2-dimethyl-	**82:** 43350z
-4-carboximidamide, 5-amino-*N*,2-dimethyl-, monohydrochloride	**82:** 43350z
-5-carboximidamide, 4-amino-*N*,1-dimethyl-	**82:** 43350z
-4-carboximidamide, 5-amino-*N*-methyl-	**82:** 43350z
-4-carboximidamide, 5-amino-1-methyl-, monohydrochloride	**82:** 43350z
-4-carboximidamide, 5-amino-*N*-(phenylmethoxy)-1-β-D-ribofuranosyl-	**76:** 135794b
-4-carboximidamide, 5-amino-1-β-D-ribofuranosyl-	**76:** 135794b
-5-carboximidamide, 4-amino-*N*,*N*,1-trimethyl-	**82:** 43350z
-4-carboximidic acid, 5-amino-1-benzyl-, ethyl ester	**51:** 2756a, **68:** 103631q
-4-carboximidic acid, 5-amino-1-benzyl-, ethyl ester, monohydrochloride	**82:** 43350z
-4-carboximidic acid, 5-amino-1-(methylthio)-, methyl ester	**71:** 13066s
-4-carboximidic acid, 5-amino-1-(methylthio)-, methyl ester, monohydrochloride	**71:** 13066s, **72:** 21645r
-4-carboximidic acid, 5-amino-1-phenyl-, ethyl ester	**68:** 103631q
-4-carboximidic acid, 5-amino-1-phenyl-, methyl ester	**74:** 125581z

TABLE 10 (*Continued*)

Compound	Reference
10.1. Amino-1,2,3-Triazolecarboxylic Acids and Their Functional Derivatives (*Continued*)	
-4-carboxylic acid, 5-acetamido-1-phenyl-, methyl ester	**74:** 125581z
-4-carboxylic acid, 1-acetyl-5-anilino-, methyl ester	**74:** 125581z
-1-carboxylic acid, 4-acetyl-5-(dimethylamino)-, ethyl ester	**72:** 31185e
-4-carboxylic acid, 5-amino-	**69:** 77238r, **72:** 21645r
-4-carboxylic acid, 5-amino-, hydrazide	**51:** 2755g, **68:** 103631q
-4-carboxylic acid, 4-amino-, methyl ester	**69:** 77238r
-4-carboxylic acid, 5-amino-1-benzyl-	**51:** 2755e, **72:** 78952t
-4-carboxylic acid, 5-amino-1-benzyl-, ethyl ester	**55:** 2662b, **68:** 103631q, **78:** 29681u, **79:** 137053v
-4-carboxylic acid, 5-amino-1-benzyl-, hydrazide	**68:** 103631q, **78:** 29681u
-4-carboxylic acid, 5-amino-1-(5-bromo-1,6-dihydro-6-oxo-1-phenyl-4-pyridazinyl)-, ethyl ester	**76:** 153691v
-4-carboxylic acid, 5-amino-1-(4-bromophenyl)-, ethyl ester	**49:** 1710h
-5-carboxylic acid, 4-[(aminocarbonyl)amino]-1-methyl-	**76:** 126928w
-4-carboxylic acid, 5-amino-1-(4-carboxylphenyl)-, diethyl ester	**49:** 1710h
-4-carboxylic acid, 5-amino-1-(3-chlorophenyl)-, ethyl ester	**73:** P98955v
-4-carboxylic acid, 5-amino-1-(4-chlorophenyl), ethyl ester	**73:** P98955v
-4-carboxylic acid, 5-amino-1-cyclohexyl-	**75:** 5801v
-4-carboxylic acid, 5-amino-1-(3,3-dimethyl-1-butenyl)-, methyl ester, (E)-	**72:** 121447w
-4-carboxylic acid, 5-amino-1-(3-fluorophenyl)-, ethyl ester	**73:** P98955v
-4-carboxylic acid, 5-amino-1-(4-fluorophenyl)-, ethyl ester	**73:** P98955v
-4-carboxylic acid, 5-amino-1-(4-methoxyphenyl)-, ethyl ester	**49:** 1710h
-4-carboxylic acid, 5-amino-1-methyl-	**72:** 21645r
-4-carboxylic acid, 5-amino-2-methyl-	**69:** 77238r
-4-carboxylic acid, 5-amino-2-methyl-, methyl ester	**79:** 137053v
-5-carboxylic acid, 4-amino-1-methyl-	**76:** 126928w, **79:** 137053v
-4-carboxylic acid, 5-amino-1-(4-methylphenyl)-, ethyl ester	**49:** 1710g
-4-carboxylic acid, 5-amino-1-phenyl-	**36:** 835[7], **56:** 12870a
-4-carboxylic acid, 5-amino-1-phenyl-, ethyl ester	**44:** 1102d, **52:** 365i, **55:** 2662a, **68:** 103631q
-4-carboxylic acid, 5-amino-1-phenyl-, hydrazide	**68:** 103631q
-4-carboxylic acid, 5-amino-1-phenyl-, methyl ester	**74:** 125581z
-4-carboxylic acid, 5-amino-1-phenyl, (2-nitroethyl) ester	**55:** 23505a
-4-carboxylic acid, 5-anilino-, ethyl ester	**55:** 2662b
-4-carboxylic acid, 5-(benzylamino)-	**72:** 78952t

TABLE 10 (*Continued*)

Compound	Reference
10.1. Amino-1,2,3-Triazolecarboxylic Acids and Their Functional Derivatives (*Continued*)	
-4-carboxylic acid, 5-(3-chloroanilino), ethyl ester	**73:** P98955v
-4-carboxylic acid, 5-(4-chloroanilino)-, ethyl ester	**73:** P98955v
-4-carboxylic acid, 1-(3-chlorophenyl)-5-[(2-hydroxyethyl)amino-, ethyl ester	**74:** P22847m
-4-carboxylic acid, 1-(4-chlorophenyl)-5-[(2-hydroxyethyl)amino]-, ethyl ester	**74:** P22847m
-4-carboxylic acid, 1-(3-chlorophenyl)-5-morpholino-, ethyl ester	**74:** P22847m
-4-carboxylic acid, 1-(4-chlorophenyl)-5-morpholino-, ethyl ester	**74:** P22847m
-4-carboxylic acid, 5-(diacetylamino)-1-phenyl-, methyl ester	**74:** 125581z
-1-carboxylic acid, 4-(diethylamino)-5-methyl-, ethyl ester	**78:** 43374c
-1-carboxylic acid, 5-(diethylamino)-4-methyl-, ethyl ester	**78:** 43374c
-2-carboxylic acid, 4-(diethylamino)-5-methyl-, ethyl ester	**78:** 43374c
-4-carboxylic acid, 5-(3-fluoroanilino)-, ethyl ester	**73:** P98955v
-4-carboxylic acid, 5-(4-fluoroanilino)-, ethyl ester	**73:** P98955v
-4-carboxylic acid, 1-(3-fluorophenyl)-5-[(2-hydroxyethyl)amino]-, ethyl ester	**74:** P22847m
-4-carboxylic acid, 1-(4-fluorophenyl)-5-[(2-hydroxyethyl)amino]-, ethyl ester	**74:** P22847m
-4-carboxylic acid, 1-(3-fluorophenyl)-5-morpholino-, ethyl ester	**74:** P22847m
-4-carboxylic acid, 1-(4-fluorophenyl)-5-morpholino-, ethyl ester	**74:** P22847m
-4-carboxylic acid, 5-formamido-, methyl ester	**69:** 77238r
-4-carboxylic acid, 5-(5-isoxazolylamino)-	**26:** 1606[6]
-4-carboxylic acid, 5-(5-isoxazolylamino)-, methyl ester	**26:** 1606[6]
-4-carboxylic acid, 5-(5-isoxazolylamino)-1-methyl-	**26:** 1606[8,9]
-4-carboxylic acid, 5-(5-isoxazolylamino)-1-methyl-, methyl ester	**26:** 1606[8,9]
-4-carboxylic acid, 2-methyl-5-[methyl[(4-methylphenyl)sulfonyl]amino]-	**84:** 135549r
-4-carboxylic acid, 2-methyl-5-[methyl[(4-methylphenyl)sulfonyl]amino]-, ethyl ester	**84:** 135549r
-4-carboxylic acid, 1-methyl-5-(phenylamino)-, ethyl ester	**76:** 140731h
-4-carboxylic acid, 5-[[(4-methylphenyl)sulfonyl]amino]-, ethyl ester	**84:** 135549r
-4-carboxylic acid, 5-(4-nitroanilino)-, ethyl ester	**49:** 1710h
5-cyano-4-[(phenylsulfonyl)amino]-	**80:** 95838t

TABLE 10 (*Continued*)

Compound	Reference
10.1. Amino-1,2,3-Triazolecarboxylic Acids and Their Functional Derivatives (*Continued*)	
5-cyano-4-(*N,N*-diethylureylenyl)-	**80:** 95838t
4-cyano-5-[(dimethylamino)methylideneamino]-1-methyl-	**73:** 77154t, **76:** 126929x
5-cyano-4-[(dimethylamino)methylideneamino]-1-methyl-	**73:** 77154t, **76:** 126929x, **79:** 146462e, **82:** 43350z
5-cyano-4-[(dimethylamino)methylideneamino]-2-methyl-	**73:** 77154t, **76:** 126929x, **79:** 146462e, **82:** 43350z
4-cyano-5-[(ethoxymethylidene)amino]-1-methyl-, ethyl ester	**80:** 82890d
5-cyano-4-[(ethoxymethylidene)amino]-1-methyl-, ethyl ester	**80:** 82890d
5-cyano-4-[(ethoxymethylidene)amino]-2-methyl-, ethyl ester	**80:** 82890d
5-cyano-4-formamido-1-methyl-	**80:** 82890d
5-cyano-4-formamido-2-methyl-	**80:** 82890d
5-cyano-4-(methylsulfonamido)-	**80:** 95838t
5-cyano-4-[[(4-methoxyphenyl)sulfonyl]amino]-	**80:** 95838t
5-cyano-4-[[(4-methylphenyl)sulfonyl]amino]-	**80:** 95838t
-1,4-dicarboxylic acid, 5-(dimethylamino)-, 1-ethyl methyl diester	**72:** 31185e
-4-thiocarboxamide, 5-amino-	**51:** 2756c
-4-thiocarboxamide, 5-amino-1-benzyl-	**51:** 2756b
-4-thiocarboxamide, 5-amino-2-phenyl-	**51:** 2811h
10.2. Miscellaneous 1,2,3-Triazoles Containing More Than One Representative Function	
4-acetamido-5-amino-1-benzyl-	**78:** 29681u
4-acetamido-2-benzoyl-	**75:** 20332m
4-acetamido-1-[(4-bromophenyl)sulfonyl]-	**83:** 192266z
4-acetamido-2-mercapto-	**72:** P67723t
1-[(4-acetamidophenyl)sulfonyl]-5-(diethylamino)-4-methyl-	**73:** 114792x, **73:** 119863f
1-acetyl-5-anilino-4-phenyl-	**74:** 125581z
1-acetyl-5-[[2,2-bis(ethoxycarbonyl)ethylidene]-amino]-4-phenyl, diethyl ester	**75:** 20332m
1-acetyl-5-(diacetylamino)-4-phenyl-	**75:** 20332m
1-acetyl-4,5-dibromo-	**54:** 4547f
1-acetyl-5-ethoxy-4-propyl-	**71:** 112865h
4-acetyl-5-hydroxy-1-phenyl-	**77:** 87530j
4-acetyl-5-hydroxy-1-(tetrazolo[5,1-*a*]phthalazin-6-yl)-, sodium salt	**81:** 37518y
4-acetyl-5-mercapto-1-phenyl-	**70:** 77874r
4-acetyl-5-mercapto-1-phenyl, 1:1 adduct with piperidine	**70:** 77874r
N-acetyl-1-methyl-5-[[(4-methylphenyl)sulfonyl]amino]-	**83:** 192266z
1-acetyl-5-(3-oxobutanamido)-4-phenyl-	**79:** 5299v
-5-amine, 1-acetyl-*N*-[[(2-ethoxycarbonyl)-1-methyl]ethenyl]-4-phenyl-	**79:** 5299v
-4-amine, 5-azido-2-phenyl-	**63:** P11749f

TABLE 10 (*Continued*)

Compound	Reference
10.2. Miscellaneous 1,2,3-Triazoles (*Continued*)	
-5-amine, 4-benzoyl-	**84:** 135549r
-5-amine, 1-benzyl-4-[(ethoxycarbonyl)amino]-, ethyl ester	**78:** 29681u
-5-amine, 1-benzyl-4-(dimethoxymethyl)-	**80:** 37075f
-5-amine, 1-[(4-bromophenyl)sulfonyl]-*N*,4-diphenyl-*N*-methyl-	**80:** 37051v
-5-amine, 1-[(4-bromophenyl)sulfonyl]-*N*-methyl-*N*-phenyl-4-(2-propenyl)-	**80:** 37051v
-5-amine, 4-butyl-*N*,*N*-diethyl-1-[[4-(dimethylamino)-phenyl]sulfonyl]-	**80:** 37051v
-5-amine, 4-butyl-*N*,*N*-diethyl-1-[(2,4,6-trimethyl-phenyl)sulfonyl]-	**80:** 37051v
-5-amine, 4-butyl-*N*-methyl-1-[(4-methylphenyl)-sulfonyl]-*N*-phenyl-	**73:** 77146s, **80:** 37051v
-5-amine, 4-butyl-*N*-methyl-1-[(2-nitrophenyl)-sulfonyl]-*N*-phenyl-	**80:** 37051v
-5-amine, 4-butyl-*N*-methyl-1-[(3-nitrophenyl)-sulfonyl]-*N*-phenyl-	**80:** 37051v
-5-amine, 4-butyl-*N*-methyl-1-[(4-nitrophenyl)-sulfonyl]-*N*-phenyl-	**73:** 77146s, **80:** 37051v
-4-amine, 2-(4-carboxyphenyl)-5-[(4-carboxy-phenyl)azo]-	**74:** P65596w
-4-amine, 5-[(4-chlorophenyl)azo]-2-phenyl-	**74:** 53665e
-4-amine, 2-(3-chlorophenyl)-5-[(3-chlorophenyl)azo]-	**74:** P65596w
-5-amine, 1-[(4-chlorophenyl)sulfonyl]-*N*,*N*-diethyl-4-methyl-	**73:** 77146s, **80:** 37051v
-1-amine, 4,5-dibenzoyl-	**84:** 31007w
-2-amine, 4,5-dibenzoyl-	**84:** 31007w
-5-amine, *N*,*N*-diethyl-1-[[4-(dimethylamino)phenyl]-sulfonyl]-4-methyl-	**80:** 37051v
-5-amine, *N*,*N*-diethyl-1-[[4-(dimethylamino)phenyl]-sulfonyl]-4-phenyl-	**80:** 37051v
-5-amine, *N*,*N*-diethyl-1-[(4-fluorophenyl)sulfonyl]-4-methyl-	**80:** 37051v
-5-amine, *N*,*N*-diethyl-1-[(4-methoxyphenyl)sulfonyl]-4-phenyl-	**73:** 77146s, **80:** 37051v
-5-amine, *N*,*N*-diethyl-4-methyl-1-(methylsulfonyl)-	**73:** 77146s, **80:** 37051v
-5-amine, *N*,*N*-diethyl-4-methyl-1-(1-naphthalenyl-sulfonyl)-	**73:** 77146s, **80:** 37051v
-5-amine, *N*,*N*-diethyl-4-methyl-1-(2-naphthalenyl-sulfonyl)-	**80:** 37051v
-5-amine, *N*,*N*-diethyl-4-methyl-1-(phenylsulfonyl)-	**73:** 77146s, **73:** 114792x, **73:** 119863f, **74:** 40817f, **80:** 37051v
-5-amine, *N*,*N*-diethyl-4-methyl-1-[(2,4,6-trimethyl-phenyl)sulfonyl]-	**73:** 77146s, **80:** 37051v
-5-amine, *N*,*N*-diethyl-4-phenyl-1-[(2,4,6-trimethyl-phenyl)sulfonyl]-	**73:** 77146s, **80:** 37051v

TABLE 10 (*Continued*)

Compound	Reference
10.2. Miscellaneous 1,2,3-Triazoles (*Continued*)	
-4-amine, 5-[[4-(dimethylamino)phenyl]azo]-2-phenyl-	**63:** P11749d
-5-amine, 1-[[4-(dimethylamino)phenyl]sulfonyl]-*N*,*N*-diphenyl-	**77:** 139556u
-5-amine, 1-[[4-(dimethylamino)phenyl]sulfonyl]-*N*,*N*-diphenyl-4-(triethylgermyl)-	**84:** 105706x
-5-amine, 1-[[4-(dimethylamino)phenyl]sulfonyl]-*N*,*N*-diphenyl-4-(triphenylgermyl)-	**84:** 105706x
-5-amine, 1-[[4-(dimethylamino)phenyl]sulfonyl]-*N*,*N*-diphenyl-4-(triphenylstannyl)-	**84:** 105706x
-5-amine, 1-[[4-(dimethylamino)phenyl]sulfonyl]-*N*-methyl-*N*-phenyl-4-(triphenylgermyl)-	**84:** 105706x
-5-amine, *N*,4-dimethyl-1-[(4-methoxyphenyl)-sulfonyl]-*N*-phenyl-	**80:** 37051v
-5-amine, *N*,4-dimethyl-1-[(4-methylphenyl)-sulfonyl]-*N*-phenyl-	**80:** 37051v
-5-amine, *N*,4-dimethyl-1-[(3-nitrophenyl)-sulfonyl]-*N*-phenyl-	**80:** 37051v
-5-amine, *N*,4-dimethyl-1-[(4-nitrophenyl)-sulfonyl]-*N*-phenyl-	**80:** 37051v
-4-amine, 2-(3,4-dimethylphenyl)-5-[(3,4-dimethylphenyl)azo]-	**74:** P65596w
-5-amine, *N*,4-dimethyl-*N*-phenyl-1-(phenylsulfonyl)-	**80:** 37051v
-5-amine, *N*,*N*-diphenyl-1-[(4-methoxyphenyl)-sulfonyl]-4-(triethylgermyl)-	**84:** 105706x
-5-amine, *N*,*N*-diphenyl-1-[(4-methoxyphenyl)-sulfonyl]-4-(triphenylgermyl)-	**84:** 105706x
-5-amine, *N*,*N*-diphenyl-1-[(4-methoxyphenyl)-sulfonyl]-4-(triphenylsilyl)-	**84:** 105706x
-5-amine, *N*,*N*-diphenyl-4-(diphenylphosphino-thioyl)-1-[(4-methylphenyl)sulfonyl]-	**81:** 136244n
-5-amine, *N*,4-diphenyl-*N*-methyl-1-[(4-methyl-phenyl)sulfonyl]-	**80:** 37051v
-5-amine, *N*,4-diphenyl-*N*-methyl-1-[(3-nitro-phenyl)sulfonyl]-	**80:** 37051v
-5-amine, *N*,4-diphenyl-*N*-methyl-1-[(4-nitro-phenyl)sulfonyl]-	**80:** 37051v
-5-amine, 4-(diphenylphosphinyl)-*N*-methyl-*N*-phenyl-1-[(2,4,6-trimethylphenyl)sulfonyl]-	**81:** 136244n
-4-amine, 2-(3-methoxyphenyl)-5[(3-methoxyphenyl)azo]-	**74:** P65596w
-5-amine, *N*-methyl-1-[(4-methylphenyl)sulfonyl]-*N*-phenyl-4-(2-propenyl)-	**80:** 37051v
-5-amine, *N*-methyl-1-(3-nitrophenyl)sulfonyl]-*N*-phenyl-4-(2-propenyl)-	**80:** 37051v
-5-amine, 5-[(4-methylphenyl)azo]-2-phenyl-	**74:** 53665e
-4-amine, 2-phenyl-5-(phenylazo)-	**74:** 53665e
-4-amine, 2-(phenylsulfonamido)-	**79:** P147417f
-4-amine, 2-(3-pyridyl)-5-(3-pyridylazo)-	**74:** P65596w
-5-amine, *N*,*N*,4-triethyl-1-[(2,4,6-trimethyl-phenyl)sulfonyl]-	**80:** 37051v

TABLE 10 (*Continued*)

Compound	Reference
10.2. Miscellaneous 1,2,3-Triazoles (*Continued*)	
4-[(aminomethylidene)amino]-5-[(hydroxyhydrazono)methyl]-	**79:** 137047w
4-azido-5-[[4-(dimethylamino)phenyl]azo]-2-phenyl-	**63:** P11749e
4-azido-2-(3,4-dimethylphenyl)-5-[(3,4-dimethylphenyl)azo]-	**74:** P65596w
4-azido-2-(3-pyridyl)-5-(3-pyridylazo)-	**74:** P65596w
4-azido-1-[(4-methylphenyl)sulfonyl]-5-phenyl-	**53:** 15065e
4-azido-2-phenyl-5-(phenylazo)-	**74:** 53665e
5-benzamido-4-benzoyl-	**57:** 5897d
1-benzamido-4-[(benzoylhydrazono)methyl]-	**71:** 49860k
1-benzamido-5-[(benzoylhydrazono)methyl]-	**71:** 49860k
1-benzamido-4-formyl-	**71:** 49860k
1-benzoyl-5-(dimethylamino)-4-phenyl-	**65:** 10524d
5-benzoyl-4-[(4,4-dimethyl-2-hydroxy-6-oxo-1-cyclohexenyl)azo]-	**84:** 135549r
1-benzoyl-5-ethoxy-4-methyl-	**72:** 21176p
2-benzoyl-4-ethoxy-5-methyl-	**79:** 31992k
5-benzoyl-4-[(2-hydroxy-1-naphthalenyl)azo]-	**84:** 135549r
4-benzoyl-5-mercapto-1-(4-methoxyphenyl)-	**70:** 77874r
4-benzoyl-5-mercapto-1-(4-methoxyphenyl)-, 1:1 adduct with piperidine	**70:** 77874r
4-benzoyl-5-mercapto-1-(4-nitrophenyl)-	**68:** 68936u
4-benzoyl-5-mercapto-1-(4-nitrophenyl)-, 1:1 adduct with piperidine	**68:** 68936u
4-benzoyl-5-mercapto-1-phenyl-	**70:** 77874r
4-benzoyl-5-mercapto-1-phenyl-, 1:1 adduct with piperidine	**70:** 77874r
5-benzoyl-4-[(4-methylphenyl)sulfonamido]-	**84:** 135549r
1-benzyl-5-[(diethoxymethylidene)amino]-4-formyl-, diethyl ester	**80:** 37075f
1-benzyl-5-[[(dimethylamino)methylidene]amino]-4-formyl-	**80:** 37075f
1-benzyl-5-[(ethoxycarbonyl)amino]-4-formyl-, ethyl ester	**80:** 37075f
1-benzyl-5-[(ethoxyethylidene)amino]-4-formyl-, ethyl ester	**80:** 37075f
1-benzyl-5-[(ethoxymethylidene)amino]-4-formyl-, ethyl ester	**80:** 37075f
1-[[(4-bromophenyl)sulfonyl]amino]-*N*,1-diacetyl-	**83:** 192266z
4-[[(4-bromophenyl)sulfonyl]amino]-*N*,2-diacetyl-	**83:** 192266z
5-[[(4-bromophenyl)sulfonyl]amino]-*N*,1-diacetyl-	**83:** 192266z
5-bromo-4-methoxy-1-methyl-	**71:** 61296p
4-butyl-1-[(4-methoxyphenyl)sulfonyl]-5-(*N*-methylanilino)-	**73:** 77146s
-4-carbonitrile, 5-(chloroazo)-	**68:** 103631q
-4-carbonyl azide, 1-(2,5-dimethylphenyl)-5-isocyanato-	**22:** 3411[3]
-4-carboxaldehyde, 5-amino-, [(5-amino-1*H*-1,2,3-triazol-4-yl)methylene]-, hydrazone	**79:** 137047w

TABLE 10 (*Continued*)

Compound	Reference
10.2. Miscellaneous 1,2,3-Triazoles (*Continued*)	
-4-carboxaldehyde, 5-amino-, oxime	**79:** 137047w
-4-carboxaldehyde, 5-amino-1-benzyl-	**79:** 137053v, **80:** 37075f
-4-carboxaldehyde, 5-amino-1-methyl-	**79:** 137053v
-4-carboxaldehyde, 5-amino-1-methyl-, conjugate monoacid	**79:** 137053v
-4-carboxaldehyde, 5-amino-2-methyl-	**79:** 137053v, **80:** 37075f
-4-carboxaldehyde, 5-amino-2-methyl-, [(5-amino-2-methyl-2*H*-1,2,3-triazol-4-yl)-methylene]hydrazone	**79:** 137053v
-4-carboxaldehyde, 5-amino-2-methyl-, conjugate monoacid	**79:** 137053v
-4-carboxaldehyde, 5-amino-2-methyl-, phenylhydrazone	**79:** 137053v
-5-carboxaldehyde, 4-amino-1-methyl-	**79:** 137053v, **80:** 37075f
-5-carboxaldehyde, 1-[[α-(benzoyloxy)benzylidene]-amino]-, benzoylhydrazone	**71:** 49860k
-5-carboxaldehyde, 1-(benzylideneamino)-	**71:** 49860k
-4-carboxaldehyde, 4-guanidino-, oxime	**79:** 137047w
-4-carboxamide, 5-(aminothio)-	**68:** 103631q
-4-carboxamide, 5-azido-	**80:** 3436j
-4-carboxamide, 5-(3-benzyl-3-methyl-1-triazeno)-	**65:** 12196h
-4-carboxamide, 5-[3-(4-bromophenyl)-1-triazeno]-	**65:** 12196h
-4-carboxamide, 5-[3-[(4-butoxyphenyl)iminomethyl]-3-butyl-1-triazenyl]-	**79:** P137166j
-4-carboxamide, 5-[3-[(4-butoxyphenyl)iminomethyl]-3-cyclohexyl-1-triazenyl]-	**79:** P137166j
-4-carboxamide, 5-[3-[(4-butoxyphenyl)iminomethyl]-3-phenyl-1-triazenyl]-	**79:** P137166j
-4-carboxamide, 5-[3-butyl-3-[(3-ethyl-4-methoxy-phenyl)iminomethyl]-1-triazenyl]-	**79:** P137166j
-4-carboxamide, 5-[3-butyl-3-(1-iminomethyl)-1-triazenyl]-	**79:** P137166j
-4-carboxamide, 5-(3-butyl-3-methyl-1-triazeno)-	**65:** 12196h
-4-carboxamide, 5-(3-butyl-1-triazeno)-	**65:** 12196h
-4-carboxamide, *N*,1-diacetyl-5-(*N*-phenylacetamido)-	**74:** 125581z
-5-carboxamide, 4-diazo-	**56:** 1445a, **65:** 12196h, **80:** 3436j
-4-carboxamide, 5-(3,3-dibutyl-1-triazeno)-	**63:** 18856g, **65:** 12196h
-4-carboxamide, 5-[[4-(dimethylamino)phenyl]azo]-	**65:** 12196h
-4-carboxamide, 5-(3,3-dimethyl-1-triazeno)-	**65:** 12196h
-4-carboxamide, 5-(3,3-dipropyl-1-triazeno)-	**63:** 18856h
-4-carboxamide, 5-hydroxy-,	**85:** P6010e
-4-carboxamide, 5-hydroxy-, monosodium salt	**85:** P6010e
-4-carboxamide, 5-[(2-hydroxy-1-naphthyl)azo]-	**65:** 12196h
-4-carboxamide, 5-hydroxy-1-β-D-ribofuranosyl-	**85:** P6010e
-4-carboxamide, 5-iodo-	**65:** 12196h
-4-carboxamide, 5-(3-isopropyl-3-methyl-1-triazeno)-	**65:** 12196h
-4-carboxamide, 5-(3-methyl-3-phenyl-1-triazeno)-	**65:** 12196h
-4-carboxamide, 5-(1-pyrrolidinylazo)-	**65:** 12196h

TABLE 10 (*Continued*)

Compound	Reference
10.2. Miscellaneous 1,2,3-Triazoles (*Continued*)	
-4-carboxamide, 5-[(trimethylsilyl)oxy]-	**85:** P6010e
-1-carboximidic acid, 4-(4-chlorophenoxy)-, 4-chlorophenyl ester	**64:** 8171e
-1-carboximidic acid, 4-(4-methoxyphenoxy)-, 4-methoxyphenyl ester	**64:** 8171e
-1-carboximidic acid, 4-(4-methylphenoxy)-, 4-methylphenyl ester	**64:** 8171e
-1-carboximidic acid, 4-phenoxy-, phenyl ester	**64:** 8171e
-4-carboxylic acid, 1-(2-benzoyl-4-chlorophenyl)-, methyl ester	**84:** 74239r
-4-carboxylic acid, 5-benzoyl-2-phenyl-	**21:** 2268[8]
-4-carboxylic acid, 1-benzyl-5-chloro-, ethyl ester	**51:** 2756e, **67:** 64305u
-4-carboxylic acid, 5-(chloroazo)-, hydrazide	**68:** 103631q
-4-carboxylic acid, 5-chloro-1-phenyl-	**53:** 18013d
-4-carboxylic acid, 5-chloro-1-phenyl-, methyl ester	**52:** 366a
-4-carboxylic acid, 1-(3-chlorophenyl)-5-hydroxy-, ethyl ester	**74:** P22847m
-4-carboxylic acid, 1-(4-chlorophenyl)-5-hydroxy-, ethyl ester	**74:** P22847m
-5-carboxylic acid, 4-[[2,6-dimethyl-4-(phenylazo)-phenyl]-1-[1-[[2,6-dimethyl-4-(phenylazo)phenyl]-oxy]formimidoyl]-, ethyl ester	**73:** 98913e
-1-carboxylic acid, 4-ethoxy, ethyl ester	**78:** 43374c
-2-carboxylic acid, 4-ethoxy, ethyl ester	**78:** 43374c
-1-carboxylic acid, 5-ethoxy-, ethyl ester	**72:** 121450s, **78:** 43374c
-1-carboxylic acid, 5-ethoxy-4-methyl-, ethyl ester	**70:** 77878v
-2-carboxylic acid, 4-ethoxy-5-methyl-, ethyl ester	**79:** 31992k
-2-carboxylic acid, 4-ethoxy-5-phenyl, ethyl ester	**78:** 111442a
-4-carboxylic acid, 5-(ethylsulfonyl)-	**82:** 58044b
-4-carboxylic acid, 1-(3-fluorophenyl)-5-hydroxy, ethyl ester	**74:** P22847m
-4-carboxylic acid, 1-(4-fluorophenyl)-5-hydroxy-, ethyl ester	**74:** P22847m
-4-carboxylic acid, 5-mercapto-1-(2-methoxyphenyl)-, ethyl ester, 1:1 adduct with piperidine	**70:** 77874r
-4-carboxylic acid, 5-mercapto-1-(2-methylphenyl)-, ethyl ester, 1:1 adduct with piperidine	**70:** 77874r
-4-carboxylic acid, 5-mercapto-1-(4-nitrophenyl)-, ethyl ester	**68:** 68936u
-4-carboxylic acid, 5-mercapto-1-(4-nitrophenyl)-, ethyl ester, 1:1 adduct with piperidine	**68:** 68936u
-4-carboxylic acid, 5-mercapto-1-phenyl-, ethyl ester	**70:** 77874r
-4-carboxylic acid, 5-mercapto-1-phenyl-, ethyl ester, 1:1 adduct with piperidine	**70:** 77874r
-1-carboxylic acid, 5-methoxy-, ethyl ester	**72:** 121450s
-4-carboxylic acid, 5-methoxy-1-(4-methoxy-5-oxo-1,3,6-cycloheptatrien-1-yl)-, ethyl ester	**79:** 5300p
-4-carboxylic acid, 5-methoxy-1-methyl-, ethyl ester	**64:** 11200d

TABLE 10 (*Continued*)

Compound	Reference
10.2. Miscellaneous 1,2,3-Triazoles (*Continued*)	
-4-carboxylic acid, 5-methoxy-1-methyl-, methyl ester	**64:** 11200c
-4-carboxylic acid, 5-methoxy-2-methyl-, ethyl ester	**64:** 11200d
-5-carboxylic acid, 4-methoxy-1-methyl-, ethyl ester	**64:** 11200d
-1-carboxylic acid, 5-methoxy-4-phenyl, ethyl ester	**78:** 29677x, **84:** 30609g
-5-carboxylic acid, 1-(2-methyl-1-oxo-2-propenyl)-, propyl ester	**85:** 47115m
-4-carboxylic acid, 1-[(4-methylphenyl)sulfonyl]-, methyl ester	**64:** 5076f
-5-carboxylic acid, 1-[(4-methylphenyl)sulfonyl]-, methyl ester	**64:** 5076f
-5(or 4)-carboxylic acid, 1-[(4-methylphenyl)-sulfonyl]-4(or 5)-phenyl-, ethyl ester	**64:** 5076g
-5-carboxylic acid, 1-(1-oxo-2-propenyl)-, propyl ester	**85:** 47115m
-4-carboxylic acid, 5-phenoxy-, ethyl ester	**64:** 8171e
-5-carboxylic acid, 4-phenoxy-1-(1-phenoxyform-imidoyl)-, ethyl ester	**64:** 8171d
4-[4-chloro-*N*-(diphenylphosphoranylidene)(phenyl-sulfonamido)]-1-[(4-methoxyphenyl)sulfonyl]-5-(methylphenylamino)-	**81:** 136244n
4-[4-chloro-*N*-(diphenylphosphoranylidene)phenyl-sulfonamido]-5-(methylphenylamino)-1-[(4-methylphenyl)sulfonyl]-	**81:** 136244n
4-[4-chloro-*N*-(diphenylphosphoranylidene)phenyl-sulfonamido]-5-(methylphenylamino)-1-[(2,4,6-trimethylphenyl)sulfonyl]-	**81:** 136244n
5-cyano-4-[[(4-nitrophenyl)sulfonyl]amino]-	**80:** 95838t
4-(diacetylamino)-2-(3,4-dimethylphenyl)-5-[(dimethylphenyl)azo]-	**74:** P65596w
4-(diacetylamino)-2-(3-pyridyl)-5-(3-pyridylazo)-	**74:** P65596w
N,1-diacetyl-4-[[(4-methylphenyl)sulfonyl]amino]-	**83:** 192266z
N,1-diacetyl-5-[[(4-methylphenyl)sulfonyl]amino]-	**83:** 192266z
N,2-diacetyl-4-[[(4-methylphenyl)sulfonyl]amino]-	**83:** 192266z
N,1-diacetyl-4-(phenylsulfonamido)-	**83:** 192266z
N,1-diacetyl-5-(phenylsulfonamido)-	**83:** 192266z
N,2-diacetyl-4-(phenylsulfonamido)-	**83:** 192266z
-4-diazonium, 5-(aminocarbonyl)-	**79:** P137166j
-4-diazonium, 5-benzoyl-, hydroxide	**84:** 135549r
-4-diazonium, 5-cyano-, chloride	**84:** 30969z
-4,5-dicarboxylic acid, 1-(2-benzoyl-4-chloro-phenyl)-, dimethyl ester	**84:** 74239r
-4,5-dicarboxylic acid, 1-hydroxy-	**70:** 19988v
-4,5-dicarboxylic acid, 1-[(4-methylphenyl)sulfonyl]-, dimethyl ester	**64:** 5076f
4-[(diethoxymethylidene)amino]-4-formyl-2-methyl-, diethyl ester	**80:** 37075f

TABLE 10 (*Continued*)

Compound	Reference
10.2. Miscellaneous 1,2,3-Triazoles (*Continued*)	
5-(diethylamino)-1-[(4-methoxyphenyl)sulfonyl]-4-methyl-	**73:** 77146s, **74:** 40817f, **74:** 114792x, **74:** 119863f, **80:** 37051v
5-(diethylamino)-4-methyl-1[(4-methoxyphenyl)sulfonyl]-	**73:** 77146s, **73:** 114792x, **73:** 119863f, **80:** 37051v
1-[[4-(dimethylamino)phenyl]sulfonyl]-4-[[N-(diphenylphosphoranylidene)-4-methylphenyl]-sulfonyl]-5-(methylphenylamino)-	**81:** 136244n
1-[[4-(dimethylamino)phenyl]sulfonyl]-5-ethoxy-4-ethyl-	**77:** 151634a
1-[[4-(dimethylamino)phenyl]sulfonyl]-5-ethoxy-4-methyl-	**77:** 151634a
1-[[4-(dimethylamino)phenyl]sulfonyl]-4-ethyl-5-methoxy-	**77:** 151634a
1-[[4-(dimethylamino)phenyl]sulfonyl]-5-methoxy-4-methyl-	**77:** 151634a
4-[[(2,2-dimethylhydrazino)methylidene]amino]-5-[(dimethylhydrazono)methyl]-	**79:** 137047w
5-[(dimethylhydrazono)methyl]-4-thioureido-	**79:** 137047w
5-[(dimethylhydrazono)methyl]-4-ureido-	**79:** 137047w
5-[(dimethylhydrazono)methyl]-4-ureido-, conjugate monoacid	**79:** 137047w
5-[(dimethylhydrazono)methyl]-4-ureido, ion (1-)	**79:** 137047w
5-(diphenylamino)-4-[N-(diphenylphosphoranylidene)-[(4-methylphenyl)sulfonyl]amino]-1-[(4-methylphenyl)sulfonyl]-	**81:** 136244n
4-[N-(diphenylphosphoranylidene) (4-methylphenyl) sulfonylamino]-1-[(4-methoxyphenyl)sulfonyl]-5-(methylphenylamino)-	**81:** 136244n
4-[N-(diphenylphosphoranylidene) (4-methylphenyl)-sulfonylamino]-5-(methylphenylamino)-1-[(4-methylphenyl)sulfonyl]-	**81:** 136244n
4-[N-(diphenylphosphoranylidene) (4-methylphenyl)-sulfonylamino]-5-(methylphenylamino)-1-[(2,4,6-trimethylphenyl)sulfonyl]-	**81:** 136244n
4-[(ethoxyethylidene)amino]-5-formyl-1-methyl-, ethyl ester	**80:** 37075f
4-[(ethoxyethylidene)amino]-5-formyl-2-methyl-, ethyl ester	**80:** 37075f
5-ethoxy-4-ethyl-1-[(2,4,6-trimethylphenyl)sulfonyl]-	**77:** 151634a
4-ethoxy-2-(4-methoxybenzoyl)-5-methyl-	**79:** 31992k
5-ethoxy-1-(4-methoxybenzoyl)-4-methyl-	**72:** 21176p
4-[(ethoxymethylidene)amino]-5-formyl-1-methyl-, ethyl ester	**80:** 37075f
4-ethoxy-5-methyl-2-(3-nitrobenzoyl)-	**79:** 31992k
4-ethoxy-5-methyl-2-(4-nitrobenzoyl)-	**79:** 31992k
5-ethoxy-4-methyl-1-(3-nitrobenzoyl)-	**72:** 21176p

TABLE 10 (*Continued*)

Compound	Reference
10.2. Miscellaneous 1,2,3-Triazoles (*Continued*)	
5-ethoxy-4-methyl-1-(4-nitrobenzoyl)-	**72:** 21176p
5-ethoxy-4-methyl-1-[(2,4,6-trimethylphenyl)sulfonyl]-	**77:** 151634a
4-ethoxy-2-(4-nitrobenzoyl)-5-phenyl-	**78:** 111442a
5-(hydrazonomethyl)-4-[(hydrazinomethylidene)amino]-	**79:** 137047w
4-[[(hydroxyamino)methylidene]amino]-5-[(hydroxyimino)methyl]-	**79:** 137047w
5-[(hydroxyimino)methyl]-4-thioureido-	**79:** 137047w
-4-methanamine, 5-amino-1-benzyl-, 3:1 adduct with carbonic acid	**79:** 146462e
4-[[(methoxyamino)methylidene]amino]-5-[(methoxyimino)methyl]-	**79:** 137047w
1-methoxy-5-(methylsulfonyl)-	**45:** 6999g
4-nitro-5-(2,4,6-trinitroanilino)-	**74:** 22769n
-4-ol, 2-benzoyl-, benzoate ester	**71:** 124340w
-4-ol, 2-benzoyl-5-phenyl-, benzoate ester	**71:** 124340w
-4-ol, 5-bromo-1-methyl-, benzoate ester	**71:** 61296p
-4-ol, 5-bromo-2-methyl-, benzoate ester	**77:** 19583d
-5-ol, 4-bromo-1-methyl-, benzoate ester	**77:** 19583d
-4-propanoic acid, α-diazo-5-hydroxy-1-(4-methoxyphenyl)-β-oxo-, ethyl ester, potassium salt	**71:** 12749e
-4-propanoic acid, α-diazo-5-hydroxy-β-oxo-1-phenyl-, ethyl ester, potassium salt	**71:** 12749e

REFERENCES

21: 2268[8]	A. Beretta, *Gazz. Chim. Ital.*, **57,** 173 (1927).
22: 423[7]	O. Dimroth and W. Michalis, *Justus Liebigs Ann. Chem.*, **459,** 39 (1927).
22: 3411[3]	A. Bertho and F. Holder, *J. Prakt. Chem.*, **119,** 189 (1928).
26: 1606[6–9]	A. Quilico, *Gazz. Chim. Ital.*, **61,** 759 (1931).
36: 835[7]	R. W. Cunningham, E. J. Fellows, and A. E. Livingston, *J. Pharmacol.*, **73,** 312 (1941).
44: 1102d	Ramart-Lucas, J. Hoch, and Grumez, *Bull. Soc. Chim. Fr.*, 447 (1949).
45: 6999g	H. J. Backer, *Rec. Trav. Chim., Pays-Bas*, **69,** 1223 (1950).
49: 1710g,h	B. R. Brown, D. L. Hammick, and S. G. Heritage, *J. Chem. Soc.*, 3820 (1953).
51: 2755–6e–e	J. R. E. Hoover and A. R. Day, *J. Amer. Chem. Soc.*, **78,** 5832 (1956).
51: 2811–2d–c	E. Richter and E. C. Taylor, *J. Amer. Chem. Soc.*, **78,** 5848 (1956).
51: 3688g	T. Noguchi, *Pharm. Bull. (Japan)*, **4,** 92 (1956).
51: 12075a	Y. Hirata, K. Iwashita, and K. Teshima, *Nagoya Sangyô Kagakû Kenkyûjho Kenkyû Hôkoku*, No. **9,** 83 (1957).
51: 14698c,d	S. Yamada, T. Mizoguchi, and A. Avata, *Yakugaku Zasshi*, **77,** 452 (1957).
52: 365–6i–c	E. Lieber, T. S. Chao, and C. N. R. Rao, *J. Org. Chem.*, **22,** 654 (1957).
52: 1997d–f	L. L. Bennett, Jr. and H. T. Baker, *J. Org. Chem.*, **22,** 707 (1957).
52: 16359f	J. Baddiley, J. G. Buchanan, and G. O. Osborne, *J. Chem. Soc.*, 1651 (1958).

53: 15065e	J. H. Boyer, C. H. Mack, N. Goebel, and L. R. Morgan, Jr., *J. Org. Chem.*, **23,** 1051 (1958).
53: 18013d	E. Lieber, T. S. Chao, and C. N. R. Rao, *Can. J. Chem.*, **37,** 118 (1959).
54: 4547f	R. Hüttel and J. Kratzer, *Chem. Ber.*, **92,** 2014 (1959).
55: 2661–2d–e	A. Dornow and J. Helberg, *Chem. Ber.*, **93,** 2001 (1960).
55: 11421–2i–a	M. A. Stevens, H. W. Smith, and G. Bosworth, *J. Amer. Chem. Soc.*, **82,** 3189 (1960).
55: 23505a	E. Profft and W. Georgi, *Justus Liebigs Ann. Chem.*, **643,** 136 (1961).
56: 1445a	Y. F. Shealy, R. F. Struck, L. B. Holum, and J. A. Montgomery, *J. Org. Chem.*, **26,** 2396 (1961).
56: 12870a	P. Bravo, G. Guadiano, A. Quilico, and A. Ricca, *Gazz. Chim. Ital.*, **91,** 47 (1961).
57: 5897d	S. Rich and J. G. Horsfall, *Conn. Agr. Expt. Sta., New Haven, Bull.*, No. **639,** 1 (1961).
63: 9943c	Y. F. Shealy and C. A. O'Dell, *J. Org. Chem.*, **30,** 2488 (1965).
63: P11749d–f	R. A. Carboni, U.S. Patent, 3,190,886 (1965).
63: 18856g	K. Hano, A. Akashi, I. Yamamoto, S. Narumi, Z. Horii, and I. Ninomiya, *Gann*, **56,** 417 (1965).
64: 5076f,g	R. Huisgen, R. Knorr, L. Möbius, and G. Szeimies, *Chem. Ber.*, **98,** 4014 (1965).
64: 11200c,d	M. Begtrup and C. Pedersen, *Acta Chem. Scand.*, **19,** 2022 (1965).
64: 8171d,e	D. Martin and A. Weise, *Chem. Ber.*, **99,** 317 (1966).
65: 713g	A. Albert and K. Tratt, *Chem. Commun.*, 243 (1966).
65: 10524d	R. Fuks, R. Buijle, and H. G. Viehe, *Angew. Chem.*, **78,** 594 (1966).
65: 12196h	Y. F. Shealy and C. A. O'Dell, *J. Med. Chem.*, **9,** 733 (1966).
67: 64305u	M. Begtrup and C. Pedersen, *Acta Chem. Scand.*, **21,** 633 (1967).
67: 90731z	A. Albert, *Chem. Commun.*, 684 (1967).
68: 68936u	M. Regitz and A. Liedhegener, *Justus Liebigs Ann. Chem.*, **710,** 118 (1967).
68: 68958c	A. Albert and K. Tratt, *J. Chem. Soc., C*, 344 (1968).
68: 103631q	D. A. Peters and P. L. McGeer, *Can. J. Physiol. Pharmacol.*, **46,** 195 (1968).
69: 19088f	T. Nozoe, H. Horino, and T. Toda, *Tetrahedron Lett.*, 5349 (1967).
69: 77238r	A. Albert, *J. Chem. Soc., C*, 2076 (1968).
69: 96637m	M. Ridi, G. Franchi, S. Mangiavacchi, and M. P. Lombardini, *Boll. Chim. Farm.*, **107,** 401 (1968).
70: 19988v	R. Huisgen and V. Weberndörfer, *Chem. Ber.*, **100,** 71 (1967).
70: 47403u	A. Albert, *J. Chem. Soc., C*, 152 (1969).
70: 77874r	M. Regitz and H. Scherer, *Chem. Ber.*, **102,** 417 (1969).
70: 77878v	G. L'abbé and H. J. Bestmann, *Tetrahedron Lett.*, 63 (1969).
70: 87690s	A. Albert, *Angew. Chem., Int. Ed. Engl.*, **8,** 132 (1969).
71: 11523q	H. Nishio, I. Yamamoto, K. Kariya, and K. Hano, *Chem. Pharm. Bull. (Tokyo)*, **17,** 539 (1969).
71: 12749e	M. Regitz and H. J. Geelhaar, *Chem. Ber.*, **102,** 1743 (1969).
71: 13066s	A. Albert, *J. Chem. Soc., D*, 500 (1969).
71: 49860k	H. El Khadem, M. A. E. Shaban, and M. A. M. Nassr. *J. Chem. Soc., C*, 1416 (1969).
71: 61296p	M. Begtrup, and C. Pedersen, *Acta Chem. Scand.*, **23,** 1091 (1969).
71: 112865h	E. Zbiral and J. Stroh, *Monatsh. Chem.*, **100,** 1438 (1969).
71: 124340w	M. Begtrup, *Acta Chem. Scand.*, **23,** 2025 (1969).
72: 21176p	G. L'abbé, P. Ykman, and G. Smets, *Tetrahedron*, **25,** 5421 (1969).
72: 21645r	A. Albert, *J. Chem. Soc., C*, 2379 (1969).
72: 31185e	H. J. Gais, K. Hafner, and M. Neuenschwander, *Helv. Chim. Acta*, **52,** 2641 (1969).

72: 55346j — J. Schawartz, M. Hornyák, and T. Süts, *Chem. Ind.* (*London*), **92,** (1970).

72: P67723t — M. Minagawa, K. Nakagawa, and M. Goto, German Patent, 1,926,547 (1970).

72: 78952t — A. Albert, *J. Chem. Soc., C,* 230 (1970).

72: 121447w — G. L'abbé and A. Hassner, *J. Heterocycl. Chem.*, **7,** 361 (1970).

72: 121450s — R. Scarpati, M. L. Graziano, and R. A. Nicolaus, *Gazz. Chim. Ital.*, **99,** 1339 (1969).

73: 77146s — M. Regitz and G. Himbert, *Tetrahedron Lett.*, 2823 (1970).

73: 77154t — A. Albert, *J. Chem. Soc., D,* 858 (1970).

73: 98913e — M. Hedayatullah, H. Iida, and L. Denivelle, *C.R. Acad. Sci., Paris, Ser. C,* **271,** 146 (1970).

73: P98955v — I. Pinter, A. Messmer, F. Ordogh, and L. Pallos, German Patent, 2,009,134 (1970).

73: 114792x — R. E. Harmon, F. Stanley, Jr., S. K. Gupta, R. A. Earl, J. Johnson, and G. Slomp, *Chem. Ind.* (*London*), 1021 (1970).

73: 119863f — R. E. Harmon, F. Stanley, Jr., S. K. Gupta, and J. Johnson, *J. Org. Chem.*, **35,** 3444 (1970).

74: 22769n — P. N. Neuman, *J. Heterocycl. Chem.*, **7,** 1159 (1970).

74: P22847m — I. Pinter, A. Messmer, F. Oerdogh, and L. Pallos, German Patent, 2,012,943 (1970).

74: 40817f — R. E. Harmon, S. K. Gupta, J. Johnson, F. Stanley, Jr., and L. J. Ladislav, *J. Pharm. Sci.*, **59,** 1694 (1970).

74: 53665e — M. Yoshida, A. Matsumoto, and O. Simamura, *Bull. Chem. Soc. Jap.*, **43,** 3587 (1970).

74: P65596w — L. A. Henderson, U.S. Patent, 3,541,107 (1970).

74: 125581z — D. R. Sutherland and G. Tennant, *J. Chem. Soc., C,* 706 (1971).

74: P125704s — A. F. Strobel and M. L. Whitehouse, German Patent, 2,041,845 (1971).

75: 5801v — N. J. Cusack, G. Shaw, and G. J. Litchfield, *J. Chem. Soc., C,* 1501 (1971).

75: 20332m — D. R. Sutherland and G. Tennant, *J. Chem. Soc., C,* 2156 (1971).

76: 126925t — A. Albert and D. Thacker, *J. Chem. Soc., Perkin Trans.*, **1,** 468 (1972).

76: 126928w — A. Albert and H. Taguchi, *J. Chem. Soc., Perkin Trans.*, **1,** 449 (1972).

76: 126929x — A. Albert, *J. Chem. Soc., Perkin Trans.*, **1,** 461 (1972).

76: 135794b — J. A. Montgomery and H. J. Thomas, *J. Med. Chem.*, **15,** 182 (1972).

76: 140731h — H. Wamhoff, *Chem. Ber.*, **105,** 743 (1972).

76: 141226j — R. L. Tolman, C. W. Smith, and R. K. Robins, *J. Amer. Chem. Soc.*, **94,** 2530 (1972).

76: 153691v — E. Gudrinece and V. V. Solov'eva, *Latv. PSR Zinat. Akad. Vestis, Kim. Ser.*, 81 (1972).

77: 19583d — M. Begtrup, *Acta Chem. Scand.*, **26,** 715 (1972).

77: 84962r — H. Iwata, I. Yamamoto, E. Gohda, K. Morita, and K. Nishino, *Biochem. Pharmacol.*, **21,** 2141 (1972).

77: 87530j — R. A. Franich, G. Lowe, and J. Parker, *J. Chem. Soc., Perkin Trans.*, **1,** 2034 (1972).

77: 113628u — D. R. Eaton, R. E. Benson, C. G. Bottomley, and A. D. Josey, *J. Amer. Chem. Soc.*, **94,** 5996 (1972).

77: 126981d — O. Makabe, S. Fukatsu, and S. Umezawa, *Bull. Chem. Soc. Jap.*, **45,** 2577 (1972).

77: 139556u — G. Himbert and M. Regitz, *Chem. Ber.*, **105,** 2963 (1972).

77: 147448a — C. W. Smith, R. W. Sidwell, R. K. Robins, and R. L. Tolman, *J. Med. Chem.*, **15,** 883 (1972).

77: 151634a — G. Himbert and M. Regitz, *Chem. Ber.*, **105,** 2975 (1972).

77: 160107h W. Hutzenlaub, R. L. Tolman, and R. K. Robins, *J. Med. Chem.*, **15,** 879 (1972).

78: 29677x R. Scarpati and M. L. Graziano, *J. Heterocycl. Chem.*, **9,** 1087 (1972).

78: 29681u C. A. Lovelette and L. Long, Jr., *J. Org. Chem.*, **37,** 4124 (1972).

78: 43374c P. Ykman, G. L'abbé, and G. Smets, *Chem. Ind. (London)*, 886 (1972).

78: 111442a P. Ykman, G. L'abbé, and G. Smets, *Tetrahedron*, **29,** 195 (1973).

79: 5299v D. R. Sutherland, G. Tennant, and R. J. S. Vevers, *J. Chem. Soc., Perkin Trans.*, **1,** 943 (1973).

79: 5300p H. Horino and T. Toda, *Bull. Chem. Soc. Jap.*, **46,** 1212 (1973).

79: 31992k G. L'abbé, *Verh. Kon. Acad. Wetensch., Lett. Schone Kunsten Belg., Kl. Wetensch.*, **34,** 49 pp (1972).

79: 137047w A. Albert and W. Pendergast, *J. Chem. Soc., Perkin Trans.*, **1,** 1625 (1973).

79: 137053v A. Albert and H. Taguchi, *J. Chem. Soc., Perkin Trans.*, **1,** 1629 (1973).

79: P137166j J. Schawarts, M. Hornyak, E. Majorszky, E. Kovacsovics, A. David, and G. Horvath, German Patent, 2,253,615 (1973).

79: 146462e A. Albert, *J. Chem. Soc., Perkin Trans.*, **1,** 1634 (1973).

79: P147417f H. J. Binte, T. Eulenberger, and H. Noack, East German Patent, 95,438 (1973).

80: 3436j Y. F. Shealy and C. A. O'Dell, *J. Heterocycl. Chem.*, **10,** 839 (1973).

80: 37051v G. Himbert and M. Regitz, *Justus Liebigs Ann. Chem.*, **9,** 1505 (1973).

80: 37075f A. Albert and H. Taguchi, *J. Chem. Soc., Perkin Trans.*, **1,** 2037 (1973).

80: 82890d A. Albert, *J. Chem. Soc., Perkin Trans.*, **1,** 2659 (1973).

80: 95838t R. Mertz, D. Van Assche, J. P. Fleury, and M. Regitz, *Bull. Soc. Chim. Fr.*, 3442 (1973).

81: 37518y A. Karlina, E. Gudriniece, and L. K. Bronfman, *Latv. PSR Zinat. Akad. Vestis, Kim. Ser.*, 202 (1974).

81: 136244n G. Himbert and M. Regitz, *Chem. Ber.*, **107,** 2513 (1974).

82: 43350z A. Albert, *J. Chem. Soc., Perkin Trans.*, **1,** 2030 (1974).

82: 58044b A. Farrington and L. Hough, *Carbohydr. Res.*, **38,** 107 (1974).

82: 118340g A. Kálmán, K. Simon, J. Schawartz, and G. Horváth, *J. Chem. Soc., Perkin Trans.*, **2,** 1849 (1974).

82: 170802y A. Albert, *J. Chem. Soc., Perkin Trans.*, **1,** 345 (1975).

83: 192266z S. Auricchio, G. Vidari, and P. Vita-Finzi, *Gazz. Chim. Ital.*, **105,** 583 (1975).

84: 30609g M. L. Graziano, M. R. Iesce, R. Palombi, and R. Scarpati, *Ann. Chim. (Rome)*, **64,** 843 (1974).

84: 30969z G. Keller, J. P. Fleury, W. Anschütz, and M. Regitz, *Bull. Soc. Chim. Fr.*, 1219 (1975).

84: 31007w T. L. Gilchrist, G. E. Gymer, and C. W. Rees, *J. Chem. Soc., Perkin Trans.*, **1,** 1747 (1975).

84: 74239r D. L. Coffen, R. I. Fryer, D. A. Katonak, and F. Wong, *J. Org. Chem.*, **40,** 894 (1975).

84: 105706x G. Himbert, D. Frank, and M. Regitz, *Chem. Ber.*, **109**, 370 (1976).

84: 135549r D. Stadler, W. Anschütz, M. Regitz, G. Keller, D. Van Assche, and J. P. Fleury, *Justus Liebigs Ann. Chem.*, 2159 (1975).

84: P136010v R. L. Tolman and C. W. Smith, U.S. Patent, 3,826,803 (1974).

85: P6010e J. T. Witkowski, R. K. Robins, and F. A. Lehmkuhl, U.S. Patent, 3,948,885 (1976).

85: 47115m K. Khuseinov, G. V. Shatalov, and B. I. Mikhantev, *Izv. Vyssh. Uchebn. Zaved. Khim. Khim. Tekhnol.*, **19,** 655 (1976).

85: 102705f A. Kálmán, L. Párkányi, J. Schawartz, and K. Simon, *Acta Crystallogr., Sect. B*, **B32,** 2245 (1976).

CHAPTER 11

Alkyl- or Aryl-Δ^2-(or Δ^4)-1,2,3-Triazolines

The most frequently used method for the preparation of the compounds that are the subject of this chapter is the addition of diazomethane to an anil. Early reports[1,2] did not show great promise, and even fair yields were exceptional (Eq. 1). More recently, great improvement has been achieved with both diaryl and certain alkyl anils (Eqs. 2,3).[3–5] A related report showed that **11.1-1** is produced in excellent yield with various heterocyclic substituents.[6] Anils of aryloxyanilines are useful in certain cases (Eq. 4).[7]

$$PhCH{=}N{-}C_6H_4{-}NO_2 + CH_2N_2 \xrightarrow{O(CH_2CH_2)_2O} \text{1-(4-nitrophenyl)-5-phenyl-}\Delta^2\text{-1,2,3-triazoline} \quad (1)$$

64% (21 hrs.)
75% (42 hrs.)

$$Ar^1CH{=}NAr^2 + CH_2N_2 \xrightarrow[H_2O]{O(CH_2CH_2)_2O} \textbf{11.1-1} \quad (2)$$

11.1-1

Ar^1	= Ph	3-NO_2Ph	4-NO_2Ph	2,4-Cl_2Ph	2,4-Cl_2Ph	2-NO_2Ph	2-ClPh
Ar^2	= Ph	Ph	Ph	Ph	3-ClPh	3,4-Cl_2Ph	3,4-Cl_2Ph
% **11.1-1**	= 75	66	70	64	92	80	98

$$(CF_3)_2C{=}N{-}CH{=}CR^1R^2 + CH_2N_2 \longrightarrow \textbf{11.1-2} \quad (3)$$

11.1-2

R^1	= Me	Ph	Me	*i*-Pr
R^2	= Me	H	H	H
% **11.1-2**	= 91	88	30*	42

* *cis-trans* mixture

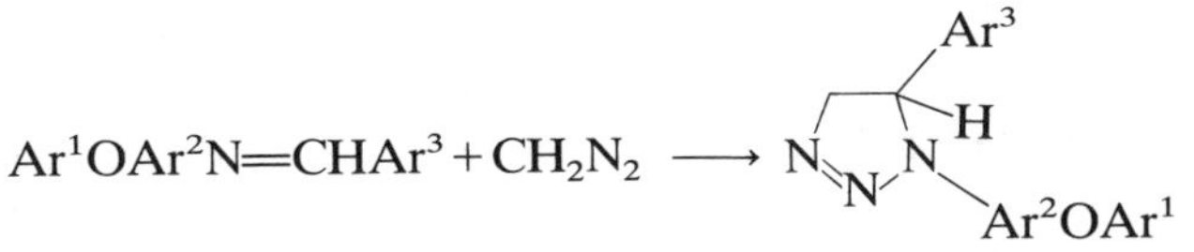

$$Ar^1OAr^2N{=}CHAr^3 + CH_2N_2 \longrightarrow \textbf{11.1-3} \quad (4)$$

Ar^1	= Ph	Ph	4-MePh	2-ClPh	2,4-Cl_2Ph
Ar^2	= Ph	Ph	Ph	3-ClPh	3-ClPh
Ar^3	= Ph	4-NO_2Ph	2,4-Cl_2Ph	4-NO_2Ph	Ph
% **11.1-3**	= 89	79	88	39	34

A second method leading to alkyl- or aryl-Δ^2-1,2,3-triazolines is the addition of azides to alkenes (Eq. 5).[8,9] These examples are somewhat exceptional and a number of other combinations produced much lower yields.[8] Scheiner has shown the stereospecificity of this reaction (Eqs. 6,7).[10]

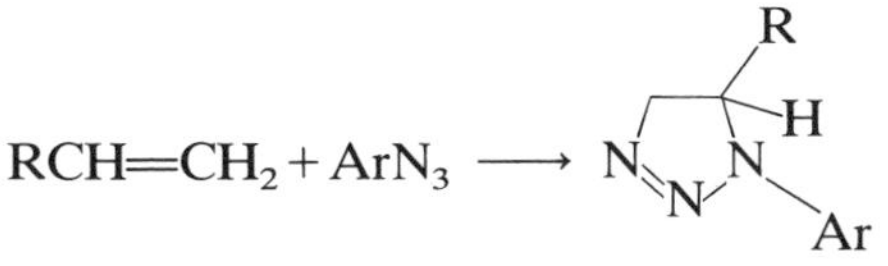

$$RCH{=}CH_2 + ArN_3 \longrightarrow \textbf{11.1-4} \quad (5)$$

R	= *n*-Bu	*n*-Bu	*n*-Bu	*n*-C_6H_{13}	2-pyridyl[9]
Ar	= 4-BrPh	4-ClPh	4-NO_2Ph	4-BrPh	4-BrPh
% **11.1-4**	= 43	89	66	79	88

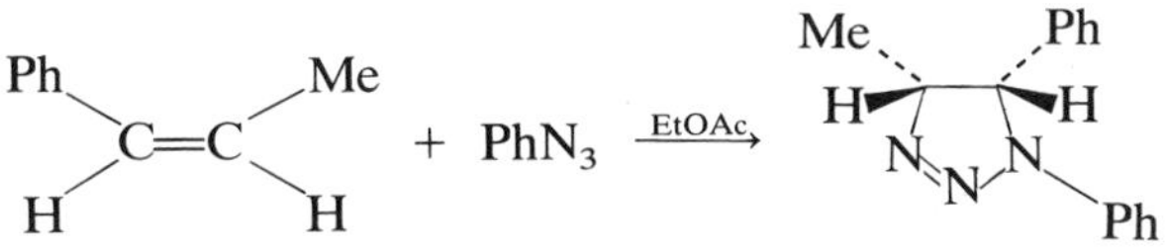

$$PhCH{=}CHMe + PhN_3 \xrightarrow{EtOAc} \text{triazoline} \quad (6)$$

22%

$$PhCH{=}CHMe + PhN_3 \xrightarrow{EtOAc} \text{triazoline} \quad (7)$$

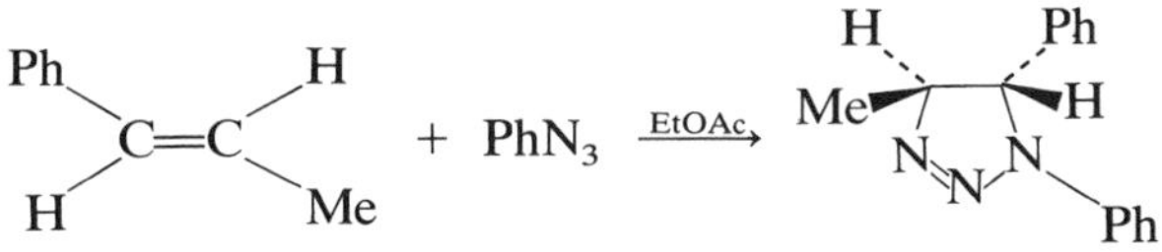

24%

An allene has been shown to react with azides (especially those bearing electron-withdrawing substituents) to produce 1,2,3-triazolines (Eq. 8).[11]

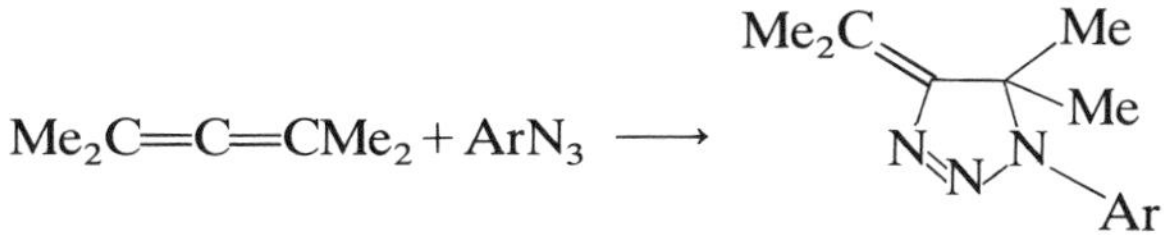

$$Me_2C{=}C{=}CMe_2 + ArN_3 \longrightarrow \quad (8)$$

Ar	%
2,4,6-$(NO_2)_3$Ph	72%
Ph	29%

The synthesis of monosubstituted Δ^2-1,2,3-triazolines from 1-arylazo-aziridines (**11.1-5**) has been reported to give excellent yields (Eq. 9).[12] The starting materials (**11.1-5**) are often unstable and were isomerized without purification. A more recent report provides a promising alternative and extension for the synthesis of these compounds from dimethyloxosulfonium methylide (**11.1-7**) (Eq. 10).[13] The use of 4-nitrophenyl- or benzoyl azide produced only triazenes.[13]

XPhN=NN(aziridine) **11.1-5** $\xrightarrow[Me_2C{=}O]{I^-}$ 1-(PhX)-Δ^2-1,2,3-triazoline **11.1-6** (9)

X	=4-Br	3-NO_2	4-NO_2	3,4-Cl_2	4-Cl	4-Me	3-Cl	3-NO_2-4-Me	Ph	4-I
% **11.1-6** =	84	64	98	83	68	76	60	90	80	85

$RN_3 + 2CH_2^-S^+OMe_2$ (**11.1-7**) $\xrightarrow{Me_2SO}$ 1-R-Δ^2-1,2,3-triazoline **11.1-8** (10)

R	=Et	*n*-C_7H_{15}	$PhCH_2$	Ph	4-EtOPh	4-BrPh
% **11.1-8** =	64	91	69	80	70	75

Finally, a related Δ^3-triazoline has been obtained in high yield (Eq. 11).[14]

1-Me-2-Ph-1,2,3-triazolium ($^+$N) FSO_3^- $\xrightarrow[MeOH]{NaBH_4}$ 1-Me-2-Ph-Δ^3-1,2,3-triazoline, 92% (11)

REFERENCES

1. **49:** 6241f
2. **56:** 1445g
3. **65:** 18576c
4. **67:** 43754z
5. **77:** 101465a
6. **83:** 9921x
7. **85:** 5558c
8. **68:** 29653e
9. **68:** 95600n
10. **68:** 113790e
11. **69:** 36040b
12. **57:** 4666a
13. **68:** 87248n
14. **78:** 159527j

TABLE 11. ALKYL- OR ARYL-Δ^2 (OR Δ^4)-1,2,3-TRIAZOLINES

Compound	Reference
-1-acetic acid, 5,5-bs(trifluoromethyl)-α-isopropyl-	**74:** 31701m
-1-acetic acid, 5,5-bis(trifluoromethyl)-α-isopropyl-, methyl ester	**74:** 31701m
1-benzyl-	**68:** 87248n
1,5-bis-(4-chlorophenyl)-	**65:** 18576c
1,5-bis-(3-nitrophenyl)-	**65:** 18576c, **81:** 100524q
1,5-bis-(4-nitrophenyl)-	**49:** 6241f, **65:** 18576c, **81:** 100524q
4,5-bis(trifluoromethyl)-	**67:** 100073c
5,5-bis(trifluoromethyl)-1-(4,5-dihydro-4-isopropyl-3*H*-pyrazol-3-yl)-	**77:** 101465a
5,5-bis(trifluoromethyl)-1-(4,5-dihydro-4-methyl-3*H*-pyrazol-3-yl)	**76:** 152837k, **77:** 101465a
5,5-bis(trifluoromethyl)-1-(4,5-dihydro-4-methyl-3*H*-pyrazol-3-yl)-, *trans*-	**78:** 76899f
5,5-bis(trifluoromethyl)-1-(3-methyl-1-butenyl)-, (E)-	**77:** 101465a
5,5-bis(trifluoromethyl)-1-(2-methyl-1-propenyl)-	**77:** 101465a
5,5-bis(trifluoromethyl)-1-(2-phenylethenyl)-	**77:** 101465a
5,5-bis(trifluoromethyl)-1-(1-propenyl)-	**77:** 101465a
1-(4-bromophenyl)-	**57:** 4666a, **68:** 87248n
1-(4-bromophenyl)-5-butyl-	**68:** 29653e
1-(4-bromophenyl)-5-(2-chlorophenyl)-	**65:** 18576c, **81:** 10524q
1-(4-bromophenyl)-5-(2,4-dichlorophenyl)-	**67:** 43754z, **81:** 100524q
1-(4-bromophenyl)-4,5-diethyl-	**68:** 29653e
1-(4-bromophenyl)-5-hexyl-	**68:** 29653e
1-(4-bromophenyl)-5-isopropenyl-	**68:** 29653e
1-(4-bromophenyl)-5-(3-nitrophenyl)-	**65:** 18576c, **81:** 100524q
1-(4-bromophenyl)-5-(4-nitrophenyl)-	**65:** 18576c, **81:** 100524q
1-(4-bromophenyl)-5-phenyl-	**65:** 18576c
1-(4-bromophenyl)-5-propenyl-, (E)-	**68:** 29653e
1-(4-bromophenyl)-5-(2-pyridyl)-	**68:** 95600m
1-(4-bromophenyl)-4,5,5-trimethyl-	**68:** 29653e
5-butyl-1-(4-chlorophenyl)-	**68:** 29653e
5-butyl-1-(4-nitrophenyl)-	**68:** 29653e
1-[3-chloro-4-(2-chlorophenoxy)phenyl]-5-(4-nitrophenyl)-	**85:** P5558c
1-[3-chloro-4-(2,4-dichlorophenoxy)phenyl]-5-phenyl-	**85:** P5558c
1-(3-chloro-4-phenoxyphenyl)-5-(3,5-dichloro-2-methoxyphenyl)-	**85:** P5558c
1-(3-chloro-4-phenoxyphenyl)-5-(4-nitrophenyl)-	**85:** P5558c
1-(3-chlorophenyl)-	**57:** 4666b
1-(4-chlorophenyl)-	**57:** 4666b
1-(3-chlorophenyl)-5-(2,4-dichlorophenyl)-	**67:** 43754z, **81:** 100524q
5-(2-chlorophenyl)-1-(3,4-dichlorophenyl)-	**67:** 43754z, **81:** 100524q
5-(4-chlorophenyl)-1-[4-(*N,N*-diethylsulfamoyl)-phenyl]-	**74:** 76377b
5-(2-chlorophenyl)-1-[4-(*N,N*-dimethylsulfamoyl)-phenyl]-	**74:** 76377b
5-(4-chlorophenyl)-1-[4-(*N,N*-dimethylsulfamoyl)-phenyl]-	**74:** 76377b

TABLE 11 (*Continued*)

Compound	Reference
5-(4-chlorophenyl)-1-[4-(ethoxycarbonyl)phenyl]-, ethyl ester	**74:** 76377b
1-(3-chlorophenyl)-5-(3-nitrophenyl)-	**65:** 18576c
1-(3-chlorophenyl)-5-(4-nitrophenyl)-	**65:** 18576c, **81:** 100524q
1-(4-chlorophenyl)-5-(3-nitrophenyl)-	**65:** 18576c, **81:** 100524q
5-(2-chlorophenyl)-1-(3-nitrophenyl)-	**65:** 18576c, **81:** 100524q
5-(4-chlorophenyl)-1-(3-nitrophenyl)-	**65:** 18576c, **81:** 100524q
1-(3-chlorophenyl)-5-phenyl-	**65:** 18576c
1-(4-chlorophenyl)-5-phenyl-	**56:** 1445h, **65:** 18576c
5-(2-chlorophenyl)-1-phenyl-	**65:** 18576c, **81:** 100524q
5-(4-chlorophenyl)-1-phenyl-	**49:** 6241h, **56:** 1446a, **65:** 18576c, **81:** 100524q
5-(4-chlorophenyl)-1-[4-(1-piperidinylsulfamoyl)-phenyl]-	**74:** 76377b
1-(4-chlorophenyl)-5-(2-pyridyl)-	**83:** 9921x
1-(4-chlorophenyl)-5-(4-pyridyl)-	**83:** 9921x
1-(4-chlorophenyl)-5-(2-quinolyl)-	**83:** 9921x
1-(4-chlorophenyl)-5-(2-thieneyl)-	**83:** 9921x
5-cyclopropyl-5-methyl-1-(4-nitrophenyl)-	**82:** 125325j
1-[4-(2,4-dichlorophenoxy)phenyl]-5-(2,4-dichloro-phenyl)-	**85:** P5558c
1-(3,4-dichlorophenyl)-	**57:** 4666a **67:** 43754a, **73:** 97689z
5-(2,4-dichlorophenyl)-1-(3,4-dichlorophenyl)-	**81:** 100524q
5-(2,4-dichlorophenyl)-1-[4-(ethoxycarbonyl)phenyl]-	**74:** 76377b
5-(2,4-dichlorophenyl)-1-[4-(4-methylphenoxy)phenyl)-	**85:** P5558c
5-(3,4-dichlorophenyl)-1-[4-(4-methylphenoxy)phenyl)-	**85:** P5558c
1-(3,4-dichlorophenyl)-5-(2-methylphenyl)-	**67:** 43754z
5-(2,4-dichlorophenyl)-1-(4-methylphenyl)-	**67:** 43754z
1-(3,4-dichlorophenyl)-5-(2-nitrophenyl)-	**67:** 43754z, **81:** 100524q
5-(2,4-dichlorophenyl)-1-(3-nitrophenyl)-	**67:** 43754z
5-(2,4-dichlorophenyl)-1-phenyl-	**67:** 43754z
5,5-dicyclopropyl-1-(4-nitrophenyl)-	**82:** 125325j
4,5-diethyl-1-(4-nitrophenyl)-	**68:** 29653e
5-[4-(dimethylamino)phenyl]-4-nitro-1-phenyl-	**66:** 10887k
5,5-dimethyl-4-isopropylidene-1-(4-nitrophenyl)-	**69:** 36040b
5,5-dimethyl-4-isopropylidene-1-phenyl-	**69:** 36040b
5,5-dimethyl-4-isopropylidene-1-(2,4,6-tri-nitrophenyl)-	**69:** 36040b
4,5-dimethyl-1-(4-nitrophenyl)-	**68:** 29653e
1,5-diphenyl-	**49:** 6241g, **56:** 1445g, **65:** 18576c, **82:** 16738a
1,5-diphenyl-4-methyl-, *cis*-	**68:** 113790e
1,5-diphenyl-4-methyl-, *trans*-	**68:** 113790e
-1-ethanesulfonic acid, 3-(3-acetamido-4-methoxyphenyl)-	**55:** P10473a
-1-ethanesulfonic acid, 3-(4-acetamido-3-methoxyphenyl)-	**55:** P10473a
-1-ethanesulfonic acid, 3-(4-acetamidophenyl)-	**55:** P10473a
-1-ethanesulfonic acid, 3-(4-biphenylyl)-	**55:** P10472i

TABLE 11 (*Continued*)

Compound	Reference
-1-ethanesulfonic acid, 3-(3-bromophenyl)-	**55:** P10472h
-1-ethanesulfonic acid, 3-(5-carbamoyl-2-methoxyphenyl)-	**55:** P10472i
-1-ethanesulfonic acid, 3-(3-chloro-1*H*-indazol-6-yl)-	**55:** P10472i
-1-ethanesulfonic acid, 3-(5-chloro-2-methoxyphenyl)-	**55:** P10472i
-1-ethanesulfonic acid, 3-(3-chloro-5-phenoxyphenyl)-	**55:** P10472i
-1-ethanesulfonic acid, 3-(2-chlorophenyl)-	**55:** P10472h
-1-ethanesulfonic acid, 3-(3-chlorophenyl)-	**55:** P10472h
-1-ethanesulfonic acid, 3-(4-chlorophenyl)-	**55:** P10472h
-1-ethanesulfonic acid, 3-(6-chloro-2-methylphenyl)-	**55:** P10472i
-1-ethanesulfonic acid, 3-[6-chloro-3-(trifluoromethyl)phenyl]-	**55:** P10472i
-1-ethanesulfonic acid, 3-(2,5-dichlorophenyl)-	**55:** P10472i
-1-ethanesulfonic acid, 3-(2,5-diethoxyphenyl)-	**55:** P10472i
-1-ethanesulfonic acid, 3-[5-(diethylsulfamoyl)-2-methoxyphenyl]-	**55:** P10472i
-1-ethanesulfonic acid, 3-(2,4-dimethoxyphenyl)-	**55:** P10472i
-1-ethanesulfonic acid, 3-(2,4-dimethylphenyl)-	**55:** P10472i
-1-ethanesulfonic acid, 3-(4-ethylphenyl)-	**55:** P10472h
-1-ethanesulfonic acid, 3-[5-(ethylsulfonyl)-2-methoxyphenyl]-	**55:** P10472i
-1-ethanesulfonic acid, 3-(2,4,6-trimethylphenyl)-	**55:** P10472i
-1-ethanesulfonic acid, 3-(3-methoxy-2-dibenzofuranyl)-	**55:** P10472i
-1-ethanesulfonic acid, 3-(2-methoxyphenyl)-	**55:** P10472i
-1-ethanesulfonic acid, 3-(4-methoxyphenyl)-	**55:** P10472i
-1-ethanesulfonic acid, 3-(5-methoxy-2-methyphenyl)-	**55:** P10472i
-1-ethanesulfonic acid, 3-(1-naphthyl)-	**55:** P10473a
-1-ethanesulfonic acid, 3-(3-nitrophenyl)-	**55:** P10472h
-1-ethanesulfonic acid, 3-(2-methyl-5-nitrophenyl)-	**55:** P10472i
-1-ethanesulfonic acid, 3-(3-sulfophenyl)-	**55:** P10472h
-1-ethanesulfonic acid, 3-(2-methylphenyl)-	**55:** P10472h
-1-ethanesulfonic acid, 3-(3-methylphenyl)-	**55:** P10472h
-1-ethanesulfonic acid, 3-(4-methylphenyl)-	**55:** P10472h
-1-ethanol, 3-(4-chlorophenyl)-	**55:** P10473a
-1-ethanol, 3-(2-chlorophenyl)-α-ethyl-	**56:** P3484h
-1-ethanol, 3-(2-chlorophenyl)-α-methyl-	**56:** P3484h
-1-ethanol, 3-(2,4-dichlorophenyl)-	**55:** P10473a
-1-ethanol, 3-(2,5-dichlorophenyl)-α,β-dimethyl-	**56:** P3484h
-1-ethanol, 3-(2,5-dichlorophenyl)-α,4-dimethyl-	**56:** P3484h
-1-ethanol, 3-(2,5-dichlorophenyl)-α-ethyl-	**56:** P3484g
-1-ethanol, 3-(2,5-dichlorophenyl)-α-methyl-	**56:** P3484h
-1-ethanol, α,4-dimethyl-3-(2,4-dimethylphenyl)-	**56:** P3484h
-1-ethanol, α,β-dimethyl-3-(2-methoxy-3-dibenzofuranyl)-	**56:** P3484i
-1-ethanol, α,β-dimethyl-3-phenyl-	**56:** P3484g
-1-ethanol, α,4-dimethyl-3-phenyl-	**56:** P3484i
-1-ethanol, α-ethyl-3-(2,4-dimethylphenyl)-	**56:** P3484h
-1-ethanol, α-ethyl-3-phenyl-	**56:** P3484h
-1-ethanol, 3-(2-methoxy-3-dibenzofuranyl)-α-methyl-	**56:** P3484i

TABLE 11 (*Continued*)

Compound	Reference
-1-ethanol, 3-(4-methoxyphenyl)-	**55:** P10473a
-1-ethanol, 3-(2-methoxyphenyl)-α-methyl-	**56:** P3484h
-1-ethanol, α-methyl-3-(2,4-dimethylphenyl)-	**56:** P3484h
-1-ethanol, α-methyl-3-phenyl-	**56:** P3484g
-1-ethanol, 3-(1-naphthyl)-	**55:** P10473a
-1-ethanol, 3-phenyl-	**55:** P10472h
1-[4-(ethoxycarbonyl)phenyl]-5-(2-furyl)-, ethyl ester	**74:** 76377b
1-[4-(ethoxycarbonyl)phenyl]-5-phenyl-, ethyl ester	**74:** 76377b
1-[4-(ethoxycarbonyl)phenyl]-5-(4-pyridyl)-, ethyl ester	**83:** 9921x
1-(4-ethoxyphenyl)-	**68:** 87248n
1-ethyl-	**68:** 87248n
2-ethyl-	**78:** 42620t
1-heptyl-	**68:** 87248n
1-(4-iodophenyl)-	**57:** 4666b
5-isopropenyl-1-(4-nitrophenyl)-	**68:** 29653e
5-isopropenyl-1-phenyl-	**27:** 2140[1]
1-(4-methoxyphenyl)-5-(2-nitrophenyl)-	**56:** 1445h
1-(4-methoxyphenyl)-5-(3-nitrophenyl)-	**65:** 18576c
5-(4-methoxyphenyl)-1-(4-nitrophenyl)-	**56:** 1445h
1-(4-methoxyphenyl)-5-(4-pyridyl)-	**83:** 9921x
4-methyl-1,5-diphenyl-	**66:** 15458s
4-methyl-1-(4-nitrophenyl)-	**57:** 4666c
1-(4-methyl-3-nitrophenyl)-	**57:** 4666b
1-(2-methyl-4-nitrophenyl)-5-phenyl-	**67:** 43754z
1-methyl-2-phenyl-	**78:** 159527j
1-(4-methylphenyl)-	**57:** 4666b
1-(3-methylphenyl)-5-phenyl-	**65:** 18576c
1-(4-methylphenyl)-5-phenyl-	**56:** 1446a, **65:** 18576c
5-(2-methylphenyl)-1-phenyl-	**65:** 18576c
1-(4-methylphenyl)-5-(4-pyridyl)-	**83:** 9921x
1-(3-nitrophenyl)-	**57:** 4666a
1-(4-nitrophenyl)-	**57:** 4666a
1-(3-nitrophenyl)-5-(2-nitrophenyl)-	**65:** 18576c, **81:** 100524q
1-(3-nitrophenyl)-5-(4-nitrophenyl)-	**65:** 18576c, **81:** 100524q
1-(2-nitrophenyl)-5-phenyl-	**67:** 43754z
1-(3-nitrophenyl)-5-phenyl-	**65:** 18576c, **81:** 100524q
1-(4-nitrophenyl)-5-phenyl-	**56:** 1445h, **65:** 18576c, **81:** 100524q
5-(2-nitrophenyl)-1-phenyl-	**56:** 1445h
5-(3-nitrophenyl)-1-phenyl-	**65:** 18576c, **81:** 100524q
5-(4-nitrophenyl)-1-phenyl-	**56:** 1445h, **65:** 18576c, **82:** 16738a
1-(4-nitrophenyl)-5-propenyl-, (E)-	**68:** 29653e
1-(3-nitrophenyl)-5-(2-pyridyl)-	**83:** 9921x
1-(3-nitrophenyl)-5-(4-pyridyl)-	**83:** 9921x
1-(3-nitrophenyl)-5-(2-quinolyl)-	**83:** 9921x
1-(4-nitrophenyl)-4,5,5-trimethyl-	**68:** 29653e

TABLE 11 (*Continued*)

Compound	Reference
1-(4-phenoxyphenyl)-5-phenyl-	**85:** P5558c
1-(4-phenoxyphenyl)-5-(trichloromethyl)-	**85:** P5558c
1-phenyl-	**50:** 3414f, **57:** 4666b, **68:** 87248n
1-phenyl-4-phosphoric acid, diethyl ester	**85:** 177551m
1-phenyl-4-phosphoric acid, dimethyl ester	**85:** 177551m
1-phenyl-5-(4-pyridyl)-	**83:** 9921x
1-[4-(1-piperidinylsulfamoyl)phenyl]-5-(2-pyridyl)-	**74:** 76377b
1-(3-pyridyl)-5-(2-quinolyl)-	**83:** 9921x

REFERENCES

27: 2140[1]	K. Alder, G. Stein and W. Friedrichsen, *Justus Liebigs Ann. Chem.*, **501,** 1 (1933).
49: 6241f–h	G. D. Buckley, *J. Chem. Soc.*, 1850 (1954).
50: 3414f	C. S. Rondestvedt, Jr. and P. K. Chang, *J. Amer. Chem. Soc.*, **77,** 6532 (1955).
55: P10472–3h–a	R. Mohr and H. Hertel, German Patent, 1,061,789 (1959).
56: 1445–6g–a	P. K. Kadaba and J. O. Edwards, *J. Org. Chem.*, **26,** 2331 (1961).
56: P3484g–i	R. Mohr and H. Hertel, German Patent, 1,075,623 (1960).
57: 4666a–c	H. W. Heine and D. A. Tomalia, *J. Amer. Chem. Soc.*, **84,** 993 (1962).
65: 18576c	P. K. Kadaba, *Tetrahedron*, **22,** 2453 (1966).
66: 10887k	G. Rembarz, B. Kirchhoff, and G. Dongowski, *J. Prakt. Chem.*, **33,** 199 (1966).
66: 15458s	P. Scheiner, *J. Amer. Chem. Soc.*, **88,** 4759 (1966).
67: 43754z	P. K. Kadaba and N. F. Fannin, *J. Heterocycl. Chem.*, **4,** 301 (1967).
67: 100073c	I. L. Knunyants and Yu. V. Zeifman, *Izv. Akad. Nauk SSSR, Ser. Khim.*, 711 (1967).
68: 29653e	P. Scheiner, *Tetrahedron*, **24,** 349 (1968).
68: 87248n	G. Gaudiano, C. Ticozzi, A. Umani-Ronchi, and P. Bravo, *Gazz. Chim. Ital.*, **97,** 1411 (1967).
68: 95600m	P. Scheiner, *Tetrahedron*, **24,** 2757 (1968).
68: 113790e	P. Scheiner, *J. Amer. Chem. Soc.*, **90,** 988 (1968).
69: 36040b	R. F. Bleiholder and H. Shechter, *J. Amer. Chem. Soc.*, **90,** 2131 (1968).
73: 97689z	P. K. Kadaba, *J. Pharm. Sci.*, **59,** 1190 (1970).
74: 31701m	K. Burger and E. Burgis, *Justus Liebigs Ann. Chem.*, **741,** 39 (1970).
74: 76377b	J. M. Stewart, R. L. Clark, and P. E. Pike, *J. Chem. Eng. Data*, **16,** 98 (1971).
76: 152837k	A. Gieren, K. Burger, and J. Fehn, *Angew. Chem., Int. Ed. Engl.*, **11,** 223 (1972).
77: 101465a	K. Burger, J. Fehn, and A. Gieren, *Justus Liebigs Ann. Chem.*, **757,** 9 (1972).
78: 42620t	C. Pigenet and H. Lumbroso, *Bull. Soc. Chim. Fr.*, 3743 (1972).
78: 76899f	A. Gieren, *Chem. Ber.*, **106,** 288 (1973).

78: 159527j	T. Isida, T. Akiyama, N. Mihara, S. Kozima, and K. Sisido, *Bull. Chem. Soc. Jap.*, **46,** 1250 (1973).
81: 100524q	P. K. Kadaba, *Pestic. Sci.*, **5,** 255 (1974).
82: 16738a	H. deSuray, G. Leroy, and J. Weiler, *Tetrahedron Lett.*, 2209 (1974).
82: 125325j	O. Gerlach, P. L. Reiter, and F. Effenberger, *Justus Liebigs Ann. Chem.*, 1895 (1974).
83: 9921x	P. K. Kadaba, *J. Heterocycl. Chem.*, **12,** 143 (1975).
85: 5558c	H. Dehne and M. Süsse, *Z. Chem.*, **16,** 102 (1976).

CHAPTER 12

Miscellaneous Δ^2-1,2,3-Triazolines

Most of the 1,2,3-triazolines bearing a functional group are amines (**12.1-1**) or ethers (**12.1-2**). These two types are most often prepared by azide addition to enamines (Eq. 1) or enol ethers (Eq. 2). Such reactions have been shown to provide excellent yields under a variety of circumstances by Huisgen and his collaborators.[1] They have also demonstrated the conversion of **12.1-1** and **12.1-2** to the corresponding 1,2,3-triazoles in high yield.[1] In a subsequent paper[2] this research group reported on the stereochemistry of the enol ether addition (Eqs. 3,4).

Ph(morpholino)C=CH$_2$ + ArN$_3$ ⟶ **12.1-1** (1)

Ar	= 4-NO_2Ph	4-MeOPh
% **12.1-1** =	99	55

R′O(R)C=CH$_2$ + 4-$N_3C_6H_4NO_2$ ⟶ **12.1-2** (2)

R	= Me	Ph	H
R′	= Et	Me	*n*-Bu
% **12.1-2** =	99	93	96

Pocar and his extremely productive collaborators have made detailed studies of the addition of azides to enamines and shown the utility of both the original method (Eqs. 5,6)[3,4] and some promising varients (Eqs. 7, 8).[5–8] The generation of the enamine *in situ* has been expanded to include ketones (Eq. 9)[7,8] and α,β-unsaturated aldehydes (Eq. 10).[9] In the latter case, the

(3) 96% 97% *cis*

(4) 70% 97% *cis*

(5) $MeCH_2CH{=}C(Me)NHPr\text{-}n$ + $4\text{-}N_3C_6H_4NO_2$ $\xrightarrow{CHCl_3}$ 85%

(6) $CH_2{=}CH{-}N$(morpholino) + $4\text{-}N_3C_6H_4NEt_2$ ⟶ 80%

(7) $RCH_2(R^1)C{=}NR^2$ + $4\text{-}N_3C_6H_4NO_2$ ⟶ **12.1-3**

R	= H	H	Me
R^1	= *i*-Pr	*n*-Pr	Et
R^2	= *n*-Pr	*n*-Bu	4-(Me_2N)Ph
% **12.1-3**	= 70	31	67

(8) RCHO + RR′NH + $4\text{-}N_3C_6H_4NO_2$ ⟶

R = Me, Et, *i*-Pr R, R′ = $O[(CH_2)_2]_2N$, Et_2, Me-Ph Excellent yields
$PhCH_2$, Ph, *n*-Pr Fair to poor yields

two possible stereoisomers are produced in equimolar quantities. Similar results are obtained using several cyclic, secondary amines. An extensive study of this addition has been reported and the yields of compounds analogous to **12.1-4** are generally good (Eq. 11).[10]

$$Me_2C{=}O + R_2NH + TiCl_4 \longrightarrow Me_2C(NR_2)_2 \xrightarrow{O_2N\text{-}C_6H_4\text{-}N_3} \qquad (9)$$

(Product: 5-Me-5-NR_2-1-(4-nitrophenyl)-Δ^2-1,2,3-triazoline)

$R_2 = O[(CH_2)_2]_2N$ 45% $PhCH_2N[(CH_2)_2]_2N$ 55%

$$MeCH{=}CHCHO + Me_2NH + N_3\text{-}C_6H_4\text{-}NO_2 \xrightarrow{CHCl_3} \qquad (10)$$

(Product: 4-[CH(Me)NMe_2]-5-NMe_2-1-(4-nitrophenyl)-Δ^2-1,2,3-triazoline, with H at C-4 and C-5)

12.1-4
about 100%

$$CH_2{=}CHCHO + R^1R^2NH + ArN_3 \longrightarrow \qquad (11)$$

(Product: 4-$R^1R^2NCH_2$-5-NR^1R^2-1-Ar-Δ^2-1,2,3-triazoline)

12.1-5

R^1 =	Me	Et	$(CH_2)_4$	$(CH_2)_5$	$(CH_2)_5$	$MeN[(CH_2)_2]_2$
R^2 =	Me	Et				
Ar =	4-NO_2Ph	4-NO_2Ph	4-NO_2Ph	4-NO_2Ph	4-ClPh	4-NO_2Ph
% **12.1-5** =	80	85	70	50	50	90
R^1 =	$O[(CH_2)_2]_2$	$O[(CH_2)_2]_2$	Me	Me	Me	Me
R^2 =			c-C_6H_{11}	Ph	Ph	2-MePh
Ar =	4-NO_2Ph	4-ClPh	4-NO_2Ph	4-NO_2Ph	4-ClPh	4-NO_2Ph
% **12.1-5** =	60	70	85	55	20	65

Some related studies deserve note; for example, when diethylacetals are employed, the reaction takes place, but the yields decrease markedly (Eq. 12).[11,12] Mixtures of isomeric enamines were found to produce very high yields of the product expected from the more stable starting material (Eq. 13).[13]

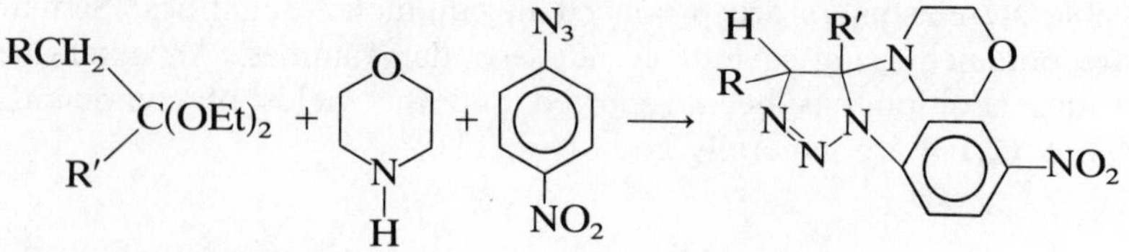

$RCH_2C(OEt)_2R'$ + morpholine + $p\text{-}N_3C_6H_4NO_2$ ⟶ triazoline (H, R, R', morpholino, N-C_6H_4-NO_2) (12)

$R,R' = Me, Et, n\text{-}Pr$

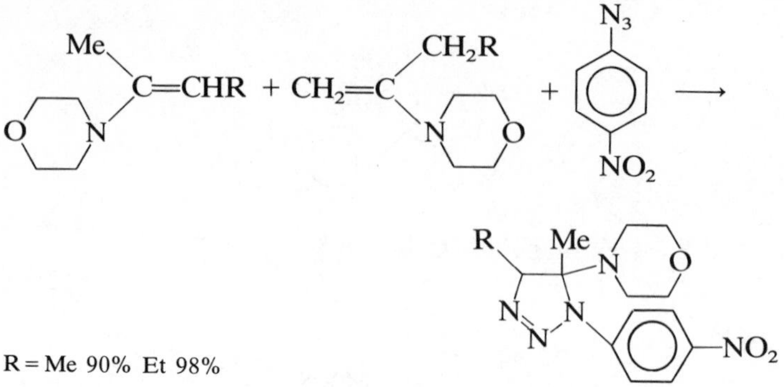

Me(morpholino)C=CHR + CH_2=C(CH_2R)(morpholino) + $p\text{-}N_3C_6H_4NO_2$ ⟶ triazoline (R, Me, morpholino, N-C_6H_4-NO_2) (13)

$R = Me$ 90% Et 98%

Finally, Pocar and his collaborators have reported a stereochemical study of the enamine–azide addition with the following conclusions:[14]

1. Kinetic control produces *trans*-triazolines.
2. High temperature or acid produces a *cis-trans* product mixture.
3. The size of the substituents does not appear to exert a significant effect.
4. It is probable that the triazoline isomerizes because that reaction produces the same product distribution as the addition reaction.

Texier and Bourgois have also studied the stereochemistry of the enamine–azide addition (Eq. 14) and have found the reaction to be thermally reversible.[15]

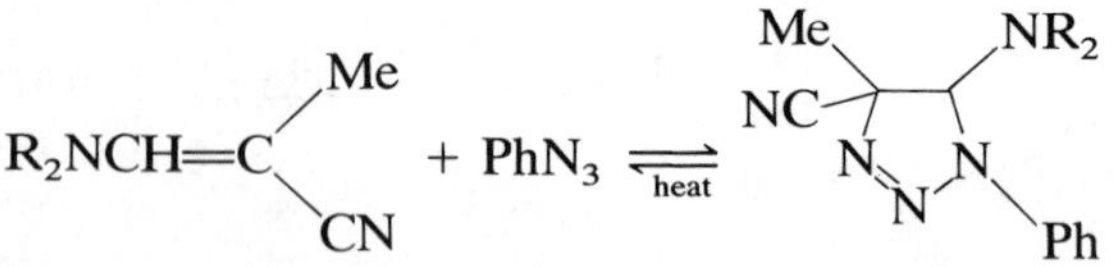

$R_2NCH{=}C(Me)CN + PhN_3 \underset{heat}{\rightleftharpoons}$ triazoline (Me, NC, NR_2, N-Ph) **12.1-6** (14)

R_2	= $(CH_2)_4$	$(CH_2)_5$	$O[(CH_2)_2]_2$
% **12.1-6**	= 30	54	72
cis/trans	= 78/22	84/16	78/22

Olsen found generally high yields with the addition of an arylazide to aldehydes or ketones in the presence of ammonia or amines (Eq. 15).[16] In this same paper he reported another example of azide-enol ether addition (Eq. 16).[16]

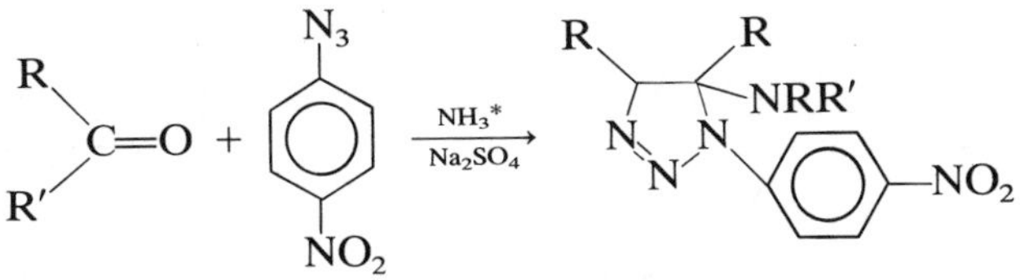

(15)

$R,R' = H, Me, Et, i\text{-}Pr$ *also $MeNH_2$ and Me_2NH

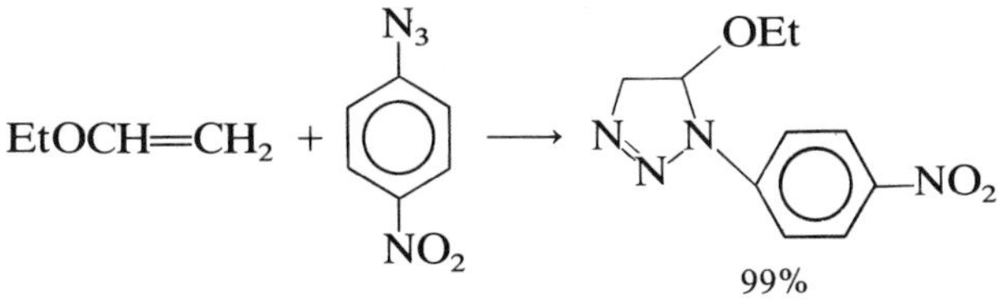

$EtOCH{=}CH_2$ + (16)

99%

In collaboration with Pedersen, Olsen has demonstrated that the 5-hydroxy-4-carboxy-1,2,3-triazoline ester (**12.1-7**), presumed to be an intermediate of the Dimroth cyclization, is stable in neutral or basic solution (Eq. 17).[17] Building on this knowledge, they have prepared an impressive array of 5-hydroxy-1,2,3-triazolines (Eq. 18) in excellent yield.[17] A series of recent publications from this laboratory has added to the synthetic utility and our understanding of this valuable method.[18–21]

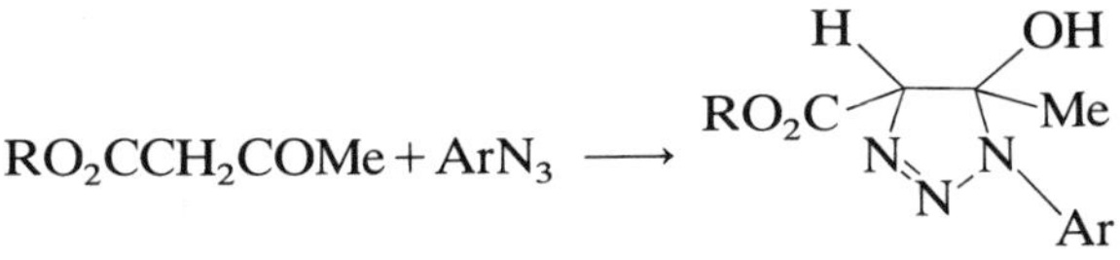

$RO_2CCH_2COMe + ArN_3 \longrightarrow$ (17)

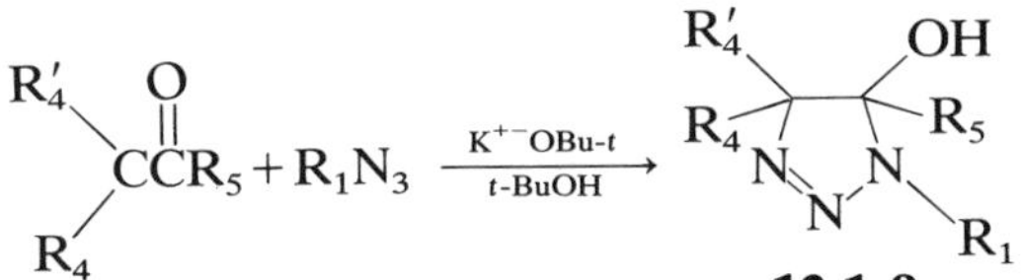

12.1-7

$R_4'R_4C{-}C(=O)R_5 + R_1N_3 \xrightarrow[t\text{-BuOH}]{K^{+-}OBu\text{-}t}$ (18)

12.1-8

R_1	= Ph	Ph	$PhCH_2$	Ph	$PhCH_2$	4-NO_2Ph
R_4	= Me	Me	Me	Me	Me	Me
R_4'	= H	H	H	Me	Me	Me
R_5	= Me	Et	Et	*i*-Pr	*i*-Pr	*i*-Pr
% **12.1-8**	= 72	85	65	71	91	84

A variety of isolated reports describe the preparation, often in high yield, of a number of 1,2,3-triazolines bearing functional groups. The following samples in Equations 19 to 22 illustrate both the range of functions possible and the need to explore more fully such syntheses.[22–24] Compared to the 1,2,3-triazoles, we find an abundance of fluoro derivatives (Eqs. 23,24).[25,26]

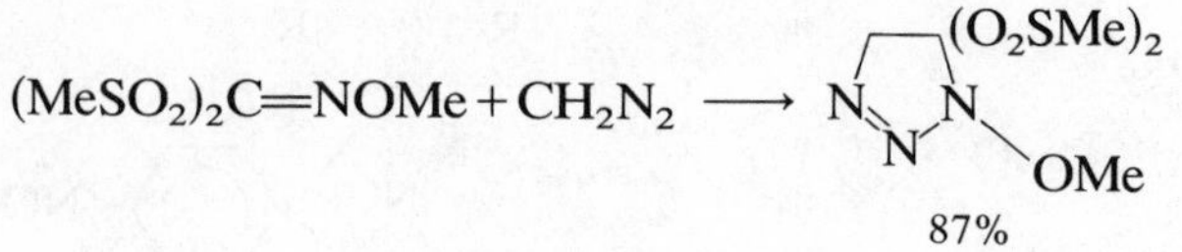

$$(MeSO_2)_2C{=}NOMe + CH_2N_2 \longrightarrow \text{[triazoline: 5,5-}(O_2SMe)_2\text{, 1-OMe]} \quad (19)$$

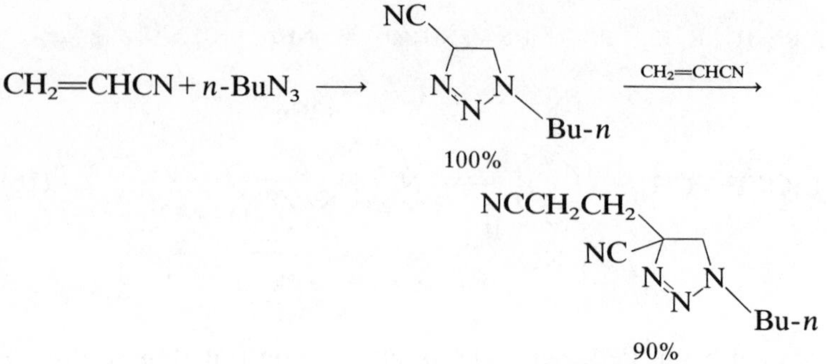

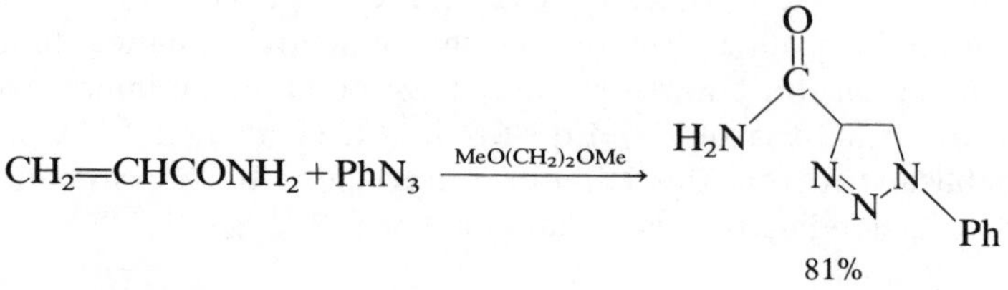

87%

$$CH_2{=}CHCN + n\text{-}BuN_3 \longrightarrow \text{[triazoline: 4-NC, 1-Bu-}n\text{]} \xrightarrow{CH_2{=}CHCN} \text{[triazoline: 4-NC, 4-}NCCH_2CH_2\text{, 1-Bu-}n\text{]} \quad (20)$$

100% 90%

$$CH_2{=}CHCONH_2 + PhN_3 \xrightarrow{MeO(CH_2)_2OMe} \text{[triazoline: 4-}H_2NC(O)\text{, 1-Ph]} \quad (21)$$

81%

$$RHC{=}C(OMe)_2 + EtO_2CN_3 \longrightarrow \text{[triazoline: 4-R, 4-H, 5,5-(OMe)}_2\text{, 1-}CO_2Et\text{]} \quad (22)$$

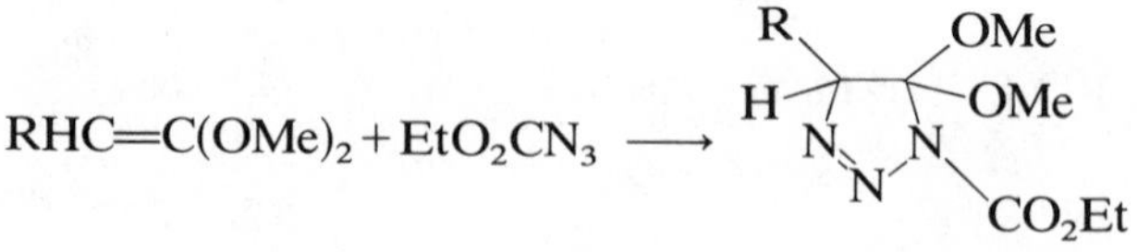

12.1-9

R	= H	Me	CH_2CN
% **12.1-9** =	60	60	100

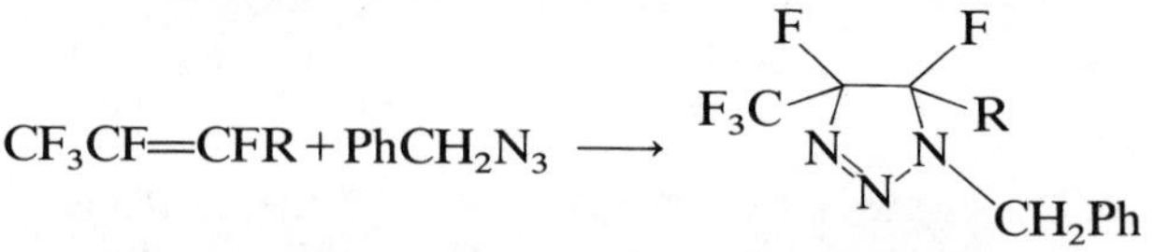

$$CF_3CF{=}CFR + PhCH_2N_3 \longrightarrow \text{[triazoline: 4-F, 4-}CF_3\text{, 5-F, 5-R, 1-}CH_2Ph\text{]} \quad (23)$$

12.1-10

R	= F	CF_3
% **12.1-10** =	85	65

$$(CF_3)_2C{=}NOC(O)O_2R + CH_2N_2 \longrightarrow \text{[triazoline: 5,5-}(CF_3)_2\text{, 1-}OC(O)O_2Et\text{]} \quad (24)$$

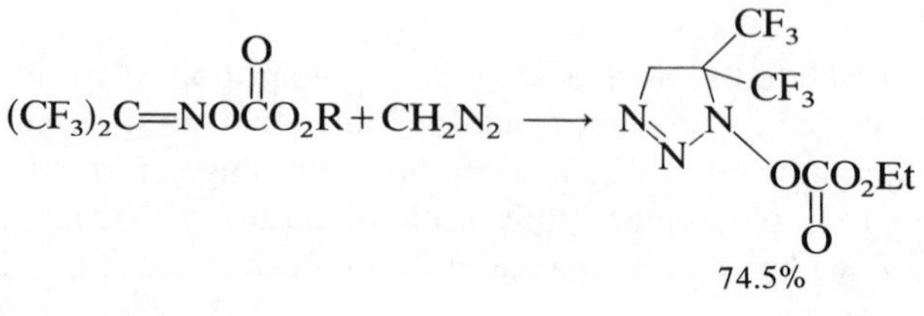

74.5%

Two rather exceptional types of compounds closely related to those being discussed in this section have been reported. The very high yield of a Δ^4-1,2,3-triazoline surely needs reexamination (Eq. 25).[27] The addition of strongly electron-withdrawing arylazides to an allene produced good yields of isopropylidene derivatives (**12.1-11**) (Eq. 26).[28]

$$CH_2{=}CHCN + PhN_3 \longrightarrow \text{(4-NC, 1-H, 2-H, 3-Ph triazoline)} \quad 97\% \tag{25}$$

$$Me_2C{=}C{=}CMe_2 + ArN_3 \longrightarrow \mathbf{12.1\text{-}11} \tag{26}$$

Ar	= Ph	4-NO_2Ph	2,4,6-$(NO_2)_3$Ph
% **12.1-11**	= 29	73	72

REFERENCES

1. **62:** 16247–8g–a
2. **62:** 16249d
3. **65:** 15367a
4. **67:** 64365p
5. **67:** 82170a
6. **67:** 82171b
7. **72:** 78995w
8. **77:** 75172f
9. **76:** 126879f
10. **76:** 126875b
11. **68:** 12925x
12. **68:** 12926y
13. **68:** 12927z
14. **76:** 139838s
15. **83:** 114301t
16. **81:** 63009c
17. **69:** 96589x
18. **80:** 59897c
19. **80:** 95834p
20. **80:** 59896b
21. **81:** 136113u
22. **45:** 6999f
23. **75:** 98501k
24. **85:** 62999w
25. **64:** 12662g
26. **77:** 151367r
27. **49:** 1048b
28. **69:** 36040b

TABLE 12. MISCELLANEOUS Δ^2 (OR Δ^4)-1,2,3-TRIAZOLINES

Compound	Reference
-1-acetic acid, 4,5-dicarboxy-	**24:** 3215[4]
4-acetyl-1-butyl-4-methyl-	**75:** 98501k
5-acetyl-1-butyl-5-methyl-	**75:** 98501k
5-acetyl-1,4-diphenyl-	**27:** 2140[1]
5-acetyl-4-methyl-1-phenyl-	**27:** 2139[9]
5-acetyl-5-methyl-1-phenyl-	**75:** 98501k
-5-amine, *N,N*-bis(isopropyl)-5-ethyl-4-methyl-1-(4-nitrophenyl)-	**76:** 139838s
-5-amine, 1-(4-bromophenyl)-*N,N*-diethyl-4-vinyl-, *trans*-	**77:** 126516f
-4-amine, 5-(4-bromophenyl)-1-phenyl-	**66:** 10887k
-4-amine, 5-(4-chlorophenyl)-1-phenyl-	**66:** 10887k
-5-amine, *N*-cyclohexyl-*N*,4-dimethyl-5-ethyl-1-(4-nitrophenyl)-	**76:** 139838s

TABLE 12 (*Continued*)

Compound	Reference
-5-amine, *N*,*N*-diethyl-4-methyl-1-(4-nitrophenyl)-5-phenyl-	**76:** 139838s
-5-amine, *N*,*N*-diethyl-5-methyl-1-(4-nitrophenyl)-4-phenyl-, *trans*-	**76:** 139838s
-5-amine, *N*,*N*-dimethyl-4-ethyl-1-(4-nitrophenyl)-, *trans*-	**77:** 75172f
-5-amine, *N*,*N*-dimethyl-5-ethyl-1-(4-nitrophenyl)-	**77:** 75172f
-5-amine, *N*,4-dimethyl-5-ethyl-1-(4-nitrophenyl)-, *cis*-	**81:** 63009c
-5-amine, *N*,4-dimethyl-5-ethyl-1-(4-nitrophenyl)-, *trans*-	**81:** 63009c
-5-amine, *N*,4-dimethyl-5-ethyl-1-(4-nitrophenyl)-*N*-phenyl-	**76:** 139838s
-5-amine, *N*,*N*-dimethyl-5-isobutyl-4-isopropyl-1-(4-nitrophenyl)-	**76:** 139838s
-5-amine, *N*,*N*-dimethyl-5-isopropyl-1-(4-nitrophenyl)-	**77:** 75172f
-5-amine, *N*,*N*-dimethyl-1-(4-nitrophenyl)-	**77:** 75172f
-5-amine, *N*,4-dimethyl-1-(4-nitrophenyl)-, *trans*-	**81:** 63009c
-5-amine, *N*,5-dimethyl-1-(4-nitrophenyl)-	**81:** 63009c
-5-amine, 4,5-dimethyl-1-(4-nitrophenyl)-, *cis*-	**81:** 63009c
-5-amine, 4,5-dimethyl-1-(4-nitrophenyl)-, *trans*-	**81:** 63009c
-4-amine, 1,5-diphenyl-	**66:** 10887k
-5-amine, 4-ethyl-*N*-methyl-1-(4-nitrophenyl)-, *trans*-	**81:** 63009c
-5-amine, 5-ethyl-*N*-methyl-1-(4-nitrophenyl)-	**81:** 63009c
-5-amine, 5-ethyl-4-methyl-1-(4-nitrophenyl)-, *cis*-	**81:** 63009c
-5-amine, 5-ethyl-4-methyl-1-(4-nitrophenyl)-, *trans*-	**81:** 63009c
-5-amine, 4-ethyl-1-(4-nitrophenyl)-, *trans*-	**81:** 63009c
-5-amine, 5-ethyl-1-(4-nitrophenyl)-	**81:** 63009c
-5-amine, 5-ethyl-1-(4-nitrophenyl)-*N*,*N*,4-trimethyl-	**76:** 139838s
-5-amine, 5-ethyl-1-(4-nitrophenyl)-*N*,*N*,4-trimethyl-, *cis*-	**77:** 75172f
-5-amine, 5-ethyl-1-(4-nitrophenyl)-*N*,*N*,4-trimethyl-, *trans*-	**77:** 75172f
-5-amine, 5-isopropyl-*N*-methyl-1-(4-nitrophenyl)-	**81:** 63009c
-5-amine, 5-isopropyl-1-(4-nitrophenyl)-	**81:** 63009c
-4-amine, 5-(4-methoxyphenyl)-1-phenyl-	**66:** 10887k
-5-amine, *N*-methyl-1-(4-nitrophenyl)-	**81:** 63009c
-5-amine, 4-methyl-1-(4-nitrophenyl)-, *trans*-	**81:** 63009c
-5-amine, 5-methyl-1-(4-nitrophenyl)-	**81:** 63009c
-5-amine, 1-(4-nitrophenyl)-	**81:** 63009c
-5-amine, 1-(4-nitrophenyl)-5-propyl-*N*,*N*,4-triethyl-	**76:** 139838s
-5-amine, 1-(4-nitrophenyl)-*N*,*N*,4,5-tetramethyl-, *cis*-	**81:** 63009c
-5-amine, 1-(4-nitrophenyl)-*N*,*N*,4,5-tetramethyl-, *trans*-	**81:** 63009c
-5-amine, 1-(4-nitrophenyl)-*N*,*N*,5-triethyl-	**76:** 139838s
-5-amine, 1-(4-nitrophenyl)-*N*,*N*,4-trimethyl-, *cis*-	**77:** 75172f
-5-amine, 1-(4-nitrophenyl)-*N*,*N*,4-trimethyl-, *trans*-	**77:** 75172f
-5-amine, 1-(4-nitrophenyl)-*N*,*N*,5-trimethyl-	**77:** 75172f
-5-amine, 1-(4-nitrophenyl)-*N*,*N*,5-trimethyl-, *cis*-	**81:** 63009c
-5-amine, 1-(4-nitrophenyl)-*N*,*N*,5-trimethyl-, *trans*-	**81:** 63009c

TABLE 12 (*Continued*)

Compound	Reference
-5-anilino-4-ethyl-1-(4-nitrophenyl)-	**72:** 78995w
4-benzoyl-5-methyl-5-(4-morpholinyl)-1-[(4-nitrophenyl)sulfonyl]-	**58:** 13943g
1-(benzoyloxy)-5,5-bis(trifluoromethyl)-	**83:** 28160q
5-(benzylamino)-5-ethyl-4-methyl-1-(4-nitrophenyl)-	**72:** 78995w
5-(benzylamino)-4-ethyl-1-(4-nitrophenyl)-	**72:** 78995w
5-(benzylamino)-4-methyl-1-(4-nitrophenyl)-	**72:** 78995w
5-(benzylamino)-5-methyl-1-(4-nitrophenyl)-	**67:** 82171b
5-(benzylamino)-1-(4-nitrophenyl)-4-pentyl-	**72:** 78955w
1-benzyl-4,5-bis(trifluoromethyl)-4,5-difluoro-	**64:** 12662g
5-(1-benzyl-4-piperazinyl)-5-methyl-1-(4-nitrophenyl)-	**74:** 31720s
1-benzyl-4,4,5-trifluoro-5-(trifluoromethyl)-	**64:** 12662g
4,4-bis(4,4′-morpholinyl)-3-(4-nitrophenyl)-	**77:** 75172f
5,5-bis(trifluoromethyl)-1-[[1-[1,1-dihydro-2,6,7-trioxa-1-phosphabicyclo[2.2.2]octyl]-2,2,2-trifluoro-1-(trifluoromethyl)ethyl]imino]-4-ethyl-	**81:** 49748m
5,5-bis(trifluoromethyl)-1-[(ethoxycarbonyl)oxy]-	**77:** 151367r, **83:** 28160q
5,5-bis(trifluoromethyl)-1-fluoro-	**74:** 111499s
5,5-bis(trifluoromethyl)-1-[(methoxycarbonyl)oxy]-	**83:** 28160q
5,5-bis(trifluoromethyl)-4-methyl-1-[[(4-methylphenyl)sulfonyl]oxy]-	**83:** 28160q
5,5-bis(trifluoromethyl)-1-[[(4-methylphenyl)-sulfonyl]oxy]-	**83:** 28160q
5,5-bis(trifluoromethyl)-1-[(phenylsulfonyl)oxy]-	**83:** 28160q
5,5-bis(trifluoromethyl)-1-[1-(triethoxyphosphor-imidoyl)-2,2,2-trifluoro-1-(trifluoromethyl)ethyl]-	**81:** 49748m
5,5-bis(trifluoromethyl)-1-[2,2,2-trifluoro-1,1-bis(trifluoromethyl)ethoxy]-	**83:** 28160q
5-(4-bromophenyl)-4-nitro-1-phenyl-	**66:** 10887k
4-[[(4-bromophenyl)sulfonyl]imino]-2,3-dimethyl-	**83:** 192266z
5-butoxy-1-(4-nitrophenyl)-	**62:** 16248a, **81:** 63009c
5-(butylamino)-4-ethyl-5-methyl-1-(4-nitrophenyl)-	**67:** 82170a
5-(butylamino)-1-(4-nitrophenyl)-5-phenyl-	**67:** 82171b
4-butyl-5-methyl-5-(4-morpholinyl)-1-(4-nitrophenyl)-	**72:** 78955w
5-butyl-5-(4-morpholinyl)-1-(4-nitrophenyl)-4-propyl-	**72:** 78955w
-4-carbonitrile, 1-butyl-	**75:** 98501k
-5-carbonitrile, 1-butyl-5-methyl-	**75:** 98501k
-4-carbonitrile, 4-methyl-5-(4-morpholinyl)-1-phenyl-, *cis*-	**83:** 114301t
-4-carbonitrile, 4-methyl-5-(4-morpholinyl)-1-phenyl-, *trans*-	**83:** 114301t
-5-carbonitrile, 5-methyl-1-phenyl-	**75:** 98501k
-4-carbonitrile, 4-methyl-1-phenyl-5-(1-piperidinyl)-, *cis*-	**83:** 114301t
-4-carbonitrile, 4-methyl-1-phenyl-5-(1-piperidinyl)-, *trans*-	**83:** 114301t
-4-carbonitrile, 4-methyl-1-phenyl-5-(1-pyrrolidinyl)-, *cis*-	**83:** 114301t
-4-carbonitrile, 4-methyl-1-phenyl-5-(1-pyrrolidinyl)-, *trans*-	**83:** 114301t
-4-carbonitrile, 1-(4-nitrophenyl)-	**64:** 11198g

TABLE 12 (*Continued*)

Compound	Reference
-4-carbonitrile, 1-phenyl-	**49:** 1048b, **64:** 11198a, **80:** 71188e
-4-carboxamide, 1-phenyl-	**68:** 95600m, **75:** 98501k
-4-carboxamide-5-thione	**51:** 2756g
-4-carboxamide-5-thione, 1-benzyl-	**51:** 2756g
-4-carboxylic acid, 4-acetyl-5-(4-chlorophenyl)-1-(4-methoxyphenyl)-, methyl ester	**76:** 113131f
-4-carboxylic acid, 4-acetyl-5-(4-chlorophenyl)-1-(4-methoxyphenyl)-, methyl ester, stereoisomer	**74:** 22744a, **76:** 113313f
-4-carboxylic acid, 4-acetyl-5-(4-chlorophenyl)-1-phenyl-, methyl ester, *trans*-	**76:** 113131f
-4-carboxylic acid, 4-acetyl-5-(4-chlorophenyl)-1-phenyl-, methyl ester, stereoisomer	**74:** 22744a, **76:** 113131f
-4-carboxylic acid, 4-acetyl-1,5-diphenyl-, ethyl ester	**70:** 114918f
-4-carboxylic acid, 4-acetyl-1,5-diphenyl-, ethyl ester, *cis*-	**74:** 22744a, **76:** 113131f
-4-carboxylic acid, 1-(4-benzoylphenyl)-, methyl ester	**64:** 11199c
-4-carboxylic acid, 1-benzyl-2-(methylcarbamoyl)-5-[(methylcarbamoyl)imino]-, ethyl ester	**72:** 21653s
-4-carboxylic acid, 1-(4-chlorophenyl)-, methyl ester	**64:** 11199c
-1-carboxylic acid, 4-(cyanomethyl)-5,5-dimethoxy-, ethyl ester	**85:** 62999w
-1-carboxylic acid, 5,5-dimethoxy-, ethyl ester	**85:** 62999w
-1-carboxylic acid, 5,5-dimethoxy-4-methyl-, ethyl ester	**85:** 62999w
-1-carboxylic acid, 5,5-dimethoxy-4-phenyl-, ethyl ester	**76:** 59076b, **78:** 29677x
-4-carboxylic acid, 1,4-dimethyl-, methyl ester	**75:** 98501k
-5-carboxylic acid, 1,5-dimethyl-, methyl ester	**75:** 98501k
-1-carboxylic acid, 5,5-dimethyl-4-isopropylidene-, ethyl ester	**69:** 36040b
-5-carboxylic acid, 1-[4-(ethoxycarbonyl)phenyl]-, 2-(9*H*-carbazol-9-yl)ethyl ester	**84:** 18157a
-1-carboxylic acid, 4-hydroxy-5-(2-oxo-2-phenylethylidene)-4-phenyl-, ethyl ester	**77:** 88403g
-1-carboxylic acid, 5-methoxy-, ethyl ester	**85:** 62999w
-4-carboxylic acid, 1-(4-methoxyphenyl)-, methyl ester	**64:** 11199c
-4-carboxylic acid, 1-(4-methylphenyl)-, methyl ester	**64:** 11198c
-5-carboxylic acid, 5-methyl-1-phenyl-, methyl ester	**64:** 11198f
-4-carboxylic acid, 1-(4-nitrobenzyl)-, methyl ester	**64:** 11198c
-4-carboxylic acid, 1-(4-nitrophenyl)-, methyl ester	**64:** 11198c
-4-carboxylic acid, 1-phenyl-, ethyl ester	**64:** 11198b
-4-carboxylic acid, 1-(2,3,4,6-tetra-*O*-acetyl-β-D-glucopyranosyl)-, methyl ester	**77:** 102129f
-4-carboxylic acid-5-thione	**65:** 2250d
-4-carboxylic acid-5-thione, 1-benzyl-	**51:** 2756e
-4-carboxylic acid-5-thione, 1-benzyl-, ethyl ester	**51:** 2756e
1-(4-chlorophenyl)-5-ethyl-4-methyl-5-(4-morpholinyl)-	**76:** 139838s

TABLE 12 (*Continued*)

Compound	Reference
1-(4-chlorophenyl)-5-ethyl-4-methyl-5-(4-morpholinyl)-, *cis*-	**77:** 75172f
1-(4-chlorophenyl)-5-ethyl-4-methyl-5-(4-morpholinyl)-, *trans*-	**77:** 75172f
1-(4-chlorophenyl)-5-ethyl-4-methyl-5-(1-piperidinyl)-	**76:** 139838s
1-(4-chlorophenyl)-4-methyl-5-(4-morpholinyl)-, *trans*-	**77:** 75172f
1-(4-chlorophenyl)-4-methyl-5-(1-pyrrolidinyl)-, *trans*-	**77:** 75172f
1-(4-chlorophenyl)-5-(4-morpholinyl)-4-[1-(4-morpholinyl)ethyl]-, [4α(R*), 5β]-	**76:** 126879f
1-(4-chlorophenyl)-5-(4-morpholinyl)-4-[1-(4-morpholinyl)ethyl]-, [4α(S*), 5β]-	**76:** 126879f
1-(4-chlorophenyl)-5-(4-morpholinyl)-4-(4-morpholinylmethyl)-	**76:** 126875b
5-(2-chlorophenyl)-4-nitro-1-phenyl-	**66:** 10887k
5-(4-chlorophenyl)-4-nitro-1-phenyl-	**66:** 10887k
1-(4-chlorophenyl)-5-(1-piperidinyl)-4-[1-(1-piperidinyl)ethyl]-, [4α(R*), 5β]-	**76:** 126879f
1-(4-chlorophenyl)-5-(1-piperidinyl)-4-[1-(1-piperidinyl)ethyl]-, [4α(S*), 5β]-, (±)-	**76:** 126879f
1-(4-chlorophenyl)-5-(1-piperidinyl)-4-(1-piperidinylmethyl)-, *trans*-	**76:** 126875b
5-(cyclohexylamino)-5-methyl-1-(4-nitrophenyl)-	**67:** 82171b
5-cyclopropyl-5-(4-morpholinyl)-1-(4-nitrophenyl)-	**72:** 78955w
5-(dibutylamino)-1-(4-nitrophenyl)-	**67:** 64365p
-4,6-dicarboxamide, 1-(1,4-diphenyl-2-oxo-3-azetidinyl)-*N*-phenyl	**68:** 58967r
-4,5-dicarboximide, 1-benzyl-*N*-(4-methoxyphenyl)-	**69:** 67261d, **71:** 49688k
-4,5-dicarboximide, 1-(4-methoxyphenyl)-*N*-methyl-	**71:** 91429b
-4,5-dicarboxylic acid, 1-benzyl-	**24:** 3231[9]
-4,5-dicarboxylic acid, 1-benzyl-, dimethyl ester	**64:** 11198e
-4,4-dicarboxylic acid, 5-(4-bromophenyl)-1-phenyl-, dimethyl ester	**76:** 113131f
-4,4-dicarboxylic acid, 5-(4-chlorophenyl)-1-phenyl-, dimethyl ester	**76:** 113131f
-4,5-dicarboxylic acid, 2-(1,2-dicarboxyethenyl)-1,3-diphenyl-, tetramethyl ester, (E)-	**68:** 68937v
-4,4-dicarboxylic acid, 1,5-diphenyl-, diethyl ester	**70:** 114918f, **76:** 113131f, **76:** 140389c
-4,4-dicarboxylic acid, 1,5-diphenyl-, dimethyl ester	**70:** 114918f, **76:** 113131f, **83:** 96896j
-4,5-dicarboxylic acid, 1-(4-methoxyphenyl)-, dimethyl ester	**64:** 11198d
-4,4-dicarboxylic acid, 1-(4-methoxylphenyl)-5-methyl-, dimethyl ester	**76:** 113131f
-4,4-dicarboxylic acid, 1-(4-methoxyphenyl)-5-phenyl-, diethyl ester	**76:** 113131f
-4,4-dicarboxylic acid, 1-(4-methoxyphenyl)-5-phenyl-, dimethyl ester	**76:** 113131f

TABLE 12 (*Continued*)

Compound	Reference
-4,4-dicarboxylic acid, 5-(4-methoxyphenyl)- 1-phenyl-, dimethyl ester	**76:** 113131f
-4,4-dicarboxylic acid, 5-methyl-1-(4-nitrophenyl)-, dimethyl ester	**76:** 113131f
-4,5-dicarboxylic acid, 2-methyl-1-(4-nitrophenyl)-, diethyl ester	**72:** 111356b
-4,4-dicarboxylic acid, 5-(4-methylphenyl)-1-phenyl-, diethyl ester	**76:** 113131f
-4,4-dicarboxylic acid, 5-methyl-1-phenyl-, dimethyl ester	**76:** 113131f
-4,4-dicarboxylic acid, 1-(4-nitrophenyl)-5-phenyl-, dimethyl ester	**76:** 113131f
-4,4-dicarboxylic acid, 5-(4-nitrophenyl)-1-phenyl-, dimethyl ester	**76:** 113131f
-4,5-dicarboxylic acid, 1-phenyl-, dihydrazide	**62:** P3568b
5-(diethylamino)-4-ethyl-1-(4-nitrophenyl)-	**68:** 12926y
5-(diethylamino)-1-(4-nitrophenyl)-	**67:** 64365p
5-[(difluoroamino)difluoromethyl]-1-fluoro- 5-(trifluoromethyl)-	**74:** 111499s
4,4-dimethoxy-5,5-dimethyl-1-phenyl-	**60:** 12005e
5-[4-(dimethylamino)anilino]-5-ethyl-4-methyl- 1-(4-nitrophenyl)-	**67:** 82170a
5-(dimethylamino)-4,4-dimethyl-1-phenyl-	**73:** 130810g
5-(dimethylamino)-4-ethyl-4-methyl-1-phenyl-	**73:** 130810g
5-[4-(dimethylamino)phenyl]-4-nitro-1-phenyl-	**66:** 10887k
2,3-dimethyl-4-[[(4-methylphenyl)sulfonyl]imino]-	**83:** 192266z
4,5-dimethyl-5-(4-morpholinyl)-1-(4-nitrophenyl)-	**68:** 12925x
4,5-dimethyl-1-(4-nitrophenyl)-5-(propylamino)-	**65:** 15367a
2,3-dimethyl-4-[[(4-nitrophenyl)sulfonyl]imino]-	**83:** 192266z
2,3-dimethyl-4-[(phenylsulfonyl)imino]-	**83:** 192266z
4,5-diphenyl-5-(4-morpholinyl)-1-(4-nitrophenyl)-, *trans*-	**76:** 139838s
1,5-diphenyl-4-nitro-	**66:** 10887k
5-ethoxy-5-methyl-1-(4-nitrophenyl)-	**62:** 16247g, **68:** 12925x, **81:** 63009c
5-ethoxy-4-methyl-1-(4-nitrophenyl)-, *trans*-	**82:** 125325j
5-ethoxy-1-(4-nitrophenyl)-	**81:** 63009c
4-ethyl-5-methoxy-1-(4-nitrophenyl)-, *trans*-	**82:** 125325j
4-ethyl-5-(*N*-methylanilino)-1-(4-nitrophenyl)-	**72:** 78955w
4-ethyl-5-methyl-5-(4-morpholinyl)-1-(4-nitrophenyl)-	**68:** 12925x, **68:** 12927z
5-ethyl-4-methyl-5-(4-morpholinyl)-1-(4-nitrophenyl)-	**76:** 139838s
5-ethyl-4-methyl-5-(4-morpholinyl)-1-(4-nitrophenyl)-, *cis*-	**77:** 75172f
5-ethyl-4-methyl-5-(4-morpholinyl)-1-(4-nitrophenyl)-, *trans*-	**77:** 75172f
5-ethyl-4-methyl-1-(4-nitrophenyl)-5-(1-piperidinyl)-	**76:** 139838s
4-ethyl-5-methyl-1-(4-nitrophenyl)-5-(propylamino)-	**65:** 15367a
5-ethyl-4-methyl-1-(4-nitrophenyl)-5-(propylamino)-	**65:** 15367a
5-ethyl-4-methyl-1-(4-nitrophenyl)-5-(1-pyrrolidinyl)-	**76:** 139838s
5-ethyl-4-methyl-1-(4-nitrophenyl)-5-(1-pyrrolidinyl)-, *cis*-	**77:** 75172f

TABLE 12 (*Continued*)

Compound	Reference
5-ethyl-4-methyl-1-(4-nitrophenyl)-5-(1-pyrrolidinyl)-, *trans*-	**77:** 75172f
4-ethyl-5-(4-morpholinyl)-1-(4-nitrophenyl)-, *trans*-	**77:** 75172f
5-ethyl-5-(4-morpholinyl)-1-(4-nitrophenyl)-	**77:** 75172f
5-ethyl-5-(4-morpholinyl)-1-(4-nitrophenyl)-4-phenyl-	**72:** 78955w
4-ethyl-5-(4-morpholinyl)-1-(4-nitrophenyl)-5-propyl-	**72:** 78955w
4-ethyl-1-(4-nitrophenyl)-5-(1-piperidinyl)-	**68:** 12927z
4-ethyl-1-(4-nitrophenyl)-5-(propylamino)-	**65:** 15367a
5-ethyl-1-(4-nitrophenyl)-5-(1-pyrrolidinyl)-	**77:** 75172f
4-ethyl-1-(4-nitrophenyl)-5-(1-pyrrolidinyl)-, *trans*-	**77:** 75172f
4-hexamido-5-thioxo-	**83:** P124032v
5-hexyl-5-(4-morpholinyl)-1-(4-nitrophenyl)-	**72:** 78955w
5-isopropyl-5-(4-morpholinyl)-1-(4-nitrophenyl)-	**77:** 75172f
4-isopropyl-1-(4-nitrophenyl)-5-(propylamino)-	**65:** 15367a
5-isopropyl-1-(4-nitrophenyl)-5-(propylamino)-	**67:** 82170a
5-isopropyl-1-(4-nitrophenyl)-5-(1-pyrrolidinyl)-	**77:** 75172f
-4-methanamine, 1-(4-chlorophenyl)-*N*-methyl-5-(methylphenylamino)-*N*-phenyl, *trans*-	**76:** 126875b
-4-methanamine, *N*-cyclohexyl-5-(cyclohexylmethylamino)-*N*-methyl-1-(4-nitrophenyl)-, *trans*-	**76:** 126875b
-4-methanamine, *N,N*-diethyl-5-(diethylamino)-1-(4-nitrophenyl)-, *trans*-	**76:** 126875b
-4-methanamine, *N,N*-dimethyl-5-(dimethylamino)-1-(4-nitrophenyl), *trans*-	**76:** 126875b
-4-methanamine, 5-(dimethylamino)-1-(4-nitrophenyl)-*N,N*,α-trimethyl-, [4α(R*), 5β]-, (±)	**76:** 126879f
-4-methanamine, *N*-methyl-5-[methyl(2-methylphenyl)amino]-*N*-(2-methylphenyl)-1-(4-nitrophenyl)- *trans*-	**76:** 126875b
-4-methanamine, *N*-methyl-5-(methylphenylamino)-1-(4-nitrophenyl)-*N*-phenyl-, *trans*-	**76:** 126875b
1-methoxy-5,5-bis(methylsulfonyl)-	**45:** 6999f
5-methoxy-4-methyl-1-(4-nitrophenyl)-, *trans*-	**82:** 125325j
5-methoxy-1-(4-nitrophenyl)-5-phenyl-	**62:** 16247h
5-(4-methoxyphenyl)-4-nitro-1-phenyl-	**66:** 10887k
5-methoxy-1-(trichloroacetyl)-	**82:** 49811x
4-methyl-5-(*N*-methylanilino)-1-(4-nitrophenyl)-	**72:** 78955w
5-methyl-1-[[(4-methylphenyl)sulfonyl]oxy]-5-(trifluoromethyl)-	**83:** 28160q
4-methyl-5-(4-morpholinyl)-1-(4-nitrophenyl)-	**72:** 78955w
4-methyl-5-(4-morpholinyl)-1-(4-nitrophenyl)-, *cis*-	**77:** 75172f
4-methyl-5-(4-morpholinyl)-1-(4-nitrophenyl)-, *trans*-	**77:** 75172f
5-methyl-5-(4-morpholinyl)-1-(4-nitrophenyl)-	**68:** 12925x, **77:** 75172f
5-methyl-5-(4-morpholinyl)-1-(4-nitrophenyl)-4-pentyl-	**72:** 78955w
4-methyl-5-(4-morpholinyl)-1-(4-nitrophenyl)-5-phenyl-	**76:** 139838s
5-methyl-5-(4-morpholinyl)-1-(4-nitrophenyl)-4-phenyl-	**72:** 78955w
5-methyl-5-(4-morpholinyl)-1-(4-nitrophenyl)-4-propyl-	**72:** 78955w
4-methyl-1-(4-nitrophenyl)-5-phenyl-5-(propylamino)-	**67:** 82171b
4-methyl-1-(4-nitrophenyl)-5-phenyl-5-(1-pyrrolidinyl)-	**76:** 139838s
5-methyl-1-(4-nitrophenyl)-4-phenyl-5-(1-pyrrolidinyl)-, *trans*-	**76:** 139838s
4-methyl-1-(4-nitrophenyl)-5-propoxy-	**62:** 16249d

TABLE 12 (*Continued*)

Compound	Reference
4-methyl-1-(4-nitrophenyl)-5-propoxy-, *cis*-	**81:** 63009c
4-methyl-1-(4-nitrophenyl)-5-propoxy-, *trans*-	**81:** 63009c
4-methyl-1-(4-nitrophenyl)-5-(propylamino)-	**63:** 11552d, **65:** 15367a
5-methyl-1-(4-nitrophenyl)-5-(propylamino)-	**72:** 78955w
4-methyl-1-(4-nitrophenyl)-5-(1-pyrrolidinyl)-, *cis*-	**77:** 75172f
4-methyl-1-(4-nitrophenyl)-5-(1-pyrrolidinyl)-, *trans*	**77:** 75172f
5-methyl-1-(4-nitrophenyl)-5-(1-pyrrolidinyl)-	**77:** 75172f
5-(4-methylphenyl)-4-nitro-1-phenyl-	**66:** 10887k
5-methyl-1-[(phenylsulfonyl)oxy]-5-(trifluoromethyl)-	**83:** 28160q
5-(4-methyl-1-piperazinyl)-4-[1-(4-methyl-1-piperazinyl)ethyl]-1-(4-nitrophenyl)-	**76:** 126879f
5-(4-methyl-1-piperazinyl)-4-[(4-methyl-1-piperazinyl)methyl]-1-(4-nitrophenyl)-, *trans*-	**76:** 126875b
5-(4-morpholinyl)-4-[1-(4-morpholinyl)ethyl]-1-(4-nitrophenyl)-, [4α(R*), 5β]-	**76:** 126879f
5-(4-morpholinyl)-4-[1-(4-morpholinyl)ethyl]-1-(4-nitrophenyl)-, [4α(S*), 5β]-, (±)-	**76:** 126879f
5-(4-morpholinyl)-4-(4-morpholinylmethyl)-1-(4-nitrophenyl)-, *trans*-	**76:** 126875b
5-(4-morpholinyl)-1-(4-nitrophenyl)-	**77:** 75172f
5-(4-morpholinyl)-1-(4-nitrophenyl)-5-pentyl-	**68:** 12925x, **72:** 78955w
5-(4-morpholinyl)-1-(4-nitrophenyl)-5-propyl-	**68:** 12927z
1-(4-nitrophenyl)-4-pentyl-5-(propylamino)-	**72:** 78955w
1-(4-nitrophenyl)-5-(1-piperidinyl)-4-(1-piperidinylmethyl)-, *trans*-	**76:** 126875b
1-(4-nitrophenyl)-5-(propylamino)-	**72:** 78955w
1-(4-nitrophenyl)-5-(1-pyrrolidinyl)-	**77:** 75172f
1-(4-nitrophenyl)-5-(1-pyrrolidinyl)-4-[1-(1-pyrrolidinyl)ethyl]-, [4α(R*), 5β]-, (±)-	**76:** 126879f
1-(4-nitrophenyl)-5-(1-pyrrolidinyl)-4-[1-(1-pyrrolidinyl)ethyl]-, [4α(S*), 5β]-, (±)-	**76:** 126879f
1-(4-nitrophenyl)-5-(1-pyrrolidinyl)-4-(1-pyrrolidinylmethyl)-, *trans*-	**76:** 126875b
-5-ol, 1-(4-aminophenyl)-5-*tert*-butyl-4-methyl-, *cis*-	**80:** 95834p
-5-ol, 1-(4-aminophenyl)-4,5-dimethyl-, *cis*-	**80:** 95834p
-5-ol, 1-(4-aminophenyl)-4,5-dimethyl-, *trans*-	**80:** 95834p
-5-ol, 1-(4-aminophenyl)-4,4-dimethyl-5-isopropyl-	**80:** 95834p, **81:** 63009c
-5-ol, 1-(4-aminophenyl)-5-isopropyl-4-methyl-, *cis*-	**80:** 95834p
-5-ol, 1-benzyl-4-(3-benzyl-1-triazenyl)-5-*tert*-butyl-	**80:** 47912r
-5-ol, 1-benzyl-5-*tert*-butyl-4-methyl-	**80:** 59897c, **84:** 30978b
-5-ol, 1-benzyl-4,5-dimethyl-	**80:** 59897c, **84:** 30978b
-5-ol, 1-benzyl-4,4-dimethyl-5-isopropyl-	**69:** 96589x, **84:** 30978b
-5-ol, 1-benzyl-5-ethyl-4-methyl-	**69:** 96589x, **80:** 59897c, **84:** 30978b
-5-ol, 1-benzyl-5-isopropyl-4-methyl-	**84:** 30978b
-5-ol, 1-benzyl-5-methyl-	**80:** 47912r, **84:** 30978b
-5-ol, 1-benzyl-4-methyl-5-phenyl-	**80:** 59897c, **84:** 30978b
-5-ol, 5-benzyl-4-methyl-1-phenyl-, *cis*-	**80:** 95834p
-5-ol, 5-benzyl-4-methyl-1-phenyl-, *trans*-	**80:** 95834p
-5-ol, 1-benzyl-5-phenyl-	**80:** 47912r, **84:** 30978b

TABLE 12 (*Continued*)

Compound	Reference
-5-ol, 1-[1,2-bis(benzoylethenyl)]-4-[(3-nitrophenyl)methylene]-5-phenyl-	**82:** 170806c
-5-ol, 1-[1,2-bis(benzoylethenyl)]-5-phenyl-4-(phenylmethylene)-	**82:** 170806c
-5-ol, 1-(4-bromophenyl)-4,4-dimethyl-5-isopropyl-	**80:** 95834p
-5-ol, 5-*tert*-butyl-1,4-dimethyl-	**80:** 59897c, **84:** 30978b
-5-ol, 5-*tert*-butyl-1,4-dimethyl-, *cis*-	**81:** 63009c
-5-ol, 5-*tert*-butyl-4,4-dimethyl-1-phenyl-	**80:** 95834p, **84:** 30978b
-5-ol, 5-*tert*-butyl-4-methyl-1-(4-nitrophenyl)-, *cis*-	**80:** 95834p
-5-ol, 5-*tert*-butyl-4-methyl-1-phenyl-	**80:** 59897c, **84:** 30978b
-5-ol, 5-*tert*-butyl-4-methyl-1-phenyl-, *cis*-	**80:** 95834p
-5-ol, 5-*tert*-butyl-1,4,4-trimethyl-	**84:** 30978b
-5-ol, 4,4-dimethyl-1,5-diphenyl-	**80:** 95834p, **84:** 30978b
-5-ol, 1,4-dimethyl-5-ethyl-	**79:** 84372p
-5-ol, 1,4-dimethyl-5-ethyl-, *cis*-	**81:** 63009c
-5-ol, 1,4-dimethyl-5-ethyl-, *cis*- (±)-	**80:** 31343v
-5-ol, 1,4-dimethyl-5-isopropyl-	**84:** 30978b
-5-ol, 1,4-dimethyl-5-isopropyl-, *cis*-	**81:** 63009c
-5-ol, 4,4-dimethyl-5-isopropyl-1-(4-methoxyphenyl)-	**80:** 95834p
-5-ol, 4,4-dimethyl-5-isopropyl-1-[(4-methylphenyl)-sulfonyl]-	**84:** 30978b
-5-ol, 4,4-dimethyl-5-isopropyl-1-(4-nitrophenyl)-	**69:** 96589x, **80:** 95834p, **84:** 30978b
-5-ol, 4,4-dimethyl-5-isopropyl-1-phenyl-	**69:** 96589x, **80:** 95834p, **84:** 30978b
-5-ol, 4,5-dimethyl-1-(4-nitrophenyl)-, *cis*-	**80:** 95834p
-5-ol, 4,5-dimethyl-1-(4-nitrophenyl)-, *trans*-	**80:** 95834p
-5-ol, 4,4-dimethyl-1-(4-nitrophenyl)-5-phenyl-	**80:** 95834p
-5-ol, 1,4-dimethyl-5-phenyl-	**80:** 59897c, **84:** 30978b
-5-ol, 4,5-dimethyl-1-phenyl-	**69:** 96589x, **80:** 59896b, **80:** 59897c, **84:** 30978b
-5-ol, 4,5-dimethyl-1-phenyl-, *cis*-	**80:** 95834p
-5-ol, 4,5-dimethyl-1-phenyl-, *trans*-	**80:** 95834p
-5-ol, 1,5-diphenyl-	**80:** 47912r
-5-ol, 1,5-diphenyl-4-ethyl-, *cis*-	**80:** 95834p
-5-ol, 1,5-diphenyl-4-methyl-	**80:** 59897c, **84:** 30978b
-5-ol, 1,5-diphenyl-4-methyl-, *cis*-	**80:** 95834p
-5-ol, 1,5-diphenyl-4-methyl-, *trans*-	**80:** 95834p
-5-ol, 5-ethyl-4-methyl-1-phenyl-	**69:** 96589x, **80:** 59897c, **84:** 30978b
-5-ol, 5-ethyl-4-methyl-1-phenyl-, *cis*-	**80:** 95834p
-5-ol, 5-ethyl-4-methyl-1-phenyl, *trans*-	**80:** 95834p
-5-ol, 5-isopropyl-4-methyl-1-(4-nitrophenyl)- *cis*-	**80:** 95834p
-5-ol, 5-isopropyl-4-methyl-1-phenyl-	**84:** 30978b
-5-ol, 5-isopropyl-4-methyl-1-phenyl-, *cis*-	**80:** 95834p
-5-ol, 5-isopropyl-1,4,4-trimethyl-	**84:** 30978b
-5-ol, 1-[3-(4-methylphenyl)-3-oxo-1-propenyl]-4-(2-oxo-2-phenylethylidene)-5-phenyl-	**82:** 170806c
-5-ol, 1-(3-oxo-3-phenyl-1-propenyl)-4-(2-oxo-2-phenylethylidene)-5-phenyl-	**82:** 170806c

TABLE 12 (*Continued*)

Compound	Reference
-5-ol, 5-phenyl-1,4,4-trimethyl-	**81:** 136113u, **84:** 30978b
-5-ol, 1,4,5-trimethyl-	**80:** 59897c, **84:** 30978b
-5-ol, 1,4,5-trimethyl-, *cis*-	**81:** 63009c
-4-propanenitrile, 1-butyl-4-cyano-	**75:** 98501k
-4-sulfonamide, *N,N*-diethyl-1-phenyl-	**50:** 3414e
-4-thiol, potassium salt	**84:** 4862q, **84:** 159507b
-4-thiol, 5-methyl-, monosodium salt	**80:** P83024m
-1-thione	**78:** 132397e
-4-thione	**65:** 2250d
-5-thione	**84:** 4862q
-5-thione, 1-(4-acetylphenyl)-	**85:** 94288e
-5-thione, 4-benzoyl-	**65:** 2250e
-4-thione, 1-benzyl-	**77:** 114317d
-5-thione, 1-benzyl-, lithium salt	**84:** 159507b
-5-thione, 1-(4-bromophenyl)-	**85:** 94288e
-5-thione, 1-(4-ethoxyphenyl)-	**85:** 94288e
-4-thione, 1-methyl-	**77:** 114317d
-4-thione, 2-methyl-	**84:** 4862q
-5-thione, 1-methyl-	**77:** 114317d
-4-thione, 3-methyl-2-phenyl-	**80:** 132667j
-5-thione, 1-(4-methylphenyl)-	**85:** 94288e
-5-thione, 1-phenyl-	**58:** 2447a

REFERENCES

24: 3215[4] T. Curtius and W. Klavehn, *J. Prakt. Chem.*, **125,** 498 (1930).

24: 3231[9] T. Curtius and K. Raschig, *J. Prakt. Chem.*, **125,** 466 (1930).

27: 2139–40[9–1] K. Alder, G. Stein, and W. Friedrichsen, *Justus Liebigs Ann. Chem.*, **501,** 1 (1933).

45: 6999f H. J. Backer, *Rec. Trav. Chim.*, **69,** 1223 (1950).

49: 1048b S. M. Gurvich and A. P. Terent'ev, *Sbornik Stateĭ Obscheĭ Khim., Akad. Naak SSSR*, **1,** 409 (1953).

50: 3414e C. S. Rondestvedt, Jr. and P. K. Chang, *J. Amer. Chem. Soc.*, **77,** 6532 (1955).

51: 2756e–g J. R. E. Hoover and A. R. Day, *J. Amer. Chem. Soc.*, **78,** 5832 (1956).

52: 2447a M. Koizumi and A. Watanabe, *Bull. Chem. Soc. Jap.*, **28,** 136 (1955).

58: 13943g R. Fusco, G. Bianchetti, D. Pocar, and R. Ugo, *Chem. Ber.*, **96,** 802 (1963).

60: 12005e R. Scarpati and D. Sica, *Gazz. Chim. Ital.*, **93,** 942 (1963).

62: P3568b J. E. Jones, U.S. Patent, 3,157,509 (1964).

62: 16247–8g–a R. Huisgen, L. Möbius, and G. Szeimies, *Chem. Ber.*, **98,** 1138 (1965).

62: 16249d R. Huisgen and G. Szeimies, *Chem. Ber.*, **98,** 1153 (1965).

63: 11552d G. Bianchetti, P. D. Croce, and D. Pocar, *Tetrahedron Lett.*, 2039 (1965).

64: 11198–9f–c G. Szeimies and R. Huisgen, *Chem. Ber.*, **99,** 491 (1966).

64: 12662g	W. R. Carpenter, A. Haymaker, and D. W. Moore, *J. Org. Chem.*, **31,** 789 (1966).
65: 2250d,e	J. Goerdeler and G. Gnad, *Chem. Ber.*, **99,** 1618 (1966).
65: 15367a	G. Bianchetti, P. D. Croce, D. Pocar, and G. G. Gallo, *Rend. Ist. Lombardo Sci. Lettere, A*, **99,** 296 (1965).
66: 10887k	G. Rembarz, B. Kirchhoff, and G. Dongowski, *J. Prakt. Chem.*, **33,** 199 (1966).
67: 64365p	P. Ferruti, D. Pocar, and G. Bianchetti, *Gazz. Chim. Ital.*, **97,** 109 (1967).
67: 82170a	G. Bianchetti, P. D. Croce, D. Pocar, and A. Vigevani, *Gazz. Chim. Ital.*, **97,** 289 (1967).
67: 82171b	G. Bianchetti, D. Pocar, P. D. Croce, and R. Stradi, *Gazz. Chim. Ital.*, **97,** 304 (1967).
68: 12925x	G. Bianchetti, P. Ferruti, and D. Pocar, *Gazz. Chim. Ital.*, **97,** 579 (1967).
68: 12926y	P. D. Croce, P. Ferruti, and R. Stradi, *Gazz. Chim. Ital.*, **97,** 589 (1967).
68: 12927z	D. Pocar, G. Bianchetti, and P. Ferruti, *Gazz. Chim. Ital.*, **97,** 597 (1967).
68: 58967r	M. S. Manhas, S. Jeng, and A. K. Bose, *Tetrahedron*, **24,** 1237 (1968).
68: 68937v	R. M. Acheson and M. W. Foxton, *J. Chem. Soc., C*, 389 (1968).
68: 95600m	P. Scheiner, *Tetrahedron*, **24,** 2757 (1968).
69: 36040b	R. F. Bleiholder and H. Schechter, *J. Amer. Chem. Soc.*, **90,** 2131 (1968).
69: 67261d	S. Oida and E. Ohki, *Chem. Pharm. Bull. (Tokyo)*, **16,** 764 (1968).
69: 96589x	C. E. Olsen and C. Pedersen, *Tetrahedron Lett.*, 3805 (1968).
70: 114918f	F. Texier and R. Carrié, *Tetrahedron Lett.*, 823 (1969).
71: 49688k	S. Oida and E. Ohki, *Chem. Pharm. Bull (Tokyo)*, **17,** 980 (1969).
71: 91429b	R. Huisgen and H. Mäder, *Angew. Chem., Int. Ed. Engl.*, **8,** 604 (1969).
72: 21653s	L. Capuano, M. Welter, and R. Zander, *Chem. Ber.*, **102,** 3698 (1969).
72: 78995w	R. Stradi and D. Pocar, *Gazz. Chim. Ital.*, **99,** 1131 (1969).
72: 111356b	Y. Hayashi, T. Watanabe, and R. Oda, *Tetrahedron Lett.*, 605 (1970).
73: 130810g	M. De Poortere and F. C. De Schryver, *Tetrahedron Lett.*, 3949 (1970).
74: 22744a	F. Texier and R. Carrié, *C.R. Acad. Sci., Ser. C*, **271,** 958 (1970).
74: 31720s	G. Bianchetti, D. Pocar, and R. Stradi, *Gazz. Chim. Ital.*, **100,** 726 (1970).
74: 111499s	B. L. Dyatkin, K. N. Makarov, and I. L. Knunyants, *Tetrahedron*, **27,** 51 (1971).
75: 98501k	W. Broeckx, N. Overbergh, C. Samyn, G. Smets, and G. L'abbé, *Tetrahedron*, **27,** 3527 (1971).
76: 59076b	R. Scarpati and M. L. Graziano, *Tetrahedron Lett.*, 4771 (1971).
76: 113131f	F. Texier and R. Carrié, *Bull. Soc. Chim. Fr.*, 4119 (1971).
76: 126875b	D. Pocar, R. Stradi, and L. M. Rossi, *J. Chem. Soc., Perkin Trans.*, **1,** 769 (1972).
76: 126879f	D. Pocar, R. Stradi, and L. M. Rossi, *J. Chem. Soc., Perkin Trans.*, **1,** 619 (1972).
76: 139838s	G. Bianchetti, R. Stradi, and D. Pocar, *J. Chem. Soc., Perkin Trans.*, **1,** 997 (1972).
76: 140389c	F. Texier and R. Carrié, *Bull. Soc. Chim. Fr.*, 258 (1972).
77: 75172f	R. Stradi, D. Pocar, and G. Bianchetti, *Org. Magn. Resonance*, **4,** 247 (1972).
77: 88403g	E. Van Loock, G. L'abbé, and G. Smets, *Tetrahedron*, **28,** 3061 (1972).
77: 102129f	M. T. García-López, G. García-Muñoz, and R. Madroñero, *J. Heterocyl. Chem.*, **9,** 717 (1972).

77: 114317d	M. Begtrup, *Acta Chem. Scand.*, **26,** 1243 (1972).
77: 126516f	H. Cardoen, S. Toppet, G. Smets, and G. L'abbé, *J. Heterocycl. Chem.*, **9,** 971 (1972).
77: 151367r	R. G. Kostyanovskii and G. K. Kadorkina, *Izv. Akad. Nauk SSSR, Ser. Khim.*, 1676 (1972).
78: 29677x	R. Scarpati and M. L. Graziano, *J. Heterocycl. Chem.*, **9,** 1087 (1972).
78: 132397e	H. M. Rauen, H. Schriewer, W. Tegtbauer, and J. E. Lasana, *Arzneim.-Forsch.*, **23,** 145 (1973).
80: 31343v	K. Kaas, *Acta Crystallogr., Sect. B*, **29,** 1458 (1973).
80: 47912r	C. E. Olsen, *Acta Chem. Scand.*, **27,** 1987 (1973).
80: 59896b	C. E. Olsen, *Acta Chem. Scand.*, **27,** 2989 (1973).
80: 59897c	C. E. Olsen, *Acta Chem. Scand.*, **27,** 2983 (1973).
80: 71188e	Y. Bando, T. Nakaya, and M. Imoto, *Makromol. Chem.*, **172,** 127 (1973).
80: 95834p	C. E. Olsen and C. Pedersen, *Acta Chem. Scand.*, **27,** 2279 (1973).
80: 132667j	M. Begtrup, *Acta Chem. Scand., Ser. B.*, **28,** 61 (1974).
81: 63009c	C. E. Olsen, *Acta Chem. Scand., Ser. B.*, **28,** 425 (1974).
81: 136113u	C. E. Olsen and C. Pedersen, *Acta Chem. Scand.*, **27,** 2271 (1973).
82: 49811x	V. P. Semenov, A. V. Studenikov, and K. A. Ogloblin, *Nov. Khim. Karbenov, Mater. Vses. Soveshch. Khim. Karbenov Ikh Analogov*, **1st,** *1972*, 254 (1973).
82: 125325j	O. Gerlach, P. L. Reiter, and F. Effenberger, *Justus Liebigs Ann. Chem.*, 1895 (1974).
82: 170806c	G. L'abbé, G. Mathys, and S. Toppet, *Chem. Ind.* (*London*), 278 (1975).
83: 28160q	R. G. Kostyanovskii, G. K. Kadorkina, G. V. Shustov, and K. S. Zakharov, *Dokl. Akad. Nauk SSSR*, **221,** 370 (1975).
83: 96896j	K. Burger, F. Manz, and A. Braun, *Synthesis*, 250 (1975).
83: 114301t	F. Texier and J. Bourgois, *J. Heterocycl. Chem.*, **12,** 505 (1975).
83: P124032v	R. Ohi, T. Shishido, T. Kitahara, and H. Shibaoka, German Patent, 2,437,353 (1975).
83: 192266z	S. Auricchio, G. Vidari, and P. Vita-Finzi, *Gazz. Chim. Ital.*, **105,** 583 (1975).
84: 4862q	J-P. Daris and J. M. Muchowski, *J. Heterocycl. Chem.*, **12,** 761 (1975).
84: 18157a	A. G. Filimoshkin, R. N. Nebedomskaya, I. P. Zherebtsov, and R. M. Livshits, *Vysokomol. Soedin., Ser. A*, **17,** 2260 (1975).
84: 30978b	C. E. Olsen, *Acta Chem. Scand., Ser. B*, **29,** 953 (1975).
84: 159507b	G. L. Dunn, J. R. E. Hoover, D. A. Berges, J. J. Taggart, L. D. Davis, E. M. Dietz, D. R. Jakas, N. Yim, P. Actor, et al., *J. Antibiot.*, **29,** 65 (1976).
85: 62999w	M. L. Graziano and R. Scarpati, *J. Heterocycl. Chem.*, **13,** 205 (1976).
85: 94288e	M. Uher, A. Rybár, A. Martvoň, and J. Leško, *Collect. Czech. Chem. Commun.*, **41,** 1551 (1976).

CHAPTER 13

Δ^2(or Δ^3)-1,2,3-Triazolin-5 (or 4)-Ones

One of the earliest studies of 1,2,3-triazolinone (**13.1-1**) or its tautomer hydroxy-1,2,3-triazole (**13.1-2**) was carried out by Pedersen (Eq. 1)[1] who, along with Begtrup, has made an impressive contribution to our understanding of the synthesis and properties of these compounds (Eqs. 2,3).[2] Most of their studies have concentrated on the reactions of these compounds and were cited in Chapter 8, but indirectly they provide access to structures of interest here (Eq. 4).[3] In most cases the yields are excellent although the analogous direct methylation leads, at best, to fair yields of product mixtures.[4,5]

In later publications Begtrup has added some excellent preparative methods (Eqs. 5,6).[6,7]

$$(EtO_2C)(ClOC)CN_2 + MeNH_2 \longrightarrow \textbf{13.1-1} \rightleftharpoons \textbf{13.1-2}\ (65\%) \quad (1)$$

13.1-1   **13.1-2**

65%

$$Ph\bar{C}HCONHMe + PhN_3 \xrightarrow{Na^+\ ^-OEt} \quad (2)$$

55%

$$R\bar{C}(CO_2NHR')_2 + PhN_3 \xrightarrow{Na^+\ ^-OEt} \textbf{13.1-3} \quad (3)$$

13.1-3

R	= Me	Ph	Me	Ph
R′	= H	H	Me	Me
% **13.1-3**	= 64	27	71	78

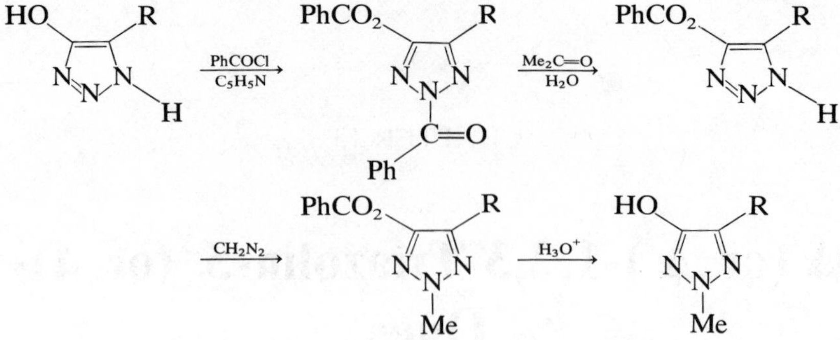

(4)

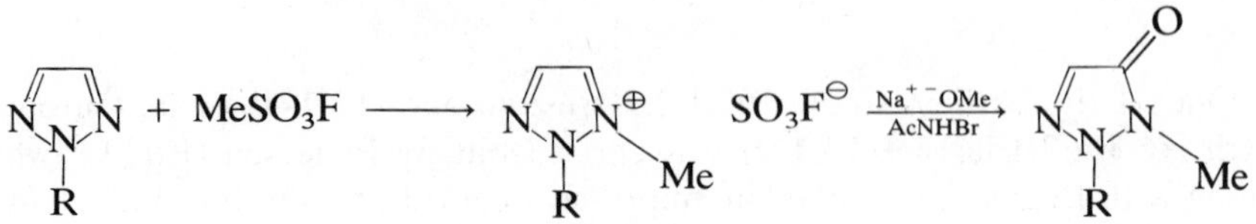

R = Me, Ph
% **13.1-4** = 79, 99
% **13.1-5** = 100, 82

(5)

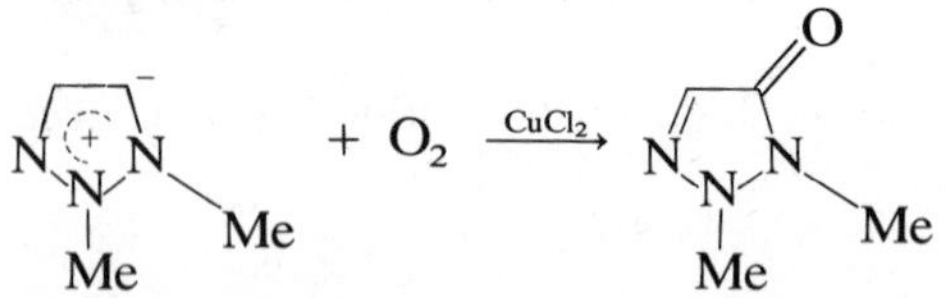

90%

(6)

Another early report that shows promise but that does not appear to have been developed, involves the basic rearrangement of syndnomimins. (Eq. 7)[8]

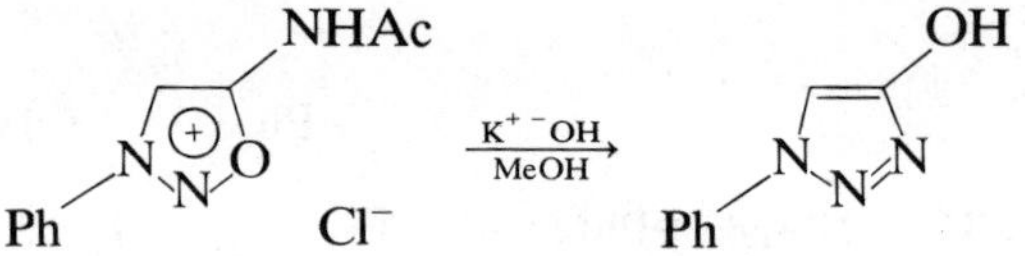

48%

(7)

Lappert and Lorberth did not report a yield for an interesting cyclization reaction, but they stated their belief that a variety of organometallic heterocyclic molecules can be obtained from reactions analogous to Equation 8.[9]

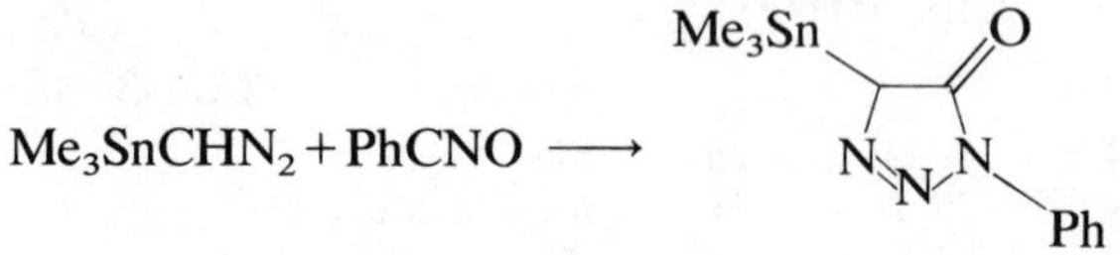

$$Me_3SnCHN_2 + PhCNO \longrightarrow$$

(8)

Activated α-dicarbonyl derivatives have been converted to *N*-oxides of hydroxy-1,2,3-triazoles and subsequently reduced in good-to-excellent yield (Eq. 9).[10]

$$O_2N\text{–}C(=NNHAr)\text{–}C(=NOH)\text{–}H \xrightarrow[H_2O]{HOAc} \textbf{13.1-6} \xrightarrow[Zn]{HOAc} \textbf{13.1-7} \quad (9)$$

Ar	= Ph	2-ClPh	3-ClPh	4-ClPh	4-MeOPh	4-EtO_2CPh
% **13.1-6**	= 52	50	50	53	58	53
% **13.1-7**	= 88		80	95		90

Phosphorous derivatives have been shown by Regitz and his collaborators to be useful in the synthesis of substituted hydroxy-1,2,3-triazoles, which probably can be hydrolyzed by analogy to many related synthetic routes (Eqs. 10,11).[11,12]

$$(EtO)_2\overset{O}{\overset{\|}{P}}CH_2CONH_2 + TsN_3 \xrightarrow{K^+\ ^-OBu\text{-}t} \text{4-HO-5-}(P(O)(OEt)_2)\text{-1H-1,2,3-triazole}\ (56\%) \quad (10)$$

$$H_2NOC\text{–}CN_2\text{–}P(O)Ph_2 \xrightarrow{K^+\ ^-OEt} \text{4-}(Ph_2P(O))\text{-5-OH-1H-1,2,3-triazole}\ (70\%) \quad (11)$$

A study of the equilibria among triazolinone, hydroxytriazole, and open-chain analogs has been reported by Buu and Edward (Eq. 12).[13] When R = EtOCO, **13.1-8** could be obtained only as the sodium salt, and R = Ph produced 40% of **13.1-8** as a crystalline compound.

Sulfur analogs of the hydroxy-1,2,3-triazoles (**13.1-10**) have been obtained from the basic rearrangement of 1,2,3-thiadiazoles (**13.1-9**) (Eq. 13)[14].

$$RCH{=}C(OEt)O^- + PhN_3 \longrightarrow \text{[4-R-5-OEt-5-}O^-\text{-1-Ph-triazoline]} \longrightarrow \text{[4-R-1-Ph-}\Delta^2\text{-1,2,3-triazolin-5-one]} \rightleftharpoons \text{13.1-8 (4-R-5-OH-1-Ph-1,2,3-triazole)} \rightleftharpoons RC(N_2)\text{–}C(=O)NHPh \quad (12)$$

13.1-8

$$\text{13.1-9 (5-amino-1,2,3-thiadiazole)} \xrightarrow[(2)\ H_3O^+]{(1)\ OH^-} \text{13.1-10 (5-mercapto-1H-1,2,3-triazole)} \quad (13)$$

13.1-9 **13.1-10** 87%

REFERENCES

1. **54:** 4547a,b
2. **62:** 1645f,g
3. **71:** 124340w
4. **64:** 11200c
5. **67:** 64305u
6. **76:** 25187y
7. **83:** 97134w
8. **58:** 7927f-h
9. **68:** 39752a
10. **80:** 108449u
11. **70:** 19987u
12. **71:** 50108c
13. **78:** 71047s
14. **65:** 2250d,e

TABLE 13. Δ^2(OR Δ^3)-1,2,3-TRIAZOLIN-5(OR 4)-ONES

Compound	Reference
-4-one	**54:** 4547a
-4-one, 2-(4-acetylphenyl)-, 1-oxide	**80:** 108449u
-4-one, 1-benzyl-	**71:** 61296p
-5-one, 1-benzyl-	**54:** 4547b
-5-one, 1-benzyl-2,4-dimethyl-	**71:** 61296p
-5-one, 1-benzyl-2-methyl-	**67:** 64305u
-5-one, 1-benzyl-4-methyl-	**71:** 61296p
-5-one, 1-benzyl-2-methyl-4-phenyl-	**71:** 61296p
-5-one, 1-benzyl-4-phenyl-	**71:** 61296p
-4-one, 2-(4-bromophenyl)-, 1-oxide	**82:** P170969h
-5-one, 1-(4-bromophenyl)-4-[2-(3-ethyl-2-benzothiazolinylidene)ethylidene]-	**47:** P6289a

TABLE 13 (*Continued*)

Compound	Reference
-5-one, 1-(4-bromophenyl)-4-[2-(3-ethyl-2-benzoxazolinylidene)ethylidene]-	**47:** P6289d
-5-one, 1-(4-bromophenyl)-4-[2-(1-methyl-2(1*H*)-quinolylidene)ethylidene]-	**47:** P6289b
-5-one, 1-(4-bromophenyl)-4-[2-(1,3,3-trimethyl-2-indolinylidene)ethylidene]-	**47:** P6288h
-4-one, 2-(3-chloro-4-fluorophenyl)-, 1-oxide	**82:** P170969h
-4-one, 2-(2-chlorophenyl)-, 1-oxide	**80:** 108449u
-4-one, 2-(3-chlorophenyl)-	**80:** 108449u
-4-one, 2-(3-chlorophenyl)-, 1-oxide	**80:** 108449u
-4-one, 2-(4-chlorophenyl)-	**80:** 108449u
-4-one, 2-(4-chlorophenyl)-, 1-oxide	**80:** 108449u
-5-one, 1-(4-chlorophenyl)-4-[2-(3-ethyl-2-benzothiazolinylidene)ethylidene]-	**47:** P6288i
-5-one, 1-(4-chlorophenyl)-4-[2-(3-ethyl-2-benzoxazolinylidene)ethylidene]-	**47:** P6289c
-5-one, 1-(4-chlorophenyl)-4-[2-(1-ethyl-2(1*H*)-quinolylidene)ethylidene]-	**47:** P6289b
-4-one, 2-[2-chloro-5-(trifluoromethyl)phenyl]-, 1-oxide	**82:** P170969
-5-one, 1-(4-chlorophenyl)-4-[2-(1,3,3-trimethyl-2-indolinylidene)ethylidene]-	**47:** P6288g
-4-one, 2-(2,5-dichlorophenyl)-, 1-oxide	**82:** P170969h
-4-one, 2-(3,4-dichlorophenyl)-, 1-oxide	**82:** P170969h
-4-one, 2-(3,5-dichlorophenyl)-, 1-oxide	**82:** P170969h
-5-one, 4-[(diethoxycarbonyl)phosphinylidyne]-, monosilver(1+) salt	**70:** 19987u
-5-one, 1,2-dimethyl-	**64:** 11200c
-4-one, 1,5-dimethyl-	**71:** 61296p
-4-one, 2,3-dimethyl-	**71:** 124340w, **76:** 25187y **83:** 97134w
-5-one, 1,2-dimethyl-4-phenyl-	**67:** 64305u
-5-one, 2,4-dimethyl-1-phenyl-	**67:** 64305u
-4-one, 1,5-diphenyl-	**58:** 7927h, **78:** 15224k
-5-one, 1,4-diphenyl-	**59:** 3913d, **62:** 1645g, **78:** 71047s
-4-one, 5,5-diphenyl-	**53:** 21902–3g-h
-5-one, 4,4-diphenyl-	**53:** 21902g
-4-one, 5,5-diphenyl-3-methyl-	**53:** 21902–3g–h
-4-one, 5-(diphenylphosphinyl)-	**71:** 50108c
-5-one, 2-[4-(ethoxycarbonyl)phenyl]-	**80:** 108449u
-5-one, 2-[4-(ethoxycarbonyl)phenyl]-, *N*-oxide	**80:** 108449u
-5-one, 4-[2-(3-ethyl-2-benzothiazolinylidene)-ethylidene]-1-phenyl-	**47:** P6289a
-5-one, 4-[2-(3-ethyl-2-benzoxazolinylidene)-ethylidene]-1-phenyl-	**47:** P6289c
-5-one, 4-[2-(1-ethyl-2(1*H*)-quinolylidene)-ethylidene]-1-phenyl-	**47:** P6289a
-4-one, 2-(4-fluorophenyl)-, 1-oxide	**82:** P170969h
-4-one, 2-(2-methoxyphenyl)-, 1-oxide	**82:** P170969h

TABLE 13 (*Continued*)

Compound	Reference
-4-one, 2-(4-methoxyphenyl)-, 1-oxide	**82:** P170969h
-4-one, 2-methyl-	**71:** 124340w
-5-one, 1-methyl-	**54:** 4547a
-5-one, 1-methyl-4,4-diphenyl-	**53:** 21899a
-5-one, 2-methyl-1,4-diphenyl-	**67:** 64305u
-5-one, 2-methyl-4-phenyl-	**62:** 1645g
-4-one, 1-methyl-5-phenyl-	**71:** 61296p
-4-one, 2-methyl-5-phenyl-	**71:** 124340w
-4-one, 3-methyl-2-phenyl-	**76:** 25187y
-5-one, 4-methyl-1-phenyl-	**62:** 1645g
-5-one, 4-methyl-2-phenyl-	**55:** 18756b
-4-one, 2-(2-methylphenyl)-, 1-oxide	**82:** P170969h
-4-one, 2-(3-methylphenyl)-, 1-oxide	**82:** P170969h
-4-one, 2-(4-methylphenyl)-, 1-oxide	**82:** P170969h
-5-one, 4-[2-(1-methyl-4(1*H*)-quinolylidene)-ethylidene]-1-phenyl-	**47:** P6289b
-4-one, 1-phenyl-	**26:** 1287[9], **58:** 7927f
-4-one, 2-phenyl-	**77:** 19583d, **80:** 108449u
-4-one, 2-phenyl-, 1-oxide	**80:** 108449u
-4-one, 5-phenyl-	**53:** 21903h, **62:** 1645f
-5-one, 1-phenyl-	**55:** 25923b, **58:** 13944a
-5-one, 1-phenyl-4-(trimethylstannyl)-	**68:** 39752a
-5-one, 2-phenyl-	**67:** 116842p
-5-one, 1-phenyl-4-[2-(1,3,3-trimethyl-2-indolinylidene)ethylidene]-	**47:** P6289a
-4-one, 2-[3-(trifluoromethyl)phenyl]-, 1-oxide	**82:** P170969h

REFERENCES

26: 1287[9]	H. Kleinfeller and G. Bönig, *J. Prakt. Chem.*, **132,** 175 (1931).
47: P6288-9h-d	J. D. Kendall, G. F. Duffin, and T. F. W. Lawrence, British Patent, 684, 707 (1952).
53: 21899a	K. Hohenlohe-Oehringen, *Monatsh. Chem.*, **89,** 588 (1958).
53: 21902-3g-h	K. Hohenlohe-Oehringen, *Monatsh. Chem.*, **89,** 557 (1958).
54: 4547a,b	C. Pedersen, *Acta Chem. Scand.*, **12,** 1236 (1958).
55: 18756b	J. Hadacek, *Spisy prirodovedecke fak. univ. Brne*, **417,** 373 (1960).
55: 25923b	F. Korte and K. Störiko, *Chem. Ber.*, **94,** 1956 (1961).
58: 7927f-h	H. U. Daeniker and J. Druey, *Helv. Chim. Acta*, **45,** 2441 (1962).
58: 13944a	R. Fusco, G. Bianchetti, D. Pocar, and R. Ugo, *Chem. Ber.*, **96,** 802 (1963).
59: 3913d	R. Scarpati, D. Sica, and A. Lionetti, *Gazz. Chim. Ital.*, **93,** 90 (1963).
62: 1645f,g	M. Begtrup and C. Pedersen, *Acta Chem. Scand.*, **18,** 1333 (1964).
64: 11200c	M. Begtrup and C. Pedersen, *Acta Chem. Scand.*, **19,** 2022 (1965).
65: 2250d,e	J. Goerdeler and G. Gnad, *Chem. Ber.*, **99,** 1618 (1966).
67: 64305u	M. Begtrup and C. Pedersen, *Acta Chem. Scand.*, **21,** 633 (1967).
67: 116842p	M. Ruccia and N. Vivona, *Ann. Chim. (Rome)*, **57,** 680 (1967).

68: 39752a M. F. Lappert and J. Lorberth, *Chem. Commun.*, 836 (1967).

70: 19987u M. Regitz, W. Anschütz, and A. Liedhegener, *Chem. Ber.*, **101,** 3734 (1968).

71: 50108c M. Regitz and W. Anschütz, *Chem. Ber.*, **102,** 2216 (1969).

71: 61296p M. Begtrup and C. Pedersen, *Acta Chem. Scand.*, **23,** 1091 (1969).

71: 124340w M. Begtrup, *Acta Chem. Scand.*, **23,** 2025 (1969).

76: 25187y M. Begtrup and K. V. Poulsen, *Acta Chem. Scand.*, **25,** 2087 (1971).

77: 19583d M. Begtrup, *Acta Chem. Scand.*, **26,** 715 (1972).

78: 15224k A. Chinone and M. Ohta, *Chem. Lett.*, 969 (1972).

78: 71047s N. T. Buu and J. T. Edward, *Can. J. Chem.*, **50,** 3719 (1972).

80: 108449u H. Lind and H. Kristinsson, *Synthesis*, 198 (1974).

82: P170969h H. Lind and H. Kristinsson, German Patent, 2,442,685 (1975).

83: 97134w M. Begtrup, *J. Chem. Soc., Chem. Commun.*, 334 (1975).

CHAPTER 14

Δ^3 (or Δ^2)-1,2,3-Triazolin-5-Ones Containing More Than One Representative Function

The synthesis of polyfunctional 1,2,3-triazolin-5-ones was investigated somewhat earlier than the alkyl and aryl derivatives discussed in Chapter 13. For example, Curtius and his students prepared a variety of such compounds from sulfonylazides and malonic esters but failed to present evidence of the yields obtained (Eq. 1).[1]

$$ArSO_2N_3 + CH_2(CO_2Et)_2 \xrightarrow{Na^+ \ ^-OEt} \text{(4-}CO_2Et\text{-5-OH-1-}SO_2Ar\text{-1,2,3-triazole)} \xrightarrow[\text{anhydrous}]{N_2H_4}$$

$$\text{(4-}ArSO_2NHC(O)\text{-5-OH-1-}NH_2\text{-1,2,3-triazole)} \xrightarrow[H_2O]{N_2H_4} \text{(4-}H_2NHNC(O)\text{-5-OH-1-}NH_2\text{-1,2,3-triazole)} \quad (1)$$

Ar = 4-ClPh, 2-$C_{10}H_7$

Another early report produced interesting results, but these diazotization reactions do not appear to be generally applicable (Eqs. 2,3).[2]

$$H_2N(EtO_2C)CHC(O)NH_2\cdot HCl \xrightarrow[Na^+ \ ^-NO_2]{HCl} \text{(4-}HO_2C\text{-5-OH-1-H-1,2,3-triazole)} \ 76\% \quad (2)$$

$$\text{(4-}EtO_2C\text{-5-}NH_2\text{-2-OH-thiazole)} \xrightarrow[Na^+ \ ^-NO_2]{HCl} \text{(4-}HO_2C\text{-5-OH-1-H-1,2,3-triazole)} \quad (3)$$

As mentioned in Chapter 13, Pedersen and Begtrup have contributed several synthetic routes to the 1,2,3-triazolinones (Eqs. 4–6).[3–5] It should be noted that Gompper carried out a reaction similar to that shown by Equation 5 some years earlier (Eq. 7).[6]

$$\text{1-R-5-hydroxy-1,2,3-triazole} + PhN_2^+ \, ^-Cl \longrightarrow \text{4-(PhN=N)-1-R-5-hydroxy-1,2,3-triazole } (\mathbf{14.1\text{-}1}) \quad (4)$$

R	= H	Me	Ph	$PhCH_2$
% **14.1-1**	= 93	59	83	68

$$CH_2(CO_2NHR)_2 + PhN_3 \xrightarrow{Na^+ \, ^-OEt} \text{4-(RHNC(=O))-1-R-5-hydroxy-1,2,3-triazole } (\mathbf{14.1\text{-}2}) \quad (5)$$

R	= H	Me
% **14.1-2**	= 72	67

$$\text{4-Br-1,2-dimethyl-1,2,3-triazolin-5-one} + PhCOCl \xrightarrow{C_5H_5N} \text{4-Br-5-}O_2CPh\text{-2-Me-1,2,3-triazole (48\%)} \xrightarrow[(2)\ H^+]{(1)\ OH^-} \text{4-Br-5-OH-2-Me-1,2,3-triazole (98\%)} \quad (6)$$

$$CH_2(CO_2Me)_2 + PhCH_2N_3 \xrightarrow{Na^+ \, ^-OMe} \text{4-}MeO_2C\text{-1-}CH_2Ph\text{-}\Delta^2\text{-1,2,3-triazolin-5-one (52\%)} \quad (7)$$

Most of the methods reported for this class of compounds show few examples and limited experimental detail. However, they should be investigated again as should the diazocyclizations in Equation 8[7] and Equation 9.[8] In Equation 9 the product structures were inferred from spectra, and the yields were said to be good.

$N_2CHCO_2Et + PhCONCO \longrightarrow$ [1-benzoyl-4-(EtO_2C)-1,2,3-triazolin-5-one] 60% (8)

$EtO_2CCH_2NHC(=O)NHN_2 + ArCOBr \xrightarrow{CH_2Cl_2}$ [4-(ArC(=O))-5-OH-1-(CH_2CO_2Et)-1,2,3-triazole] (9)

Ar = Ph, 3-BrPh

The synthesis of sulfur analogs has been achieved through basic rearrangement of 1,2,3-thiadiazoles (Eq. 10).[9]

[4-R-5-NH_2-1,2,3-thiadiazole] $\xrightarrow[(2)\ H_3O^+]{(1)\ OH^-}$ [4-R-5-SH-1H-1,2,3-triazole] **14.1-3** (10)

R = CO_2H COPh

% **14.1-3** = 100 92

REFERENCES

1. **24:** 3230[1–9]
2. **44:** 1965h
3. **54:** 4546–7i–b
4. **62:** 1645f
5. **77:** 19583d
6. **51:** 13855b
7. **62:** 6468a
8. **67:** 53855x
9. **65:** 2250d,e

TABLE 14. Δ^3(OR Δ^2)-1,2,3-TRIAZOLIN-5(OR 4)-ONES CONTAINING MORE THAN ONE REPRESENTATIVE FUNCTION

Compound	Reference
4-acetamido-2-phenyl-	**67:** 116842p
-1-acetic acid, 4-(3-bromobenzoyl)-	**67:** 53855x
-1-acetic acid, 4-benzoyl-	**67:** 53855x
4-acetyl-2-(2-hydroxyphenyl)-1-phenyl-	**48:** 4221i
4-acetyl-2-(4-hydroxyphenyl)-1-phenyl-	**48:** 4221i
4-acetyl-2-(2-hydroxy-4-sulfo-1-naphthyl)-1-phenyl-	**48:** 4221h
4-amino-	**67:** 116842p
4-amino-2-phenyl-	**67:** 116842p
1-(2-benzothiazolyl)-4-benzoyl-	**75:** 48991z
1-benzyl-(4-phenylazo)-	**54:** 4547b
5-bromo-2,3-dimethyl-	**76:** 25187y
-4-one, 5-bromo-1-methyl-	**71:** 61296p
-4-one, 5-bromo-2-methyl-	**77:** 19583d
-4-one, 5-bromo-2-phenyl-	**82:** P170969h
-4-carboxamide	**51:** 2756d, **62:** 1645f
-4-carboxamide, 1-amino-*N*-(1-naphthylsulfonyl)-	**24:** 3230[1]
-4-carboxamide, 1-amino-*N*-(2-naphthylsulfonyl)-	**24:** 3230[5]
-4-carboxamide, 1-benzyl-	**51:** 2756d
-4-carboxamide, 1-[(benzylidene)amino]-*N*-(1-naphthylsulfonyl)-	**24:** 3230[1]
-4-carboxamide, 1-[(benzylidene)amino]-*N*-(2-naphthylsulfonyl)-	**24:** 3230[5]
-4-carboxamide, *N*,1-dimethyl-	**62:** 1645f
-4-carboxamide, 1-(isopropylideneamino)-*N*-(2-naphthylsulfonyl)-	**24:** 3230[5]
-4-carboxamide, 1-(1-naphthylsulfonyl)-	**24:** 3229[9]
-4-carboxamide, 1-(2-naphthylsulfonyl)-	**24:** 3230[4]
-4-carboxamide, 5-oxo-2-β-D-ribofuranosyl-	**85:** P6010e
-4-carboxamide, 1-phenyl-(?)	**51:** 14698c
-4-carboxamide, 4-phenyl-	**53:** 21903i
-4-carboxamidoxime	**55:** 11422c
-4-carboxylic acid, ethyl ester	**44:** 1965h
-4-carboxylic acid, 1-(*N*-acetylsulfanyl)-, ethyl ester	**54:** 18409a
-4-carboxylic acid, 1-amino	**24:** 3230[1]
-4-carboxylic acid, 1-(2-anthraquinonylsulfonyl)-, ethyl ester	**24:** 3230[9]
-4-carboxylic acid, 1-benzoyl-, ethyl ester	**62:** 6468a
-4-carboxylic acid, 1-benzyl-	**51:** 13855b
-4-carboxylic acid, 1-benzyl-, ethyl ester	**51:** 2756d
-4-carboxylic acid, 1-benzyl-, methyl ester	**51:** 13855b
-4-carboxylic acid, 1-(benzylideneamino)-, benzalhydrazide	**24:** 3230[1]
-4-carboxylic acid, 1-(4-bromophenyl)-, ethyl ester	**50:** 9838e
-4-carboxylic acid, 1-(*N*-carboxysulfanilyl)-, diethyl ester	**54:** 18409a, **56:** 7931f
-4-carboxylic acid, 1-(4-chlorophenylsulfonyl)-, ethyl ester	**24:** 3229[6]

TABLE 14 (*Continued*)

Compound	Reference
-4-carboxylic acid, 1,2-dimethyl-, methyl ester	**64:** 11200c
-4-carboxylic acid, 1-(4-methoxyphenyl)-, ethyl ester	**50:** 9838e
-4-carboxylic acid, 1-(4-methylphenyl)-, ethyl ester	**50:** 9838e
-4-carboxylic acid, 1-(1-naphthylsulfonyl)-	**24:** 3229[8]
-4-carboxylic acid, 1-(2-naphthylsulfonyl)-, ethyl ester	**24:** 3230[4]
-4-carboxylic acid, 1-[(2-nitrobenzylidene)amino]-, (2-nitrobenzylidene)hydrazide	**24:** 3230[1]
-4-carboxylic acid, 1-phenyl-, ethyl ester	**50:** 9838e
-4-carboxylic acid, 4-phenyl-, ethyl ester	**53:** 21903h
-4-carboxylic acid, 1-(4-quinolyl)-, oxide	**58:** 4545d
-4-carboxylic acid, 1-[(salicylal)amino]-,	**23:** 3230[1]
(salicylal)hydrazide	**23:** 3230[1]
-4-carboxylic acid, 1-sulfanilyl-, ethyl ester	**38:** 5004[9], **54:** 18409a, **56:** 7931f
-4-one, 5-chloro-2-phenyl-	**82:** P170969h
-4-one, 2,3-dimethyl-5-methoxy-	**76:** 25187y
-4-one, 5-methoxy-3-methyl-2-phenyl-	**76:** 25187y
1-methyl-4-(phenylazo)-	**54:** 4547b
4-(phenylacetamido)-	**67:** 116842p
4-(phenylazo)-	**54:** 4546i
2-phenyl-4-(phenylacetamido)-	**67:** 116842p
1-phenyl-4-(phenylazo)-	**54:** 4547b

REFERENCES

24: 3229[8,9] T. Curtius, H. Bottler, and G. Hasse, *J. Prakt. Chem.*, **125,** 366 (1930).

24: 3230[1–9] T. Curtius, H. Bottler, W. Randenbusch, R. Tüxen, and H. Derlon, *J. Prakt. Chem.*, **125,** 380 (1930).

38: 5004[9] T. Ekstrand, *Svensk Kem. Tid.*, **54,** 257 (1942).

44: 1965h A. H. Cook, I. Heilbron, and G. D. Hunter, *J. Chem. Soc.*, 1443 (1949).

48: 4221h,i J. Poskočil and Z. J. Allan, *Chem. Listy*, **47,** 1801 (1953).

50: 9838e J. E. Leffler and S-K Liu, *J. Amer. Chem. Soc.*, **78,** 1949 (1956).

51: 2756d S. J. Davis and C. S. Rondestvedt, Jr., *Chem. Ind.* (*London*), 845 (1956).

51: 13855b R. Gompper, *Chem. Ber.*, **90,** 382 (1957).

51: 14698c M. Haring and T. Wagner-Jauregg, *Helv. Chim. Acta*, **40,** 852 (1957).

53: 21903h,i K. Hohenlohe-Oehringen, *Monatsh. Chem.*, **89,** 557 (1958).

54: 4546–7i–b C. Pedersen, *Acta Chem. Scand.*, **12,** 1236 (1958).

54: 18409a A. Lespagnol, D. Bar, Mme. Erb-Debruyne, and Mme. Delhomenie-Sauvage, *Bull. Soc. Chim. Fr.*, 490 (1960).

55: 11422c M. A. Stevens, H. W. Smith, and G. B. Brown, *J. Amer. Chem. Soc.*, **82,** 3189 (1960).

56: 7931f A. Lespagnol, R. Osteux, and D. Bar, *Bull. Soc. Chim. Biol.*, **43,** 789 (1961).

58: 4545d S. Kamiya, *Chem. Pharm. Bull.* (*Tokyo*), **10,** 471 (1962).

62: 1645f M. Begtrup and C. Pedersen, *Acta Chem. Scand.*, **18,** 1333 (1964).

62: 6468a R. Neidlein, *Chem. Ber.*, **97,** 3476 (1964).
64: 11200c M. Begtrup and C. Pedersen, *Acta Chem. Scand.*, **19,** 2022 (1965).
65: 2250d,e J. Goerdeler and G. Gnad, *Chem. Ber.*, **99,** 1618 (1966).
67: 53855x J. H. Looker and J. W. Carpenter, *Can. J. Chem.*, **45,** 1727 (1967).
67: 116842p M. Ruccia and N. Vivona, *Ann. Chim.* (*Rome*), **57,** 680 (1967).
71: 61296p M. Begtrup and C. Pedersen, *Acta Chem. Scand.*, **23,** 1091 (1969).
75: 48991z S. Maiorana and G. Pagani, *Chim. Ind.* (*Milan*), **53,** 470 (1971).
76: 25187y M. Begtrup and K. V. Poulsen, *Acta Chem. Scand.*, **25,** 2087 (1971).
77: 19583d M. Begtrup, *Acta Chem. Scand.*, **26,** 715 (1972).
82: P170969h H. Lind and H. Kristinsson, German Patent, 2,442,685 (1975).
85: P6010e J. T. Witkowski, R. K. Robins, and F. A. Lehmkuhl, U.S. Patent, 3,948,885 (1976).

CHAPTER 15

Bi- and Bis[1,2,3-Triazoles and 1,2,3-Triazolines]

The principal synthetic methods described in previous sections have been modified and employed for the preparation of symmetrical bis[1,2,3-triazoles]. Very few examples of bis[1,2,3-triazolines] are known, and these too are prepared using methods closely related to those of the mono-substituted cases.

Of all methods certainly azide-acetylene addition is the most common; for example, Sasaki and his collaborators have studied medium ring diazides and their 1,2,3-triazole products (Eqs. 1,2).[1] In an earlier paper they described a number of analogous reactions that occur in generally excellent yield.[2]

$$+ \; 2MeO_2CC{\equiv}CCO_2Me \longrightarrow \qquad (1)$$

15.1-1

A much earlier report involves one of the rare examples of fluorine-containing systems prepared in good yield (Eq. 3).[3]

The addition of azide ion, arylazides, and trimethylsilylazide to acetylenes have all been carried out in good-to-excellent yield (Eqs. 4–7).[4–7] The synthesis of **15.1-2** can be carried out step-wise and the mono-addition product isolated.[4] The arylazide case (Eq. 6) was also applied to a symmetrical biphenyl with comparable results.[6]

The base catalyzed addition of azides to active methylene compounds has been shown to be applicable to bis-analogs (Eq. 8).[8]

N_3 N_3 ⇌ N_3 N_3 + $2MeO_2CC{\equiv}CCO_2Me$

MeO_2C CO_2Me MeO_2C CO_2Me N N N 90% N N N MeO_2C CO_2Me $\xrightarrow{\text{heat}}$ **15.1-1** (2)

$F_3CC{\equiv}CC{\equiv}CCF_3 + 2$ (N_3, NMe_2) ⟶

F_3C CF_3 Me_2N N N N N N N NMe_2 (3)

$RC{\equiv}CC(=O)C{\equiv}CR + Na^+ \ {}^-N_3$ ⟶ R O C R N N N H H N N N **15.1-2** (4)

R	= Me	Ph
% **15.1-2** =	76	77

PhO_2S C=CH CH=C SO_2Ph R R $\xrightarrow[\text{HCONMe}_2]{Na^+ \ {}^-N_3}$

R R N N N H N N N H **15.1-3** (5)

R	= CN	CO_2Me
% **15.1-3** =	91	89

$$o\text{-}C_6H_4(N_3)_2 + 2MeO_2CC{\equiv}CCO_2Me \longrightarrow o\text{-}C_6H_4[\text{1-(4,5-bis(}CO_2Me\text{)-1,2,3-triazolyl)}]_2 \quad 65\% \tag{6}$$

$$HC{\equiv}C(CH_2)_2C{\equiv}CH + 2Me_3SiN_3 \longrightarrow \text{bis(2-}Me_3Si\text{-1,2,3-triazol-4-yl)}(CH_2)_2 \quad 77\%$$

$$\xrightarrow{H_2O} \text{bis(1}H\text{-1,2,3-triazol-4-yl)}(CH_2)_2 \quad 100\% \tag{7}$$

$$NCCH_2CO_2Et + N_3\text{—}C_6H_4\text{—}SO_2\text{—}C_6H_4\text{—}N_3 \xrightarrow{Na^{+}\,{}^{-}OEt}$$

$$[\text{4-}EtO_2C\text{-5-}H_2N\text{-1,2,3-triazol-1-yl—}C_6H_4]_2SO_2 \quad 30\% \tag{8}$$

Two examples of nucleophilic substitution have been demonstrated in the synthesis of bis[1,2,3-triazoles] (Eq. 9).[9]

$$2\ \text{4-PhC(O)-5-Me-1}H\text{-1,2,3-triazole} + Br(CH_2)_{5\ \text{or}\ 6}Br \xrightarrow[Na_2CO_3]{Me_2C{=}O}$$

$$\text{[4-PhC(O)-5-Me-1,2,3-triazol-1-yl]}_2(CH_2)_{5\ \text{or}\ 6} \quad 85\% \tag{9}$$

The extensive work of L'abbé and his collaborators with phosphorous ylides and related compounds shows great promise in this synthetic application (Eqs. 10–13).[10–13]

$$2\ (Me)(EtO_2C)C{=}PPh_3 + N_3CO{-}C_6H_4{-}OCN_3 \longrightarrow \qquad (10)$$

(or 1,3-phenylene) 73%

$$2RCCH{=}PPh_3 + N_3{-}C_6H_4{-}N_3 \longrightarrow \textbf{15.1-4} \qquad (11)$$

R	= Me	Ph
% **15.1-4**	= 83	91

$$2\ (Z)(^{-}O)C{=}CHP^{+}Ph_3 + N_3CO{-}C_6H_4{-}CON_3 \longrightarrow \textbf{15.1-5} \qquad (12)$$

1,4-phenylene	Z = Me	Ph	1,3-phenylene	Z = Ph
% **15.1-5**	= 60	76		62

Smith and his students have extended their productive studies to include bis[1,2,3-triazole] examples (Eqs. 14,15).[14,15] The first reaction appears to be an exception and related systems provide much lower yields.[14] The diazo compound (**15.1-6**) can be reduced in high yield using hydrazine and palladium.[15] The members of Smith's group have also made a detailed study of the thermolysis of diazo-1,2,3-triazoles in various aromatic solvents and have found useful amounts of bis-products (Eq. 16).[16]

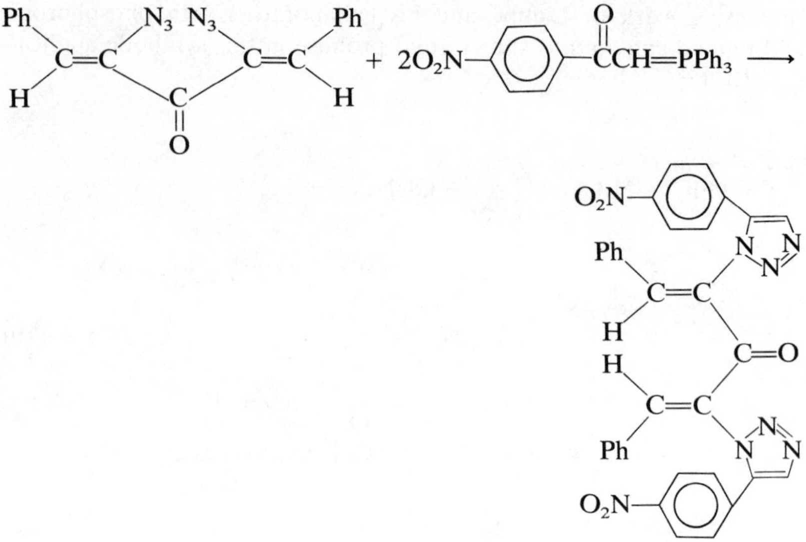

Ph N$_3$ N$_3$ Ph

C=C C=C + 2O$_2$N–C$_6$H$_4$–C(=O)CH=PPh$_3$ ⟶

H C H

O

O$_2$N

Ph N N

C=C N

H

H C=O (13)

C=C N

Ph N N

O$_2$N

70%

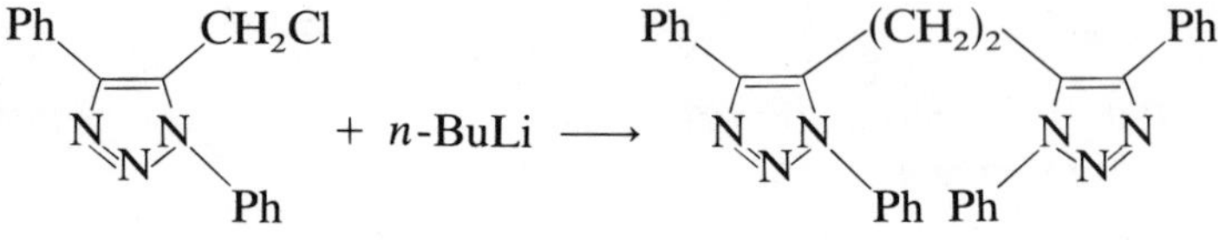

Ph CH$_2$Cl Ph (CH$_2$)$_2$ Ph

N N + *n*-BuLi ⟶ N N N N (14)

N N N

Ph Ph Ph

97%

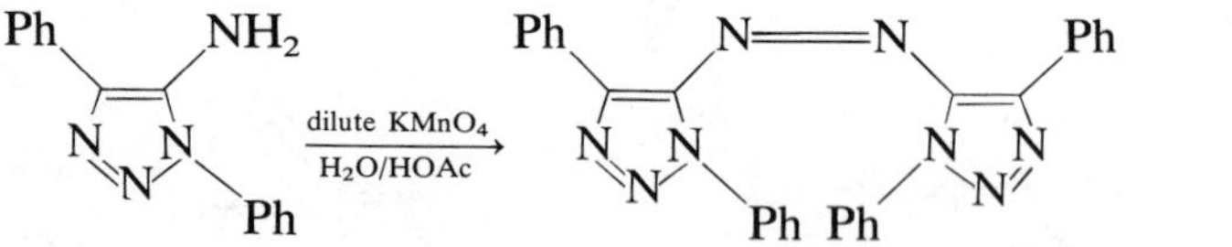

Ph NH$_2$ Ph N=N Ph

N N dilute KMnO$_4$ / H$_2$O/HOAc ⟶ N N N N (15)

N N N

Ph Ph Ph

15.1-6

30%

Albert and his collaborators, in their studies of 8-azapurines, have obtained several interesting examples with nitrogen derivatives connecting two 1,2,3-triazoles (Eqs. 17,18).[17,18] In the absence of ammonia the yield of **15.1-11** rose to 85%! In a later paper Albert reported an intermediate of this reaction (**15.1-12**) formed in excellent yield (Eq. 19).[19]

The preparation of bis[1,2,3-triazole] disulfides has been accomplished both by direct oxidation of the thiol (Eq. 20)[20] and preliminary rearrangement of 1,2,3-thiadiazoles (Eq. 21).[21]

An early report of a crossed aldol condensation shows promise (Eq. 22),[22] as does an example of synthesis from a sugar derivative (Eq. 23).[23]

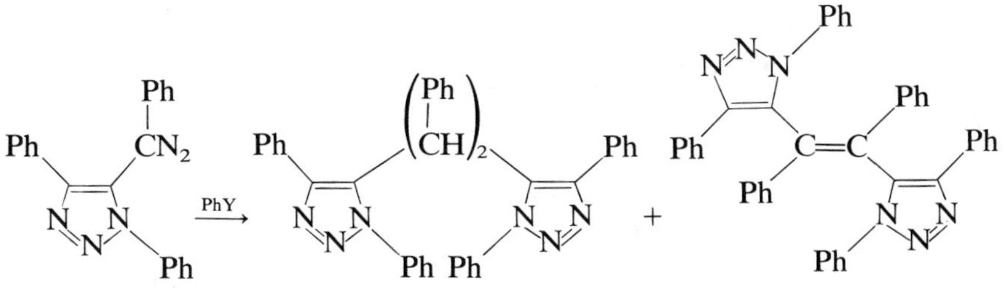

Ph Ph CN_2 N N N Ph $\xrightarrow{PhY}$ Ph (Ph CH)$_2$ Ph N N N Ph Ph N N N + Ph N N N Ph Ph C=C Ph Ph N N N Ph

15.1-7 **15.1-8**

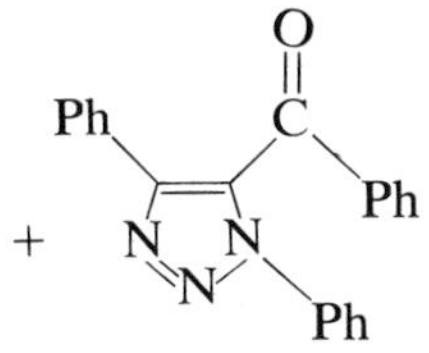

+ Ph C(=O) Ph N N N Ph (16)

15.1-9

Y	% **15.1-7**	% **15.1-8**	% **15.1-9**
H	42.0	5.5	6.7
Me	38.0	12.3	12.5
OMe	55.2	5.5	4.7
OPr-*i*	46.2	3.0	10.9
NO_2	39.3	0	25.6

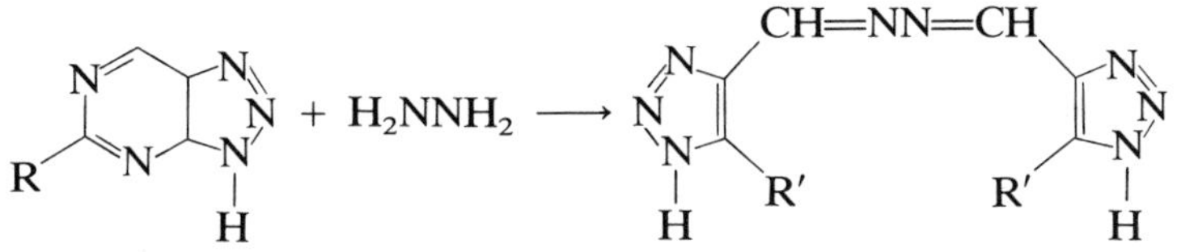

R N N N N N H + H_2NNH_2 ⟶ CH=NN=CH N N N H R′ R′ N N N H (17)

15.1-10

R	= H	OH	SH
R′	= NH_2	$NHCONH_2$	$NHCSNH_2$
% **15.1-10**	= 73	88	72

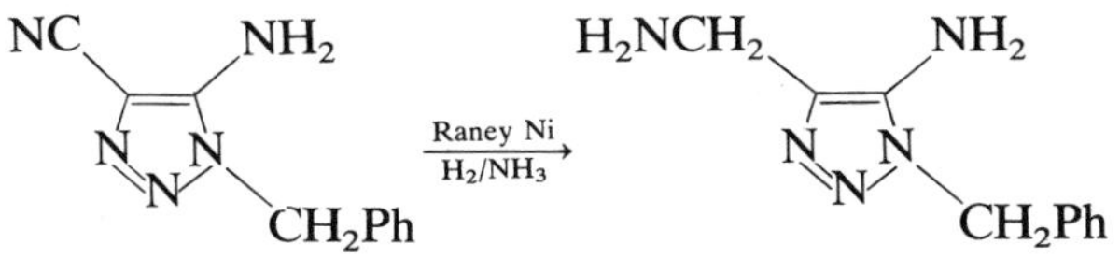

NC NH_2 N N N CH_2Ph $\xrightarrow[H_2/NH_3]{\text{Raney Ni}}$ H_2NCH_2 NH_2 N N N CH_2Ph

66%

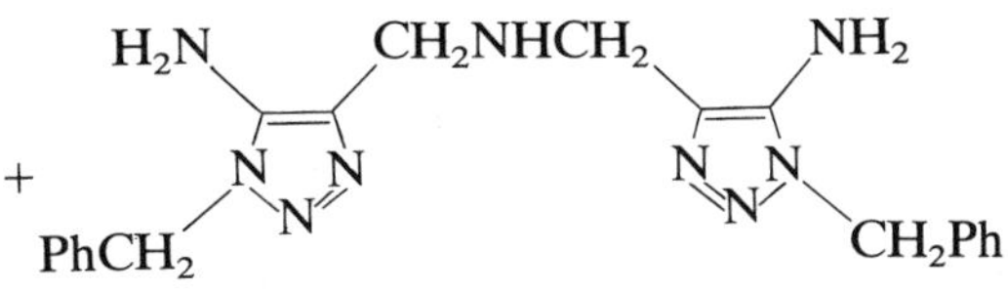

+ H_2N CH_2NHCH_2 NH_2 N N N $PhCH_2$ N N N CH_2Ph (18)

15.1-11

15%

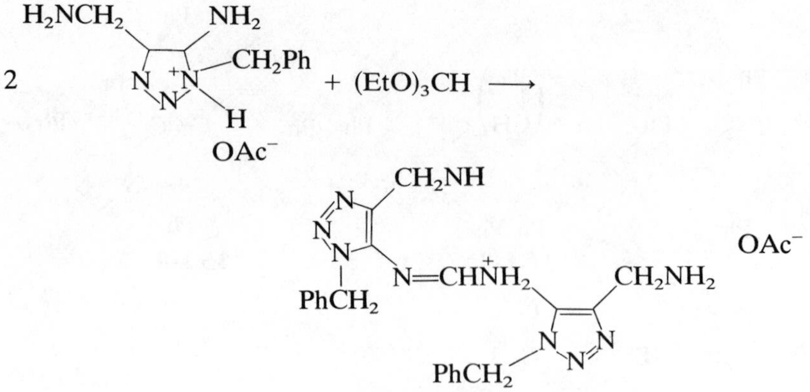

2 [H_2NCH_2, NH_2, CH_2Ph, H, triazolium] OAc^- + $(EtO)_3CH$ ⟶

[CH_2NH, $PhCH_2$, $N{=}CH\overset{+}{N}H_2$, CH_2NH_2, $PhCH_2$] OAc^- (19)

15.1-12
90%

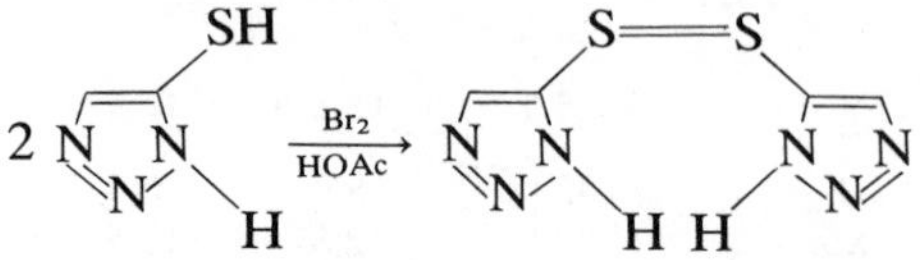

2 [SH, H] $\xrightarrow[HOAc]{Br_2}$ [S═S, H H] (20)
75%

[O, C, R^1, NHR^2, N, S, N] + Me_3COCl $\xrightarrow{HOAc}$

[O, C, R^1, S═S, C, O, R^1, R^2 R^2] (21)

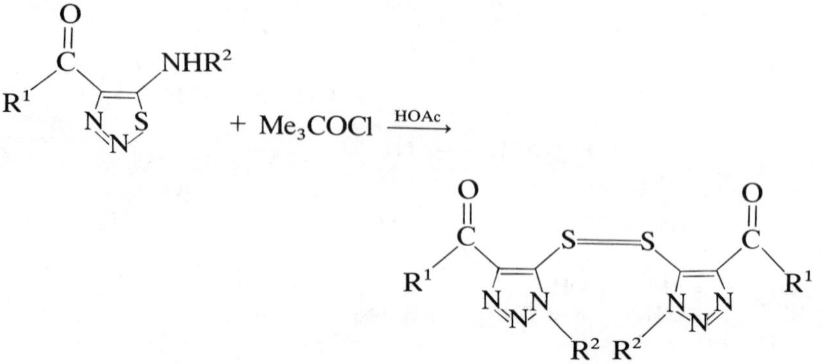

15.1-13

R^1 =	Ph	Me	OEt	OEt	Ph	EtO	EtO	Ph
R^2 =	Ph	Ph	Ph	2-MePh	4-MeOPh	2-MeOPh	4-NO_2Ph	4-NO_2Ph
% **15.1-13** =	86	82	55	63	62	61	74	92

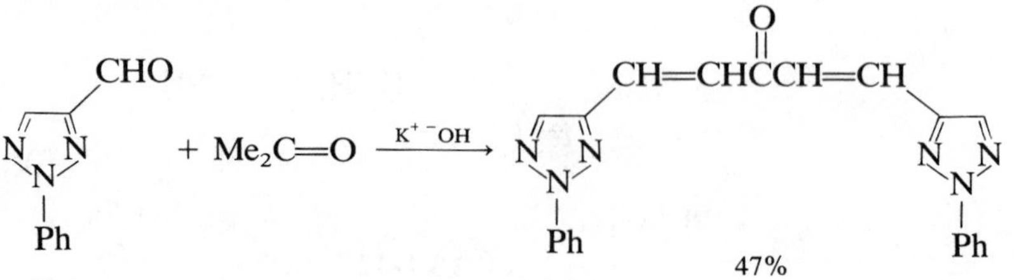

[CHO, Ph] + $Me_2C{=}O$ $\xrightarrow{K^+\ ^-OH}$ [$CH{=}CH\overset{O}{\overset{\|}{C}}CH{=}CH$, Ph, Ph] (22)
47%

$$\xrightarrow{CuSO_4} \qquad (23)$$

69%

Recently, an important first step toward understanding metal ion-azapurine interactions was taken with the synthetic and crystallographic study of a copper(II) complex of a bis[1,2,3-triazole], one resonance contributor of which (**15.1-14**) is shown (Eq. 24).[24]

$$+ CuCl_2 \xrightarrow{HCl} \qquad (24)$$

15.1-14

REFERENCES

1. **78:** 83889f	2. **76:** 126021v	3. **65:** 18484h	4. **80:** 108452q
5. **80:** 3439n	6. **67:** 90739h	7. **65:** 15414f	8. **47:** 10525h,i
9. **82:** 171481e	10. **72:** 21176p	11. **75:** 141988t	12. **76:** 25192w
13. **77:** 164609w	14. **68:** 78204t	15. **73:** 44674j	16. **80:** 145179b
17. **79:** 137047w	18. **79:** 146462e	19. **84:** 121778c	20. **65:** 2250e
21. **70:** 77874r	22. **46:** 6123–4g–a	23. **69:** 27660b	24. **83:** 21202w

TABLE 15. BI- AND BIS[1,2,3-TRIAZOLES AND 1,2,3-TRIAZOLINES]

Compound	Reference
15.1. Bi- and Bis[1,2,3-Triazoles]	
-1-acetic acid, 4-(1-hydroxy-1-methylethyl)-, 2-[4-(1-hydroxy-1-methylethyl)-1-*H*-1,2,3-triazol-1-yl], ethyl ester	**62:** P10443g
-4-acrolein, 1-phenyl-, azine	**64:** 3523e
-5-amine, *N*-[[5-amino-1-benzyl-1*H*-1,2,3-triazol-4-yl]methylene]-1-benzyl-4-(dimethoxymethyl)-	**80:** 37075f
-5-amine, 1,1′-bis(4-bromophenyl)-*N*,*N*-diethyl-4,4′,5,5′-tetrahydro-, [4,5′-bi-	**77:** 126516f
5,5′-azobis[1,4-diphenyl-	**73:** 44674j
4,4′-([1,1′-biphenyl]-4,4′-diyldi-3,2-oxiranediyl)-bis[2-phenyl-	**80:** P134946y
2,2′-(4,4′-biphenylylene)bis[4-nitro-	**61:** P1873h
1,1′-bis[4-(dimethylamino)phenyl]-5,5′-bis(trifluoromethyl)-4,4′-bi	**65:** 18484h
bis(4-benzoyl-, ketonato- mercury	**69:** 92399n
5,4-bis(2,2′-benzoxazolyl)-1,4-phenylene[bi-	**80:** P72087h
1,1′-bis[4,4′-bis[(carboxyamino)phenyl]-, 1-(2-hydroxyethyl)-3-(4-1*H*-1,2,3-triazol-1-ylphenyl)urea ester	**72:** P100715s
1,1′-bis[4,4′-bis[(carboxyamino)phenyl]-, 2,2′-(4-phenylenedioxy)diethanol diester	**72:** P100715s
1,1′-bis[4,4′-bis[(carboxyamino)phenyl]-, tetramethylene diester	**72:** P100715s
2,2′-bis(2-hydroxyphenyl)-5,5′-diphenyl[4,4′bi-	**71:** P124445j
4,4′-[2,2′-bis(2-methoxyphenyl)-5,5′-dimethylbi-	**71:** P124445j
2,2′-[1,2-bis[5-(2-sulfophenyl)]ethenediyl]bis-	**82:** 93787t
2,2′-[1,2-bis[5-(2-sulfophenyl)]ethenediyl]bis-[4-methyl-, ion(2-)	**76:** 101063x
2,2′-[1,2-bis(2-sulfophenyl)ethenediyl]-4,4′-bis[4-(4-sulfophenyl)-, tetrasodium salt	**71:** P126001k
2,2′-[1,2-bis(2-sulfophenyl)ethenediyl]-4,4′-bis[5-(4-phenyl-, disodium salt	**72:** P100716t
5,4-[*N*,*N*″-bis(thioureidoazino)bis(methylidyne-	**79:** 137047w
5,4-[*N*,*N*″-bis(ureyleneazino)bis(methylidyne-	**79:** 137047w
-4-carbonitrile, 5,5′-(1,4-phenylene)bis-	**80:** 3439n
-4-carbonyl azide, 5,5′-ureidobis[1-(2,5-dimethylphenyl)-,	**22:** 3411[3]
1,1′-carbonylbis-	**79:** 78696g
-4-carboxaldehyde, 5-amino-, [(5-amino-1*H*-1,2,3-triazol-4-yl)methylene]hydrazone	**79:** 137047w
-4-carboxaldehyde, 5-amino-2-methyl-, [(5-amino-2-methyl-2*H*-1,2,3-triazol-4-yl)methylene]hydrazone	**79:** 137053v
-4-carboxaldehyde, 1,5-diphenyl-, azine	**68:** 78204t
-5-carboxaldehyde, 1,4-diphenyl-, azine	**68:** 78204t
-4-carboxamide, *N*,*N*′-[1,3,4-oxadiazole-2,5-diylbis(4-amino-3,1-anthraquinonylene)]bis[1-phenyl-	**63:** P15025a, **68:** P88191a
-4-carboxamide, *N*,*N*′-1,4-phenylenebis[2-[4-[(4,5-dihydro-3-methyl-5-oxo-1-phenyl-1*H*-pyrazol-4-yl)azo]phenyl]-5-methyl-	**80:** P122396f

TABLE 15 (*Continued*)

Compound	Reference
15.1. Bi- and Bis[1,2,3-Triazoles] (*Continued*)	
-4-carboxamide, *N,N'*-1,4-phenylenebis[2-[4-[(2-hydroxy-1-naphthalenyl)azo]phenyl]-5-methyl-	**80:** P122396f
-4-carboxamide, *N,N'*-1,4-phenylenebis[5-methyl-2-[4-[[2-oxo-1-[(phenylamino)carbonyl]propyl]-azo]phenyl]-	**80:** P122396f
-4-carboxanilide, 2-phenyl-4'-(2*H*-1,2,3-triazol-2-yl)-,	**46:** 6123h
-4-carboxanilide, 4'-(4-methyl-2*H*-1,2,3-triazol-2-yl)-2-phenyl-,	**46:** 6124a
-4-carboxanilide, 2-(4-carboxyphenyl), (2-phenyl-2*H*-1,2,3-triazol-4-yl)-, methyl ester	**46:** 6123g
-4-carboximidamide, N^2, bis[5-amino-, tetrachloro-, dihydrogen, (OC-6-11)-cuprate(2-),	**83:** 21202w
-4-carboxylic acid, 2,2'-(1,4-biphenylene)bis[5-(2-carboxyphenyl)-,	**22:** 782[6]
-4-carboxylic acid, 2,2'-(3,7-dibenzothiophenediyl)-bis[5-phenyl-, S,S-dioxide	**71:** P92665n
-4-carboxylic acid, 5,5'-dithiobis[1-(2-methoxyphenyl)-, diethyl ester	**70:** 77874r
-4-carboxylic acid, 5,5'-dithiobis[1-(2-methylphenyl)-, diethyl ester	**70:** 77874r
-4-carboxylic acid, 5,5'-dithiobis[1-(4-nitrophenyl)-, diethyl ester	**70:** 77874r
-4-carboxylic acid, 5,5'-dithiobis[1-phenyl-, diethyl ester	**70:** 77874r
-4-carboxylic acid, 5,5'-(1,4-phenylene)bis-, dimethyl ester	**80:** 3439n
-4-carboxylic acid, 5,5'-(1,3-phenylene)bis-[1-benzyl-, diethyl ester	**85:** 123460n
-4-carboxylic acid, 5,5'-(1,4-phenylene)bis[1-benzyl-, diethyl ester	**85:** 123460n
-4-carboxylic acid, 5,5'-(1,4-phenylene)bis[2-(2-methoxy-5-methylphenyl)-, dimethyl ester	**77:** P7313c
-4-carboxylic acid, 1-phenyl-, (1-phenyl-1*H*-1,2,3-triazol-4-yl)-, methyl ester	**42:** 2600c
-4-carboxylic acid, 2-phenyl-, (2-phenyl-2*H*-1,2,3-triazol-4-yl-), methyl ester	**46:** 6123g
-4-carboxylic acid, 1,1'-(sulfonyldi-1,4-phenylene)-bis[5-amino-,	**47:** 10525h
-4-carboxylic acid, 1,1'-(sulfonyldi-1,4-phenylene)-bis[5-amino-, diethyl ester	**47:** 10525h
2,2'-[3,7-dibenzothiophenediyl-4,4'-(sulfophenyl)]bis-, 5,5-dioxide, disodium salt	**71:** P92665n
4,1'-(1,4-dibenzoyl-1,6-hexandiyl)bis[5-methyl-	**82:** 171481e
-4,5-dicarboxylic acid, 1,1'-bicyclo[4.2.0]octa-2,4-diene-7,8)-diylbis-, tetramethyl ether	**76:** 126021v
-4,5-dicarboxylic acid, 1,1'-bicyclo[4.2.0]octane-7,8-diylbis-, tetramethyl ester	**76:** 126021v
-4,5-dicarboxylic acid, 1,1'-(2,2'-biphenylene)bis-, tetramethyl ester	**67:** 90739h

TABLE 15 (*Continued*)

Compound	Reference
15.1. Bi- and Bis[1,2,3-Triazoles] (*Continued*)	
-5,5′-dicarboxylic acid, 4,4′-[2,2′-bis(2-hydroxyphenyl)bi-	**71:** P124445j
-5,5′-dicarboxylic acid, 4,4′-[2,2′-bis(2-hydroxyphenyl)bi-, dimethyl ester	**71:** P124445j
-5,5′-dicarboxylic acid, 4,4′-[2,2′-bis(2-hydroxyphenyl)bi-, dioctyl ester	**71:** P124445j
-5,5′-dicarboxylic acid, 4,4′-[2,2′-bis(2-methoxyphenyl)bi-	**71:** P124445j
-4,5-dicarboxylic acid, 1,1′-[2,3-bis(methylene)-1,4-butanediyl]bis-, tetramethyl ester	**79:** 126209z
-4,5-dicarboxylic acid, 1,1′-[5-(2-butynyl)-6-phenyl-2,4-pyrimidinediyl]bis-, tetramethyl ester	**72:** P55493e
-4,5-dicarboxylic acid, 1,1′-(1,2-cyclooctanediyl)bis-, tetramethyl ester, *cis*-	**76:** 126021v
-4,5-dicarboxylic acid, 1,1′-(3-cyclooctene-1,2-diyl)bis-, tetramethyl ester, (1α, 2α, 3Z)-	**76:** 126021v
-4,5-dicarboxylic acid, 1,1′-(5-cyclooctene-1,2-diyl)bis-, tetramethyl ester, (1α, 2α, 5Z)-	**76:** 126021v
-4,5-dicarboxylic acid, 1,1′-(4-cyclopenten-1,3-diyl)bis-, tetramethyl ester, *cis*-	**78:** 83889f
-4,5-dicarboxylic acid, 1,1′-(decahydro-1,3-dioxo-2-phenyl-4,7-ethano-1*H*-cyclobut[*f*]isoindole-5,6-diyl)bis-, tetramethyl ester	**76:** 126021v
-4,5-dicarboxylic acid, 1,1′-(decahydro-1,3-dioxo-2-phenyl-4,7-etheno-1*H*-cyclobut[*f*]isoindole-5,6-diyl)bis-, tetramethyl ester, (3aα, 4α, 4aβ, 5β, 6β, 6aβ, 7α7aα)-	**76:** 126021v
-4,5-dicarboxylic acid, 1,1′-[3,4-dibromo-5-(diphenylmethylene)-1,2-cyclohexanediyl]bis-, tetramethyl ester	**76:** 72184e
-4,5-dicarboxylic acid, 1,1′-(8,9-dihydro-1,4-dioxaspiro[4.6]undeca-6,10-diene-8,9-diyl)bis-, tetramethyl ester, *cis*-	**76:** 126021v
-4,5-dicarboxylic acid, 1,1′-[(2,5-dihydroxy-1,4-phenylene)dimethylene]bis-, tetramethyl ester, diacetate ester	**64:** 14184g
-4,5-dicarboxylic acid, 1,1′-[3-(diphenylmethylene)-1,2-cyclopentanediyl]bis-, tetramethyl ester	**76:** 72184e
-4,5-dicarboxylic acid, 1,1′-[5-(diphenylmethylene)-3-cyclopentene-1,2-diyl]bis-, tetramethyl ester, *cis*-	**76:** 72184e
-4,5-dicarboxylic acid, 1,1′-[5-(diphenylmethylene)-3-cyclopentene-1,2-diyl]bis-, tetramethyl ester, *trans*-	**76:** 72184e
-4,5-dicarboxylic acid, 1,1′-[1-(ethoxycarbonyl)-2,3-dihydro-1*H*-azepine-2,3-diyl]bis-, tetramethyl ester, *cis*-	**76:** 126021v

TABLE 15 (*Continued*)

Compound	Reference
15.1. Bi- and Bis[1,2,3-Triazoles] (*Continued*)	
-4,5-dicarboxylic acid, 1,1′-[1-(ethoxycarbonyl)-4,5-dihydro-1*H*-azepine-4,5-diyl]bis-, tetramethyl ester	**76:** 126021v
-4,5-dicarboxylic acid, 1,1′-[5-(2-ethoxyethyl)-6-phenyl-2,4-pyrimidinediyl]bis-, tetramethyl ester	**72:** P55493e
-4,5-dicarboxylic acid, 1,1′-[2,3,3a,4,7,7a-hexahydro-5,6-bis(methoxycarbonyl)-4,7-ethano-1*H*-indene-1,3-diyl]bis-, tetramethyl ester, (1α, 3α, 3aα, 4β, 7β, 7aα)-	**78:** 4007r, **78:** 83889f
-4,5-dicarboxylic acid, 1,1′-(5-methyl-6-phenyl-2,4-pyrimidineyl)bis-, tetramethyl ester	**72:** P55493e
-4,5-dicarboxylic acid, 1,1′-(1,2-phenylene)bis-, tetramethyl ester	**67:** 90739h
-4,5-dicarboxylic acid, 1,1′-[6-phenyl-5-(2-propynyl)-2,4-pyrimidinediyl]bis-, tetramethyl ester	**72:** P55493e
5,5′-(1,2-dichloroethylene)bis[1,4-diphenyl-	**68:** 78204t
4,4′-(1,2-dihydroxy-1,2-diphenylethylene)-1,2-bis[1-ethyl-5-methyl-	**82:** 171481e
4,4′-(1,2-dihydroxy-1,2-diphenylethylene)-1,2-bis[5-methyl-	**75:** 109587w
4,4′-(1,2-dihydroxyethylene)-1,2-bis[2-phenyl-, *threo*-	**69:** 27660b
4,4′-(1,2-dihydroxyethylene)-1,2-bis[2-phenyl-, diacetate ester, *threo*-	**69:** 27660b
2,2′-(2,5-dihydroxy-1,4-phenylene)bis[4,5-diphenyl-	**83:** P132619g
-4,5-dimethanol, 1,1′-(oxydiethylene)bis-,	**62:** P10443h
4,4′-(dioxoethyl)bis[2-phenyl-	**69:** 27660b
5,5′-(1,2-diphenyl-1,2-ethanediyl)bis[1,4-diphenyl-	**80:** 145179b
5,5′-(1,2-diphenyl-1,2-ethenediyl)bis[1,4-diphenyl-	**80:** 145179b
1,1′-(1,5-diphenyl-3-oxo-1,4-pentadiene-2,4-diyl)bis[5-phenyl-, (Z,Z)-	**77:** 164609w
4,4′-dithiodi-,	**65:** 2250e
5,5′-dithiobis[4-acetyl-1-phenyl-	**70:** 77874r
5,5′-dithiobis[4-benzoyl-1-(4-methoxyphenyl)-	**70:** 77874r
5,5′-dithiobis[4-benzoyl-1-(4-nitrophenyl)-	**70:** 77874r
5,5′-dithiobis[4-benzoyl-1-phenyl-	**70:** 77874r
[(1,2-ethandiyl)azo]carbonic acid, bis[1-phenyl-5,4-, diethyl ester	**76:** 34177x
4,4′-ethylenebis-	**65:** 15414f
5,5′-ethylenebis 1,4-diphenyl-	**68:** 78204t
4,4′-ethylenebis[2-(trimethylsilyl)-	**65:** 15414f
4-[2′-hydroxy-2′-(2-phenyl-2*H*-1,2,3-triazol-4-yl)-acetyl]-2-phenyl-	**63:** 11540h
4-[2′-hydroxy-2′-[[2-(4-methylphenyl)-2*H*-1,2,3-triazol-4-yl]acetyl]-2-(4-methylphenyl)-	**63:** 1850h, **63:** 11540f
1,1′-(iminodiethylene)bis-	**64:** 2082g
-4-methanamine, 5-amino-*N*-[[5-amino-1-benzyl-1*H*-1,2,3-triazol-4-yl]methyl]-1-benzyl-	**79:** 146462e

TABLE 15 (*Continued*)

Compound	Reference
15.1. Bi- and Bis[1,2,3-Triazoles] (*Continued*)	
-4-methanamine, 5-[[[[4-(aminomethyl)-1-benzyl-1*H*-1,2,3-triazol-5-yl]amino]methylene]amino]-1-benzyl-	**84:** 121778c
-4-methanamine, 5-[[[[4-(aminomethyl)-1-benzyl-1*H*-1,2,3-triazol-5-yl]amino]methylene]amino]-1-benzyl-, monoacetate ester	**84:** 121778c
-4-methanamine, 5-[[[[4-(aminomethyl)-1-benzyl-1*H*-1,2,3-triazol-5-yl]amino]methylene]amino]-1-benzyl-, mononitrate ester	**84:** 121778c
-4-methanol, 1,1′-(oxydiethylene)bis-	**62:** P10443h
-4-methanol, 1,1′-(oxydiethylene)bis[α,α-dimethyl-	**62:** P10443h
-4-methanol, 1,1′-(oxydiethylene)bis[α-methyl-	**62:** P10443g
-4-methanol, 1,1′-pentamethylenebis[α,α-dimethyl-	**62:** P10443h
-4-methanol, 1,1′-pentamethylenebis[α-methyl-	**62:** P10443h
-4-methanol, 1,1′-(1,4-phenylenedimethylene)-bis[α,α-dimethyl-	**62:** P10444a
-4-methanol, 1-phenyl-α-[(1-phenyl-1*H*-1,2,3-triazol-4-yl)methyleneamino]-	**49:** 6242c
-4-methanol, 1,1′-tetramethylenebis-	**62:** P10443g
-4-methanol, 1,1′-tetramethylenebis[α,α-dimethyl-	**62:** P10443g
[methylenebis[(1,4-phenylene)2-cyano-1-(3-methoxy-1,4-phenylene)ethyl]-2,2′-bis[4-methyl-5-phenyl-	**82:** P172618s
1,4,5,8-naphthalenetetracarboxylic 1,8:4,5-diimide, *N,N*′-bis[(2-ethylamino)-3-propyl]-4,4′-	**68:** P68786v
oxalylbis[2-(4-methylphenyl)-	**63:** 11540h
oxalybis[2-phenyl-	**63:** 11540g
oxalylbis[2-phenyl-], bis(phenylhydrazone)	**63:** 11541a
oxalylbis[2-phenyl-], phenylhydrazone	**63:** 11541a
oxomethyl-4,4′-bis[5-methyl-	**80:** 108452q
oxomethyl-4,4′-bis[5-phenyl-	**80:** 108452q
5,5′-(oxydimethylene)bis[1,4-diphenyl-	**68:** 78204t
1,1′-(1,4-phenylene)bis[5-methyl-	**74:** 141988t, **79:** 31992k
1,1′-(1,4-phenylene)bis[5-phenyl-	**74:** 141988t, **79:** 31992k
2,2′-(1,4-phenylene)bis[4-nitro-	**61:** P1873h
1,1′-(1,3-phenylenedicarbonyl)bis[5-ethoxy-4-methyl-	**72:** 21176p
1,1′-(1,4-phenylenedicarbonyl)bis[5-ethoxy-4-methyl-	**72:** 21176p
2,2′-(1,4-phenylenedicarbonyl)bis[4-ethoxy-5-methyl-	**79:** 31992k
1,1′-(1,4-phenylenedicarbonyl)bis[5-methyl-	**76:** 25192w
2,2′-(1,3-phenylenedicarbonyl)bis[4-ethoxy-5-methyl-	**79:** 31992k
2,2′-(1,4-phenylenedicarbonyl)bis[4-methyl-	**76:** 25192w, **79:** 31992k
1,1′-(1,3-phenylenedicarbonyl)bis[5-phenyl-	**76:** 25192w
1,1′-(1,4-phenylenedicarbonyl)bis[5-phenyl-	**76:** 25192w
2,2′-(1,3-phenylenedicarbonyl)bis[4-phenyl-	**76:** 25192w, **79:** 31992k
2,2′-(1,4-phenylenedicarbonyl)bis[4-phenyl-	**76:** 25192w, **79:** 31992k
1,1′-(1,4-phenylenedimethylene)bi-	**73:** 3865d
-4-propanoic acid, α,α′-[1,2-ethenediylbis[(3-sulfo-4,1-phenylene)-2-hydrazinyl-1-ylidene]]bis[5-methyl-β-oxo-, diethyl ester	**84:** P181615g

TABLE 15 (*Continued*)

Compound	Reference
15.1. Bi and Bis[1,2,3-Triazoles] (*Continued*)	
-4-propanoic acid, α,α′-[1,2-ethenediylbis[(3-sulfo-4,1-phenylene)-2-hydrazinyl-1-ylidene]] bis[5-methyl-β-oxo-2-phenyl-, diethyl ester	**84:** P181615g
succinic acid, 2,3-bi-, dimethyl ester	**68:** 68937v
succinic acid, bis[4′,4″-(1H-1,2,3-triazol-1-yl) anilide-	**72:** P100715s
-4-sulfonic acid, 2,2-[1,2-ethenediylbis(3-sulfo-4,1-phenylene)bis[5-[5-methyl-2-(sulfophenyl)-2H-1,2,3-triazol-4-yl]-, hexasodium salt	**84:** P181615g
1,1′-[sulfonyldi-(1,4-phenylene)]bis[5-amino-	**47:** 10525i
terephthalic acid, [bis 4′,4″-(1*H*-1,2,3-triazol-1-yl)anilide-	**72:** P100715s
4,4′,5,5′-tetrahydro-5,5,5′,5′-tetrakis(trifluoromethyl)[1,1′-bi-	**85:** 5148u
5,5′-thiobis[1-phenyl-	**58:** 2447a
4,4′-[[3,3′-thiobis(propanamido)]bi-	**72:** P67723t
4,4′-[*N,N*′-(thioureylene)bi-	**72:** P67723t
4,4′-(*N,N*′-ureylene)bis[5-(4-chlorophenyl)-1-phenyl-	**81:** 3850k
4,4′-(*N,N*′-ureylene)bis[1,5-diphenyl-	**81:** 3850k
5,5′-vinylenebis[1,4-diphenyl-	**68:** 78204t
15.2. Bi- and Bis[1,2,3-Triazolines]	
-4-carboxamide, *N,N*-(1,5-naphthylenedisulfonyl)-bis[5-oxo-	**24:** 3230[8]
-4-carboxamide, 1,1′-(1,5-naphthylenedisulfonyl)-bis[5-oxo-	**24:** 3230[8]
-4-carboxylic acid, 1,1′-(1,5-naphthylenedisulfonyl)-bis[5-oxo-	**24:** 3230[7]
-4,5-dicarboximide, *N,N*′-hexamethylenebis[3-butyl-	**72:** 3846c
-4,5-dicarboximide, 3,3′-pentamethylenebis(*N*-phenyl-	**72:** 3846c
-1-ethanesulfonic acid, 3,3′-(4,4′-biphenylene)bis-	**55:** P10472i
-1-ethanesulfonic acid, 3,3′-(2,2′-dimethoxy-4,4′-biphenylene)bis	**55:** P10473a
-1-ethanesulfonic acid, 3,3′-[vinylenebis(sulfophenylene)]bis-	**55:** P10473a
-1-ethanol, 3,3′-(4,4′-biphenylene)bis[α,β-dimethyl-	**56:** P3484i
-1-ethanol, 3,3′-(4,4′-biphenylene)bis[α-methyl-	**56:** P3484h

REFERENCES

22: 3411[3]	A. Bertho and F. Hölder, *J. Prakt. Chem.*, **119,** 173 (1928).
24: 3230[7,8]	T. Curtius and R. Tüxen, *J. Prakt. Chem.*, **125,** 401 (1930).
42: 2600c	R. Hüttel and A. Gebhardt, *Justus Liebigs Ann. Chem.*, **558,** 34 (1947).
46: 6123–4g–a	J. L. Riebsomer and D. A. Stauffer, *J. Org. Chem.*, **16,** 1643 (1951).
47: 10525h,i	D. Liberman, M. Moyeux, and A. Rouaix, *Bull. Soc. Chim. Fr.*, **19,** 719 (1952).
49: 6242c	R. Hüttel, T. Schneiderhan, H. Hertwig, A. Leuchs, V. Reincke, and J. Miller, *Justus Liebigs Ann. Chem.*, **585,** 115 (1954).
55: P10472–3i–a	R. Mohr and H. Hertel, German Patent, 1,061,789 (1959).
56: P3484h,i	R. Mohr and H. Hertel, German Patent, 1,075,623 (1960).
58: 2447a	M. Koizumi and A. Watanabe, *Bull. Chem. Soc. Jap.*, **28,** 136 (1955).
61: P1873h	R. Mohr and M. Zimmermann, German Patent, 1,168,437 (1964).
62: P10443–4g–a	H. A. Stansbury, Jr., J. A. Durden, Jr., and W. H. Catlette, U.S. Patent, 3,161,651 (1964).
63: 1850h	H. El Khadem, Z. M. El-Shafei, and M. H. Meshreki, *J. Chem. Soc.*, 1524 (1965).
63: 11540–1f–a	H. J. Binte and G. Henseke, *Z. Chem.*, **5,** 268 (1965).
63: P15025a	CIBA Ltd., Belgian Patent, 650,257 (1965).
64: 2082g	H. Gold, *Justus Liebigs Ann. Chem.*, **688,** 205 (1965).
64: 3523e	W. König, M. Coenen, W. Lorenz, F. Bahr, and A. Bassl, *J. Prakt. Chem.*, **30,** 96 (1965).
64: 14184g	J. O. Fournier and J. B. Miller, *J. Heterocycl. Chem.*, **2,** 488 (1965).
65: 2250e	J. Goerdeler and G. Gnad, *Chem. Ber.*, **99,** 1618 (1966).
65: 15414f	L. Birkofer and P. Wegner, *Chem. Ber.*, **99,** 2512 (1966).
65: 18484h	W. P. Norris and W. G. Finnegan, *J. Org. Chem.*, **31,** 3292 (1966).
67: 90739h	S. K. Khetan and M. V. George, *Can. J. Chem.*, **45,** 1993 (1967).
68: P68786v	S. Schutz, M. Bock, and H. Otten, British Patent, 1,081,124 (1967).
68: 68937v	R. M. Acheson and M. W. Foxton, *J. Chem. Soc., C*, 389 (1968).
68: 78204t	P. A. S. Smith and J. G. Wirth, *J. Org. Chem.*, **33,** 1145 (1968).
68: P88191a	K. Weber and P. Hugelshofer, Swiss Patent, 439,534 (1967).
69: 27660b	G. Hanisch and G. Henseke, *Chem. Ber.*, **101,** 2074 (1968).
69: 92399n	L. D. Gavrilova and S. I. Zhdanov, *Elektrokhimiya*, **4,** 841 (1968).
70: 77874r	M. Regitz and H. Scherer, *Chem. Ber.*, **102,** 417 (1969).
71: P92665n	A. Dorlars and O. Neuner, British Patent, 1,155,229 (1969).
71: P124445j	Geigy, A.-G., French Patent, 1,559,131 (1969).
71: P126001k	A. Dorlars and O. Neuner, South African Patent, 68 02,800 (1968).
72: 3846c	Y. Gilliams and G. Smets, *Makromol. Chem.*, **128,** 263 (1969).
72: 21176p	G. L'abbé, P. Ykman, and G. Smets, *Tetrahedron*, **25,** 5421 (1969).
72: P55493e	H. A. Wagner, U.S. Patent, 3,479,357 (1969).
72: P67723t	M. Minagawa, K. Nakagawa, and M. Goto, German Patent, 1,926,547 (1970).
72: P100715s	S. Petersen, H. Striegler, H. Gold, and A. Haberkorn, South African Patent, 69 01,244 (1969).
72: P100716t	Farbenfabriken Bayer A.-G., French Patent, 1,577,760 (1969).
73: 3865d	A. J. Hubert, *Bull. Soc. Chim. Belg.*, **79,** 195 (1970).
73: 44674j	P. A. S. Smith, G. J. W. Breen, M. K. Hajek, and D. V. C. Awang, *J. Org. Chem.*, **35,** 2215 (1970).
74: 141988t	P. Ykman, G. L'abbé, and G. Smets, *Tetrahedron*, **27,** 845 (1971).
75: 109587w	J. Van Thielen, T. van Thien, and F. C. De Schryver, *Tetrahedron Lett.*, 3031 (1971).

76: 25192w	P. Ykman, G. L'abbé, and G. Smets, *Tetrahedron*, **27,** 5623 (1971).
76: 34177x	H. Wamhoff and P. Sohár, *Chem. Ber.*, **104,** 3510 (1971).
76: 72184e	T. Sasaki, K. Minamoto, and K. Usami, *Bull. Chem. Soc. Jap.*, **44,** 3471 (1971).
76: 101063x	C. Eckhardt and H. Hefti, *J. Soc. Dyers Colour.*, **87,** 365 (1971).
76: 126021v	T. Sasaki, K. Kanematsu, and Y. Yukimoto, *J. Org. Chem.*, **37,** 890 (1972).
77: P7313c	H. Lind, German Patent, 2,133,012 (1972).
77: 126516f	H. Cardoen, S. Toppet, G. Smets, and G. L'abbé, *J. Heterocycl. Chem.*, **9,** 971 (1972).
77: 164609w	P. Ykman, G. Mathys, G. L'abbé, and G. Smets, *J. Org. Chem.*, **37,** 3213 (1972).
78: 4007r	T. Sasaki, K. Kanematsu, and Y. Yukimoto, *Chem. Lett.*, 1005 (1972).
78: 83889f	T. Sasaki, K. Kanematsu, and Y. Yukimoto, *J. Chem. Soc., Perkin Trans.*, **1,** 375 (1973).
79: 31992k	G. L'abbé, *Verh. Kon. Acad. Wetensch., Lett. Schone Kunsten Belg., Kl. Wetensch.*, **34,** 49 pp (1972).
79: 78696g	W. Walter and M. Radke, *Justus Liebigs Ann. Chem.*, 636 (1973).
79: 126209z	Y. Gaoni, *Tetrahedron Lett.*, 2361 (1973).
79: 137047w	A. Albert and W. Pendergast, *J. Chem. Soc., Perkin Trans.*, **1,** 1625 (1973).
79: 137053v	A. Albert and H. Taguchi, *J. Chem. Soc., Perkin Trans.*, **1,** 1629 (1973).
79: 146462e	A. Albert, *J. Chem. Soc., Perkin Trans.*, **1,** 1634 (1973).
80: 3439n	G. Beck and D. Günther, *Chem. Ber.*, **106,** 2758 (1973).
80: 37075f	A. Albert and H. Taguchi, *J. Chem. Soc., Perkin Trans.*, **1,** 2037 (1973).
80: P72087h	G. Beck and D. Günther, German Patent, 2,212,694 (1973).
80: 108452q	Y. Tanaka, S. R. Velen and S. I. Miller, *Tetrahedron*, **29,** 3271 (1973).
80: P122396f	B. Hirsch, H. J. Heckemann, and K. P. Thi, East German Patent, 99,998 (1973).
80: P134946y	H. Rempfler, H. Bosshard, and K. Weber, German Patent, 2,332,098 (1974).
80: 145179b	P. A. S. Smith and E. M. Bruckmann, *J. Org. Chem.*, **39,** 1047 (1974).
81: 3850k	F. G. Badder, M. N. Basyouni, F. A. Fouli, and W. I. Awad, *J. Indian Chem. Soc.*, **50,** 589 (1973).
82: 93787t	D. Lorke, *MVC-Rep., Miljoevardscentrum, Stockholm*, **2,** 109 (1973).
82: 171481e	F. C. De Schryver, T. V. Thien, S. Toppet, and G. Smets, *J. Polym. Sci., Polym. Chem. Ed.*, **13,** 227 (1975).
82: P172618s	J. Schroeder, German Patent, 2,335,218 (1975).
83: 21202w	L. G. Purnell, J. C. Shepherd, and D. J. Hodgson, *J. Amer. Chem. Soc.*, **97,** 2376 (1975).
83: P132619g	M. Matsuo, K. Koga, and S. Hotta, Japanese Patent, 74 61,070 (1974).
84: 121778c	A. Albert, *J. Chem. Soc., Perkin Trans.*, **1,** 291 (1976).
84: P181615g	F. Fleck, A. V. Mercer, and R. Paver, German Patent, 2,535,069 (1976).
85: 5148u	K. Burger, S. Tremmel, and H. Schickaneder, *J. Fluorine Chem.*, **7,** 471 (1976).
85: 123460n	R. A. Henry, *J. Chem. Eng. Data*, **21,** 503 (1976).

CHAPTER 16

1,2,3-Triazolium, Triazolinium and Meso-Ionic Compounds

Early attempts to prepare these interesting and useful salts (**16.1-1**) involved both alkylation of the 1,2,3-triazole ring (Eqs. 1–3)[1,2] and cyclization methods (Eqs. 4,5).[2,3] Uniformly high yields have been realized from these methods as well as the subsequent conversion of **16.1-2** to the alcohol **16.1-3**.[3]

A potentially exciting cycloaddition substrate was first prepared as a most probable intermediate (Eq. 6)[4] and later isolated in good yield (Eq. 7).[5]

1-Benzyl-1,2,3-triazole (CH_2Ph) + MeI → (1)

1-Methyl-1,2,3-triazole (Me) + $PhCH_2I$ → 1-Me-3-CH_2Ph-1,2,3-triazolium I^- (2)

16.1-1
about 80%

1-Phenyl-1,2,3-triazole (Ph) + MeI ⟶ 1-Me-3-Ph-1,2,3-triazolium I^- (3)

16.1-1
84%

$CH{=}C(R)SO_2Cl$ + PhN_3 ⟶ 1-Ph-3-[$CH_2CH(R)SO_2Cl$]-1,2,3-triazolium (4)

16.1-1

R	= H	Me
% **16.1-1**	= 99	98

(5)

16.1-2

16.1-3

R′	= hydrogen					
Ar	= 2-MePh	4-ClPh	3-BrPh	3-NO_2Ph	4-MeOPh	1,1′(4, 4′)$(C_6H_4)_2$
% **16.1-2** =	65	70	65	40	60	50

(6)

85%

(7)

73%

In several papers Albert and his collaborators have explored the synthesis of **16.1-4;** an important intermediate in their syntheses of 8-azapurines (Eq. 8).[6,7]

(8)

16.1-4

75%

By far the most important contributions to our understanding of the synthesis and utility of these compounds have been made by Pedersen and Begtrup. They noticed an apparent exception to the general trend of low yields (Eq. 9)[8] and developed the method carefully and in detail (Eqs. 10 to 12).[9–11] Building on this, Begtrup has found a number of analogous reactions that take place in exceptional yield (Eq. 13).[12–16] A limited number of closely related compounds were also prepared by similar means (Eqs. 14,15).[13,17]

OMe, CH_2Ph + MeI $\xrightarrow{CDCl_3}$ O^-, Me, CH_2Ph 91% (9)

OC(O)Ph, CH_2Ph + MeI $\xrightarrow{MeCN}$ O_2CPh, Me, CH_2Ph, I^- 81%

$\xrightarrow[H_2O]{Ba(OH)_2}$ O^-, Me, CH_2Ph 94% (10)

R, OH, CH_2Ph $\xrightarrow[Na^{+-}OH]{MeI}$ R, O^-, Me, CH_2Ph **16.1-5** (11)

R	= H	Me	Ph	Br
% **16.1-5**	= 100	77	68	90

Br, Me + MeI ⟶ Br, Me, Me, I^- $\xrightarrow[Na^{+-}I]{Na^{+-}OH}$ O^-, Me, Me 96% (12)

$$\text{1-}R^1\text{-4-}R^4\text{-5-}R^5\text{-1,2,3-triazole} + TsOMe \longrightarrow \textbf{16.1-6}\ (TsO^-) \quad (13)$$

R^1	= Me	Me	Me	$PhCH_2$	Ph	Ph	Ph	$PhCH_2$	$PhCH_2$	Ph	$PhCH_2$
R^4	= H	Br	Br	H	H	Me	H	Br	Br	H	H
R^5	= H	H	Br	H	H	H	Me	H	Br	SMe	SMe
% **16.1-6**	= 98	91	87	97	100	81	100	86	84	100	99

$$\text{1-benzyl-1,2,3-triazole} + PhCH_2Br \longrightarrow \text{1,3-bis(}CH_2Ph\text{)triazolium } Br^- \ (97\%) \quad (14)$$

$$\text{2-}R^2\text{-4-}R^4\text{-5-}R^5\text{-1,2,3-triazole} + MeSO_3F \longrightarrow \textbf{16.1-7}\ (FSO_3^-) \quad (15)$$

R^2	= Me	Ph	Me
R^4, R^5	= H	H	Br
% **16.1-7**	= 79	99	99

Finally, Begtrup has prepared some interesting sulfur derivatives by two novel reactions (Eqs. 16,17).[18,19] Other examples of **16.1-9** were obtained as mixtures and in lower yield.[19]

$$\text{1-}R'\text{-3-}R\text{-4-X-triazolium } A^- \xrightarrow[HCONMe_2]{Na_2S} \textbf{16.1-8}\ (S^-) \quad (16)$$

R	= Me	Me	$PhCH_2$	Ph
R′	= Me	$PhCh_2$	Me	Me
X	= Br	Br	Br	Cl
% **16.1-8**	= 55	43	30	93

$$\text{1,3-R,R'-4-R''-1,2,3-triazolium} + S_8 \xrightarrow[\text{HCONMe}_2]{\text{Na}^{+-}\text{H}} \textbf{16.1-9} \quad (17)$$

R	= Me	$PhCH_2$.	Me	$PhCH_2$	Me
R′	= Me	$PhCH_2$	Me	Me	Ph
R″	= H	H	Br	Cl	Me
% **16.1-9**	= 100	97	56	71	87

REFERENCES

1. **49:** 10274a
2. **50:** 3414g,h
3. **58:** 10191a–g
4. **74:** 141656h
5. **77:** 75175j
6. **65:** 713g
7. **68:** 68958c
8. **67:** 64305u
9. **67:** 100074d
10. **71:** 61296p
11. **66:** 37847v
12. **72:** 43570q
13. **74:** 141647f
14. **75:** 87745s
15. **75:** 109510r
16. **76:** 72454t
17. **76:** 25187y
18. **75:** 63704b
19. **83:** 58717u

TABLE 16. 1,2,3-TRIAZOLIUM, TRIAZOLINIUM AND MESO-IONIC COMPOUNDS

Compound	Reference
16.1. Alkyl- or Aryl-1,2,3-Triazolium Compounds	
1-(4-acetamidophenyl)-3-benzyl-, chloride	**72:** P100715s
1-(3-acetamidophenyl)-3-methyl-, methyl sulfate	**67:** P100995e
1-(4-acetamidophenyl)-3-methyl-, methyl sulfate	**67:** P100995e
1-(4-amino-3-chlorophenyl)-3-methyl-, methyl sulfate	**72:** P100715s
1-[3-[(4-amino-8-hydroxy-1-naphthyl)azo]phenyl]-3-methyl-, methyl sulfate	**67:** P100995e
1-[4-[(4-amino-8-hydroxy-1-naphthyl)azo]phenyl]-3-methyl-, methyl sulfate	**67:** P100995e
1-(4-aminophenyl)-3-benzyl-, chloride	**72:** P100715s
1-(4-aminophenyl)-3-benzyl-, chloride, hydrochloride	**72:** P100715s
1-(4-aminophenyl)-3-methyl-, methyl sulfate	**72:** P100715s
3-benzyl-1-didodecyl-, bromide	**42:** 5059e
1-benzyl-3-ethyl-4-phenyl-, iodide	**51:** 13855a
1-benzyl-3-ethyl-5-phenyl-, iodide	**51:** 13855a
3-benzyl-1-ethyl-4-phenyl-, iodide	**51:** 13855a
3-benzyl-1-ethyl-5-phenyl-, iodide	**51:** 13855a
1-benzyl-3-methyl-, bromide	**74:** 141647f
1-benzyl-3-methyl-, iodide	**49:** 10274a
1-benzyl-3-methyl-, 4-methylphenyl sulfonate	**74:** 141647f, **75:** 87745s **76:** 72454t, **83:** 58717u
3-benzyl-1-phenyl-, bromide	**83:** 27081c
3-benzyl-1-phenyl-, 4-methylphenyl sulfonate	**74:** 141647f
1-[4-(carboxyamino)phenyl]-3-methyl-, chloride, 1-[4-[3-(2-hydroxyethyl)ureido]phenyl]-3-methyl-1*H*-1,2,3-triazolium chloride ester	**72:** P100715s

TABLE 16 (*Continued*)

Compound	Reference
16.1. Alkyl- or Aryl-1,2,3-Triazolium Compounds (*Continued*)	
1-[4-(carboxyamino)phenyl]-3-methyl-, methyl sulfate,-2,2′-(1,4-phenylenedioxy)diethanol diester	**72:** P100715s
1-[4-(carboxyamino)phenyl]-3-methyl-, methyl sulfate, oxydiethylene ester	**72:** P100715s
1-[4-(carboxyamino)phenyl]-3-methyl-, methyl sulfate, tetramethylene ester	**72:** P100715s
1-[3-chloro-4-[[4-[(2-cyanoethyl)ethylamino]-(2-methylphenyl)]azo]phenyl]-3-methyl-, chloride	**67:** P100995e
1-[3-chloro-4-[[4-(dimethylamino)-2-(2-methylphenyl)]azo]phenyl]-3-methyl-, methyl sulfate	**67:** P100995e
1-[3-chloro-4-[[4-[[4-(*N*-methylanilino)phenyl]azo](2-methylphenyl)]azo]phenyl]-3-methyl-, methyl sulfate	**67:** P100995e
1-[2-[4-[(2-chloro-4-nitrophenyl)azo]-*N*-ethylanilino]ethyl]-3-methyl-, chloride	**69:** P28576r
1-[2-[4-[(2-chloro-4-nitrophenyl)azo]-*N*-ethylanilino]ethyl]-3-methyl-, methyl sulfate	**69:** P28576r
1-[2-[[4-[3-(4-chlorophenyl)-2-pyrazolin-1-yl]phenyl]sulfonyl]ethyl]-3-methyl-, methyl sulfate	**72:** P91491m
1-[2-(chlorosulfonyl)ethyl]-3-phenyl-, chloride	**50:** 3414g
1-[3-[[4-[2-cyanoethyl)ethylamino]phenyl]azo]-4-ethoxyphenyl]-3-ethyl-, 4-methylphenyl sulfonate	**67:** P100995e
1-[4-[[4-[(2-cyanoethyl)ethylamino]phenyl]azo]-phenyl]-3-methyl-, methyl sulfate	**67:** P100995e
1,3-dibenzyl-, bromide	**74:** 141647f
1,3-dibenzyl-, 4-methylphenyl sulfonate	**74:** 141647f, **83:** 58717u
1-[2-[[4-(diethylamino)phenyl]azo]phenyl]-3,5-dimethyl-, trichlorozincate (1-)	**68:** P22709u
1,3-diethyl-4(or 5)-phenyl-, iodide	**51:** 13855b
1,2-dimethyl-, chloride	**76:** 25187y, **83:** 97134w
1,2-dimethyl-, fluorosulfate	**76:** 25187y
1,2-dimethyl-, hydroxide, inner salt	**83:** 97134w
1,3-dimethyl-,′ bromide	**72:** 43570q
1,3-dimethyl-, hydroxide, inner salt	**83:** 97134w
1,3-dimethyl-, 4-methylphenyl sulfonate	**72:** 43570q, **74:** 141647f, **83:** 58717u, **83:** 97134w
1,3-dimethyl-4-ethyl-	**83:** 97134w
1,4-dimethyl-3-phenyl-, bromide	**80:** 132667j
3,4-dimethyl-1-phenyl-, bromide	**80:** 132667j
3,4-dimethyl-1-phenyl-, 4-methylphenyl sulfonate	**74:** 141647f, **83:** 58717u
3,5-dimethyl-1-phenyl-, 4-methylphenyl sulfonate	**74:** 141647f
1,3-dimethyl-4-(2-formylethenyl)-, iodide	**75:** 63702z
1-[4-[[4-[ethyl(2-hydroxyethyl)amino](2-methylphenyl)]-azo](3-methylphenyl)]-3-methyl-, methyl sulfate	**67:** P100995e
1,1′-[hexamethylenebis(thioureylene-1,4-phenylene]bis-[3-methyl-, bis(methylsulfate	**72:** P100715s
1,1′-[hexamethylenebis[ureylene(2-methyl-1,4-phenylene)]]bis[3-methyl-, bis(methyl sulfate)	**72:** P100715s

TABLE 16 (*Continued*)

Compound	Reference
16.1. Alkyl- or Aryl-1,2,3-Triazolium Compounds (*Continued*)	
1,1′-[hexamethylenebis(ureylene-1,4-phenylene)]bis-[3-methyl-, bis(methyl sulfate)	**72:** P100715s
4-(1-hydroxy-1-methylethyl)-1-methyl-1-phenyl-, iodide	**46:** 8652a
3-methyl-1-[4-[[4-[[4-(*N*-methylanilino)phenyl]-azo](2-methylphenyl)]azo]phenyl]-, methyl sulfate	**67:** P100995e
1-methyl-2-[4-[(3-methyl-5-oxo-2-pyrazolin-4-yl)azo]phenyl]-, chloride	**63:** P3082d
3-methyl-1-(4-nitrophenyl)-, methyl sulfate	**72:** P100715s
1-methyl-2-phenyl-, bromide	**80:** 132667j
1-methyl-2-phenyl-, fluorosulfate	**76:** 25187y, **78:** 159527j
1-methyl-2-phenyl-, tetrafluoroborate (1-)	**83:** P19056h
1-methyl-3-phenyl-, bromide	**80:** 132667j
3-methyl-1-phenyl-, 4-methylphenyl sulfonate	**74:** 141647f, **76:** 72454t **83:** 58717u
5-methyl-3-phenyl-1-(2-sulfopropyl)-, hydroxide, inner salt	**50:** 3414h
3-methyl-1-[4-[(1,2,3,4-tetrahydro-2,7-dihydroxy-benzo[*h*]quinolin-6-yl)azo]phenyl]-, methyl sulfate	**67:** P100995e
1,1′-[1,4-phenylenebis(iminocarbonylimino-4,1-phenylene)]bis[3-methyl-, bis(methyl sulfate)	**79:** 100462b
1,1′-[1,4-phenylenebis[ureylene-3-methyl-1,4-phenylene)]]bis[3-methyl-, bis(methyl sulfate)	**72:** P100715s
1,1′-[1,4-phenylenebis(ureylene-1,4-phenylene)]bis]3-benzyl-, dichloride	**72:** P100715s
3-phenyl-1-(2-sulfoethyl)-, hydroxide, inner salt	**50:** 3414h
1,1′-[succinylbis(imino-1,4-phenylene)]bis[3-ethyl-, dibromide	**72:** P100715s
1,1′-[terephthaloylbis(imino-1,4-phenylene)]bis[3-ethyl-, dibromide	**72:** P100715s
1,3,4,5-tetramethyl-, (?)	**83:** 97134w
1,1′-(ureylenedi-1,4-phenylene)bis[3-benzyl-, dichloride	**72:** P100715s
1,1′-(ureylenedi-1,4-phenylene)bis[3-methyl-, dichloride	**72:** P100715s
16.2. Miscellaneous 1,2,3-Triazolium Compounds	
5-acetamido-3-methyl-1-(4-methylphenyl)-, hydroxide, inner salt	**73:** 120571r
5-acetamido-3-methyl-1-(4-methylphenyl)-, hydroxide, inner salt, monopicrate	**73:** 120571r
5-acetamido-3-methyl-1-phenyl-, hydroxide, inner salt	**73:** 120571r
5-acetamido-3-methyl-1-phenyl-, picrate	**73:** 120571r
5-acetyl-1,2-dimethyl-, (?)	**83:** 97134w
4-acetyl-5-[4-(dimethylamino)styryl]-1,2-diphenyl-, perchlorate	**59:** 15410f
4-acetyl-1,2-diphenyl-5-methyl-, perchlorate	**59:** 15410f
5-amino-4-(aminocarbonyl)-1-benzyl-3-methyl-, 4-methylphenyl sulfonate	**68:** 68958c, **80:** 82890d

TABLE 16 (*Continued*)

Compound	Reference
16.2. Miscellaneous 1,2,3-Triazolium Compounds (*Continued*)	
5-amino-1-benzyl-4-carbamoyl-3-methyl-, 4-methylphenyl sulfonate	**65:** 713g
5-amino-3-methyl-1-(4-methylphenyl)-, picrate	**73:** 120571r
5-amino-3-methyl-1-phenyl-, chloride	**73:** 120571r
1-benzamido-2-benzoyl-4,5-dimethyl-, hydroxide, inner salt	**72:** 66869h
4-benzoyl-1,2-diphenyl-5-methyl-, perchlorate	**59:** 15410f
5-benzoyloxy-1-benzyl-3-methyl-, iodide	**67:** 100074d
1-benzyl-4-[(2-bromocyclohexyl)thio]-3-methyl-, bromide	**77:** 114317d
1-benzyl-4-bromo-5-hydroxy-3-methyl-, hydroxide, inner salt	**71:** 61296p
1-benzyl-4-bromo-3-methyl-, 4-methylphenyl sulfonate	**74:** 141647f, **75:** 87745s
1-benzyl-5-bromo-3-methyl-, 4-methylphenyl sulfonate	**74:** 141647f, **75:** 87745s
1-benzyl-4-(bromothio)-3-methyl-, bromide	**77:** 114317d
1-benzyl-4-chloro-5-mercapto-3-methyl-, hydroxide, inner salt	**83:** 58717u
3-benzyl-4-chloro-5-mercapto-1-methyl-, hydroxide, inner salt	**83:** 58717u
1-benzyl-4-chloro-3-methyl-, 4-methylphenyl sulfonate	**83:** 58717u
3-benzyl-4-chloro-1-methyl-, 4-methylphenyl sulfonate	**83:** 58717u
1-benzyl-4,5-dibromo-3-methyl-, 4-methylphenyl sulfonate	**74:** 141647f, **75:** 87745s
1-benzyl-3,4-dimethyl-5-hydroxy-, hydroxide, inner salt	**71:** 61296p
1-benzyl-5-hydroxy-3-methyl-, hydroxide, inner salt	**67:** 100074d
3-benzyl-4-hydroxy-1-methyl-, hydroxide, inner salt	**67:** 64305u
3-benzyl-5-hydroxy-1-methyl-, hydroxide, inner salt	**71:** 61296p, **76:** 72454t
1-benzyl-5-hydroxy-3-methyl-4-phenyl-, hydroxide, inner salt	**71:** 61296p
3-benzyl-4-mercapto-1-methyl-, hydroxide, inner salt	**75:** 63704b, **77:** 114317d, **83:** 58717u
3-benzyl-5-mercapto-1-methyl-, hydroxide, inner salt	**75:** 63704b, **76:** 72454t, **77:** 114317d, **83:** 58717u
1-benzyl-5-methoxy-3-methyl-, iodide	**67:** 64305u
1-benzyl-3-methyl-4-(methylthio)-, 4-methylphenyl sulfonate	**76:** 72454t
3-benzyl-1-methyl-5-[(2-oxopropyl)thio]-, bromide	**77:** 114317d
4,5-bis(methylthio)-1,3-dimethyl-, (?)	**83:** 97134w
3-(4-bromo-2-chlorophenyl)-1-butyl-4-hydroxy-, hydroxide, inner salt	**84:** P121847z
3-(4-bromo-2-chlorophenyl)-1-*sec*-butyl-4-hydroxy-, hydroxide, inner salt	**84:** P121847z
3-(4-bromo-2-chlorophenyl)-1-*tert*-butyl-4-hydroxy-, hydroxide, inner salt	**84:** P121847z
3-(4-bromo-2-chlorophenyl)-1-cyclohexyl-4-hydroxy-, hydroxide, inner salt	**84:** P121847z

TABLE 16 (*Continued*)

Compound	Reference
16.2. Miscellaneous 1,2,3-Triazolium Compounds (*Continued*)	
3-(4-bromo-2-chlorophenyl)-1-ethyl-4-hydroxy-, hydroxide, inner salt	**84:** P121847z
3-(4-bromo-2-chlorophenyl)-4-hydroxy-1-isobutyl-, hydroxide, inner salt	**84:** P121847z
3-(4-bromo-2-chlorophenyl)-4-hydroxy-1-isopentyl-, hydroxide, inner salt	**84:** P121847z
3-(4-bromo-2-chlorophenyl)-4-hydroxy-1-isopropyl-, hydroxide, inner salt	**84:** P121847z
3-(4-bromo-2-chlorophenyl)-4-hydroxy-1-pentyl-, hydroxide, inner salt	**84:** P121847z
N-bromo-4,5-dimethyl-, bromide	**50:** 9393c
4-bromo-1,2-dimethyl-, fluorosulfate	**76:** 25187y
4-bromo-1,3-dimethyl-, 4-methylphenyl sulfonate	**72:** 43570q, **74:** 141647f, **75:** 109510r, **83:** 85717u
5-bromo-1,2-dimethyl-, (?)	**83:** 97134w
5-bromo-1,2-dimethyl-, fluorosulfate	**76:** 25187y
5-bromo-1,3-dimethyl-, chloride	**66:** 37847v
4-bromo-1,3-dimethyl-5-hydroxy-, hydroxide, inner salt	**66:** 37847v, **72:** 43570q, **75:** 109510r
4-bromo-1,3-dimethyl-5-mercapto-, hydroxide, inner salt	**83:** 58717u
5-bromo-1,3-dimethyl-4-methoxy-, 4-methylphenyl sulfonate	**75:** 109510r
4-bromo-5-hydroxy-3-methyl-1-phenyl-, hydroxide inner salt	**74:** 141647f, **76:** 72454t
5-bromo-4-hydroxy-3-methyl-1-phenyl-, hydroxide, inner salt	**74:** 141647f
4-bromo-5-hydroxy-3-methyl-1-(4-methylphenyl)-, hydroxide, inner salt	**73:** 120571r
3-(4-bromo-2-methylphenyl)-4-hydroxy-1-isopropyl-, hydroxide, inner salt	**84:** P121847z
1-*tert*-butyl-3-(4-chloro-2-nitrophenyl)-4-hydroxy-, hydroxide, inner salt	**84:** P121847z
1-butyl-3-(2,4-dichlorophenyl)-4-hydroxy-, hydroxide, inner salt	**84:** P121847z
1-*sec*-butyl-3-(2,4-dichlorophenyl)-4-hydroxy-, hydroxide, inner salt	**84:** P121847z
1-*tert*-butyl-3-(2,4-dichlorophenyl)-4-hydroxy-, hydroxide, inner salt	**84:** P121847z
1-butyl-4-hydroxy-3-(2,4,5-trichlorophenyl)-, hydroxide, inner salt	**84:** P121847z
1-*sec*-butyl-4-hydroxy-3-(2,4,5-trichlorophenyl)-, hydroxide, inner salt	**84:** P121847z
1-*tert*-butyl-4-hydroxy-3-(2,4,5-trichlorophenyl)-, hydroxide, inner salt	**84:** P121847z
1-butyl-4-hydroxy-3-(2,4,6-trichlorophenyl)-, hydroxide, inner salt	**84:** P121847z

TABLE 16 (*Continued*)

Compound	Reference
16.2. Miscellaneous 1,2,3-Triazolium Compounds (*Continued*)	
4-carboxy-1-[2-[[4-(dimethylamino)phenyl]azo]phenyl]-3,5-dimethyl-, trichlorozincate (1-), methyl ester	**67:** P44797c, **68:** P22709u
4-carboxy-3-ethyl-1-[2-[3-(3-ethyl-2-benzothiazolinylidene)propenyl]-1-methylbenzimidazolium-6-yl]-5-methyl-, diiodide	**74:** 48048b
4-carboxy-1-[2-[2-(3-ethyl-4-oxo-2-thioxo-5-thiazolidene)ethylidene]-3-methyl-5-benzothiazolidinyl]-3,5-dimethyl-, iodide	**74:** 48048b
5-chloro-1,2-dimethyl-, (?)	**83:** 97134w
4-chloro-1-methyl-3-phenyl-, bromide	**80:** 132667j
4-chloro-1-methyl-3-phenyl-, 4-methylphenyl sulfonate	**77:** 114317d
4-chloro-3-methyl-1-phenyl-, bromide	**80:** 132667j
3-(4-chloro-2-methylphenyl)-1-cyclopentyl-4-hydroxy-, hydroxide, inner salt	**84:** P121847z
3-(4-chloro-2-nitrophenyl)-4-hydroxy-1-isopropyl-, hydroxide, inner salt	**84:** P121847z
1-[2-[[4-[(2-cyanoethyl)methylamino]phenyl]azo]phenyl]-4-(methoxycarbonyl)-3,5-dimethyl-, (T-4)-tetrachlorozincate (2-)	**76:** P128683m
4-cyano-1,3-dimethyl-5-[[(4-methylphenyl)-sulfonyl]amino]-, hydroxide, inner salt	**84:** 135549r
1-cyclohexyl-3-(2,4-dichlorophenyl)-4-hydroxy-, hydroxide, inner salt	**84:** P121847z
1-cyclopentyl-3-(2,4-dichlorophenyl)-4-hydroxy-, hydroxide, inner salt	**84:** P121847z
1,3-dibenzyl-5-hydroxy-, hydroxide, inner salt	**71:** 61296p
1,3-dibenzyl-5-mercapto-, hydroxide, inner salt	**83:** 58717u
4,5-dibromo-1,2-dimethyl-, fluorosulfate	**76:** 25187y
4,5-dibromo-1,3-dimethyl-, (?)	**83:** 97134w
4,5-dibromo-1,3-dimethyl-, 4-methylphenyl sulfonate	**72:** 43570q, **75:** 109510r
3-(2,4-dibromophenyl)-1-ethyl-4-hydroxy-, hydroxide, inner salt	**84:** P121847z
4,5-dichloro-1,3-dimethyl-, (?)	**83:** 97134w
3-(2,4-dichlorophenyl)-1-ethyl-4-hydroxy-, hydroxide, inner salt	**84:** P121847z
3-(2,4-dichlorophenyl)-4-hydroxy-1-isobutyl-, hydroxide, inner salt	**84:** P121847z
3-(2,4-dichlorophenyl)-4-hydroxy-1-isopentyl-, hydroxide, inner salt	**84:** P121847z
3-(2,4-dichlorophenyl)-4-hydroxy-1-isopropyl-, hydroxide, inner salt	**84:** P121847z
3-(2,4-dichlorophenyl)-4-hydroxy-1-propyl-, hydroxide, inner salt	**84:** P121847z
1,3-dimethyl-5-hydroxy-, hydroxide, inner salt	**66:** 37847v, **77:** 61894f, **83:** 97134w
1,3-dimethyl-4-hydroxy-5-methoxy-, hydroxide, inner salt	**75:** 109510r

TABLE 16 (*Continued*)

Compound	Reference
16.2. Miscellaneous 1,2,3-Triazolium Compounds (*Continued*)	
1,3-dimethyl-5-hydroxy-4-phenyl-, hydroxide, inner salt	**67:** 64305u, **80:** 132667j
3,4-dimethyl-5-hydroxy-1-phenyl-, hydroxide, inner salt	**67:** 64305u, **74:** 141647f
3,5-dimethyl-4-hydroxy-1-phenyl-, hydroxide, inner salt	**74:** 141647f
1,3-dimethyl-4-mercapto-, hydroxide, inner salt	**75:** 63704b, **77:** 114317d, **83:** 58717u
1,5-dimethyl-4-mercapto-3-phenyl-, hydroxide, inner salt	**83:** 58717u
1,3-dimethyl-5-methoxy-, iodide	**64:** 11200c
1,3-dimethyl-4-methoxy-, 4-methylphenyl sulfonate	**74:** 141647f, **75:** 109510r
1,3-dimethyl-5-methoxy-, trichlorozincate	**66:** P19822d
1,3-dimethyl-4-[methyl-[(4-methylphenyl)-sulfonyl]amino]-, iodide	**83:** 192266z
1,4-dimethyl-5-[(2-methyl-4*H*-pyrazolo[1,5-α]-benzimidazol-3-yl]azo]-, methyl sulfate	**66:** P105886j
1,2-dimethyl-5-(methylthio)-, (?)	**83:** 97134w
5,5′-[(2,5-dimethyl-1,4-phenylene)bis[methylene-(ethylimino)-1,4-phenyleneazo]]bis[1-benzyl-4-methyl-, bis(methyl sulfate)	**66:** P47321h
5,5′-[(2,5-dimethyl-1,4-phenylene)bis[methylene-(ethylimino)-1,4-phenyleneazo]]bis[1,4-diethyl-, bis(ethyl sulfate)	**66:** P47321h
5,5′-[(2,5-dimethyl-1,4-phenylene)bis[methylene[(2-hydroxy-3-methoxypropyl)imino](2-methyl-1,4-phenylene)azo]]-bis[1,4-dimethyl-bis(methyl sulfate)	**66:** P47321h
5-[(1,2-dimethyl-1*H*-pyrazolo[1,5-*α*]benzimidazol-3-yl)azo]-1,4-dimethyl-, methyl sulfate	**66:** P105886j
1,4-diphenyl-5-hydroxy-3-methyl-, hydroxide, inner salt	**67:** 64305u
1,4-diphenyl-5-imino-, hydroxide, inner salt	**61:** 654f
4,4′-dithiobis[1-benzyl-3-methyl-, dibromide	**77:** 114317d
5-ethoxy-1,3-dimethyl-, chloride	**66:** 37847v
5-ethoxy-3-methyl-1-(4-methylphenyl)-, tetrafluoroborate (1-)	**73:** 120571r
5-ethoxy-3-methyl-1-(4-methylphenyl)-, picrate	**73:** 120571r
1,1′-ethylenebis[4-acetyl-2-(4-chlorophenyl)-5-phenyl-, diperchlorate	**73:** 66510e
1,1′-ethylenebis[4-acetyl-2,5-diphenyl, diperchlorate	**73:** 66510e
1,1′-ethylenebis[4-acetyl-2-(4-ethoxyphenyl)-5-methyl-, diperchlorate	**73:** 66510e
1,1′-ethylenebis[4-acetyl-2-(4-methoxyphenyl)-5-methyl-, diperchlorate	**73:** 66510e
1,1′-ethylenebis[4-acetyl-2-(4-methylphenyl)-5-phenyl-, diperchlorate	**73:** 66510e
5,5′-[ethylenebis[oxyethylene(ethylimino)-1,4-phenyleneazo]]bis[1,4-dimethyl-, bis(methyl sulfate)	**66:** P47321h

TABLE 16 (*Continued*)

Compound	Reference
16.2. Miscellaneous 1,2,3-Triazolium Compounds (*Continued*)	
5-[[4-(1-ethylhydrazino)phenyl]azo]-1,4-dimethyl-, trichlorozincate	**66:** P19822d
3-ethyl-5-hydroxy-1-phenyl-, hydroxide, inner salt	**67:** 64305u
1-ethyl-4-hydroxy-3-(2,4,5-trichlorophenyl)-, hydroxide, inner salt	**84:** P121847z
3-ethyl-5-methoxy-1-phenyl-, iodide	**67:** 64305u
1,1′-hexamethylenebis[4-acetyl-2-(4-chlorophenyl)-5-methyl-, diperchlorate	**73:** 66510e
1,1′-hexamethylenebis[4-acetyl-2,5-diphenyl-, diperchlorate	**73:** 66510e
1,1′-hexamethylenebis[4-acetyl-2-(4-methoxyphenyl)-5-methyl-, diperchlorate	**73:** 66510e
1,1′-hexamethylenebis[4-acetyl-5-methyl-2-(4-methylphenyl)-, diperchlorate	**73:** 66510e
1,1′-hexamethylenebis[4-acetyl-5-methyl-2-phenyl-, diperchlorate	**73:** 66510e
1,1′-hexamethylenebis[4-acetyl;-2-(4-methylphenyl)-5-phenyl-, diperchlorate	**73:** 66510e
4-hydroxy-1-isobutyl-3-(2,4,5-trichlorophenyl)-, hydroxide, inner salt	**84:** P121847z
4-hydroxy-1-isopropyl-3-(2,4,5-trichlorophenyl)-, hydroxide, inner salt	**84:** P121847z
4-hydroxy-1-isopropyl-3-(2,4,6-trichlorophenyl)-, hydroxide, inner salt	**84:** P121847z
5-hydroxy-3-methyl-1-(4-methylphenyl)-, chloride	**73:** 120571r
5-hydroxy-3-methyl-1-(4-methylphenyl)-, hydroxide, inner salt	**73:** 120571r
5-hydroxy-3-methyl-1-(4-methylphenyl)-4-(tricyanoethenyl)-, hydroxide, inner salt	**73:** 120571r
4-hydroxy-1-methyl-5-(methylthio)-1-phenyl-, hydroxide, inner salt	**76:** 72454t
4-hydroxy-3-methyl-5-(methylthio)-1-phenyl-, hydroxide, inner salt	**76:** 72454t
4-hydroxy-3-methyl-1-phenyl-, hydrozide, inner salt	**74:** 141647f, **76:** 72454t
5-hydroxy-3-methyl-1-phenyl-, hydroxide, inner salt	**67:** 64305u, **73:** 120571r, **76:** 72454t
5-hydroxy-1,3,4-trimethyl-, hydroxide, inner salt	**71:** 61296p
5-imino-1-(4-methylphenyl)-4-phenyl-, hydroxide, inner salt	**61:** 654f
4-mercapto-3-methyl-1-phenyl-, hydroxide, inner salt	**76:** 72454t, **83:** 58717u
5-mercapto-3-methyl-1-phenyl-, hydroxide, inner salt	**75:** 63704b, **77:** 114317d, **83:** 58717u
1-methyl-4-(methylthio)-3-phenyl-, 4-methylphenyl sulfonate	**76:** 72454t
5,5′-[(4-methylphenylimino)bis[ethylene-(ethylimino)-1,4-phenyleneazo]]bis[1,4-diethyl-, bis(ethyl sulfate)	**66:** P47321h
1,1′-pentamethylenebis[4-acetyl-2-(4-methoxyphenyl)-5-methyl-, diperchlorate	**73:** 66510e

TABLE 16 (*Continued*)

Compound	Reference
16.2. Miscellaneous 1,2,3-Triazolium Compounds (*Continued*)	
1,1′-pentamethylenebis[4-acetyl-5-methyl-2-(4-methylphenyl)-, diperchlorate	**73:** 66510e
1,1′-pentamethylenebis[4-acetyl-5-methyl-2-phenyl-, diperchlorate	**73:** 66510e
1-(phenylamino)-2,4,5-triphenyl-, hydroxide, inner salt	**74:** 141656h, **77:** 75175j
5,5′-[1,4-phenylenebis[methylene(ethylimino)-1,4-phenyleneazo]]bis[1,4-dimethyl-, bis(methyl sulfate)	**66:** P47321h
4,4′-sulfinylbis[5-hydroxy-3-methyl-1-(4-methylphenyl)-, dihydroxide, bis(inner salt)	**73:** 120571r
1,1′-tetramethylenebis[4-acetyl-2-(4-chlorophenyl)-5-methyl-, diperchlorate	**73:** 66510e
1,1′-tetramethylenebis[4-acetyl-2-(4-methoxyphenyl)-5-methyl-, diperchlorate	**73:** 66510e
1,1′-tetramethylenebis[4-acetyl-2-(4-methoxyphenyl)-5-phenyl-, diperchlorate	**73:** 66510e
1,1′-tetramethylenebis[4-acetyl-5-methyl-2-(4-methylphenyl)-, diperchlorate	**73:** 66510e
1,1′-tetramethylenebis[4-acetyl-5-methyl-2-phenyl-, diperchlorate	**73:** 66510e
4,4′-thiobis[5-hydroxy-3-methyl-1-(4-methylphenyl)-, dihydroxide, bis(inner salt)	**73:** 120571r
4,4′-thiobis[5-hydroxy-3-methyl-1-phenyl-, dihydroxide, bis(inner salt)	**73:** 120571r
1,1′-trimethylenebis[4-acetyl-2-(4-methoxyphenyl)-5-phenyl-, diperchlorate	**73:** 66510e
16.3. Δ^2-1,2,3-Triazolinium Compounds	
3-(4-acetamidophenyl)-1-(2-hydroxyethyl)-, hydroxide, sulfate, inner salt	**58:** 10191c
3-[4-(aminocarbonyl)-1*H*-pyrazol-3-yl]-1-(2-chloroethyl)-, chloride	**78:** 34925f
3-(4-aminophenyl)-1-(2-hydroxyethyl)-, hydroxide, sulfate, inner salt	**58:** 10191d
3,3′-(4,4′-biphenylene)bis[1-(2-hydroxyethyl)-, hydroxide, sulfate, inner salt	**58:** 10191d
3,3′-(4,4′-biphenylene)bis[1-(2-hydroxy-1-methylethyl)-, hydroxide, sulfate, inner salt	**58:** 10191e
3,3′-(4,4′-biphenylene)bis[1-(2-hydroxypropyl)-, chloride	**58:** 10191g
3,3′-(4,4′-biphenylene)bis[1-(2-hydroxypropyl)-, hydroxide, sulfate, inner salt	**58:** 10191d
3-(4-bromo-2-chlorophenyl)-1-butyl-4-hydroxy-, hydroxide, inner salt	**84:** P164788y
3-(4-bromo-2-chlorophenyl)-1-*sec*-butyl-4-hydroxy-, hydroxide, inner salt	**84:** P164788y
3-(4-bromo-2-chlorophenyl)-1-*tert*-butyl-4-hydroxy-, hydroxide, inner salt	**84:** P164788y

TABLE 16 (*Continued*)

Compound	Reference
16.3. Δ^2-1,2,3-Triazolium Compounds (*Continued*)	
1-(4-bromo-2-chlorophenyl)-3-butyl-5-oxo-, hydroxide, inner salt	**78:** P136300r
1-(4-bromo-2-chlorophenyl)-3-*sec*-butyl-5-oxo-, hydroxide, inner salt	**78:** P136300r
1-(4-bromo-2-chlorophenyl)-3-*tert*-butyl-5-oxo-, hydroxide, inner salt	**78:** P136300r
3-(4-bromo-2-chlorophenyl)-1-cyclohexyl-4-hydroxy-, hydroxide, inner salt	**84:** P164788y
1-(4-bromo-2-chlorophenyl)-3-cyclohexyl-5-oxo-, hydroxide, inner salt	**78:** P136300r
3-(4-bromo-2-chlorophenyl)-1-ethyl-4-hydroxy-, hydroxide, inner salt	**84:** P164788y
1-(4-bromo-2-chlorophenyl)-3-ethyl-5-oxo-, hydroxide, inner salt	**78:** P136300r
3-(4-bromo-2-chlorophenyl)-4-hydroxy-1-isobutyl-, hydroxide, inner salt	**84:** P164788y
3-(4-bromo-2-chlorophenyl)-4-hydroxy-1-isopentyl-, hydroxide, inner salt	**84:** P164788y
3-(4-bromo-2-chlorophenyl)-4-hydroxy-1-isopropyl-, hydroxide, inner salt	**84:** P164788y
3-(4-bromo-2-chlorophenyl)-4-hydroxy-1-pentyl-, hydroxide, inner salt	**84:** P164788y
1-(4-bromo-2-chlorophenyl)-3-isobutyl-5-oxo-, hydroxide, inner salt	**78:** P136300r
1-(4-bromo-2-chlorophenyl)-3-isopentyl-5-oxo-, hydroxide, inner salt	**78:** P136300r
1-(4-bromo-2-chlorophenyl)-3-isopropyl-5-oxo-, hydroxide, inner salt	**78:** P136300r
1-(4-bromo-2-chlorophenyl)-5-oxo-3-pentyl-, hydroxide, inner salt	**78:** P136300r
3-(4-bromo-2-methylphenyl)-4-hydroxy-1-isopropyl-, hydroxide, inner salt	**84:** P164788y
1-(4-bromo-2-methylphenyl)-3-isopropyl-5-oxo-, hydroxide, inner salt	**78:** P136300r
3-(3-bromophenyl)-1-(2-hydroxyethyl)-, hydroxide, sulfate, inner salt	**58:** 10191b
1-*tert*-butyl-3-(4-chloro-2-nitrophenyl)-4-hydroxy-, hydroxide, inner salt	**84:** P164788y
3-*tert*-butyl-1-(4-chloro-2-nitrophenyl)-5-oxo-, hydroxide, inner salt	**78:** P136300r
1-butyl-3-(2,4-dichlorophenyl)-4-hydroxy-, hydroxide, inner salt	**84:** P164788y
1-*sec*-butyl-3-(2,4-dichlorophenyl)-4-hydroxy-, hydroxide, inner salt	**84:** P164788y
1-*tert*-butyl-3-(2,4-dichlorophenyl)-4-hydroxy-, hydroxide, inner salt	**84:** P164788y
3-butyl-1-(2,4-dichlorophenyl)-5-oxo-, hydroxide, inner salt	**78:** P136300r

TABLE 16 (*Continued*)

Compound	Reference
16.3. Δ^2-1,2,3-Triazolium Compounds (*Continued*)	
3-*sec*-butyl-1-(2,4-dichlorophenyl)-5-oxo-, hydroxide, inner salt	**78:** P136300r
1-butyl-4-hydroxy-3-(2,4,5-trichlorophenyl)-, hydroxide, inner salt	**84:** P164788y
1-*sec*-butyl-4-hydroxy-3-(2,4,5-trichlorophenyl)-, hydroxide, inner salt	**84:** P164788y
1-*tert*-butyl-4-hydroxy-3-(2,4,5-trichlorophenyl)-, hydroxide, inner salt	**84:** P164788y
1-butyl-4-hydroxy-3-(2,4,6-trichlorophenyl)-, hydroxide, inner salt	**84:** P164788y
3-butyl-5-oxo-1-(2,4,5-trichlorophenyl)-, hydroxide, inner salt	**78:** P136300r
3-*sec*-butyl-5-oxo-1-(2,4,5-trichlorophenyl)-, hydroxide, inner salt	**78:** P136300r
3-*tert*-butyl-5-oxo-1-(2,4,5-trichlorophenyl)-, hydroxide, inner salt	**78:** P136300r
3-butyl-5-oxo-1-(2,4,6-trichlorophenyl)-, hydroxide, inner salt	**78:** P136300r
1-(5-carbamoylimidazol-4-yl)-3-(2-chloroethyl)-, chloride	**70:** 51776k, **73:** 28983d
3-(5-carbamoyl-2-methoxyphenyl)-1-(2-hydroxyethyl)-, hydroxide, sulfate, inner salt	**58:** 10191c
3-(3-carboxyphenyl)-1-(2-hydroxyethyl)-, hydroxide, sulfate, inner salt, ethyl ester	**58:** 10191c
1-(2-chloroethyl)-3-(2-chlorophenyl)-, chloride	**54:** 9900f
1-(2-chloroethyl)-3-(4-chlorophenyl)-, chloride	**54:** 9900f
1-(2-chloroethyl)-3-(5-cyano-1*H*-imidazol-4-yl)-, chloride	**83:** 178922d
1-(2-chloroethyl)-3-(2-methyl-6-nitrophenyl)-, chloride	**65:** 13584b
1-(2-chloroethyl)-3-(2-nitrophenyl)-, chloride	**54:** 9900f
1-(2-chloroethyl)-3-(2-methylphenyl)-, chloride	**54:** 9900f
1-(2-chloroethyl)-3-(4-methylphenyl)-, chloride	**54:** 9900f
1-(2-chloroethyl)-3-(4-nitrophenyl)-, chloride	**54:** 9900f
1-(2-chloroethyl)-3-phenyl-, chloride	**54:** 9900f
3-(3-chloro-1*H*-indazol-6-yl)-1-(2-hydroxyethyl)-, hydroxide, sulfate, inner salt	**58:** 10191c
3-(4-chloro-2-methylphenyl)-1-cyclopentyl-4-hydroxy-, hydroxide, inner salt	**84:** P164788y
1-(4-chloro-2-methylphenyl)-3-cyclopentyl-5-oxo-, hydroxide, inner salt	**78:** P136300r
3-(6-chloro-2-methylphenyl)-1-(2-hydroxyethyl)-, hydroxide, sulfate, inner salt	**58:** 10191c
3-(4-chloro-2-nitrophenyl)-4-hydroxy-1-isopropyl-, hydroxide, inner salt	**84:** P164788y
1-(4-chloro-2-nitrophenyl)-3-isopropyl-5-oxo-, hydroxide, inner salt	**78:** P136300r
3-(5-chloro-2-phenoxyphenyl)-1-(2-hydroxyethyl)-, hydroxide, sulfate, inner salt	**58:** 10191c

TABLE 16 (*Continued*)

Compound	Reference
16.3. Δ^2-1,2,3-Triazolium Compounds (*Continued*)	
3-(2-chlorophenyl)-1-(2-hydroxybutyl)-, hydroxide, sulfate, inner salt	**58:** 10191d
3-(2,5-dichlorophenyl)-1-(2-hydroxyethyl)-, chloride	**58:** 10191g
3-(2,5-dichlorophenyl)-1-(2-hydroxyethyl)-, hydroxide, sulfate, inner salt	**58:** 10191b
3-(2,4-dichlorophenyl)-4-hydroxy-1-isobutyl-, hydroxide, inner salt	**84:** P164788y
3-(2,4-dichlorophenyl)-4-hydroxy-1-isopentyl-, hydroxide, inner salt	**84:** P164788y
3-(2,4-dichlorophenyl)-4-hydroxy-1-isopropyl-, hydroxide, inner salt	**84:** P164788y
3-(2,5-dichlorophenyl)-1-(2-hydroxy-1-methylpropyl)-, hydroxide, sulfate, inner salt	**58:** 10191d
3-(2,5-dichlorophenyl)-1-(2-hydroxypropyl)-, hydroxide, sulfate, inner salt	**58:** 10191d
3-(2,4-dichlorophenyl)-4-hydroxy-1-propyl-, hydroxide, inner salt	**84:** P164788y
3-(2,5-dichlorophenyl)-1-(2-hydroxypropyl)-4-methyl-, hydroxide, sulfate, inner salt	**58:** 10191e
1-(2,4-dichlorophenyl)-3-isobutyl-5-oxo-, hydroxide, inner salt	**78:** P136300r
1-(2,4-dichlorophenyl)-3-isopentyl-5-oxo-, hydroxide, inner salt	**78:** P136300r
1-(2,4-dichlorophenyl)-3-isopropyl-5-oxo-, hydroxide, inner salt	**78:** P136300r
1-(2,4-dichlorophenyl)-5-oxo-3-propyl-, hydroxide, inner salt	**78:** P136300r
3-(2,4-diethoxyphenyl)-1-(2-hydroxyethyl)-, hydroxide, sulfate, inner salt	**58:** 10191c
3-[5′-(diethylsulfamoyl)-2-methoxyphenyl]-1-(2-hydroxyethyl)-, hydroxide, sulfate, inner salt	**58:** 10191c
3-(2,4-dimethoxyphenyl)-1-(2-hydroxyethyl)-, hydroxide, sulfate, inner salt	**58:** 10191c
2,2-dimethyl-1-isopropyl-5-(isopropylimino)-4-oxo-, hydroxide, inner salt	**75:** 20308h
3-(2,4-dimethylphenyl)-1-(2-hydroxyethyl)-, hydroxide, sulfate, inner salt	**58:** 10191c
3-(2,5-dimethylphenyl)-1-(2-hydroxybutyl)-, hydroxide, sulfate, inner salt	**58:** 10191d
3-(2,4-dimethylphenyl)-1-(2-hydroxypropyl)-, hydroxide, sulfate, inner salt	**58:** 10191e
1-ethyl-4-hydroxy-3-(2,4,5-trichlorophenyl)-, hydroxide, inner salt	**84:** P164788y
3-ethyl-5-oxo-1-(2,4,5-trichlorophenyl)-, hydroxide, inner salt	**78:** P136300r
3-(4-ethylphenyl)-1-(2-hydroxyethyl)-, hydroxide, sulfate, inner salt	**58:** 10191b
3-[5-(ethylsulfonyl)-2-methoxyphenyl]-1-(2-hydroxyethyl)-, hydroxide, sulfate, inner salt	**58:** 10191c

TABLE 16 (*Continued*)

Compound	Reference
16.3. Δ^2-1,2,3-Triazolium Compounds (*Continued*)	
1-(2-hydroxybutyl)-3-phenyl-, hydroxide, sulfate, inner salt	**58:** 10191d
1-(2-hydroxyethyl)-3-[5-[(2-hydroxyethyl)sulfonyl]-2-methoxyphenyl]-, chloride	**64:** P6800b
1-(2-hydroxyethyl)-3-[5-[(2-hydroxyethyl)sulfonyl]-2-methoxyphenyl]-, hydroxide, inner salt, hydrogen sulfate ester	**64:** P6800b
1-(2-hydroxyethyl)-3-(2-methoxy-3-dibenzofuranyl)-, hydroxide, sulfate, inner salt	**58:** 10191c
1-(2-hydroxyethyl)-3-(2-methoxyphenyl)-, hydroxide, sulfate, inner salt	**58:** 10191b
1-(2-hydroxyethyl)-3-(4-methoxyphenyl)-, chloride	**58:** 10191g
1-(2-hydroxyethyl)-3-(4-methoxyphenyl)-, hydroxide, sulfate, inner salt	**58:** 10191c
1-(2-hydroxyethyl)-3-(5-methoxy-2-methylphenyl)-, hydroxide, sulfate, inner salt	**58:** 10191c
1-(2-hydroxyethyl)-3-(2-methyl-5-nitrophenyl)-, hydroxide, sulfate, inner salt	**58:** 10191c
1-(2-hydroxyethyl)-3-(2-methylphenyl)-, hydroxide, sulfate, inner salt	**58:** 10191b
1-(2-hydroxyethyl)-3-(3-methylphenyl)-, hydroxide, sulfate, inner salt	**58:** 10191b
1-(2-hydroxyethyl)-3-(4-methylphenyl)-, hydroxide, sulfate, inner salt	**58:** 10191b
1-(2-hydroxyethyl)-3-(1-naphthyl)-, chloride	**58:** 10191g
1-(2-hydroxyethyl)-3-(1-naphthyl)-, hydroxide, sulfate, inner salt	**58:** 10191d
1-(2-hydroxyethyl)-3-(2-nitrophenyl)-, hydroxide, sulfate, inner salt	**58:** 10191c
1-(2-hydroxyethyl)-3-phenyl-, chloride	**58:** 10191f
1-(2-hydroxyethyl)-3-phenyl-, hydroxide, sulfate, inner salt	**58:** 10191a
1-(2-hydroxyethyl)-3-(3-sulfophenyl)-, hydroxide, sulfate, inner salt	**58:** 10191c
1-(2-hydroxyethyl)-3-(2,4,6-trimethylphenyl)-, hydroxide, sulfate, inner salt	**58:** 10191c
4-hydroxy-1-isobutyl-3-(2,4,5-trichlorophenyl)-, hydroxide, inner salt	**84:** P164788y
4-hydroxy-1-isopropyl-3-(2,4,5-trichlorophenyl)-, hydroxide, inner salt	**84:** P164788y
4-hydroxy-1-isopropyl-3-(2,4,6-trichlorophenyl)-, hydroxide, inner salt	**84:** P164788y
1-(2-hydroxy-1-methylpropyl)-3-(2-methoxy-3-dibenzofuranyl)-, chloride	**58:** 10191g
1-(2-hydroxy-1-methylpropyl)-3-(2-methoxy-3-dibenzofuranyl)-, hydroxide, sulfate, inner salt	**58:** 10191e

TABLE 16 (*Continued*)

Compound	Reference
16.3. Δ^2-1,2,3-Triazolium Compounds (*Continued*)	
1-(2-hydroxy-1-methylpropyl)-3-phenyl-, hydroxide, sulfate, inner salt	**58:** 10191e
1-(2-hydroxypropyl)-3-(2-methoxy-3-dibenzofuranyl)-, chloride	**58:** 10191g
1-(2-hydroxypropyl)-3-(2-methoxy-3-dibenzofuranyl)-, hydroxide, sulfate, inner salt	**58:** 10191d
1-(2-hydroxypropyl)-3-(4-methoxyphenyl)-, hydroxide, sulfate, inner salt	**58:** 10191d
1-(2-hydroxypropyl)-4-methyl-3-phenyl-, hydroxide, sulfate, inner salt	**58:** 10191e
1-(2-hydroxypropyl)-3-phenyl-, chloride	**58:** 10191f
1-(2-hydroxypropyl)-3-phenyl-, hydroxide, sulfate, inner salt	**58:** 10191d
3-isobutyl-5-oxo-1-(2,4,5-trichlorophenyl)-, hydroxide, inner salt	**78:** P136300r
3-isopropyl-5-oxo-1-(2,4,5-trichlorophenyl)-, hydroxide, inner salt	**78:** P136300r
3-isopropyl-5-oxo-1-(2,4,6-trichlorophenyl)-, hydroxide, inner salt	**78:** P136300r

REFERENCES

42: 5059e D. Jerchel, *Fiat Rev. German Sci. Biochemistry*, Pt I, 59 (1939–1946).

46: 8652a Fr. Moulin, *Helv. Chim. Acta*, **35,** 167 (1952).

49: 10274*a* R. H. Wiley and J. Moffat, *J. Amer. Chem. Soc.*, **77,** 1703 (1955).

50: 3414g,h C. S. Rondestvedt, Jr. and P. K. Chang, *J. Amer. Chem. Soc.*, **77,** 6532 (1955).

50: 9393c R. Hüttel and G. Welzel, *Justus Liebigs Ann. Chem.*, **593,** 207 (1955).

51: 13855a,b R. Gompper, *Chem. Ber.*, **90,** 382 (1957).

54: 9900f K. A. Kornev and K. K. Khomenkova, *Ukrain. Khim. Zhur.*, **25,** 484 (1959).

58: 10191a-g R. Mohr and H. Hertel, *Chem. Ber.*, **96,** 114 (1963).

59: 15410f B. Hirsch and A. Bassl, *Z. Chem.*, **3,** 351 (1963).

61: 654f P. A. S. Smith, L. O. Krbechek, and W. Resemann, *J. Amer. Chem. Soc.*, **86,** 2025 (1964).

63: P3082d R. Sureau, G. R. H. Mingasson, A. Golla, and V. Dupre, French Patent, 1,391,676 (1965).

64: P6800b Farberwerke Hoechst A.-G., Dutch Patent, 6,503,892 (1965).

64: 11200c M. Begtrup and C. Pedersen, *Acta Chem. Scand.*, **19,** 2022 (1965).

65: 713g A. Albert and K. Tratt, *Chem. Commun.*, 243 (1966).

65: 13584*b* A. Kasahara and T. Hongu, *Nippon Kagaku Zasshi*, **86,** 1343 (1965).

66: P19822d J. Voltz, U.S. Patent, 3,280,100 (1966).

66: 37847v M. Begtrup and C. Pedersen, *Acta Chem. Scand.*, **20,** 1555 (1966).

66: P47321h	W. Yamatani and K. Matsui, U.S. Patent, 3,291,788 (1966).
66: P105886j	G. Wolfrum, R. Puetter, and K. H. Menzel, German Patent, 1,234,891 (1967).
67: P44797c	Farbwerke Hoechst A.-G., Dutch Patent, 6,608,828 (1966).
67: 64305u	M. Begtrup and C. Pedersen, *Acta Chem. Scand.*, **21,** 633 (1967).
67: 100074d	M. Begtrup, K. Hansen, and C. Pedersen, *Acta Chem. Scand.*, **21,** 1234 (1967).
67: P100995e	Farbenfabriken Bayer A.-G., Dutch Patent, 6,610,038 (1967).
68: P22709u	Farbwerke Hoechst A.-G., French Patent, 1,484,099 (1967).
68: 68958c	A. Albert and K. Tratt, *J. Chem. Soc., C,* 344 (1968).
69: P28576r	G. Wolfrum and H. Gold, British Patent, 1,111,236 (1968).
70: 51776k	D. J. Abraham, J. S. Rutherford, and R. D. Rosenstein, *J. Med. Chem.*, **12,** 189 (1969).
71: 61296*p*	M. Begtrup and C. Pedersen, *Acta Chem. Scand.*, **23,** 1091 (1969).
72: 43570q	M. Begtrup and P. A. Kristensen, *Acta Chem. Scand.*, **23,** 2733 (1969).
72: 66869h	S. Petersen and H. Heitzer, *Angew. Chem., Int. Ed. Engl.*, **9,** 67 (1970).
72: P91491m	Farbenfabriken Bayer A.-G., French Patent, 2,001,069 (1969).
72: P100715s	S. Petersen, H. Striegler, H. Gold, and A. Haberkorn, South African Patent, 69 01,244 (1969).
73: 28983d	P. D. Sternglanz, *J. Pharm. Sci.*, **59,** 708 (1970).
73: 66510e	A. Singh and B. Hirsch, *Indian J. Chem.*, **8,** 514 (1970).
73: 120571r	K. T. Potts and S. Husain, *J. Org. Chem.*, **35,** 3451 (1970).
74: 48048b	L. F. Avramenko, Yu. B. Vilenskii, B. M. Ivanov, N. V. Kudryavskaya, I. A. Ol'shevskaya, V. Ya. Pochinok, L. I. Skripnik, L. N. Fedorova, and I. P. Fedorova, *Usp. Nauch. Fotogr.*, **14,** 12 (1970).
74: 141647f	M. Begtrup, *Acta Chem. Scand.*, **25,** 249 (1971).
74: 141656h	C. S. Angadiyavar, K. B. Sukumaran, and M. V. George, *Tetrahedron Lett.*, 633 (1971).
75: 20308h	W. Lwowski, R. A. de Mauriac, R. A. Murray, and L. Lunow, *Tetrahedron Lett.*, 425 (1971).
75: 63702z	L. S. Davies and G. Jones, *J. Chem. Soc., C,* 2572 (1971).
75: 63704b	M. Begtrup, *Tetrahedron Lett.*, 1577 (1971).
75: 87745s	M. Begtrup, *Acta Chem. Scand.*, **25,** 803 (1971).
75: 109510r	M. Begtrup, *Acta Chem. Scand.*, **25,** 795 (1971).
76: 25187y	M. Begtrup and K. V. Poulsen, *Acta Chem. Scand.*, **25,** 2087 (1971).
76: 72454t	M. Begtrup, *Acta Chem. Scand.*, **25,** 3500 (1971).
76: P128683m	R. Mohr, E. Mundlos, K. Hohmann, and J. Ostermeier, German Patent, 2,029,314 (1971).
77: 61894f	K. T. Potts and S. Husain, *J. Org. Chem.*, **37,** 2049 (1972).
77: 75175j	K. B. Sukumaran, C. S. Angadiyavar, and M. V. George, *Tetrahedron,* **28,** 3987 (1972).
77: 114317d	M. Begtrup, *Acta Chem. Scand.*, **26,** 1243 (1972).
78: P136300r	M. J. Abu-EL-Haj and J. W. McFarland, German Patent, 2,239,400 (1973).
78: 159527j	T. Isida, T. Akiyama, N. Mihara, S. Kozima, and K. Sisido, *Bull. Chem. Soc. Jap.*, **46,** 1250 (1973).
79: 100462b	A. Haberkorn and S. Petersen, *Advan. Antimicrob. Antineoplastic Chemother., Proc. Int. Congr. Chemother., 7th 1971* **1,** 409 (1972).
80: 82890d	A. Albert, *J. Chem. Soc., Perkin Trans.*, **1,** 2659 (1973).
80: 132667j	M. Begtrup, *Acta Chem. Scand., Ser. B,* **28,** 61 (1974).
83: P19056h	D. S. Bailey, M. D. Shea, R. T. Brongo, J. C. Fleming, and J. W. Manthey, German Patent, 2,437,382 (1975).
83: 27081c	A. Jart, *Acta Polytech. Scand., Chem. Incl. Metall. Ser.*, **121,** 118 pp (1974).

83: 58717u	M. Begtrup, *Acta Chem. Scand., Ser. B*, **B29,** 141 (1975).
83: 97134w	M. Begtrup, *J. Chem. Soc., Chem. Commun.*, 334 (1975).
83: 178922z	Y. F. Shealy and C. A. O'Dell, *J. Pharm. Sci.*, **64,** 954 (1975).
83: 192266z	S. Auricchio, G. Vidari, and P. Vita-Finzi, *Gazz. Chim. Ital.*, **105,** 583 (1975).
84: P121847z	M. J. Abu-El-Haj and J. W. McFarland, U.S. Patent, 3,933,843 (1976).
84: 135549r	D. Stadler, W. Anschütz, M. Regitz, G. Keller, D. Van Assche, and J. P. Fleury, *Justus Liebigs Ann. Chem.*, 2159 (1975).
84: P164788y	M. J. Abu-El-Haj and J. W. McFarland, U.S. Patent, 3,939,174 (1976).

CHAPTER 17

Complex 1,2,3-Triazoles Related to Practical Applications

A great many 1,2,3-triazoles have been examined by industrial laboratories in, for example, textiles and medicine. The synthetic methods employed have been illustrated at length in the preceding chapters. The practical problems related to the application of these compounds result in rather elaborate and often indefinite structure assignment. The organization of such information for simplicity and for the reliability of searching seems best approached by using the formula index, which achieves the economy of space by eliminating a great deal of repetitious nomenclature. The following systems make up the largest portion of compounds included in the formula index:

1. Optical brightners (extended conjugated systems).
2. Optical brightners (coumarin ring systems).
3. Penicillin related systems.
4. Polymers and copolymers.

The following general rules are observed in the arrangement of the tables:

1. Polymers are listed by the monomer formula and following the isomeric small molecules.
2. Copolymers and adducts are listed under the triazole component.
3. Isomers are arranged by the registry number used in the *Chemical Abstracts.*

The literature has been examined through *Chemical Abstracts,* Vol. **83** (1975).

TABLE 17.1 MISCELLANEOUS 1,2,3-TRIAZOLES AND RELATED COMPOUNDS

Molecular formula						Salt, adduct, or	Registry	
C	H	Cl	N	O	S	copolymer	number	Reference
4	1	2	3	2		$(C_6H_{14}N_2)_x$	55184-85-5	**83:** P11149v
4	1	2	3	2		$(C_{15}H_{16}O_2)_x$	51982-34-4	**81:** 26334j
4	1	2	3	2		$(C_{17}H_{20}O_2)_x$	52002-64-9	**81:** 26334j
4	3		3	4		$(C_6H_{16}N_2)_x$	39050-67-4	**78:** 98054d
5	3	2	3	2		$(C_6H_8N_2)_x$	52319-13-8	**82:** P31753v, **82:** 73727v
5	3	2	3	2		$(C_6H_{14}N_2)_x$	52319-12-7	**82:** 73727v
5	3	2	3	2		$(C_6H_{14}N_2)_x$	52319-23-0	**82:** P31753v, **82:** 73727v
5	3	2	3	2		$(C_4Cl_2N_2O_2S)_x$	53808-23-4	**82:** P31753v
5	3	2	3	2		$(C_6H_{14}N_2 \cdot C_5H_3Cl_2N_3O_2)_x$	52319-24-1	**82:** 73727v
5	9		3			$(C_8H_8)_x$	28408-75-5	**70:** P48020d
6	7		3	2			55988-94-8	**83:** P115706r
7	9		3	2			55988-96-0	**83:** P115706r
8	5		3				26967-15-7	**68:** 87784j, **71:** 13464v
8	11		3	2			55988-98-2	**83:** P115706r
8	11		3	2			55989-01-0	**83:** P115706r
8	13		3	2			55988-99-3	**83:** P115706r
9	13		3	2			55989-03-2	**83:** P115706r
10	11		3			$(C_8H_8)_x$	28408-76-6	**70:** P48020d
10	11		5	3	2		37539-03-0	**77:** P126654z, **78:** P159644v, **81:** P136161h
10	13		5	2			55155-31-2	**83:** P11149v
10	15		3	2			56597-55-8	**83:** 147900r
10	15		5	2			37383-07-6	**78:** 98054d
11	7	2	3	2		$(C_6H_{14}N_2)_x$	55155-22-1	**83:** 179654e
11	9		5	2			37157-31-6	**77:** P7315e
11	9		5	2			52303-55-6	**82:** P31753v, **82:** 73727v
11	12		6	4	2		43005-17-0	**79:** P53343b
11	13		5	3	2		52115-96-5	**80:** P120970q
11	13		5	3	2		53374-77-9	**81:** P49694r
11	15		5	2			52303-54-5	**82:** 73727v
11	15		5	2			52357-59-2	**82:** 73727v
11	15		5	2			53858-40-5	**82:** P31753v
12	12		4	2			29207-83-8	**73:** 130944d
12	13		5	4	2		54470-13-2	**82:** P43446k
12	14		4	2			21217-57-2	**71:** 2032b
12	15		5	4	1		36042-05-4	**76:** 149496t
13	10	1	5	2			5036-42-0	**71:** 2032b
13	11		5	2			5036-41-9	**71:** 2032b
13	12		3	3		CH_3O_4S	52688-43-4	**81:** 19043k
13	12		6				41909-46-0	**79:** 19154f
13	12		6				41909-47-1	**79:** 19154f
13	15		5	5	2		53689-34-2	**81:** P136161h, **83:** P58856p

TABLE 17.1 (*Continued*)

Molecular Formula						Salt, adduct, or	Registry	
C	H	Cl	N	O	S	copolymer	number	Reference
13	15		5	6	3	Na	51245-09-1	**80:** P48019s
14	8		6	2			38957-34-5	**77:** P164713a, **78:** 84321b
14	8		6	2			39736-54-4	**78:** 84321b
14	8		6	4			26485-36-9	**72:** P112806k, **83:** P116987v
14	10		6	2			26485-37-0	**72:** P112806k, **83:** P116987v
14	11	1	4	2			5036-33-9	**71:** 2032b
14	11		3	1			41581-20-8	**79:** 66882b
14	12		4	2			5036-22-6	**71:** 2032b
14	12		4	2			5036-45-3	**71:** 2032b
14	13		5	3			6012-30-2	**71:** 2032b
14	13		5	4	1		24149-51-7	**71:** 2032b
14	14		6				25190-20-9	**72:** 3846c
14	14		6				25190-21-0	**72:** 3846c
14	15		5	7	1		29126-08-7	**73:** P120648w
14	16		4	2			24149-36-8	**71:** 2032b
15	10	1	5	2			37157-40-7	**77:** P7315e
15	10	1	5	2			56003-45-3	**83:** P50607n
15	10		4	2			23454-66-2	**71:** P22927a
15	11		5	2			34791-90-7	**77:** P7315e
15	11		5	3			27802-54-6	**74:** P14197j
15	14		4	2			5036-28-2	**71:** 2032b
15	14		4	2			5036-29-3	**71:** 2032b
15	14		4	2			5036-32-8	**71:** 2032b
15	14		4	2			24145-37-7	**71:** 2032b
15	14		4	3			5036-34-0	**71:** 2032b
15	14		6	6	1	Na	29126-03-2	**73:** P120648w
15	16		6				25190-18-5	**72:** 3846c
15	16		6				25190-19-6	**72:** 3846c
15	17		5	1			57021-93-9	**83:** P207574z
15	17		5	6	2		54470-14-3	**82:** P43446k
16	10		6				31783-05-8	**69:** 107160d
16	11	1	4	2			41428-27-7	**78:** P137966z
16	11	1	4	2			52048-47-2	**81:** P122801e
16	12		4	2			23454-68-4	**71:** P22927a
16	12		4	2			41113-12-6	**78:** P31421c, **78:** P85944u, **78:** P99060q **78:** P137966z
16	12		4	2			52048-63-2	**81:** P122801e
16	12		6	3			26485-38-1	**72:** P112806k, **83:** P116987v
16	12		7	2		CH_3O_4S	29641-54-1	**73:** P100073b
16	13	1	4				41500-46-3	**78:** P85944u, **78:** P137966z
16	13	1	4				52048-48-3	**81:** P122801e
16	14		4				41113-10-4	**78:** P31421c, **78:** P99060q

TABLE 17.1 (*Continued*)

Molecular Formula						Salt, adduct, or	Registry	
C	H	Cl	N	O	S	copolymer	number	Reference
16	14		4				52048-64-3	**81:** P122801e
16	14		6	2			26867-94-7	**82:** P172621n
16	14		6	2			34791-89-4	**77:** P7315e
16	14		6	3			27802-55-7	**74:** P14197j
16	15		5			HCl	41344-74-5	**78:** P99060q
16	15		5	3			21217-46-9	**71:** 2032b
16	16		6	3			37157-36-1	**77:** P7315e
16	16		6	4	3		37539-01-8	**77:** P126654z
16	16		6	4	3		37539-02-9	**77:** P126654z
16	18		6				41909-49-3	**79:** 19154f
16	18		6				41909-50-6	**79:** 19154f
16	21		5	5	2		56206-40-7	**83:** P58856p
17	10		4	4			19778-44-0	**69:** P52937e, **82:** P32468t
17	11	1	3	1			41426-28-2	**83:** P61732a
17	11	2	3	1			52455-39-7	**83:** P61732a
17	11	2	3	1			52455-46-6	**83:** P61732a
17	11		5	2			23918-37-8	**72:** P45038w
17	12	1	3	1			41113-15-9	**78:** P31421c
17	12	1	3	1			52455-41-1	**83:** P61732a
17	12	1	3	1			52455-43-3	**83:** P61732a
17	12	1	3	1			52455-47-7	**83:** P61732a
17	12	1	3	2			41427-82-1	**78:** P85944u, **83:** P61732a
17	12	1	3	2			52455-35-3	**83:** P61732a
17	12		4	2			19695-29-5	**69:** P52937e, **82:** P32468t
17	12		4	2			37157-32-7	**77:** P7315e
17	12		4	2			52085-04-8	**81:** P51143d
17	13	1	4	2			51395-15-4	**80:** P122419r
17	13	2	5	1			35793-11-4	**76:** P128833k
17	13		3	1			30818-73-6	**74:** P65594u
17	13		3	2			30818-75-8	**78:** P31421c, **78:** P85944u, **78:** P137966z, **83:** P61732a
17	13		3	3			30818-74-7 39385-89-2	**78:** P31421c **83:** P61732a
17	13		5				25730-38-5	**72:** P45038w
17	13		5				42962-69-6	**79:** P80310v
17	13		5	2			37157-33-8	**77:** P7315e
17	14	1	5	2			37157-30-5	**77:** P7314d
17	14	2	6	5	2	K	56187-51-0	**83:** 43359e
17	14		4	1			41113-24-0	**78:** P31421c
17	14		4	2			23454-55-9	**71:** P22927a
17	14		4	2			36510-77-7	**76:** 140659r
17	14		4	2			37739-17-6	**77:** P90079z
17	14		4	2			52042-54-3	**80:** 132664f
17	14		4	3			23454-64-0	**71:** P22927a
17	14		4	3			51419-18-2	**80:** P122419r

TABLE 17.1 (*Continued*)

C	H	Cl	N	O	S	Salt, adduct, or copolymer	Registry number	Reference
17	14		6				41532-49-4	**78:** P99065v
17	14		6	4			37157-38-3	**77:** P7315e
17	15		5	1			34570-15-5	**76:** P128833k
17	15		5	2			29641-52-9	**73:** P100073b
17	15		5	4	1		35793-13-6	**76:** P128833k
17	15		5	5	2	K	39049-40-6	**77:** 147455a
17	15		5	6	2		38889-02-0	**77:** 147455a
17	17		5	4			5036-39-5	**71:** 2032b
17	17		5	5	2		38889-03-1	**77:** 147455a
17	17		6	2		CH_3O_4S	29641-53-0	**73:** P100073b
17	17		6	2		CH_3O_4S	37157-41-8	**77:** P7315e
17	17		6	3		CH_3O_4S	30194-47-9	**74:** P14197j
17	19		5	2			55465-57-1	**83:** 179654e
17	19		5	8	2		29125-90-4	**73:** P120648w
18	11	2	3	2			53185-39-0	**82:** P17894k
18	11		5				53185-38-9	**82:** P17894k
18	12	1	3	2			31521-42-3	**74:** P100617c, **74:** P100618d, **81:** P79389n
18	12	1	3	3			52084-93-2	**81:** P51143d
18	12		4	4			19778-45-1	**69:** P52937e, **82:** P32468t
18	12		6				42962-63-0	**79:** P80310v
18	13		3	2			1965-74-8	**81:** 19043k, **83:** 105984r
18	13		3	2			25054-84-6	**72:** P101860x, **73:** P121566e, **77:** P7317g
18	13		3	3			26759-72-8	**72:** P122925n, **77:** P7317g, **81:** P137597s
18	13		3	3			52084-90-9	**81:** P51143d
18	13		3	3			52570-45-3	**81:** P79389n
18	14	1	3	1			52455-38-6	**83:** P61732a
18	14	1	3	2			41426-29-3	**78:** P85944u, **78:** P137966z
18	14	1	5	1			16761-45-8	**68:** P106086b
18	14	2	4	2			34502-50-6	**77:** 75174h
18	14	2	4	2			36280-35-0	**76:** 140659r
18	14	2	4	2			36357-70-7	**76:** 140659r
18	14		4	2			19686-28-3	**69:** P52937e, **82:** P32468t
18	14		4	2			37157-34-9	**77:** P7315e
18	14		4	2			52085-06-0	**81:** P51143d, **81:** P171366a
18	14		4		1		18013-54-2	**68:** P106086b
18	14		6	6			34502-44-8	**77:** 75174h
18	14		6	6			52042-52-1	**80:** 132664f
18	14		6	6			52042-53-2	**80:** 132664f

TABLE 17.1 (*Continued*)

Molecular Formula						Salt, adduct, or copolymer	Registry number	Reference
C	H	Cl	N	O	S			
18	15		3	1			52455-42-2	**83:** P61732a
18	15		3	2			52455-34-2	**83:** P61732a
18	15		5				18028-73-4	**68:** P106086b
18	15		5	2			37157-35-0	**77:** P7315e
18	15		5	2			52497-86-6	**81:** P171366a
18	15		5	8	2		38921-34-5	**77:** 147455a
18	16		4	2			19226-31-4	**69:** 36042d, **74:** 141654f
18	16		4	2			20220-70-6	**71:** P22927a
18	16		4	2			23454-57-1	**71:** P22927a
18	16		4	2			36289-49-3	**76:** 140659r
18	16		4	4	1		25730-37-4	**72:** P45038w
18	16		6	8	2		56834-89-0	**83:** P179082s
18	17		5	4	1		25730-36-3	**72:** P45038w
18	17		5	4	1		38921-35-6	**77:** 147455a
18	17		5	5	2	K	39049-41-7	**77:** 147455a
18	17		5	6	2		36905-17-6	**77:** 147455a
18	17		7	6	2		51647-18-8	**80:** P83012f
18	18		6	2			29641-48-3	**73:** P100073b
18	18		6	4	1		38921-39-0	**77:** 147455a
18	18		6	4	2		37539-00-7	**77:** P126654z
18	18		6	4	2	Na	37539-05-2	**77:** P126654z
18	18		6	5	2		51627-14-6	**80:** P83024m, **81:** P105538g, **83:** P193352m
18	18		6	5	2	Na	53781-48-9	**81:** P105538g
18	18		6	5	2	$xC_3H_8O_2$	57235-40-2	**83:** P193352m
18	19		5	10	1	Na	29126-02-1	**73:** P120648w
18	19		7	4	3		51440-49-4	**80:** 66979j
19	13		5				41532-75-6	**78:** P99065v, **83:** P99223y
19	13		5	1			34180-59-1	**75:** P152985m
19	14	1	3	2			35123-44-5	**76:** P60964w
19	14	1	3	3			52084-94-3	**81:** P51143d
19	14		4	2	1		18013-53-1	**68:** P106086b
19	14		4	3			52085-05-9	**81:** P51143d
19	15	1	4	1			38880-44-3	**77:** P141496t
19	15	1	4	2			16761-44-7	**68:** P106086b
19	15	1	4	3			18028-74-5	**68:** P106086b
19	15		3	2			18146-59-3	**69:** P52937e, **73:** P121566e, **77:** P7317g, **81:** P51153g, **81:** P137597s, **82:** P32468t
19	15		3	3			26726-37-4	**72:** P122925n, **77:** P7317g
19	15		3	3			35123-35-4	**76:** P60964w
19	15		3	3			50995-76-1	**80:** P49275j

TABLE 17.1 (*Continued*)

Molecular Formula						Salt, adduct, or	Registry	
C	H	Cl	N	O	S	copolymer	number	Reference
19	15		3	3			51542-03-1	**81:** P51143d
19	15		3	3			52084-95-4	**81:** P51143d
19	15		3	4			52002-49-0	**81:** 26334j
19	16	1	5	1			16761-43-6	**68:** P106086b
19	16		4	1			18013-51-9	**68:** P106086b
19	16		4	2			18013-64-4	**68:** P106086b
19	16		4	2			18028-71-2	**68:** P106086b
19	16		4	4			23454-65-1	**71:** P22927a
19	16		6	1			41909-48-2	**79:** 19154f
19	16		6	4	1		29125-97-1	**73:** P120648w
19	17		5	1			16761-42-5	**68:** P106086b
19	18		4	1			41113-23-9	**78:** P31421c
19	18		4	2			23454-61-7	**71:** P22927a, **78:** P137850g
19	18		4	2			23454-77-5	**71:** P22927a
19	18		4	2			36541-16-9	**76:** 140659r
19	18		4	2			38880-47-6	**77:** P141496t
19	18		4	2			52042-55-4	**80:** 132664f
19	18		4	4			52042-56-5	**80:** 132664f
19	18		6	5	2		53683-30-0	**82:** 11679x
19	18		6	6	1		38889-04-2	**77:** 147455a
19	20		6	4	2		40901-40-4	**78:** P159644v
19	20		6	4	2		53374-78-0	**81:** P49694r
19	20		6	4	2	Na	53374-79-1	**81:** P49694r
19	20		6	4	3		40851-78-3	**78:** P159644v
19	20		6	5	1		38889-06-4	**77:** 147455a
19	20		6	5	2		51627-15-7	**80:** P83024m
19	20		6	5	2		52438-32-1	**80:** P83013g
19	20		6	5	2		54172-63-3	**83:** P58856p
19	20		6	5	2	$C_2HF_3O_2$	52353-32-9	**80:** P83013g
19	20		6	5	2	$C_2HF_3O_2$	52484-88-5	**80:** P83013g
19	20		6	5	2	HCl	54172-65-5	**81:** P136161h, **83:** P58856p
19	20		6	5	2	Na	54172-64-4	**81:** P136161h, **83:** P58856p
19	20		6	6	2		54172-69-9	**81:** P136161h, **83:** P58856p
19	20		6	6	2	HCl	54172-72-4	**81:** P136161h, **83:** P58856p
19	20		6	6	2	Na	54172-70-2	**81:** P136161h, **83:** P58856p
19	20		8	5	2		54895-44-2	**82:** P112086g
19	20		8	5	2		54895-45-3	**82:** P112086g
19	22		6	4	2		56198-92-6	**83:** P28250u
19	23		5	8	1	Na	29125-85-7	**73:** P120648w
19	24		6	4	2		56168-18-4	**83:** P28250u
20	14		6	2			52725-14-1	**82:** P172621n
20	14		6	5	1	Na	37838-20-3	**77:** P7315e
20	15		7	4	1	Na	29126-00-9	**73:** P120648w
20	16	2	6	6	2		52570-54-4	**81:** P79389n

TABLE 17.1 (*Continued*)

Molecular Formula						Salt, adduct, or	Registry	
C	H	Cl	N	O	S	copolymer	number	Reference
20	16	2	6	6	2	2Na	29633-69-0	**74:** P100617c
20	16		4	3			18013-50-8	**68:** P106086b
20	16		4	3			19695-34-2	**69:** P52937e, **82:** P32468t
20	16		4	3			19695-39-7	**69:** P52937e, **82:** P32468t
20	16		4	4			16781-21-8	**68:** P106086b
20	16		6	1			18028-72-3	**68:** P106086b
20	17	1	4	3			16761-48-1	**68:** P106086b
20	17		3	2			19683-12-6	**69:** P52937e, **81:** P51153g, **82:** P32468t
20	17		3	3			26726-38-5	**72:** P122925n, **77:** P7317g, **81:** P137597s
20	17		5	2			16761-46-9	**68:** P106086b
20	17		5	2			52085-22-0	**81:** P51143d
20	17		5	3			18013-56-4	**68:** P106086b
20	18	1	5	1			16786-41-7	**68:** P106086b
20	18	1	5	2			16781-25-2	**68:** P106086b
20	18	1	5	2			18028-67-6	**68:** P106086b
20	18		6				41909-45-9	**79:** 19154f
20	18		6	4	1		29125-82-4	**73:** P120648w
20	18		6	8	1		38921-38-9	**77:** 147455a
20	18		6	8	2		52570-53-3	**81:** P79389n
20	19	1	4	2			51419-15-9	**80:** P122419r
20	19		5	8	2		38921-33-4	**77:** 147455a
20	19		5	8	2	Na	29125-84-6	**73:** P120648w
20	20		4	2			23454-56-0	**71:** P22927a
20	20		4	2			23454-62-8	**71:** P22927a
20	20		4	2			23454-69-5	**71:** P22927a
20	20		4	2			34519-95-4	**77:** 75174h
20	20		4	2			34599-20-7	**77:** 75174h
20	20		4	2			52042-51-0	**80:** 132664f
20	20		4	4			19226-32-5	**69:** 36042d
20	20		6	5	1		38889-05-3	**77:** 147455a
20	20		6	5	2		53683-34-4	**82:** P43446k
20	20		6	6	1		36905-18-7	**77:** 147455a
20	20		6	6	3	$C_2HF_3O_2$	55767-21-0	**83:** P10111q
20	20		6	7	2		55767-31-2	**83:** P10111q
20	21		5	5	1		38889-29-1	**77:** 147455a
20	21		7	6	2	$C_2HF_3O_2$	55767-38-9	**83:** P10111q
20	22		6	4	3		40851-75-0	**78:** P159644v
20	22		6	5	1		38889-07-5	**77:** 147455a
20	22		6	6	2		54172-75-7	**83:** P58856p
21	11	1	6	2			37157-28-1	**77:** P7314d
21	14	1	5	2			34771-66-9	**77:** P7314d, **81:** P25679p, **83:** P193339n
21	14	1	5	2			52468-64-1	**81:** P93326w

TABLE 17.1 (*Continued*)

C	H	Cl	N	O	S	Salt, adduct, or copolymer	Registry number	Reference
21	14	1	5	5	1	Na	37157-29-2	**77:** P7314d
21	14		4	2			37699-20-0	**77:** P90079z
21	14		6	2			38583-33-4	**77:** P128099w
21	15	2	3				31652-30-9	**74:** P112048f, **74:** P125697s, **74:** P125698t, **83:** P206290s
21	15	2	3				57381-87-0	**83:** P206290s
21	15		5	2			26485-34-7	**83:** P116987v
21	15		5	2			37157-39-4	**77:** P7315e
21	15		5	3			27802-56-8	**74:** P14197j
21	16	1	3				57381-80-3	**83:** P206290s
21	16	1	3				57381-81-4	**83:** P206290s
21	16	1	3				57381-82-5	**83:** P206290s
21	16		4	2			57381-85-8	**83:** P206290s
21	17		3				31216-81-6	**74:** P112048f, **74:** P125697s, **74:** P125698t, **83:** P206290s
21	17		3	4			19778-49-5	**69:** P52937e, **77:** P7313c **81:** P51152f **82:** P32468t
21	17		3	4			35123-33-2	**76:** P60964w
21	18	1	3	2			19695-52-4	**69:** P52937e, **81:** P51153g
21	18	1	5	2			16781-22-9	**68:** P106086b
21	18		4	2	1		18013-52-0	**68:** P106086b
21	18		4	3			19695-35-3	**69:** P52937e, **82:** P32468t
21	18		4	3			34180-58-0	**75:** P152985m
21	18		4	4			19778-51-9	**69:** P52937e, **82:** P32468t
21	18		4	4			19778-54-2	**69:** P52937e, **82:** P32469t
21	18		4	4			52085-08-2	**81:** P51143d
21	19		3	2			19778-56-4	**69:** P52937e
21	19		3	2			29597-03-3	**73:** P121566e
21	19		3	3			52084-89-6	**81:** P51143d
21	19		3	3			52084-91-0	**81:** P51143d
21	19		3	4			52002-50-3	**81:** 26334j
21	19		5	2			16761-47-0	**68:** P106086b
21	19		5	2			52497-90-2	**81:** P171366a
21	19		5	8	1		37651-14-2	**77:** P88521u
21	20	1	3	3			35793-07-8	**76:** P128832j
21	20		4	1			34184-15-1	**75:** P152985m
21	21	1	4	2			51419-14-8	**80:** P122419r
21	21		3	3			34549-36-5	**76:** P128832j, **77:** P7313c, **77:** P141496t

TABLE 17.1 (*Continued*)

C	H	Cl	N	O	S	Salt, adduct, or copolymer	Registry number	Reference
21	21		5				34180-65-9	**75:** P152985m
21	21		5	6	1	Na	29125-98-2	**73:** P120648w
21	21		5	6	1	Na	29125-99-3	**73:** P120648w
21	22		4	2			23454-87-7	**71:** P22927a
21	22		4	2			23531-65-9	**71:** P22927a
21	22		6	7	2		53689-35-3	**81:** P105538g
21	22		6	7	2	$0.5C_2HF_3O_2$	55767-44-7	**83:** P10111q
21	24		6	10	4	3Na	54743-39-4	**82:** P43437h
22	14	1	5	2			38583-31-2	**77:** P128099w
22	14	2	4	2			29207-79-2	**73:** 130944d
22	14	2	4	2			34502-42-6	**77:** 75174h
22	15		3	3			25054-83-5	**72:** P101860x, **77:** P7317g
22	15		3	4			26726-41-0	**72:** P122925n, **77:** P7317g, **81:** P137597s
22	15		5	2			36528-32-2	**77:** P128099w
22	16		4				57381-84-7	**83:** P206290s
22	16		4	1			18013-59-7	**68:** P106086b
22	16		4	2			23454-58-2	**71:** P22927a, **77:** P90079z
22	16		4	2			29207-75-8	**73:** 130944d
22	16		4	2			29207-84-9	**73:** 130944d
22	16		4	5	1		25730-35-2	**72:** P45038w
22	16		6				33793-17-8	**75:** P110997m
22	16		6				33793-18-9	**75:** P110997m
22	17		6	2		CH_3O_4S	38583-34-5	**77:** P128099w
22	18		4	3			19778-55-3	**69:** P52937e, **82:** P32468t
22	18		6	10	2	2Na	22092-04-2	**70:** P97963x, **73:** P78559r
22	19		3				57381-86-9	**83:** P206290s
22	19		3	4			19695-33-1	**69:** P52937e, **82:** P32468t
22	20		4	3			16761-40-3	**68:** P106086b
22	20		4	3			19778-50-8	**69:** P52937e, **82:** P32468t
22	20		4	4			18013-62-2	**68:** P106086b
22	20		4	4			19695-40-0	**69:** P52937e, **82:** P32468t
22	20		6	9	1		37060-07-4	**77:** P126662a
22	21		3	2			19695-49-9	**69:** P52937e
22	21		3	2			29597-05-5	**73:** P121566e
22	21		5	2			18013-60-0	**68:** P106086b
22	21		5	3			18013-55-3	**68:** P106086b
22	21		5	8	1		29125-83-5	**73:** P120648w
22	21		5	9	1		29125-89-1	**73:** P120648w
22	22		4	1			34180-51-3	**75:** P152985m
22	22		6	1			41909-51-7	**79:** 19154f
22	22		6	8	1		39031-39-5	**77:** 147455a

TABLE 17.1 (*Continued*)

C	H	Cl	N	O	S	Salt, adduct, or copolymer	Registry number	Reference
22	22		10	8	2	2Na	54673-45-9	**82:** P126614q
22	23		5	1			16781-24-1	**68:** P106086b
22	24		4	2			23454-75-3	**71:** P22927a
22	24		4	2			23454-80-0	**71:** P22927a
22	24		6	7	2		54172-77-9	**81:** P136161h
22	24		6	7	2		56258-67-4	**83:** P58856p
22	24		6	8	2		54172-80-4	**81:** P136161h, **83:** P58856p
22	25		5	1			41113-25-1	**78:** P31421c
23	13	2	3	2			31521-41-2	**74:** P100617c, **74:** P100618d, **81:** P79389n
23	13	2	3	2			31521-44-5	**74:** P100617c, **74:** P100618d, **81:** P79389n
23	13	2	3	2			31616-60-1	**74:** P100617c, **74:** P100618d, **81:** P79389n
23	14	1	3	2			29818-24-4	**74:** P100617c, **74:** P100618d **81:** P79389n
23	14	1	3	3			31616-61-2	**74:** P100618d
23	14	1	3	3			52570-44-2	**81:** P79389n
23	14	1	3	3			52570-47-5	**81:** P79389n
23	14	1	3	3			52673-70-8	**81:** P79389n
23	15	1	6	2			52048-55-2	**81:** P122801e
23	15	3	4	2			34502-41-5	**77:** 75174h
23	15	3	4	2			52042-57-6	**80:** 132664f
23	15		3	2			19683-08-0	**69:** P52937e, **73:** P121566e, **81:** P51153g, **82:** P32468t
23	15		3	2	1		29818-23-3	**74:** P100613y
23	15		3	3			29920-34-1	**74:** P100613y
23	15		3	3			52673-68-4	**81:** P79389n
23	16	1	3	1			49744-03-8	**80:** P16460n
23	16	1	5	2			38583-32-3	**77:** P128099w
23	16	1	7	4	1		52048-70-1	**81:** P51154h
23	16		4	1			31573-83-8	**74:** P88677n
23	16		4	1			34184-18-4	**75:** P152985m
23	16		4	1			41427-88-7	**78:** P85944u, **78:** P137966z
23	16		4	3			18013-58-6	**68:** P106086b
23	16		4	4			37699-17-5	**77:** P90079z
23	16		6	2			52048-51-8	**81:** P122801e
23	16		6	5	1	Na	52048-69-8	**81:** P51154h
23	17		3	2			49744-09-4	**80:** P16460n
23	17		5				37936-24-6	**78:** P85944u
23	17		5	3			41426-26-0	**78:** P137966z

TABLE 17.1 (*Continued*)

Molecular Formula C	H	Cl	N	O	S	Salt, adduct, or copolymer	Registry number	Reference
23	17		5	3			52371-67-2	**81:** P93093t
23	17		5	4			26485-35-8	**72:** P112806k, **83:** P116987v
23	18		4	2			23454-59-3	**71:** P22927a, **77:** P90079z
23	18		4	2			23454-60-6	**71:** P22927a
23	18		4	2			23454-70-8	**71:** P22927a
23	18		4	2			23454-73-1	**71:** P22927a
23	18		4	2			49738-41-2	**79:** 126403h
23	19		5	1			37757-58-7	**78:** P31421c
23	20	1	5	4	1		57339-36-3	**83:** P206291t
23	20	1	5	4	1		57339-37-4	**83:** P206291t
23	21		5	4	1		57339-34-1	**83:** P206291t
23	21		5	4	1	K	57339-35-2	**83:** P206291t
23	21		5	10	1		29125-91-5	**73:** P120648w
23	22		4	3			19778-53-1	**69:** P52937e, **82:** P32468t
23	22		4	3			35126-03-5	**76:** P60964w
23	23	1	8				41113-13-7	**78:** P31421c
23	23		3	3			26726-39-6	**72:** P122925n, **77:** P7317g, **81:** P137597s
23	23		5	2			18013-65-5	**68:** P106086b
23	23		5	8	1		29129-64-4	**73:** P120648w
23	23		5	8	2	Na	29125-86-8	**73:** P120648w
23	24		6	7	2		54470-15-4	**82:** P43446k
23	25		5	8	1		29126-04-3	**73:** P120648w
23	26		4	2			23454-71-9	**71:** P22927a
23	26		4	2			23454-76-4	**71:** P22927a
23	26		4	2			23454-78-6	**71:** P22927a
23	26		4	2			23454-86-6	**71:** P22927a
23	26		6	7	2		51627-16-8	**81:** P105538g, **83:** P193352m
23	26		6	7	2	Na	53689-33-1	**81:** P105538g
23	28		6	2			35793-04-5	**76:** P128832j
23	28		6	5	2		56206-50-9	**83:** P58856p
24	15	2	3	2			28094-63-5	**73:** P26625q
24	15	2	3	2			35125-97-4	**76:** P60954w
24	15	2	3	3			26759-65-9	**72:** P122925n, **77:** P7317g, **81:** P137597s
24	15		5	1			34180-56-8	**75:** P152985m
24	15		7	2			52048-76-7	**81:** P51154h
24	16	1	3	1			37936-25-7	**78:** P85944u, **78:** P137966z
24	16	1	3	2			19683-10-4	**69:** P52937e, **81:** P51153g, **82:** P32468t
24	16	1	3	2			28094-45-3	**73:** P26625q

TABLE 17.1 (*Continued*)

Molecular Formula						Salt, adduct, or copolymer	Registry number	Reference
C	H	Cl	N	O	S			
24	16	1	3	2			28094-46-4	**73:** P26625q
24	16	1	3	2			28094-50-0	**73:** P26625q
24	16	1	3	2			31521-40-1	**74:** P100617c, **74:** P100618d, **81:** P79389n
24	16	1	3	2			31521-43-4	**74:** P100617c, **74:** P100618d, **81:** P79389n
24	16	1	3	2			35123-43-4	**76:** P60964w
24	16	1	3	3			26726-46-5	**72:** P122925n, **77:** P7317g, **81:** P137597s
24	16	1	3	3			31521-45-6	**74:** P100617c, **74:** P100618d, **81:** P79389n
24	16	1	3	3			35123-37-6	**76:** P60964w
24	16	1	5	1			41427-87-6	**78:** P85944u, **78:** P137966z
24	16		4	3			19695-31-9	**69:** P52937e, **81:** P51152f, **82:** P32468t
24	16		4	5			26726-50-1	**72:** P122925n, **77:** P7317g, **81:** P137597s
24	17	1	4	1			29399-92-6	**73:** P67701e
24	17	1	4	1			34180-50-2	**75:** P152985m
24	17	1	4	1			41428-19-7	**78:** P85944u
24	17	1	4		1		29363-25-5	**73:** P67701e
24	17		3	1			41428-11-9	**78:** P137966z
24	17		3	2			19683-09-1	**77:** P7317g, **82:** P32468t, **82:** P172621n
24	17		3	2			19683-13-7	**69:** P52937e, **81:** P51153g, **82:** P32468t
24	17		3	2			19695-45-5	**69:** P52937e
24	17		3	2			25826-28-2	**73:** P26625q **77:** P153929z
24	17		3	3			19695-47-7	**69:** P52937e
24	17		3	3			26726-45-4	**72:** P122925n, **77:** P7317g, **81:** P137597s
24	17		3	3			26726-48-7	**72:** P122925n, **77:** P7317g, **81:** P137597s
24	17		3	3			35123-34-3	**76:** P60964w
24	17		3	3			52084-92-1	**81:** P51143d
24	17		3	3			52570-46-4	**81:** P79389n
24	17		3	3			52673-69-5	**81:** P79389n
24	17		3	4			52570-48-6	**81:** P79389n

TABLE 17.1 (*Continued*)

C	H	Cl	N	O	S	Salt, adduct, or copolymer	Registry number	Reference
24	17		3	5	1		26759-73-9	**72:** P122925n, **77:** P7317g, **81:** P137597s
24	17		3	5	1		26759-74-0	**72:** P122925n, **77:** P7317g, **81:** P137597s
24	17		5	1			41427-86-5	**78:** P85944u, **78:** P137966z
24	17		5	2			34180-57-9	**75:** P152985m
24	18	1	3	2	1		41427-64-9	**78:** P85945v
24	18		4	1			34184-17-3	**75:** P152985m
24	18		4		1		29363-22-2	**73:** P67701e
24	18		6	2			52048-62-1	**81:** P122801e
24	18		6	3	1	Na	41532-60-9	**78:** P99065v
24	19	1	6	1			41428-23-3	**78:** P85944u, **78:** P137966z
24	19	1	6	1			52048-54-1	**81:** P122801e
24	19	1	6	2			41428-21-1	**78:** P85944u, **78:** P137966z
24	19		3	2			49744-08-3	**80:** P16460n
24	19		3	3	1	Na	41500-44-1	**78:** P85945v
24	19		5				34180-64-8	**75:** P152985m
24	19		5				41426-27-1	**78:** P85944u, **78:** P137966z
24	19		5	2			41427-85-4	**78:** P85944u, **78:** P137966z
24	19		5	3	1		34180-70-6	**75:** P152985m
24	19		7	4	1	Na	25730-29-4	**72:** P45038w
24	20	1	7	3			19695-44-4	**69:** P52937e **82:** P32468t
24	20		4	2			23454-63-9	**71:** P22927a
24	20		4	2			23454-74-2	**71:** P22927a
24	20		4	2			29207-76-9	**73:** 130944d
24	20		4	2			29207-77-0	**73:** 130944d
24	20		4	2			34502-43-7	**77:** 75174h
24	20		4	4			29207-78-1	**73:** 130944d
24	20		6	1			41427-99-0	**78:** P85944u, **78:** P137966z
24	20		6	1			52048-49-4	**81:** P122801e
24	20		6	2			41427-94-5	**78:** P85944u, **78:** P137966z
24	20		6	2			52048-50-7	**81:** P122801e
24	20		6	4	1	Na	26593-38-4	**72:** P45038w
24	20		6	4	1	Na	41532-59-6	**78:** P99065v
24	21	1	6	2			41428-20-0	**78:** P85944u, **78:** P137966z
24	21		5			HCl	41345-02-2	**78:** P99059w
24	22		6	2			41427-93-4	**78:** P85944u, **78:** P137966z
24	23		3				57381-91-6	**83:** P206290s

TABLE 17.1 (*Continued*)

C	H	Cl	N	O	S	Salt, adduct, or copolymer	Registry number	Reference
24	23		5	10	1		29125-88-0	**73:** P120648w
24	23		5	10	1		29125-95-9	**73:** P120648w
24	28	1	5	2			35793-05-6	**76:** P128832j
24	28		4	2			23454-83-3	**71:** P22927a
24	28		6	7	2		51627-19-1	**80:** P83024m
24	28		6	7	2		52353-30-7	**80:** P83013g
24	28		6	7	2		52438-35-4	**80:** P83013g
24	28		6	7	2		54172-62-2	**81:** P136161h, **83:** P58856p
24	28		6	8	2		54172-71-3	**81:** P136161h
24	29		5	2			34549-35-4	**76:** P128832j
24	32		6	6	2		56168-17-3	**83:** P28250u
25	16	1	3	2			31773-56-5	**74:** P113262h, **77:** P36415t
25	16	1	3	2			34627-93-5	**76:** P128835n, **77:** P153927x, **79:** P67828a, **79:** P80350h
25	16	1	5	1			34184-16-2	**75:** P152985m
25	16	2	4				41427-50-3	**78:** P85945v
25	16		4	3			32903-63-2	**75:** P37932e
25	17	1	4				41425-68-7	**78:** P85943t
25	17	1	4				41427-28-5	**78:** P85945v
25	17	1	4				41427-29-6	**78:** P85945v
25	17	1	4				41427-30-9	**78:** P85945v
25	17	1	4				41427-46-7	**78:** P85945v
25	17	1	4				41428-33-5	**78:** P85945v
25	17	1	4				56617-19-7	**83:** 105984r
25	17		3	3			31773-55-4	**74:** P113262h, **77:** P36415t
25	17		3	4			32896-21-2	**75:** P37932e
25	17		3	5			26759-64-8	**72:** P122925n, **77:** P7317g, **81:** P137597s
25	17		5	1			34180-53-5	**75:** P152985m
25	17		7	2			52048-74-5	**81:** P51154h
25	18	1	3	2			19778-47-3	**69:** P52937e, **81:** P51153g, **82:** P32468t
25	18	1	3	2			19778-48-4	**69:** P52937e, **81:** P51153g, **82:** P32468t
25	18	1	3	2			35123-45-6	**76:** P60964w
25	18	1	3	3			35125-96-3	**76:** P60964w
25	18		4				37936-26-8	**78:** P85945v
25	18		4				41425-64-3	**78:** P85943t
25	18		4				41425-73-4	**78:** P85943t
25	18		4				41425-77-8	**78:** P85943t
25	18		4				52455-56-8	**83:** P61732a

TABLE 17.1 (*Continued*)

C	H	Cl	N	O	S	Salt, adduct, or copolymer	Registry number	Reference
25	18		4	1			41427-45-6	**78:** P85945v
25	18		4	3			19695-32-0	**69:** P52937e, **81:** P51152f, **82:** P32468t
25	18		4	4			19695-43-3	**69:** P52937e, **82:** P32468t
25	18		4	4			19778-52-0	**69:** P52937e, **82:** P32468t
25	19	1	4	1			34184-14-0	**75:** P152985m
25	19	1	4	1			41427-53-6	**78:** P85945v
25	19		3	2			19683-11-5	**69:** P52937e, **81:** P51153g, **82:** P32468t
25	19		3	2			19683-14-8	**69:** P52937e, **81:** P51153g, **82:** P32468t
25	19		3	2			19695-46-6	**69:** P52937e
25	19		3	2			19695-48-8	**69:** P52937e
25	19		3	2			19695-50-2	**69:** P52937e
25	19		3	2			25054-85-7	**72:** P101860x, **73:** P121566e, **77:** P7317g
25	19		3	2			28094-47-5	**73:** P26625q
25	19		3	2			29702-95-2	**73:** P110809k
25	19		3	2			32061-94-2	**73:** P26625q **77:** P153929z
25	19		3	3			19778-57-5	**69:** P52937e
25	19		3	3			25781-33-3	**73:** P26625q
25	19		3	3			26726-42-1	**72:** P122925n, **77:** P7317g **81:** P137597s
25	19		3	3			26726-49-8	**72:** P122925n, **77:** P7317g, **81:** P137597s
25	19		3	3			26759-69-3	**72:** P122925n, **77:** P7317g, **81:** P137597s
25	19		3	3			28094-43-1	**73:** P26625q
25	19		3	3			35125-98-5	**76:** P60964w
25	19		3	3			35178-70-2	**76:** P60964w
25	19		3	4			26759-66-0	**72:** P122925n, **77:** P7317g, **81:** P137597s
25	19		3	4			35123-36-5	**76:** P60964w
25	19		3	4	1		25781-34-4	**73:** P26625q
25	19		7	1			25751-47-7	**72:** P45038w
25	20	1	3	2	1		41500-43-0	**78:** P85945v
25	20		4	1			26513-14-4	**73:** P67701e, **75:** P152985m

TABLE 17.1 (*Continued*)

Molecular Formula C	H	Cl	N	O	S	Salt, adduct, or copolymer	Registry number	Reference
25	20		4	1			37936-45-1	**78:** P85943t
25	20		4	1			41428-36-8	**78:** P85945v
25	20		4	2			29363-27-7	**73:** P67701e
25	20		4	3			18013-57-5	**68:** P106086b
25	20		4	3	1		29363-28-8	**73:** P67701e
25	21	1	4	2			38880-48-7	**77:** P141496t
25	21		3	2	1		41428-38-0	**78:** P85945v
25	21		5				34180-63-7	**75:** P152985m
25	21		5	1			34180-67-1	**75:** P152985m
25	21		7	2			25730-12-5	**72:** P45038w
25	21		7	4	1		52048-65-4	**81:** P51154h
25	22		4	2			37699-19-7	**77:** P90079z
25	24		5	4		$C_2H_3O_2$	38583-61-8	**77:** P90077x
25	25		5	8	1	Na	29125-87-9	**73:** P120648w
25	26		4	3			34180-55-7	**75:** P152985m
25	26		6	2			54969-65-2	**82:** 171481e
25	26		6	2		$(C_{29}H_{28}O_2)_x$	54969-74-3	**82:** 171481e
25	28		6	2			54983-70-9	**82:** 171481e
25	30		4	2			23454-67-3	**71:** P22927a
25	30		4	2			23454-79-7	**71:** P22927a
26	17	1	6				52048-56-3	**81:** P122801e
26	17	1	6				52048-58-5	**81:** P122801e
26	17	1	6	3	1	Na	25730-28-3	**72:** P45038w
26	17		5				41425-74-5	**78:** P85943t
26	17		5				41425-78-9	**78:** P85943t
26	18	1	3	4			35123-40-1	**76:** P60964w
26	18	1	3	4			52028-98-5	**81:** P51152f
26	18	1	7	2	1		52123-23-6	**81:** P122801e
26	18		4	3			32903-26-7	**75:** P37932e
26	18		6				31573-82-7	**74:** P88677n, **81:** P122801e
26	18		6				41428-05-1	**78:** P85944u, **78:** P137966z
26	18		6	3	1	Na	25730-26-1	**72:** P45038w
26	18		6	3	1	Na	41428-06-2	**78:** P85944u, **78:** P137966z
26	18		6	3	1	Na	41428-08-4	**78:** P85944u, **78:** P137966z
26	18		6	3	1	Na	52048-60-9	**81:** P122801e
26	18		6	6	2	$2C_{25}H_{53}N_3$	31773-52-1	**74:** P113261g
26	18		6	6	2	2Na	26798-30-1	**72:** P45038w
26	19	1	4				41427-49-0	**78:** P85945v
26	19	1	4				41427-61-6	**78:** P85945v
26	19	1	4	1			41427-47-8	**78:** P85945v
26	19	1	4	1			41427-62-7	**78:** P85945v
26	19	1	6				56634-76-5	**83:** P133399x
26	19	2	3	1			41428-17-5	**78:** P85944u, **78:** P137966z
26	19		3	3			41428-13-1	**78:** P85944u, **78:** P137966z

TABLE 17.1 (*Continued*)

Molecular Formula C	H	Cl	N	O	S	Salt, adduct, or copolymer	Registry number	Reference
26	19		3	4			19695-53-5	**69:** P52937e
26	19		3	4			35123-29-6	**76:** P60964w
26	19		3	4			35123-39-8	**76:** P60964w
26	20	1	3	1			41428-12-0	**78:** P85944u, **78:** P137966z
26	20		4				41425-75-6	**78:** P85943t
26	20		4				41427-32-1	**78:** P85945v
26	20		4				41428-32-4	**78:** P85945v
26	20		4	1			41427-33-2	**78:** P85945v
26	20		4	1			41427-90-1	**78:** P85944u, **78:** P137966z
26	20		4	1			41428-30-2	**78:** P85945v
26	20		4	3			19695-42-2	**69:** P52937e, **82:** P32468t
26	20		4	4			19695-41-1	**69:** P52937e, **82:** P32468t
26	20		4	4			19695-55-7	**69:** P52937e, **82:** P32468t
26	20		6				31573-80-5	**74:** P88677n
26	20		6				37717-27-4	**78:** P99060q
26	20		6	6	2	2Na	52174-71-7	**81:** P65249e
26	20		6	9	3	3Na	52174-70-6	**81:** P65249e
26	20		8	4	2		25730-24-9	**72:** P45038w
26	21		3	2			19683-15-9	**69:** P52937e, **81:** P51153g, **82:** P32468t
26	21		3	2			19695-51-3	**69:** P52937e
26	21		3	3			19783-69-8	**69:** P52937e
26	21		3	3			35123-38-7	**76:** P60964w
26	21		7	1			25751-48-8	**72:** P45038w
26	22	1	3	2	1		41427-56-9	**78:** P85945v
26	22		4	1			29363-24-4	**73:** P67701e
26	22		4	2			35126-02-4	**76:** P60964w
26	22		4	5	1		26792-16-5	**72:** P122925n
26	22		6	1			41344-76-7	**78:** P99060q
26	23	1	4	2	1		41427-54-7	**78:** P85945v
26	23		3	2	1		41428-39-1	**78:** P85945v
26	23		3	2	1		41428-40-4	**78:** P85945v
26	23		5				34180-61-5	**75:** P152985m
26	23		5				34180-69-3	**75:** P152985m
26	24		4	2	1		41428-37-9	**78:** P85945v
26	25	1	8	2			19695-63-7	**69:** P52937e, **82:** P32468t
26	26		6	10	2	$(C_6H_{13}NO_2$ $C_6H_{11}NO)_x$	25853-69-4	**72:** P68221q
26	28		6	2			54969-66-3	**82:** 171481e
26	28		6	2		$(C_{26}H_{18}O_3)_x$	54969-42-5	**82:** 171481e
26	28		6	2		$(C_{29}H_{24}O_2)_x$	54969-41-4	**82:** 171481e
26	28		6	2		$(C_{29}H_{28}O_2)_x$	54969-75-4	**82:** 171481e
26	30		6	2			54983-71-0	**82:** 171481e

TABLE 17.1 (*Continued*)

Molecular Formula						Salt, adduct, or	Registry	
C	H	Cl	N	O	S	copolymer	number	Reference
26	33		5	2			35793-03-4	**76:** P128832j
27	16	1	3	2			31521-46-7	**74:** P100617c, **74:** P100618d, **81:** P79389n
27	16	1	7				25730-08-9	**72:** P45038w
27	16	1	7				25730-09-0	**72:** P45038w
27	16		6	2			52497-95-7	**81:** P171366a
27	17		3	3			52673-71-9	**81:** P79389n
27	17		7				23918-34-5	**72:** P45038w
27	17		7				42962-70-9	**83:** P99222x
27	17		7	3	1	Na	25730-25-0	**72:** P45038w
27	18	1	7				41532-65-4	**78:** P99065v
27	19	1	6	1			52048-57-4	**81:** P122801e
27	19	1	6	1			52048-59-6	**81:** P122801e
27	19	1	8	2			19695-59-1	**69:** P52937e, **82:** P32468t
27	19		7				31573-84-9	**74:** P88677n
27	19		7				41532-63-2	**78:** P99065v
27	19		7	1			41532-64-3	**83:** P195247e
27	20		4	2			52455-31-9	**83:** P61732a
27	20		4	2			52455-55-7	**83:** P61732a
27	20		6	1			52048-52-9	**81:** P122801e
27	20		6	2			56634-75-4	**83:** P133399x
27	20		6	3	1	Na	25730-27-2	**72:** P45038w
27	21	1	4	2			41427-48-9	**78:** P85945v
27	21		3	4			19695-54-6	**69:** P52937e
27	21		3	4			35178-71-3	**76:** P60964w
27	21		7	1			41532-62-1	**78:** P99065v, **83:** P195247e
27	21		7	1			56634-74-3	**83:** P133399x
27	21		7	2	1		41428-07-3	**78:** P85944u, **78:** P137966z
27	21		7	3			41532-54-1	**78:** P99065v
27	22		4	2			41428-31-3	**78:** P85945v
27	22		4	3			34180-52-4	**75:** P152985m
27	23	1	6	4			41428-26-6	**78:** P85944u, **78:** P137966z
27	24	1	3	2	1		41427-55-8	**78:** P85945v
27	24		4	2			35126-05-7	**76:** P60964w
27	25	1	6	2			41428-24-4	**78:** P85944u, **78:** P137966z
27	25		7	6	1		52048-71-2	**81:** P51154h
27	26		6	2			41428-00-6	**78:** P85944u, **78:** P137966z
27	29		9	2			37757-56-5	**78:** P31421c
27	34		4	2			23454-72-0	**71:** P22927a
28	16		8				41538-03-8	**79:** P80310v
28	18	2	4	2			34502-47-1	**77:** 75174h
28	18	2	4	2			34502-49-3	**77:** 75174h

TABLE 17.1 (*Continued*)

Molecular Formula C	H	Cl	N	O	S	Salt, adduct, or copolymer	Registry number	Reference
28	18		6	2			52497-94-6	**81:** P171366a
28	18		6	6			34502-40-4	**77:** 75174h
28	18		6	6			34502-45-9	**77:** 75174h
28	18		8				41532-70-1	**78:** P99065v, **83:** P99223y
28	19	2	3	3	1	Na	50745-51-2	**79:** P127405x
28	19		3	2			25781-35-5	**73:** P26625q
28	19		3	2			42242-33-1	**79:** P116278g
28	19		5	2			52497-88-8	**81:** P171366a
28	19		7				25730-07-8	**72:** P45038w
28	19		7				42962-67-4	**83:** P99222x
28	19		7				42962-71-0	**79:** P80310v
28	19		7	1			25730-10-3	**72:** P45038w
28	20	1	3	3	1	Na	50745-49-8	**79:** P127405x
28	20	1	7				41532-67-6	**83:** P195247e
28	20	2	4	2			41427-57-0	**78:** P85945v
28	20	2	4	2			41427-58-1	**78:** P85945v
28	20	2	4	2			41696-69-9	**78:** P85945v
28	20		4	2			19226-33-6	**69:** 36042d
28	20		6				52048-53-0	**81:** P122801e
28	20		6	2			52497-96-8	**81:** P171366a
28	20		6	3	1	Na	25751-51-3	**72:** P45038w
28	20		6	11	3	3Na	22092-06-4	**70:** P97963x
28	20		8	6	2	2Na	54673-47-1	**82:** P126614q
28	21	1	4	2			41427-23-0	**78:** P85945v
28	21	1	4	2			41427-24-1	**78:** P85945v
28	21	1	4	2			41427-25-2	**78:** P85945v
28	21	1	4	2			41427-52-5	**78:** P85945v
28	21	1	8	2			19695-58-0	**69:** P52937e, **82:** P32468t
28	21	1	8	2			19695-60-4	**69:** P52937e, **82:** P32468t
28	21	2	3	2			42446-57-1	**79:** P116278g
28	21		3				30818-76-9	**74:** P65594u
28	21		3	3	1	Na	27546-11-8	**74:** P65594u
28	21		3	4	1	Na	50745-59-0	**79:** P127405x
28	21		7	1			41532-66-5	**78:** P99065v
28	22		4	2			41428-35-7	**78:** P85945v
28	22		6	2	1		25730-18-1	**72:** P45038w
28	22		6	3			52497-87-7	**81:** P171366a
28	23		3	4			35123-41-2	**76:** P60964w
28	23		3	4			52455-49-9	**83:** P61732a
28	23		5	1			34180-54-6	**75:** P152985m
28	23		7	2	1		25729-98-0	**72:** P45038w
28	23		7	2	1		25730-15-8	**72:** P45038w
28	23		7	3	1		25730-17-0	**72:** P45038w
28	24		4	3			35125-99-6	**76:** P60964w
28	24		6				56634-78-7	**83:** P133399x
28	24		6	6	2	2Na	52174-68-2	**81:** P65249e

TABLE 17.1 (*Continued*)

Molecular Formula						Salt, adduct, or	Registry	
C	H	Cl	N	O	S	copolymer	number	Reference
28	25		3	3			19971-35-8	**69:** P52937e
28	26	2	4	10			20640-45-3	**69:** 36375q
28	26	2	4	10			23259-11-2	**70:** 58163s
28	26		4	5	1		37150-82-6	**81:** P137597s
28	27		9	2			19695-36-4	**69:** P52937e, **82:** P32468t
28	28		4	10			13032-88-7	**70:** 58163s
28	28		4	10			23259-10-1	**70:** 58163s
28	28		8	4	1		52048-68-7	**81:** P51154h
29	17		9				41532-72-3	**83:** P99223y
29	18	1	3	2			29702-98-5	**73:** P110809k
29	18	1	3	2			31521-47-8	**74:** P100617c, **74:** P100618d, **81:** P79389n
29	18	2	4				41935-54-0	**83:** P81213k
29	19	1	4				41935-53-9	**83:** P81213k
29	19	1	4	1			49743-73-9	**80:** P16460n
29	19	1	4	1			49743-75-1	**80:** P16460n
29	19	1	4	1			49743-77-3	**80:** P16460n
29	19		3	2			25054-82-4	**72:** P101860x, **72:** P122925n, **73:** P121566e, **81:** P137597s
29	19		3	2			28094-51-1	**73:** P26625q, **77:** P153929z
29	19		3	3			25080-04-0	**72:** P122925n, **77:** P7317g, **81:** P137697s
29	19		3	3			52570-49-7	**81:** P79389n
29	19		3	5	1		26792-17-6	**72:** P122925n, **77:** P7317g, **81:** P137597s
29	19		7				25730-11-4	**72:** P45038w
29	20	1	5				41427-84-3	**78:** P85944u, **78:** P137966z
29	20		4	1			29636-55-3	**74:** P88677n
29	20		4	1			41934-50-3	**79:** P127405x
29	20		4	1			49743-72-8	**80:** P16460n
29	20		4	4	1	Na	49743-81-9	**80:** P16460n
29	20		6	4	1		52048-66-5	**81:** P51154h
29	20		6	5	1		52048-75-6	**81:** P51154h
29	21	1	6	1	1		41500-47-4	**78:** P85944u, **78:** P137966z
29	21	1	6	2			41428-25-5	**78:** P85944u, **78:** P137966z
29	21		3	5	1	x Na	30818-71-4	**74:** P65594u
29	21		5				34180-68-2	**75:** P152985m
29	21		5				41427-83-2	**78:** P85944u, **78:** P137966z

TABLE 17.1 *(Continued)*

Molecular Formula						Salt, adduct, or	Registry	
C	H	Cl	N	O	S	copolymer	number	Reference
29	21		5				49743-82-0	**80:** P16460n
29	21		5	3			49744-04-9	**80:** P16460n
29	22	1	3	4	1	Na	50745-50-1	**79:** P127405x
29	22		4	4	1	Na	30818-72-5	**74:** P65594u
29	22		6	1	1		41428-02-8	**78:** P85944u, **78:** P137966z
29	22		6	2			41428-01-7	**78:** P85944u, **78:** P137966z
29	23	1	4	3			41427-59-2	**78:** P85945v
29	24		4	2			41428-41-5	**78:** P85945v
29	24		4	3			41427-26-3	**78:** P85945v
29	25	1	4				41427-60-5	**78:** P85945v
29	25		3	4			35123-42-3	**76:** P60964w
29	26		4				41427-31-0	**78:** P85945v
29	26		4	2			35126-01-3	**76:** P60964w
29	26		4	3			35126-04-6	**76:** P60964w
29	26		4	4			35126-06-8	**76:** P60964w
29	27		5	8	1		29125-81-3	**73:** P120648w
29	27		9	7	2	2Na	41113-14-8	**78:** P31421c
29	28		6	3			25730-13-6	**72:** P45038w
29	29	1	4	2			23454-84-4	**71:** P22927a
29	29		3	3			54969-68-5	**82:** 171481e
29	31		8	4	1	CH_3O_4S	52048-73-4	**81:** P51154h
29	33		9	2			19695-56-8	**69:** P52937e, **82:** P32468t
29	38		4	2			23454-81-1	**71:** P22927a
30	19		5	2			53199-84-1	**82:** P17894k
30	20	2	4				50670-47-8	**79:** P147435k
30	20	2	6	4	2		34564-10-8	**76:** P128834m
30	20	2	6	6	2		52570-51-1	**81:** P79389n
30	20	2	6	6	2	2Na	31521-48-9	**74:** P100617c
30	20		6				35102-59-1	**76:** P73762d
30	20		6	2			32903-27-8	**75:** P37932e
30	20		6	3	1	Na	35102-60-4	**76:** P73762d
30	21	1	4	1			50745-48-7	**83:** P81213k
30	21		3	2			19695-30-8	**69:** P52937e, **81:** P51153g, **82:** P32468t
30	21		3	2			42446-62-8	**79:** P116278g
30	21		3	3			28094-49-7	**73:** P26625q
30	21		3	3			42446-59-3	**79:** P116278g
30	22	1	3	3	1	Na	49795-66-6	**79:** P106146y
30	22	1	5				34180-62-6	**75:** P152985m
30	22	1	5				41426-30-6	**78:** P85944u, **78:** P137966z
30	22		4				50670-44-5	**79:** P147435k
30	22		4	1			29363-23-3	**73:** P67701e
30	22		4	1			49743-79-5	**80:** P16460n
30	22		4	3	1		49743-74-0	**80:** P16460n

TABLE 17.1 (*Continued*)

C	H	Cl	N	O	S	Salt, adduct, or copolymer	Registry number	Reference
30	22		6	5	1		52048-67-6	**81:** P51154h
30	22		6	6	2		37069-54-8	**79:** P137160c
30	22		6	6	2	$2CH_5N_3$	31773-50-9	**74:** P113261g
30	22		6	6	2	$2C_{13}H_{13}N_3$	31773-51-0	**74:** P113261g
30	22		6	6	2	$2C_{14}H_{31}N$	37432-50-1	**77:** P7222x
30	22		6	6	2	$2C_{25}H_{53}N_3$	31773-53-2	**74:** P113261g
30	22		6	6	2	2K	52237-03-3	**81:** P93085s, **81:** P107815u
30	22		6	6	2	2Na	23743-28-4	**79:** P137160c, **81:** P25679p, **81:** P38957c, **83:** P193339n
30	22		6	8	2		52570-50-0	**81:** P79389n
30	23		3				29081-46-7	**74:** P65594u
30	23		3				41425-71-2	**78:** P85943t
30	23		3	3	1	Na	42181-64-6	**79:** P106146y
30	23		5				34180-66-0	**75:** P152985m
30	24		4	2			23454-82-2	**71:** P22927a
30	24		4	2			34502-46-0	**77:** 75174h
30	24		4	2			34502-48-2	**77:** 75174h
30	24		4	4			19226-34-7	**69:** 36042d
30	24		8	6	2	2Na	54673-48-2	**82:** P126614q
30	27		3	2			42407-16-9	**79:** P116278g
30	28		4	3			35126-00-2	**76:** P60964w
30	28		8	2	1		25730-23-8	**72:** P45038w
30	28		8	4	2		25730-19-2	**72:** P45038w
30	31		9	2			19695-37-5	**69:** P52937e, **82:** P32468t
30	32		4	10			13400-19-6	**70:** 58163s
30	32		4	10			13400-20-9	**70:** 58163s
30	32		4	12			20640-44-2	**69:** 36375q
31	19	3	6				41113-19-3	**78:** P31421c
31	19		7				34234-31-6	**76:** P73762d
31	19		7				51202-88-1	**81:** P38962a
31	20	1	3	4			35123-31-0	**76:** P60964w
31	20		4	3			32896-22-3	**75:** P37932e
31	20		4	3			32903-28-9	**75:** P37932e
31	20		4	3			32903-29-0	**75:** P37932e
31	21	1	4				41427-51-4	**78:** P85945v
31	21		3	4			28094-41-9	**73:** P26625q
31	21		3	4			33936-22-0	**76:** P60964w
31	21		5	2			32896-26-7	**75:** P37932e
31	21		5	2			53185-55-0	**82:** P17894k
31	22		4				41425-66-5	**78:** P85943t
31	22		4				41428-34-6	**78:** P85945v
31	22		6				37757-57-6	**78:** P31421c
31	22		6	3	1	Na	52596-37-9	**81:** P51142c
31	22		8	3			41532-56-3	**78:** P99065v
31	23		3	2			42446-65-1	**79:** P116278g

TABLE 17.1 (*Continued*)

Molecular Formula						Salt, adduct, or copolymer	Registry number	Reference
C	H	Cl	N	O	S			
31	23		3	3			42446-60-6	**79:** P116278g
31	23		3	3			42446-61-7	**79:** P116278g
31	23		3	3			42446-63-9	**79:** P116278g
31	23		7	2	1		51732-33-3	**81:** P51142c
31	25		5				34206-90-1	**75:** P152985m
31	25		5	1			41500-36-1	**78:** P85944u, **78:** P137966z
31	26		4	2			42446-70-8	**79:** P116278g
31	26		5			CH_3O_4S	49743-83-1	**80:** P16460n
31	26		8	2	1		25730-20-5	**72:** P45038w
31	29		3	2			42446-68-4	**79:** P116278g
31	30		8	3			41532-55-2	**78:** P99065v
32	20	2	6	1			41344-85-8	**78:** P99060q
32	20	2	6	1			41344-86-9	**78:** P99060q
32	20	2	6	2	1		41345-00-0	**78:** P99059w
32	21		5	3			32903-25-6	**75:** P37932e
32	22	2	6				41344-80-3	**78:** P99060q
32	22	2	6				41344-92-7	**78:** P99060q
32	22	2	6	6	2	2Na	49694-01-1	**79:** P106146y
32	22		4	2	1		33014-37-8	**75:** P37932e
32	22		4	3			31676-24-1	**75:** P37932e
32	22		6	1			41344-82-5	**78:** P99060q
32	22		6	2	1		41344-99-4	**78:** P99059w
32	22		6	2	1		51131-76-1	**80:** P84255z
32	22		6	3	1		25730-14-7	**72:** P45038w
32	22		6	10	2	2Na	21163-10-0	**70:** P97963x
32	23		3	4			35178-69-9	**76:** P60964w
32	23		3	5			35123-30-9	**76:** P60964w
32	23		5	2			53199-83-0	**82:** P17894k
32	23		5	4			53185-40-3	**82:** P17894k
32	23		7	2	1		25730-16-9	**72:** P45038w
32	24	2	6	6	2	2Na	33753-84-3	**75:** P99255v
							32892-85-6	**82:** P74470t
32	24		6				37936-44-0	**78:** P85943t, **80:** P16457s, **80:** P134946y
32	24		6				41344-79-0	**78:** P99060q
32	24		6	2			41425-80-3	**78:** P85943t
32	24		6	2	1		56634-77-6	**83:** P133399x
32	24		6	6	2	2Na	49795-67-7	**79:** P106146y
32	25		5	6	1	Na	29125-96-0	**73:** P120648w
32	26		6	12	4	4Na	33799-88-1	**75:** P99255v
32	27		5				34180-60-4	**75:** P152985m
32	29		7	3	1		25751-52-4	**72:** P45038w
32	30		4	2			41427-27-4	**78:** P85945v
32	35		9	2			19695-38-6	**69:** P52937e, **82:** P32468t
32	36		4	2			23454-85-5	**71:** P22927a
33	22		4	1			49743-70-6	**80:** P16460n

TABLE 17.1 (*Continued*)

C	H	Cl	N	O	S	Salt, adduct, or copolymer	Registry number	Reference
33	25		3	2			42446-58-2	**79:** P116278g
33	25		3	4			35123-32-1	**76:** P60964w
33	25		7	2	1		25751-49-9	**72:** P45038w
33	27	1	4	1			49743-76-2	**80:** P16460n
33	27	1	4	1			49743-78-4	**80:** P16460n
33	27		5	6	1		29129-65-5	**73:** P120648w
33	28		4	1			49743-71-7	**80:** P16460n
33	31		3	4			42446-69-5	**79:** P116278g
33	31		5	8	2		29126-07-6	**73:** P120648w
33	33	1	8	2	1		52048-61-0	**81:** P122801e
33	33		3	2			42407-17-0	**79:** P116278g
33	33		9	2			19695-61-5	**69:** P52937e, **82:** P32468t
33	34		8	2	1		25751-50-2	**72:** P45038w
33	35		9	4	2		25730-22-7	**72:** P45038w
34	15		9	2			53185-58-3	**82:** P17894k
34	22		6	2			32896-25-6	**75:** P37932e
34	22		8				41425-50-7	**78:** P85943t
34	22		8				41500-35-0	**78:** P85943t
34	24		8	6	2	2Na	33753-87-6	**75:** P99255v, **82:** P74470t
34	24		8	8	2	2Na	53305-34-3	**82:** P74470t
34	25	1	6				41345-03-3	**78:** P99059w
34	25		3				41973-71-1	**79:** P147435k
34	26		6				42222-28-6	**79:** P106145x
34	26		6	6	2	2Na	41345-01-1	**78:** P99059w
34	27		3	5			42446-64-0	**79:** P116278g
34	28		6				41425-52-9	**78:** P85943t
34	28		6	6	2	2Na	49795-68-8, 52174-69-3	**79:** P106146y **81:** P65249e
34	28		8	8	2	2Na	21163-11-1	**70:** P38909m
34	28		8	8	2	2Na	33753-88-7	**75:** P99255v
34	30		4	1			49743-80-8	**80:** P16460n
34	37		8	2	1	CH_3O_4S	25730-33-0	**72:** P45038w
35	23		3	2			42446-66-2	**79:** P116278g
35	24		4	1			49743-68-2	**80:** P16460n
35	24		4	1			49743-69-3	**80:** P16460n
35	25		7				37717-28-5	**78:** P99059w
35	28		4	3			32896-24-5	**75:** P37932e
35	29		3	2			42446-67-3	**79:** P116278g
36	24		6	3	1		35102-58-0	**76:** P73762d
36	26	2	12	6	2	2Na	54641-95-1	**82:** P126614q
36	26	2	12	6	2	2Na	54641-96-2	**82:** P126614q
36	26		4				50670-38-7	**79:** P147435k
36	26		10	8	2	2Na	54641-97-3	**82:** P126614q
36	28		12	6	2	2Na	54641-93-9	**82:** P126614q
36	28		12	12	4	4Na	54673-50-6	**82:** P126614q
36	30	2	12	8	2	2Na	54673-51-7	**82:** P126614q
36	32		12	8	2	2Na	54673-49-3	**82:** P126614q

TABLE 17.1 (*Continued*)

C	H	Cl	N	O	S	Salt, adduct, or copolymer	Registry number	Reference
36	36		8	6	2		34564-09-5	**76:** P128834m
36	42		10	4	2		25730-21-6	**72:** P45038w
36	47		9	2			19695-57-9	**69:** P52937e, **82:** P32468t
37	26		6	3			32896-23-4	**75:** P37932e
37	30	2	6	10	1		37132-88-0	**77:** P126662a
37	30		8	1			41532-53-0	**78:** P99065v
37	54		4	2			38880-49-8	**77:** P141496t
38	23		5	2			53185-41-4	**82:** P17894k
38	26		6	6	2		25730-32-9	**72:** P45038w
38	28		6				41425-56-3	**78:** P85943t
38	30		10	8	2	2Na	54673-46-0	**82:** P126614q
38	30		10	10	2	2Na	54641-90-6	**82:** P126614q
38	32		12	6	2	2Na	54641-94-0	**82:** P126614q
39	28		10	1			41532-58-5	**78:** P99065v
39	36		4	3			32844-73-8	**75:** P37932e
40	26		8				41425-59-6	**78:** P85943t
40	30		6	12	4	2Na	53359-79-8	**82:** P74470t
40	30		6	12	4	4Na	33753-86-5	**75:** P99255v
44	42		10	8	2	2Na	54641-89-3	**82:** P126614q
56	40		6				37421-60-6	**77:** P63358h

TABLE 17.2. MISCELLANEOUS 1,2,3-TRIAZOLES AND RELATED COMPOUNDS

Molecular formula											Registry	
C	H	Br	F	I	N	O	P	Si	S	M+	number	Reference
13	12		3		5	4			3		52115-91-0	**80:** P120970q
13	19				5	3		1	2		56254-37-6	**83:** P58856p
14	14		3		5	4			3	Na	52115-92-1	**80:** P120970q
17	11			1	4						41425-69-8	**78:** P85943t
17	14	1			5	2					37157-27-0	**77:** P7314d
17	14	2			6	5			2	K	56187-52-1	**83:** 43359e
17	14	2			6	5			2	K	56187-52-1	**83:** 43359e
17	14			2	6	5			2	K	56187-53-2	**83:** 43359e
18	14	2			4	2					35474-96-1	**76:** 140659r
18	14	2			4	2					36474-97-2	**76:** 140659r
18	17		1		6	5			2		51746-35-1	**80:** P83009k
20	15		6		5	4			1		38921-36-7	**77:** 147455a
20	15		6		5	4			1		38921-37-8	**77:** 147455a
20	16		6		6	4			1		38921-40-3	**77:** 147455a
21	16	1			3						57381-88-1	**83:** P206290s
21	16		1		3						57381-83-6	**83:** P206290s
21	16			1	3						57381-90-5	**83:** P206290s
21	19		3		6	6			2		56206-54-3	**83:** P58856p
22	16		3		3						57381-79-0	**83:** P206290s
22	16		3		3						57381-89-2	**83:** P206290s

TABLE 17.2 (*Continued*)

Molecular formula-											Registry	
C	H	Br	F	I	N	O	P	Si	S	M+	number	Reference
22	20		1		5	8			1		29125-92-6	**73:** P120648w
22	20		1		5	8			1		29125-93-7	**73:** P120648w
22	20		1		5	8			1		29125-94-8	**73:** P120648w
23	14	1			3	2					29633-67-8	**74:** P100617c, **81:** P79389n
23	22				5	3	1				41532-68-7	**78:** P99065v
25	16		3		3	3					26759-67-1	**72:** P122925n, **77:** P7317g, **81:** P137597s
26	27		3		6	9			2		56206-53-2	**83:** P58856p
27	32				3	5	1				35126-08-0	**76:** P60964w
28	18	2			4	2					34519-96-5	**77:** 75174h
28	18	2			4	2					34519-97-6	**77:** 75174h
28	18	2			4	2					34566-67-1	**77:** 75174h
28	18	2			4	2					56428-74-1	**83:** 106649x
28	26	2			4	10					20640-46-4	**69:** 36375q
28	26	2			4	10					20640-47-5	**69:** 36375q
28	26			2	4	10					20640-48-6	**69:** 36375q
30	15	4			5	2					53185-57-2	**82:** P17894k
30	20		2		6	6			2	2Na	35542-80-4	**77:** P63261w
32	34				3	5	1				35126-07-9	**76:** P60964w
33	36				3	5	1				35126-09-1	**76:** P60964w
33	36				3	6	1				35126-10-4	**76:** P60964w

REFERENCES

68: 87784j — V. V. Korshak, E. S. Krongauz, A. P. Travnikova, A. V. D'yachenko, A. A. Askadskii, and V. P. Sidorova, *Dokl. Akad. Nauk USSR*, **178,** 607 (1968).

68: P106086b — Hickson and Welch, Netherlands Patent, 6,613,363 (1967).

69: 36042d — N. E. Alexandrou and E. M. Micromastoras, *Tetrahedron Lett.*, 231 (1968).

69: 36375q — H. El Khadem, M. A. M. Nassr, and M. A. E. Shaban, *J. Chem. Soc., C,* 1465 (1968).

69: P52737e — O. Neuner, A. Dorlars, C. W. Schellhammer, and O. Berendes, British Patent, 1,113,918 (1968).

69: 107160d — Y. Gilliams and G. Smets, *Makromol. Chem.*, **117,** 1 (1968).

70: P38909m — Farbenfabriken Bayer A.-G., French Patent, 1,508,550 (1968).

70: P48020d — R. K. Gavurina, P. A. Medvedeva, G. A. Koval'chuk, G. S. Akimova, V. N. Chistokletov, and A. A. Petrov, Russian Patent, 224,066 (1968).

70: 58163s — H. El Khadem and M. A. E. Shaban, *J. Chem. Soc.*, **C,** 519 (1967).

70: P97963x — A. Dorlars, O. Neuner, and R. Puetter, German Patent, 1,287,550 (1969).

71: 2032b — S. Petersen, W. Gauss, H. Kiehne, and L. Juehling, *Z. Krebsforsch.*, **72,** 162 (1969).

71: 13464v — V. V. Rode, E. M. Bondarenko, A. V. D'yachenko, E. S. Krongauz, and V. V. Korshak, *Vysokomol. Soedin., Ser. A*, **11,** 828 (1969).

71: P22927a	A. Dorlars and C. W. Schellhammer, South African Patent 68 01,094 (1968).
72: 3246c	Y. Gilliams and G. Smets, *Makromol. Chem.*, **128,** 263 (1969).
72: P45038w	I. Okubo and M. Tsujimoto, German Patent, 1,805,371 (1969).
72: P68221q	A. Dorlars and E. Istel, British Patent, 1,172,328 (1969).
72: P101860x	Geigy, A.-G., French Patent, 1,556,530 (1969).
72: P112806k	A. Dorlars and W. D. Wirth, South African Patent, 68 08,154 (1969).
72: P129925n	Geigy, A.-G., French Patent, 1,556,529 (1969).
73: P26625q	R. Kirchmayr and J. Rody, German Patent, 1,936,760 (1970).
73: P67701e	G. G. Di Giovanoel and R. Zweidler, German Patent, 1,958,589 (1970).
73: P78559r	Farbenfabriken Bayer A.-G., French Patent, 1,565,741 (1969).
73: P100073b	Farbenfabriken Bayer A.-G., French Patent, 2,004,074 (1969).
73: P110809k	Geigy, A.-G., French Patent, 1,563,902 (1969).
73: P120648w	Glaxo Laboratories Ltd., French Patent, 1,572,190 (1969).
73: P121566e	F. Fleck, H. Balzer and H. Aebli, French Patent, 1,568,007 (1969).
73: 130944d	H. El Khadem, M. A. E. Shaban, and M. A. M. Nassr, *J. Chem. Soc., C*, 2167 (1970).
74: P14197j	A. Dorlars, German Patent, 1,906,662 (1970).
74: P65594u	A. Dorlars and O. Neuner, German Patent, 1,917,740 (1970).
74: P88677n	F. Fleck and H. R. Schmid, German Patent, 2,025,792 (1970).
74: P100613y	H. Schlaepfer, German Patent, 2,029,122 (1970).
74: P100617c	R. Kirchmayr, German Patent, 2,029,157 (1970).
74: P100618d	R. Kirchmayr, German Patent, 2,029,157 (1970).
74: P112048f	G. Jaeger and K. H. Buechel, German Patent, 1,940,626 (1971).
74: P113261g	W. Horstmann, German Patent, 1,929,664 (1970).
74: P113262h	H. Schlaepfer, German Patent, 2,029,096 (1970).
74: P125697s	W. Draber and K. H. Buechel, German Patent, 1,940,627 (1971).
74: P125698t	K. H. Buechel and W. Draber, German Patent, 1,940,628 (1971).
74: 141654f	H. Bauer, A. J. Boulton, W. Fedeli, A. R. Katritzky, A. Majik-Hamid, F. Mazza, and A. Vaciago, *Angew. Chem., Int. Ed. Engl.*, **10,** 129 (1971).
75: P37932e	H. Aebli, H. Balzer, and F. Fleck, German Patent, 2,047,547 (1971).
75: P99255v	H. Balzer, F. Fleck, A. V. Mercer, and R. Paver, German Patent, 2,062,383 (1971).
75: P110997m	J. Schroeder and C. W. Schellhammer, German Patent, 1,955,068 (1971).
75: P152985m	O. Neuner, A. Dorlars, and P. Schnegg, German Patent, 1,962,353 (1971).
76: P60964w	R. Kirchmayr, German Patent, 2,112,198 (1971).
76: P73762d	F. Fleck and H. Schmid, German Patent, 2,117,567 (1971).
76: P128832j	A. F. Strobel and M. L. Whitehouse, German Patent, 2,129,863 (1971).
76: P128833k	A. F. Strobel and M. L. Whitehouse, German Patent, 2,129,855 (1971).
76: P128834m	A. Dorlars and O. Neuner, German Patent, 2,032,172 (1972).
76: P128835n	H. Schlaepfer, German Patent, 2,032,088 (1972).
76: 140659r	H. Bauer, A. J. Boulton, W. Fedeli, A. R. Katritzky, A. Majid-Hamid, F. Mazza, and A. Vaciago, *J. Chem. Soc., Perkin Trans.*, **2,** 662 (1972).
76: 149496t	R. G. Micetich, R. Raap, J. Howard, and I. Pushkas, *J. Med. Chem.*, **15,** 333 (1972).
77: P7222x	Geigy, A.-G., French Patent, 2,067,301 (1971).
77: P7313c	H. Lind, German Patent, 2,133,012 (1972).
77: P7314d	A. Dorlars and C. W. Schellhammer, German Patent, 2,037,854 (1972).
77: P7315e	A. Dorlars and H. Gold, German Patent, 2,040,189 (1972).
77: P7317g	R. Kirchmayr, H. Heller, and J. Rody, U.S. Patent, 3,646,054 (1972).
77: P36415t	Geigy, A.-G., French Patent, 2,051,162 (1971).
77: P63261w	K. D. Bode, R. Germann, and U. Schuessler, German Patent, 2,051,302 (1972).

77: P63358h A. E. Siegrist, German Patent, 2,148,015 (1972).

77: 75174h N. E. Alexandrou and E. D. Micromastoras, *J. Org. Chem.*, **37,** 2345 (1972).

77: P88521u D. A. Berges, G. L. Dunn, and J. R. E. Hoover, German Patent, 2,158,330 (1972).

77: P90077x G. Boehmke and H. Theidel, German Patent, 2,050,726 (1972).

77: P90079z T. Noguchi, K. Tsukamoto, K. Isogami, and H. Hojo, Japanese Patent, 71 33,148 (1971).

77: P126654z L. B. Crast, Jr., German Patent, 2,202,274 (1972).

77: P126662a M. Gregson, M. C. Cook, and G. I. Gregory, German Patent, 2,204,060 (1972).

77: P128099w R. Zweidler, G. Kabas, H. Schlaepfer, and I. J. Fletcher, German Patent, 2,159,797 (1972).

77: P141496t A. F. Strobel and M. L. Whitehouse, French Patent, 2,095,314 (1972).

77: 147455a D. Willner, A. M. Jelenevsky, and L. C. Cheney, *J. Med. Chem.*, **15,** 948 (1972).

77: P153927x Geigy, A.-G. French Patent, 2,097,448 (1972).

77: P153929z R. Kirchmayr and J. Rody, U.S. Patent, 3,686,202 (1972).

77: P164713a W. Gauss and S. Petersen, German Patent, 2,106,845 (1972).

78: P31421c H. Schlaepfer, G. Kabas, and I. J. Fletcher, German Patent, 2,213,840 (1972).

78: 84321b W. Gauss, H. Heitzer, and S. Petersen, *Justus Liebigs Ann. Chem.*, **764,** 131 (1972).

78: P85943t F. Fleck, H. Kittl, H. R. Schmid, H. Schmid, and S. Valenti, German Patent, 2,212,480 (1972).

78: P85944u H. Schlaepfer, G. Kabas, and I. J. Fletcher, German Patent, 2,213,895 (1972).

78: P85945v G. Kabas, H. Schlaepfer, and I. J. Fletcher, German Patent, 2,213,839 (1972).

78: 98054d B. I. Mikhant'ev, G. V. Shatalov, and V. D. Galkin, *Tr. Voronezh. Univ.*, **95,** 8 (1972).

78: P99059w G. Kabas, H. Schaepfer, and I. J. Fletcher, German Patent, 2,213,754 (1972).

78: P99060q G. Kabas, H. Schlaepfer, and I. J. Fletcher, German Patent, 2,213,753 (1972).

78: P99065v H. Aebli, F. Fleck, and H. Schmid, German Patent, 2,225,075 (1972).

78: P137850g K. D. Bode, A. Brueggemann, and V. Weber, German Patent, 2,126,929 (1973).

78: P137966z Ciba-Geigy, A.-G., French Patent, 2,131,661 (1972).

78: P159644v R. U. Lemieux and R. Raap, German Patent, 2,222,954 (1973).

79: 19154f P. Ykman, G. L'abbé, and G. Smets, *J. Indian Chem. Soc.*, **49,** 1245 (1972).

79: P53343b J. R. E. Hoover, German Patent, 2,259,776 (1973).

79: 66882b K. Ikeda, G. Smets, and G. L'abbé, *J. Polym. Sci., Polym. Chem. Ed.*, **11,** 1177 (1973).

79: P67828a H. Schlaepfer, U.S. Patent, 3,743,639 (1973).

79: P80310v F. Fleck and H. Schmid, German Patent, 2,256,354 (1973).

79: P80350h H. Schlaepfer, British Patent, 1,317,307 (1973).

79: P106145x K. Weber and C. Luethi, German Patent, 2,262,632 (1973).

79: P106146y K. Weber and H. Schlaepfer, German Patent, 2,262,633 (1973).

79: P116278g O. Neuner and A. Dorlars, German Patent, 2,161,343 (1973).

79: 126403h K. S. Balachandran and M. V. George, *Tetrahedron*, **29,** 2119 (1973).

79: P127405x F. Fleck, A. V. Mercer, R. Paver, and H. Schmid, German Patent, 2,261,062 (1973).

79: P137160c U. Claussen, H. Gold, and J. Schroeder, German Patent, 2,210,261 (1973).
79: P147435k F. Fleck, H. Kittl, and S. Valenti, German Patent, 2,262,340 (1973).
80: P16457s K. Weber and H. Schlaepfer, German Patent, 2,262,578 (1973).
80: P16460n P. Liechti and H. Schlaepfer, German Patent, 2,309,614 (1973).
80: P48019s R. M. De Marinis and J. R. E. Hoover, German Patent, 2,322,127 (1973).
80: P49275j A. Domergue and R. F. M. Sureau, German Patent, 2,319,828 (1973).
80: 66979j M. Misiek, A. J. Moses, T. A. Pursiano, F. Leitner, and K. E. Price, *J. Antibiot.*, **26,** 737 (1973).
80: P83009k G. L. Dunn and J. R. E. Hoover, German Patent, 2,330,307 (1974).
80: P83012f P. Crooij and A. Colinet, German Patent, 2,331,599 (1974).
80: P83013g G. L. Dunn and J. R. E. Hoover, German Patent, 2,330,296 (1974).
80: P83024m G. L. Dunn and J. R. E. Hoover, German Patent, 2,316,866 (1974).
80: P84255z I. Ohkubo and M. Tsujimoto, Japanese Patent, 73 20,406 (1973).
80: P120970q R. M. De Marinis and J. R. E. Hoover, German Patent, 2,336,345 (1974).
80: P122419r A. Dorlars, A. Vogel, and C. W. Schellhammer, German Patent, 2,226,524 (1973).
80: 132664f N. E. Alexandrou, N. A. Rodios, and C. P. Hadjiantonrou-Lonizou, *Org. Magn. Resonance*, **5,** 579 (1973).
80: P134946y H. Rempfler, H. Bosshard, and K. Weber, German Patent, 2,332,098 (1974).
81: 19043k D. Basting, F. P. Schaefer, and B. Steyer, *Appl. Phys.*, **3,** 81 (1974).
81: P25679p U. Claussen, German Patent, 2,254,300 (1974).
81: 26334j B. I. Mikhant'ev, G. V. Shatalov, V. D. Galkin, and V. K. Vornova, *Monomery. Vysokomol. Soedin*, 100 (1973).
81: P38957c H. Gold and U. Claussen, German Patent, 2,242,784 (1974).
81: P38962a F. Fleck and H. Schmid, Swiss Patent, 543,521 (1973).
81: P49694r L. B. Crast, Jr., U.S. Patent, 3,813,388 (1974).
81: P51142c F. Fleck and H. Schmid, Swiss Patent, 544, 103 (1973).
81: P51143d H. Schlaepfer, German Patent, 2,329,991 (1974).
81: P51152f E. Asaumi and E. Kobayashi, Japanese Patent, 73 37,325 (1973).
81: P51153g T. Yanagisawa, Japanese Patent, 73 37,324 (1973).
81: P51154h I. Ohkubo and M. Tsujimoto, Japanese Patent, 73 41,117 (1973).
81: P65249e F. Fleck, A. V. Mercer, R. Paver, and H. Schmid, German Patent, 2,345,159 (1974).
81: P79389n R. Kirchmayr, U.S. Patent, 3,816,413 (1974).
81: P93085s A. Dorlars, H. Gold, and W. Horstmann, German Patent, 2,248,820 (1974).
81: P93093t B. Hirsch and H. J. Heckemann, East German Patent, 101,892 (1973).
81: P93326w H. Dierkes, K. Schoenol, J. Walter, and F. Mueller, German Patent, 2,242,597 (1974).
81: P105538g D. Willner and L. B. Crast, Jr., German Patent, 2,364,192 (1974).
81: P107815u K. D. Bode and R. Mueiller, German Patent, 2,312,064 (1974).
81: P122801e I. Ohkubo, M. Tsujimoto, and R. Tsukahara, Japanese Patent, 73 41,119 (1973).
81: P136161h S. R. Baker, L. C. Cheney, and C. T. Holdrege, German Patent, 2,404,592 (1974).
81: P137597s R. Kirchmayr, H. Heller, and J. Rody, Swiss Patent, 547,821 (1974).
81: P171366a H. Schlaepfer, German Patent, 2,355,116 (1974).
82: 11679x J. M. Essery, U. Corbin, V. Spancmanis, L. B. Crast, Jr., R. G. Graham, P. F. Misco, Jr., D. Willner, D. N. McGregor, and L. C. Cheney, *J. Antibiot.*, **27,** 573 (1974).

82: P17894k F. Fleck, H. Schmid, and S. Valenti, German Patent, 2,358,005 (1974).
82: P31753v A. Cairncross, U.S. Patent, 3,838,112 (1974).
82: P32468t O. Neuner, A. Dorlars, C. W. Schellhammer, and O. Berendes, German Patent, 1,794,396 (1974).
82: P43437h M. A. Kaplan and A. P. Granatek, U.S. Patent, 3,840,535 (1974).
82: P43446k J. M. Essery and L. C. Cheney, U.S. Patent, 3,780,032 (1974).
82: P43445j Y. Kodama and T. Ishimaru, Japanese Patent, 74 48,691 (1974).
82: 73727v R. N. Macdonald, A. Cairncross, J. B. Sieja, and W. H. Sharkey, *J. Polym. Sci., Polym. Chem. Ed.*, **12,** 663 (1974).
82: P74470t H. Balzer, F. Fleck, A. V. Mercer, and R. Paver, Swiss Patent, 551,987 (1974).
82: P112086g J. R. Hoover and J. A. Weisbach, German Patent, 2,422,068 (1974).
82: P126614q F. Fleck, H. Schmid, A. V. Mercer, and R. E. Paver, German Patent, 2,423,091 (1974).
82: 171481e F. C. De.Schryver, T. V. Thien, S. Toppet, and G. Smets, *J. Polym. Sci., Polym. Chem. Ed.*, **13,** 227 (1975).
82: P172621n A. Dorlars, O. Neuner, and U. Claussen, German Patent, 2,340,237 (1975).
83: P10111q J. G. Gleason, German Patent, 2,437,143 (1975).
83: P11149v M. Russo, V. Guidotti, and F. Grippa, French Patent, 2,221,477 (1974).
83: P28250u T. Naito, J. Okumura, H. Hoshi, and H. Kamachi, German Patent, 2,442,302 (1975).
83: P43359e R. Gericke, W. Rogalski, R. Bergmann, H. Wahlig and W. Hameister, German Patent, 2,345,402 (1975).
83: P50607n Bayer A.-G., French Patent, 2,220,104 (1974).
83: P58856p S. R. Baker, C. T. Holdrege, and L. C. Cheney, South African Patent, 74 00,665 (1974).
83: P61732a A. Dorlars and O. Neuner, German Patent, 2,258,276 (1974).
83: P81213k F. Fleck, H. Schmid, A. V. Mercer, and R. Paver, Swiss Patent, 561,709 (1975).
83: P99222x F. Fleck and H. Schmid, Swiss Patent, 561,708 (1975).
83: P99223y H. Aebli, F. Fleck, and H. Schmid, Swiss Patent, 562,228 (1975).
83: 105984r B. Steyer and F. P. Schaefer, *Appl. Phys.*, **7,** 113 (1975).
83: 106649x S. C. Kokkou and P. J. Rentzeperis, *Acta Crystallogr., Sect. B*, **B31,** 1564 (1975).
83: P115706r B. I. Mikhant'ev, G. V. Shatalov, V. D. Galkin, V. S. Voishchev, and O. V. Voishcheva, Russian Patent, 460,281 (1975).
83: P116987v A. Dorlars and W. D. Wirth, U.S. Patent, 3,839,333 (1974).
83: P133399x F. Fleck and H. R. Schmid, Swiss Patent, 562,812 (1975).
83: 147900r B. I. Mikhant'ev, G. V. Shatalov, V. D. Galkin, V. S. Voishchev, and O. V. Voishcheva, *Vysokomol. Soedin., Ser. B,* **17,** 467 (1975).
83: P179082s J. Bradshaw, M. C. Cook, G. I. Gregory, J. D. Crocker, and D. R. Sutherland, German Patent, 2,460,537 (1975).
83: 179654e M. Russo, *Kunststoffe*, **65,** 346 (1975).
83: P193339n U. Claussen, German Patent, 2,338,881 (1975).
83: P193352m M. A. Kaplan and A. P. Granatek, German Patent, 2,500,386 (1975).
83: P195247e H. Aebli, F. Fleck, and H. Schmid, Swiss Patent, 566,326 (1975).
83: P206290s K. H. Buechel, H. Gold, P. E. Frohberger, and H. Kaspers, German Patent, 2,407,305 (1975).
83: P206291t H. Arie, A. Saika, M. Yamanaka, I. Saito, Y. Inai, T. Sato, K. Ema and S. Nomoto, Japanese Patent, 75 96,593 (1975).
83: P207574z H. Schenermann, W. Mach, and D. Angort, German Patent, 2,406,220 (1975).

Index

All authors have been indexed. Numbers in parentheses are used to indicate the reference involved, since not all authors are mentioned specifically.